Jonathan Lawry, Enrique Miranda, Alberto Bugarin, Shoume
Przemys aw Grzegorzewski, Olgierd Hyrniewicz (Eds.)

Soft Methods for Integrated Uncertainty Modelling

Advances in Soft Computing

Editor-in-chief
Prof. Janusz Kacprzyk
Systems Research Institute
Polish Academy of Sciences
ul. Newelska 6
01-447 Warsaw
Poland
E-mail: kacprzyk@ibspan.waw.pl

Further volumes of this series
can be found on our homepage:
springer.com

Tetsuzo Tanino, Tamaki Tanaka, Masahiro
Inuiguchi
*Multi-Objective Programming and Goal
Programming,* 2003
ISBN 3-540-00653-2

Mieczys aw K opotek, S awomir T.
Wierzchoń, Krzysztof Trojanowski (Eds.)
*Intelligent Information Processing and Web
Mining,* 2003
ISBN 3-540-00843-8

Ajith Abraham, Katrin Franke, Mario
Kˆppen (Eds.)
Intelligent Systems Design and Applications,
2003
ISBN 3-540-40426-0

Ahmad LotÝ, Jonathan M. Garibaldi (Eds.)
Applications and Science in Soft-Computing,
2004
ISBN 3-540-40856-8

Mieczys aw K opotek, S awomir T.
Wierzchoń, Krzysztof Trojanowski (Eds.)
*Intelligent Information Processing and Web
Mining,* 2004
ISBN 3-540-21331-7

Miguel LÛpez-DÌaz, MarÌaÁ. Gil,
Przemys aw Grzegorzewski, Olgierd
Hryniewicz, Jonathan Lawry
*Soft Methodology and Random Information
Systems,* 2004
ISBN 3-540-22264-2

Kwang H. Lee
*First Course on Fuzzy Theory and
Applications,* 2005
ISBN 3-540-22988-4

Barbara Dunin-Keplicz, Andrzej Jankowski,
Andrzej Skowron, Marcin Szczuka
*Monitoring, Security, and Rescue
Techniques in Multiagent Systems,* 2005
ISBN 3-540-23245-1

Bernd Reusch (Ed.)
*Computational Intelligence, Theory and
Applications: International Conference 8th
Fuzzy Days in Dortmund, Germany,
Sept. 29 – Oct. 01, 2004 Proceedings,* 2005
ISBN 3-540-2280-1

Frank Hoffmann, Mario Kˆppen, Frank
Klawonn, Rajkumar Roy (Eds.)
*Soft Computing: Methodologies and
Applications,* 2005
ISBN 3-540-25726-8

Ajith Abraham, Bernard de Baets, Mario
Kˆppen, Bertram Nickolay (Eds.)
*Applied Soft Computing Technologies: The
Challenge of Complexity,* 2006
ISBN 3-540-31649-3

Ashutosh Tiwari, Joshua Knowles, Erel
Avineri, Keshav Dahal, Rajkumar Roy (Eds.)
Applications of Soft Computing, 2006
ISBN 3-540-29123-7

Mieczys aw A. K opotek, S awomir T.
Wierzchoń, Krzysztof Trojanowski (Eds.)
*Intelligent Information Processing and Web
Mining,* 2006
ISBN 3-540-33520-X

Jonathan Lawry, Enrique Miranda, Alberto
Bugarin, Shoumei Li, Maria Angeles Gil,
Przemys aw Grzegorzewski, Olgierd
Hyrniewicz (Eds.)
*Soft Methods for Integrated Uncertainty
Modelling,* 2006
ISBN 3-540-34776-3

Jonathan Lawry
Enrique Miranda
Alberto Bugarin
Shoumei Li
Maria Angeles Gil
Przemys aw Grzegorzewski
Olgierd Hyrniewicz
(Eds.)

Soft Methods for Integrated Uncertainty Modelling

 Springer

Jonathan Lawry
AI Group
Department of Engineering
Mathematics
University of Bristol
Bristol, BS8 1TR, UK

Shoumei Li
Department of Applied Mathematics
Beijing University of Technology
Beijing 100022, P.R. China

Enrique Miranda
Rey Juan Carlos University
Statistics and Operations Research
C-Tulip·n s/n
MÛstoles28933, Spain

Maria Angeles Gil
Universiad de Oviedo
Fac. Ciencias
Dpto. Estadistica e I.O y D.M.
Calle Calvo Sotelo s/n
33071 Oviedo, Spain

Alberto Bugarin
Intelligent Systems Group
Department of Electronics
& Computer Science
University of Santiago de Compostela
Santiago de Compostela, Spain

Przemys aw Grzegorzewski
Olgierd Hyrniewicz
Systems Research Institute
Polish Academy of Sciences
Newelska 6
01-447 Warsaw, Poland

Library of Congress Control Number: 2006928309

ISSN print edition: 1615-3871
ISSN electronic edition: 1860-0794
ISBN-10 3-540-34776-3 Springer Berlin Heidelberg New York
ISBN-13 978-3-540-34776-7 Springer Berlin Heidelberg New York

Springer is a part of Springer Science+Business Media
springer.com
© Springer-Verlag Berlin Heidelberg 2006
Printed in The Netherlands

Typesetting: by the author and techbooks using a Springer LATEX macro package
Cover design: *Erich Kirchner*, Heidelberg

Printed on acid-free paper   SPIN: 11757825   89/techbooks   5 4 3 2 1 0

Preface

The idea of soft computing emerged in the early 1990s from the fuzzy systems community, and refers to an understanding that the uncertainty, imprecision and ignorance present in a problem should be explicitly represented and possibly even exploited rather than either eliminated or ignored in computations. For instance, Zadeh defined 'Soft Computing' as follows:

> Soft computing differs from conventional (hard) computing in that, unlike hard computing, it is tolerant of imprecision, uncertainty and partial truth. In effect, the role model for soft computing is the human mind.

Recently soft computing has, to some extent, become synonymous with a hybrid approach combining AI techniques including fuzzy systems, neural networks, and biologically inspired methods such as genetic algorithms. Here, however, we adopt a more straightforward definition consistent with the original concept. Hence, soft methods are understood as those uncertainty formalisms not part of mainstream statistics and probability theory which have typically been developed within the AI and decision analysis community. These are mathematically sound uncertainty modelling methodologies which are complementary to conventional statistics and probability theory.

In addition to probabilistic factors such as measurement error and other random effects, the modelling process often requires us to make qualitative and subject judgements that cannot easily be translated into precise probability values. Such judgements give rise to a number of different types of uncertainty including; fuzziness if they are based on linguistic information; epistemic uncertainty when their reliability is in question; ignorance when they are insufficient to identify or restrict key modelling parameters; imprecision when parameters and probability distributions can only be estimated within certain bounds. Statistical theory has not traditionally been concerned with modelling uncertainty arising in this manner but soft methods, a range of powerful techniques developed within AI, attempt to address those problems where the encoding of subjective information is unavoidable. Therefore, a more realistic modelling process providing decision makers with an accurate reflection of the true current state of our knowledge (and ignorance) requires an integrated framework incorporating both probability theory, statistics and soft methods. This fusion motivates innovative research at the interface between computer science (AI), mathematics and systems engineering.

This edited volume is the proceedings of the 2006 International Workshop on Soft Methods in Probability and Statistics (SMPS 2006) hosted by the Artificial Intelligence Group at the University of Bristol, between 5-7 September 2006. This is the third of a series of biennial meetings organized in 2002 by the Systems Research Institute from the Polish Academy of Sciences in Warsaw, and in 2004 by the Department of Statistics and Operational Research at the University of Oviedo in Spain. These conferences provide a forum for discussion and research into the fusion of soft methods with probability and statistics, with the ultimate goal of integrated uncertainty modelling in complex systems involving human factors.

The papers in the volume are organized into a number of key themes each addressing a different aspect of the integration of soft methods with probability and statistics. These are identified both as being longstanding foundational problems, as well as promising avenues of research with the potential of providing significant advances in the modelling and representation of knowledge and uncertainty. Also vital to the development of any academic discipline is the identification and exploration of challenging new application areas. It is only through the application of existing tools and methodologies to the analysis of uncertainty in large-scale complex systems that fundamental research issues can be identified and new capabilities developed.

Part I presents abstracts of four keynote presentations by Lotfi Zadeh, Gert de Cooman, Jim Hall and Vladik Kreinovich. Prof. Zadeh's talk provides details on the latest developments in his theory of generalised uncertainty. Prof. de Cooman's talk describes a theory of linguistic probabilities based on imprecise probabilities. Prof. Hall gives an overview of the application of soft methods in Earth Systems Engineering. Prof. Kreinovich describes algorithms for statistical data processing under interval uncertainty and investigates their complexity. Part II on Soft Methods in Statistics and Random Information Systems presents current research leading to the development of new statistical tools incorporating fuzziness. Part III on Probability of Imprecisely-Valued Random Elements With Applications focusses on aspects of probability theory incorporating imprecision. Part IV on Applications and Modelling of Imprecise Operators considers how linguistic quantifiers can be used to describe uncertainty. Part V on Imprecise Probability theory concerns the uncertainty measures corresponding to upper and lower probabilities and previsions. Part VI on Possibility, Evidence and Interval Methods contains papers on possibility and evidence theory as well as interval methods. Finally, part VII presents a range of challenging applications requiring the integration of uncertainty, fuzziness and imprecision.

Bristol,   *Jonathan Lawry*
May 2006

Contents

Part I Keynote Papers

**Generalized Theory of Uncertainty (GTU) – Principal Concepts
and Ideas**
Lotfi A. Zadeh 3

Reasoning with Vague Probability Assessments
Gert de Cooman 5

Soft Methods in Earth Systems Engineering
Jim W. Hall 7

**Statistical Data Processing under Interval Uncertainty: Algorithms and
Computational Complexity**
Vladik Kreinovich 11

Part II Soft Methods in Statistics and Random Information Systems

On Testing Fuzzy Independence
Olgierd Hryniewicz 29

Variance Decomposition of Fuzzy Random Variables
Andreas Wünsche, Wolfgang Näther 37

Fuzzy Histograms and Density Estimation
Kevin Loquin, Olivier Strauss 45

**Graded Stochastic Dominance as a Tool for Ranking the Elements
of a Poset**
Karel De Loof, Hans De Meyer, Bernard De Baets 53

On Neyman-Pearson Lemma for Crisp, Random and Fuzzy Hypotheses
Adel Mohammadpour, Ali Mohammad-Djafari 61

Fuzzy Probability Distributions Induced by Fuzzy Random Vectors
Wolfgang Trutschnig 71

On the Identifiability of TSK Additive Fuzzy Rule-Based Models
José Luis Aznarte M., José Manuel Benítez 79

**An Asymptotic Test for Symmetry of Random Variables Based
on Fuzzy Tools**
González-Rodríguez. G., Colubi, A., D'Urso P., Giordani, P. 87

Exploratory Analysis of Random Variables Based on Fuzzifications
Colubi, A., González-Rodríguez. G., Lubiano, M.A., Montenegro, M. 95

A Method to Simulate Fuzzy Random Variables
González-Rodríguez. G., Colubi, A., Gil, M.A., Coppi, R. 103

Friedman's Test for Ambiguous and Missing Data
Edyta Mrówka, Przemysław Grzegorzewski 111

**Part III Probability of Imprecisely-Valued Random Elements
with Applications**

Measure-Free Martingales with Application to Classical Martingales
S.F. Cullender, W.-C. Kuo, C.C.A. Labuschagne and B.A. Watson 121

A Note on Random Upper Semicontinuous Functions
Hung T. Nguyen, Yukio Ogura, Santi Tasena and Hien Tran 129

Optional Sampling Theorem and Representation of Set-Valued Amart
Shoumei Li, Li Guan 137

On a Choquet Theorem for Random Upper Semicontinuous Functions
Yukio Ogura 145

A General Law of Large Numbers, with Applications
Pedro Terán, Ilya Molchanov 153

Part IV Applications and Modelling of Imprecise Operators

Fuzzy Production Planning Model for Automobile Seat Assembling
J. Mula, R. Poler, J.P. Garcia-Sabater 163

Optimal Selection of Proportional Bounding Quantifiers in Linguistic Data Summarization
Ingo Glöckner 173

A Linguistic Quantifier Based Aggregation for a Human Consistent Summarization of Time Series
Janusz Kacprzyk, Anna Wilbik, Sławomir Zadrożny 183

Efficient Evaluation of Similarity Quantified Expressions in the Temporal Domain
F. Díaz-Hermida, P. Cariñena, A. Bugarín 191

Part V Imprecise Probability Theory

Conditional Lower Previsions for Unbounded Random Quantities
Matthias C. M. Troffaes 201

Extreme Lower Probabilities
Erik Quaeghebeur, Gert de Cooman 211

Equivalence Between Bayesian and Credal Nets on an Updating Problem
Alessandro Antonucci, Marco Zaffalon 223

Varying Parameter in Classification Based on Imprecise Probabilities
Joaquín Abellán, Serafín Moral, Manuel Gómez and Andrés Masegosa 231

Comparing Proportions Data with Few Successes
F.P.A. Coolen, P. Coolen-Schrijner 241

A Unified View of Some Representations of Imprecise Probabilities
S. Destercke, D. Dubois 249

Part VI Possibility, Evidence and Interval Methods

Estimating an Uncertain Probability Density
Yakov Ben-Haim 261

Theory of Evidence with Imperfect Information
J. Recasens 267

Conditional IF-probability
Katarína Lendelová 275

On Two Ways for the Probability Theory on IF-sets
Beloslav Riečan 285

A Stratification of Possibilistic Partial Explanations
Sara Boutouhami, Aicha Mokhtari 291

Finite Discrete Time Markov Chains with Interval Probabilities
Damjan Škulj 299

Evidence and Compositionality
Wagner Borges, Julio Michael Stern 307

High Level Fuzzy Labels for Vague Concepts
Zengchang Qin and Jonathan Lawry 317

Part VII Integrated Uncertainty Modelling in Applications

Possibilistic Channels for DNA Word Design
Luca Bortolussi, Andrea Sgarro 327

**Transformation of Possibility Functions in a Climate Model
of Intermediate Complexity**
Hermann Held, Thomas Schneider von Deimling 337

Fuzzy Logic for Stochastic Modeling
Özer Ciftcioglu and I. Sevil Sariyildiz 347

A CUSUM Control Chart for Fuzzy Quality Data
Dabuxilatu Wang 357

**A Fuzzy Synset-Based Hidden Markov Model
for Automatic Text Segmentation**
Viet Ha-Thuc, Quang-Anh Nguyen-Van, Tru Hoang Cao and Jonathan Lawry .. 365

**Applying Fuzzy Measures for Considering Interaction Effects
in Fine Root Dispersal Models**
Wolfgang Näther, Konrad Wälder 373

**Scoring Feature Subsets for Separation Power in Supervised Bayes
Classification**
Tatjana Pavlenko, Hakan Fridén 383

**Interval Random Variables and Their Application in Queueing Systems
with Long–Tailed Service Times**
Bartłomiej Jacek Kubica, Krzysztof Malinowski 393

Online Learning for Fuzzy Bayesian Prediction
N.J. Randon, J. Lawry, I.D. Cluckie 405

Index 413

Part I

Keynote Papers

Generalized Theory of Uncertainty (GTU) – Principal Concepts and Ideas

Lotfi A. Zadeh*

Department of EECS, University of California, Berkeley, CA 94720-1776
zadeh@eecs.berkeley.edu

Uncertainty is an attribute of information. The path-breaking work of Shannon has led to a universal acceptance of the thesis that information is statistical in nature. Concomitantly, existing theories of uncertainty are based on probability theory. The generalized theory of uncertainty (GTU) departs from existing theories in essential ways. First, the thesis that information is statistical in nature is replaced by a much more general thesis that information is a generalized constraint, with statistical uncertainty being a special, albeit important case. Equating information to a generalized constraint is the fundamental thesis of GTU.

Second, bivalence is abandoned throughout GTU, and the foundation of GTU is shifted from bivalent logic to fuzzy logic. As a consequence, in GTU everything is or is allowed to be a matter of degree or, equivalently, fuzzy. Concomitantly, all variables are, or are allowed to be granular, with a granule being a clump of values drawn together by a generalized constraint.

And third, one of the principal objectives of GTU is achievement of NL-capability, that is, the capability to operate on information described in natural language. NL-capability has high importance because much of human knowledge, including knowledge about probabilities, is described in natural language. NL-capability is the focus of attention in the present paper.

The centerpiece of GTU is the concept of a generalized constraint. The concept of a generalized constraint is motivated by the fact that most real-world constraints are elastic rather than rigid, and have a complex structure even when simple in appearance. Briefly, if X is a variable taking values in a universe of discourse, U, then a generalized constraint on X, $GC(X)$, is an expression of the form X isr R, where R is a constraining relation, and r is an indexical variable which defines the modality of the constraint, that is, its semantics. The principal constraints are possibilistic (r = blank); veristic ($r = v$); probabilistic ($r = p$); random set ($r = r$); fuzzy graph ($r = fg$); usuality ($r = u$); bimodal ($r = bm$); and group ($r = g$). Generalized constraints may be combined, qualified, propagated and counterpropagated. A gener-

* Research supported in part by ONR N00014-02-1-0294, BT Grant CT1080028046, Omron Grant, Tekes Grant, Chevron Texaco Grant and the BISC Program of UC Berkeley.

L.A. Zadeh: *Generalized Theory of Uncertainty (GTU) – Principal Concepts and Ideas*, Advances in Soft Computing **6**, 3–4 (2006)
www.springerlink.com

alized constraint may be a system of generalized constraints. The collection of all generalized constraints constitutes the generalized constraint language, GCL.

The fundamental theses of GTU may be expressed as the symbolic equality $I(X) = GC(X)$, when $I(X)$ is the information about X. In GTU, a proposition is viewed as an answer to a question of the form "What is the value of X?" and thus is a carrier of information about X. In this perspective, the meaning of p, $M(p)$, is the information which it carries about X. An important consequence of the fundamental thesis of GTU is what is referred to as the meaning postulate: $M(I) = GC(X(p))$. This symbolic equality plays a pivotal role in GTU's NL-capability.

A prerequisite to computation with information described in natural language is precisiation of meaning. More specifically, if p is a proposition or a system of propositions drawn from a natural language, then the meaning of p is precisiated by expressing p as a generalized constraint, that is, translating p into the generalized constraint language GCL. The object of precisiation, p, and the result of precisiation, p^*, are referred to as the precisiend and precisiand, respectively. The degree to which the intension, that is, the attribute-based meaning of p^* matches the intension of p is referred to as the cointension of p^* and p. A precisiend, p^*, is cointensive if cointension of p^* and p is in some specified sense, high.

In GTU, deduction of an answer: ans(q), to a query, q, involves these modules: (a) Precisiation module, P; (b) Protoform module, Pr; and (c) Deduction/Computation module, D/C. The Precisiation module operates on the initial information set, p, expressed as INL, and results in a cointensive precisiend, p^*. The Protoform module serves as an interface between the Precisiation module and the Deduction/Computation module. The input to Pr is a generalized constraint, p^*, and its output is a protoform of p^*, that is, its abstracted summary, p^{**}. The Deduction/Computation module is basically a database (catalog) of rules of deduction which are, for the most part, rules which govern generalized constraint propagation and counterpropagation. The principal deduction rule is the Extension Principle. The rules are protoformal, with each rule having a symbolic part and a computational part. The protoformal rules are grasped into modules, with each module comprising rules which are associated with a particular class of generalized constraints, that is, possibilistic constraints, probabilistic constraints, veristic constraints, usuality constraints, etc.

The paper concludes with examples of computation with uncertain information described in natural language.

Reasoning with Vague Probability Assessments

Gert de Cooman

Ghent University, Research Group SYSTeMS, Technologiepark – Zwijnaarde 914, 9052 Zwijnaarde, Belgium
gert.decooman@UGent.be

In this lecture, I expound and comment on a model, or even more ambitiously, a theory, for representing, and drawing inferences from, vague probability assessments. The details of this theory have been published in two papers, the first [3] dealing with its behavioural underpinnings, and the second [1, 2] with its deeper mathematical aspects.

In a first part, I intend to discuss the basic features of the model, and explain how we can use so-called *possibilistic previsions* to mathematically represent vague probability assessments. I shall then discuss a number of requirements, or axioms, that can be imposed on such previsions, and the inference method, called *natural extension*, that such requirements generate. This inference method allows possibilistic previsions to be used as a basis for decision making, or as a prior in statistical reasoning. In addition, I shall discuss the connections between the theory of possibilistic previsions, Zadeh's theory of fuzzy probability [5, 6], and Walley's theory of coherent lower previsions [4].

In a second part, the emphasis will lie on providing possibilistic previsions with a behavioural and operationalisable interpretation. I shall discuss how the notion of a buying function, which is a mathematical object describing a modeller's beliefs about whether a subject will (not) accept to buy a given gamble for a given price, leads in certain well-defined circumstances to a hierarchical model that is mathematically completely equivalent to that of a special class of possibilistic previsions.

References

[1] G. de Cooman. A behavioural model for vague probability assessments. *Fuzzy Sets and Systems*, 154:305–358, 2005. With discussion.

[2] G. de Cooman. Further thoughts on possibilistic previsions: a rejoinder. *Fuzzy Sets and Systems*, 154:375–385, 2005.

[3] G. de Cooman and P. Walley. A possibilistic hierarchical model for behaviour under uncertainty. *Theory and Decision*, 52:327–374, 2002.

G. de Cooman: *Reasoning with Vague Probability Assessments*, Advances in Soft Computing **6**, 5–6 (2006)
www.springerlink.com   © Springer-Verlag Berlin Heidelberg 2006

[4] P. Walley. *Statistical Reasoning with Imprecise Probabilities*. Chapman and Hall, London, 1991.

[5] L. A. Zadeh. Fuzzy probabilities. *Information Processing and Management*, 20:363–372, 1984.

[6] L. A. Zadeh. Toward a perception-based theory of probabilistic reasoning with imprecise probabilities. *Journal of Statistical Planning and Inference*, 105:233–264, 2002.

Soft Methods in Earth Systems Engineering

Jim W. Hall

Tyndall Centre for Climate Change Research, School of Civil Engineering and Geosciences,
University of Newcastle upon Tyne NE1 7RU, UK
jim.hall@ncl.ac.uk

The narrowly defined technical problems that occupied civil engineers during the last century and a half, such as the mechanics of the materials steel, concrete and water, have for most practical purposes been solved. The outstanding challenges relate to interactions between technological systems, the natural environment and human society, at a range of scales up to the global. Management of these coupled systems is obviously a problem of decision making under uncertainty, informed by, on the one hand, sometimes quite dense datasets but, on the other, perhaps only the vaguest of intuitions about the behaviour of the systems in question. An extension of the scope of engineering from a narrowly focussed technical activity to one that more consciously engages with society and the natural environment means that approaches based upon the strictures of individual decision rationality may have to be modified as part of collective and perhaps highly contested decision processes.

The territory of Earth Systems Engineering outlined above seems to be fertile ground for soft methods in probability and statistics. There are severe uncertainties, often associated with human interaction with the technical and environmental systems in question. Information appears in a range of formats, including imprecise measurements and vague linguistic statements. Decision makers and citizens may legitimately be averse to ambiguities in the information at their disposal, particularly if the decisions impact upon future generations as much or more than upon our own. In highly contested decision processes the arrival of a technical expert with a solution that they claim to be 'optimal' according to some narrowly defined criteria of rationality is unlikely to be helpful.

The motivation for the use of soft methods in Earth Systems Engineering may be clear, yet their adoption in situations of practical significance is still quite limited. A brief review of some practical applications will reveal some successes and some important outstanding challenges. Even though the studies described are far more applied than much of what is published in the technical literature of fuzzy set theory, possibility theory, imprecise probability theory and Dempster-Shafer theory, these are nonetheless studies that have taken place in university Civil Engineering departments in partnership with enlightened individuals from industry and government. The step into 'unsupervised' industry practice will be achieved with the help of

J.W. Hall: *Soft Methods in Earth Systems Engineering*, Advances in Soft Computing **6**, 7–10 (2006)
www.springerlink.com

convenient software tools of the standard, for example, now widely used for analysis of (precise) Bayesian networks.

The applications

Slope stability analysis: Analysis of slope hydrology and soil mechanics is limited by scarcity of data and limitations in constitutive models of soils [1][2]. Information on soil properties is sometimes reported in the literature as intervals rather than probability distributions. Using slopes in Hong Kong as an example, we have demonstrated how imprecise input data, together with probabilistic information on hydrological properties, can be propagated through a numerical model of slope response. The analysis raises questions about the representation of dependency using random relations.

Condition assessment of flood defence systems: Routine inspection of flood defence infrastructure in the UK provides a linguistic classification of the condition of the infrastructure. These condition grades are approximately related to the variables that are input to reliability analysis. The linguistic classification naturally lends itself to fuzzy representation and we have demonstrated how this information can then be used to generate fuzzy probabilities of system failure, which can be used as a basis for prioritisation of maintenance [3].

Model and regionalisation uncertainties in flood risk analysis: We have explored the use of Info-Gap analysis [4] for analysis of flood risk management decisions under severe model and statistical uncertainties [5]. The analysis has illustrated the relative robustness of alternative flood management options.

Uncertainties in projections of global climate change: Analysis of uncertainties in projections of global mean temperature is attracting considerable attention in the global climate modelling community, but, despite severe uncertainties, is being addressed within conventional probabilistic paradigms, with a few notable exceptions [6]. We have illustrated how uncertainties in climate sensitivity can be represented with sets of probability measures [7] and how non-additive measures may be used to represent the vagueness associated with socio-economic scenarios [8].

Imprecise probabilities of abrupt climate change: Working jointly with the Potsdam Institute for Climate Impact Research we have conducted an elicitation exercise to obtain imprecise probabilities of a set of critical 'tipping points' in the Earth System. The probabilities of abrupt climate change were obtained conditional upon global mean temperature increasing within specified corridors. The work illustrates the severe uncertainties surrounding these critical aspects of the global climate.

Imprecise network models: Bayesian belief networks are being increasingly used in systems analysis and quantified risk analysis. Applications of interval version of network models in attribution of climate change [9] and environmental risk analysis have provided new insights for decision-makers into the sources and implications of uncertainty.

Some reflections

Application of soft methods to realistic practical examples has seldom been straight-forward. Some testing challenges are etched on our memories:

Elicitation: Imprecise representations have provided an expressive framework for representation of expert beliefs. However, it is doubtful that even experts who are well versed in probability are fully informed about the commitments that are being made in their statements. Whilst imprecise representation is intuitively attractive, experts are often doubtful about the exact location of the bounds that they provide. It is worrying that the bounding judgements experts have most difficulty making have a pivotal influence upon the output of the analysis.

Computation of imprecise probabilities: The optimisation problems associated with propagating sets of probability measures through computationally expensive, non-monotonic numerical models are of course, not trivial. Whilst bounding cases of concentrations of probability mass may be identified, justifying the opposite bound can be far more problematic in practice. Our experience with interval-valued belief networks has been favourable, but we have restricted ourselves to simple network structures. The arrival of algorithms for computing credal networks of a general structure is awaited with anticipation.

Proliferation of uncertainty: Bounding analyses have yielded disappointing amounts of information, and, in the limit, vacuous bounds. Whilst this may be a legitimate reflection of the state of knowledge it leaves decision-makers at a loss as to how to proceed. The dilation of intervals in the application of Generalized Bayes Rule is particularly disappointing.

Choice of uncertainty representation: Some situations obviously suggest a particular uncertainty representation, for example when information appears in interval format or when decision-makers are interested in the robustness of well specified options. In other situations the choice of an appropriate representation is contentious.

Aggregation of evidence: We have tested a variety of approaches to aggregation of evidence from different sources [10] with disappointing outcomes. Disjunctive approaches have provided uninformatively wide bounds whereas the conjunction of evidence has yielded null sets.

Concluding remarks

Of course 'hard' probability and statistics have a longer and richer pedigree than the methods discussed at the SMPS conference. It is hardly fair to expect the same extent of elaboration from soft methods, given their relative immaturity and the small size of their research community. Yet probability and statistics are remarkable not only in the richness of theory and methodology that has been developed through the centuries but also in the remarkable and expanding number of fields of application where the theories have become completely assimilated into human endeavour. The reliability-based Eurocodes and the used of geostatistics are just two examples that come to mind. Meanwhile soft methods are still the realm of cranks and enthusiasts. The severe uncertainties associated with managing coupled technological and natural systems and the heterogeneity of information that is obtained from these systems provides ample justification for departure from the conventional probabilistic paradigm. Yet to do so requires attention both to outstanding theoretical challenges and also to the process of applying soft methods in practice.

References

[1] J.W. Hall, E. Rubio, and M.J. Anderson. Random sets of probability measures in slope hydrology and stability analysis. *ZAMM: Journal of Applied Mathematics and Mechanics*, 84(10-11):710–720, 2004.

[2] E. Rubio, J.W. Hall, and M.G. Anderson. Uncertainty analysis in a slope hydrology and stability model using probabilistic and imprecise information. *Computers and Geotechnics*, 31:529–536, 2004.

[3] R.J. Dawson and J.W. Hall. Probabilistic condition characterisation of coastal structures using imprecise information. In J. McKee Smith, editor, *Coastal Engineering 2002, Proc. 28th Int. Conf.*, volume 2, pages 2348–2359. World Scientific, 2003.

[4] Y. Ben-Haim. *Information-Gap Decision Theory: Decisions Under Severe Uncertainty*. Academic Press, 2001.

[5] D. Hine and J.W. Hall. Info-gap analysis of flood model calibration. In *Hydroinformatics 2006: Proc. 6th Int. Conf. on Hydroinformatics*, in press.

[6] E. Kriegler and H. Held. Utilizing belief functions for the estimation of future climate change. *International Journal of Approximate Reasoning*, 39(2-3):185–209, 2005.

[7] J.W. Hall, G. Fu, and J. Lawry. Imprecise probabilities of climate change: aggregation of fuzzy scenarios and model uncertainties. *Climatic Change*, in press.

[8] J. Lawry, J.W. Hall, and G. Fu. A granular semantics for fuzzy measures and its application to climate change scenarios. In F.G. Cozman, R. Nau, and E. Seidenfeld, editors, *ISIPTA '05, Proc. 4th Int. Symp. on Imprecise Probabilities and Their Applications*, pages 213–221, 2005.

[9] J.W. Hall, C. Twyman, and A. Kay. Influence diagrams for representing uncertainty in climate-related propositions. *Climatic Change*, 69:343–365, 2005.

[10] K. Sentz and S. Ferson. Combination of evidence in dempster-shafer theory. Technical Report SAND2002-0835, Sandia National Laboratories, Albuquerque, New Mexico, 2002.

Statistical Data Processing under Interval Uncertainty: Algorithms and Computational Complexity

Vladik Kreinovich

Department of Computer Science, University of Texas at El Paso,
El Paso, TX 79968, USA
vladik@utep.edu

1 Main Problem

Why indirect measurements? In many real-life situations, we are interested in the value of a physical quantity y that is difficult or impossible to measure directly. Examples of such quantities are the distance to a star and the amount of oil in a given well. Since we cannot measure y directly, a natural idea is to measure y *indirectly*. Specifically, we find some easier-to-measure quantities $x_1, \ldots, x_n$ which are related to y by a known relation $y = f(x_1, \ldots, x_n)$; this relation may be a simple functional transformation, or complex algorithm (e.g., for the amount of oil, numerical solution to an inverse problem). Then, to estimate y, we first measure the values of the quantities $x_1, \ldots, x_n$, and then we use the results $\tilde{x}_1, \ldots, \tilde{x}_n$ of these measurements to to compute an estimate $\tilde{y}$ for y as $\tilde{y} = f(\tilde{x}_1, \ldots, \tilde{x}_n)$:

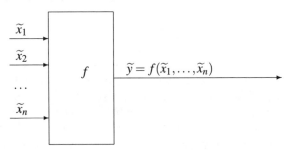

For example, to find the resistance R, we measure current I and voltage V, and then use the known relation $R = V/I$ to estimate resistance as $\tilde{R} = \tilde{V}/\tilde{I}$.

Computing an estimate for y based on the results of direct measurements is called *data processing*; data processing is the main reason why computers were invented in the first place, and data processing is still one of the main uses of computers as number crunching devices.

V. Kreinovich: *Statistical Data Processing under Interval Uncertainty: Algorithms and Computational Complexity*, Advances in Soft Computing **6**, 11–26 (2006)
www.springerlink.com

Comment. In this paper, for simplicity, we consider the case when the relation between x_i and y is known exactly; in some practical situations, we only known an approximate relation between x_i and y.

Why interval computations? From computing to probabilities to intervals. Measurement are never 100% accurate, so in reality, the actual value x_i of i-th measured quantity can differ from the measurement result $\widetilde{x}_i$. Because of these *measurement errors* $\Delta x_i \stackrel{\text{def}}{=} \widetilde{x}_i - x_i$, the result $\widetilde{y} = f(\widetilde{x}_1, \ldots, \widetilde{x}_n)$ of data processing is, in general, different from the actual value $y = f(x_1, \ldots, x_n)$ of the desired quantity y.

It is desirable to describe the error $\Delta y \stackrel{\text{def}}{=} \widetilde{y} - y$ of the result of data processing. To do that, we must have some information about the errors of direct measurements.

What do we know about the errors Δx_i of direct measurements? First, the manufacturer of the measuring instrument must supply us with an upper bound Δ_i on the measurement error. If no such upper bound is supplied, this means that no accuracy is guaranteed, and the corresponding "measuring instrument" is practically useless. In this case, once we performed a measurement and got a measurement result $\widetilde{x}_i$, we know that the actual (unknown) value x_i of the measured quantity belongs to the interval $\mathbf{x}_i = [\underline{x}_i, \overline{x}_i]$, where $\underline{x}_i = \widetilde{x}_i - \Delta_i$ and $\overline{x}_i = \widetilde{x}_i + \Delta_i$.

In many practical situations, we not only know the interval $[-\Delta_i, \Delta_i]$ of possible values of the measurement error; we also know the probability of different values Δx_i within this interval. This knowledge underlies the traditional engineering approach to estimating the error of indirect measurement, in which we assume that we know the probability distributions for measurement errors Δx_i.

In practice, we can determine the desired probabilities of different values of Δx_i by comparing the results of measuring with this instrument with the results of measuring the same quantity by a standard (much more accurate) measuring instrument. Since the standard measuring instrument is much more accurate than the one use, the difference between these two measurement results is practically equal to the measurement error; thus, the empirical distribution of this difference is close to the desired probability distribution for measurement error. There are two cases, however, when this determination is not done:

- First is the case of cutting-edge measurements, e.g., measurements in fundamental science. When a Hubble telescope detects the light from a distant galaxy, there is no "standard" (much more accurate) telescope floating nearby that we can use to calibrate the Hubble: the Hubble telescope is the best we have.
- The second case is the case of measurements on the shop floor. In this case, in principle, every sensor can be thoroughly calibrated, but sensor calibration is so costly – usually costing ten times more than the sensor itself – that manufacturers rarely do it.

In both cases, we have no information about the probabilities of Δx_i; the only information we have is the upper bound on the measurement error.

In this case, after we performed a measurement and got a measurement result $\widetilde{x}_i$, the only information that we have about the actual value x_i of the measured quantity is that it belongs to the interval $\mathbf{x}_i = [\widetilde{x}_i - \Delta_i, \widetilde{x}_i + \Delta_i]$. In such situations, the only information that we have about the (unknown) actual value of $y = f(x_1, \ldots, x_n)$ is that y belongs to the range $\mathbf{y} = [\underline{y}, \overline{y}]$ of the function f over the box $\mathbf{x}_1 \times \ldots \times \mathbf{x}_n$:

$$\mathbf{y} = [\underline{y}, \overline{y}] = \{f(x_1, \ldots, x_n) \mid x_1 \in \mathbf{x}_1, \ldots, x_n \in \mathbf{x}_n\}.$$

The process of computing this interval range based on the input intervals $\mathbf{x}_i$ is called *interval computations*; see, e.g., [19, 37].

Interval computations techniques: brief reminder. Historically the first method for computing the enclosure for the range is the method which is sometimes called "straightforward" interval computations. This method is based on the fact that inside the computer, every algorithm consists of elementary operations (arithmetic operations, min, max, etc.). For each elementary operation $f(a,b)$, if we know the intervals $\mathbf{a}$ and $\mathbf{b}$ for a and b, we can compute the exact range $f(\mathbf{a}, \mathbf{b})$. The corresponding formulas form the so-called *interval arithmetic*. For example,

$$[\underline{a}, \overline{a}] + [\underline{b}, \overline{b}] = [\underline{a} + \underline{b}, \overline{a} + \overline{b}]; \quad [\underline{a}, \overline{a}] - [\underline{b}, \overline{b}] = [\underline{a} - \overline{b}, \overline{a} - \underline{b}];$$

$$[\underline{a}, \overline{a}] \cdot [\underline{b}, \overline{b}] = [\min(\underline{a} \cdot \underline{b}, \underline{a} \cdot \overline{b}, \overline{a} \cdot \underline{b}, \overline{a} \cdot \overline{b}), \max(\underline{a} \cdot \underline{b}, \underline{a} \cdot \overline{b}, \overline{a} \cdot \underline{b}, \overline{a} \cdot \overline{b})].$$

In straightforward interval computations, we repeat the computations forming the program f step-by-step, replacing each operation with real numbers by the corresponding operation of interval arithmetic. It is known that, as a result, we get an enclosure $\mathbf{Y} \supseteq \mathbf{y}$ for the desired range.

In some cases, this enclosure is exact. In more complex cases (see examples below), the enclosure has excess width.

Example. Let us illustrate the above idea on the example of estimating the range of the function $f(x) = (x - 2) \cdot (x + 2)$ on the interval $x \in [1, 2]$.

We start with parsing the expression for the function, i.e., describing how a computer will compute this expression; it will implement the following sequence of elementary operation:

$$r_1 := x - 2; \quad r_2 := x + 2; \quad r_3 := r_1 \cdot r_2.$$

The main idea behind straightforward interval computations is to perform the same operations, but with *intervals* instead of *numbers*:

$$\mathbf{r}_1 := [1, 2] - [2, 2] = [-1, 0]; \quad \mathbf{r}_2 := [1, 2] + [2, 2] = [3, 4];$$

$$\mathbf{r}_3 := [-1, 0] \cdot [3, 4] = [-4, 0].$$

For this function, the actual range is $f(\mathbf{x}) = [-3, 0]$.

Comment: This is just a toy example, there are more efficient ways of computing an enclosure $\mathbf{Y} \supseteq \mathbf{y}$.

There exist more sophisticated techniques for producing a narrower enclosure, e.g., a centered form method. However, for each of these techniques, there are cases when we get an excess width. Reason: as shown in [25], the problem of computing the exact range is known to be NP-hard even for polynomial functions $f(x_1, \ldots, x_n)$ (actually, even for quadratic functions f).

Practical problem. In some practical situations, in addition to the lower and upper bounds on each random variable x_i, we have some additional information about x_i.

So, we arrive at the following problem:

- we have a data processing algorithm $f(x_1, \ldots, x_n)$, and
- we have some information about the uncertainty with which we know x_i (e.g., measurement errors).

We want to know the resulting uncertainty in the result $y = f(x_1, \ldots, x_n)$ of data processing.

In interval computations, we assume that the uncertainty in x_i can be described by the interval of possible values. In real life, in addition to the intervals, we often have some information about the probabilities of different values within this interval. What can we then do?

2 What is the Best Way to Describe Probabilistic Uncertainty?

In order to describe how uncertainty in x_i affects y, we need to know what is the best way to represent the corresponding probabilistic uncertainty in x_i.

In probability theory, there are many different ways of representing a probability distribution. For example, one can use a probability density function (pdf), or a cumulative distribution function (CDF), or a probability measure, i.e., a function which maps different sets into a probability that the corresponding random variable belongs to this set. The reason why there are many different representations is that in different problems, different representations turned out to be the most useful.

We would like to select a representation which is the most useful for problems related to risk analysis. To make this selection, we must recall where the information about probabilities provided by risk analysis is normally used.

How is the partial information about probabilities used in risk analysis? The main objective of risk analysis is to make decisions. A standard way of making a decision is to select the action a for which the expected utility (gain) is the largest possible. This is where probabilities are used: in computing, for every possible action a, the corresponding expected utility. To be more precise, we usually know, for each action a and for each actual value of the (unknown) quantity x, the corresponding value of the utility $u_a(x)$. We must use the probability distribution for x to compute the expected value $E[u_a(x)]$ of this utility.

In view of this application, the most useful characteristics of a probability distribution would be the ones which would enable us to compute the expected value $E[u_a(x)]$ of different functions $u_a(x)$.

Which representations are the most useful for this intended usage? General idea. Which characteristics of a probability distribution are the most useful for computing mathematical expectations of different functions $u_a(x)$? The answer to this question depends on the type of the function, i.e., on how the utility value u depends on the value x of the analyzed parameter.

Smooth utility functions naturally lead to moments. One natural case is when the utility function $u_a(x)$ is smooth. We have already mentioned, in Section I, that we usually know a (reasonably narrow) interval of possible values of x. So, to compute the expected value of $u_a(x)$, all we need to know is how the function $u_a(x)$ behaves on this narrow interval. Because the function is smooth, we can expand it into Taylor series. Because the interval is narrow, we can safely consider only linear and quadratic terms in this expansion and ignore higher-order terms: $u_a(x) \approx c_0 + c_1 \cdot (x - x_0) + c_2 \cdot (x - x_0)^2$, where x_0 is a point inside the interval. Thus, we can approximate the expectation of this function by the expectation of the corresponding quadratic expression: $E[u_a(x)] \approx E[c_0 + c_1 \cdot (x - x_0) + c_2 \cdot (x - x_0)^2]$, i.e., by the following expression: $E[u_a(x)] \approx c_0 + c_1 \cdot E[x - x_0] + c_2 \cdot E[(x - x_0)^2]$. So, to compute the expectations of such utility functions, it is sufficient to know the first and second moments of the probability distribution.

In particular, if we use, as the point x_0, the average $E[x]$, the second moment turns into the variance of the original probability distribution. So, instead of the first and the second moments, we can use the mean E and the variance V.

In risk analysis, non-smooth utility functions are common. In engineering applications, most functions are smooth, so usually the Taylor expansion works pretty well. In risk analysis, however, not all dependencies are smooth. There is often a threshold x_0 after which, say, a concentration of a certain chemical becomes dangerous.

This threshold sometimes comes from the detailed chemical and/or physical analysis. In this case, when we increase the value of this parameter, we see the drastic increase in effect and hence, the drastic change in utility value. Sometimes, this threshold simply comes from regulations. In this case, when we increase the value of this parameter past the threshold, there is no drastic increase in effects, but there is a drastic decrease of utility due to the necessity to pay fines, change technology, etc. In both cases, we have a utility function which experiences an abrupt decrease at a certain threshold value x_0.

Non-smooth utility functions naturally lead to CDFs. We want to be able to compute the expected value $E[u_a(x)]$ of a function $u_a(x)$ which changes smoothly until a certain value x_0, then drops it value and continues smoothly for $x > x_0$. We usually know the (reasonably narrow) interval which contains all possible values of x. Because the interval is narrow and the dependence before and after the threshold is smooth, the resulting change in $u_a(x)$ before x_0 and after x_0 is much smaller than the change at x_0. Thus, with a reasonable accuracy, we can ignore the small changes before and after x_0, and assume that the function $u_a(x)$ is equal to a constant u^+ for $x < x_0$, and to some other constant $u^- < u^+$ for $x > x_0$.

The simplest case is when $u^+ = 1$ and $u^- = 0$. In this case, the desired expected value $E[u_a^{(0)}(x)]$ coincides with the probability that $x < x_0$, i.e., with the corresponding value $F(x_0)$ of the cumulative distribution function (CDF). A generic function $u_a(x)$ of this type, with arbitrary values u^- and u^+, can be easily reduced to this simplest case, because, as one can easily check, $u_a(x) = u^- + (u^+ - u^-) \cdot u^{(0)}(x)$ and hence, $E[u_a(x)] = u^- + (u^+ - u^-) \cdot F(x_0)$.

Thus, to be able to easily compute the expected values of all possible non-smooth utility functions, it is sufficient to know the values of the CDF $F(x_0)$ for all possible x_0.

3 How to Represent Partial Information about Probabilities

General idea. In many cases, we have a complete information about the probability distributions that describe the uncertainty of each of n inputs.

However, a practically interesting case is how to deal with situations when we only have partial information about the probability distributions. How can we represent this partial information?

Case of cdf. If we use cdf $F(x)$ to represent a distribution, then full information corresponds to the case when we know the exact value of $F(x)$ for every x. Partial information means:

- either that we only know approximate values of $F(x)$ for all x, i.e., that for every x, we only know the interval that contains $F(x)$; in this case, we get a *p-box*;
- or that we only know the values of $F(x)$ for some x, i.e, that we only know the values $F(x_1), \ldots, F(x_n)$ for finitely many values $x = x_1, \ldots, x_n$; in this case, we have a *histogram*.

It is also possible that we know only approximate values of $F(x)$ for some x; in this case, we have an *interval-valued histogram*.

Case of moments. If we use moments to represent a distribution, then partial information means that we either know the exact values of finitely many moments, or that we know intervals of possible values of several moments.

4 Resulting Problems

This discussion leads to a natural classification of possible problems:

- If we have complete information about the distributions of x_i, then, to get validated estimates on uncertainty of y, we have to use Monte-Carlo-type techniques; see, in particular, papers by D. Lodwick et al. [33, 34]
- If we have p-boxes, we can use methods proposed by S. Ferson et al. [13, 14, 15, 23, 43, 46].

- If we have histograms, we can use methods proposed by D. Berleant et al. [6, 7, 8, 9, 10, 44, 53].
- If we have moments, then we can use methods proposed by S. Ferson, V. Kreinovich, M. Orshansky, et al. [18, 22, 30, 41, 42].

There are also additional issues, including:

- how we get these bounds for x_i?
- specific practical applications, like the appearance of histogram-type distributions in problems related to privacy in statistical databases,
- etc.

5 Case Study

Practical problem. In some practical situations, in addition to the lower and upper bounds on each random variable x_i, we know the bounds $\mathbf{E}_i = [\underline{E}_i, \overline{E}_i]$ on its mean E_i.

Indeed, in measurement practice (see, e.g., [11]), the overall measurement error Δx is usually represented as a sum of two components:

- a *systematic* error component $\Delta_s x$ which is defined as the expected value $E[\Delta x]$, and
- a *random* error component $\Delta_r x$ which is defined as the difference between the overall measurement error and the systematic error component: $\Delta_r x \overset{\text{def}}{=} \Delta x - \Delta_s x$.

In addition to the bound Δ on the overall measurement error, the manufacturers of the measuring instrument often provide an upper bound Δ_s on the systematic error component: $|\Delta_s x| \leq \Delta_s$.

This additional information is provided because, with this additional information, we not only get a bound on the accuracy of a single measurement, but we also get an idea of what accuracy we can attain if we use repeated measurements to increase the measurement accuracy. Indeed, the very idea that repeated measurements can improve the measurement accuracy is natural: we measure the same quantity by using the same measurement instrument several (N) times, and then take, e.g., an arithmetic average $\bar{x} = \frac{\tilde{x}^{(1)} + \ldots + \tilde{x}^{(N)}}{N}$ of the corresponding measurement results $\tilde{x}^{(1)} = x + \Delta x^{(1)}, \ldots, \tilde{x}^{(N)} = x + \Delta x^{(N)}$.

- If systematic error is the only error component, then all the measurements lead to exactly the same value $\tilde{x}^{(1)} = \ldots = \tilde{x}^{(N)}$, and averaging does not change the value – hence does not improve the accuracy.
- On the other hand, if we know that the systematic error component is 0, i.e., $E[\Delta x] = 0$ and $E[\tilde{x}] = x$, then, as $N \to \infty$, the arithmetic average tends to the actual value x. In this case, by repeating the measurements sufficiently many times, we can determine the actual value of x with an arbitrary given accuracy.

In general, by repeating measurements sufficiently many times, we can arbitrarily decrease the random error component and thus attain accuracy as close to Δ_s as we want.

When this additional information is given, then, after we performed a measurement and got a measurement result $\tilde{x}$, then not only we get the information that the actual value x of the measured quantity belongs to the interval $\mathbf{x} = [\tilde{x} - \Delta, \tilde{x} + \Delta]$, but we can also conclude that the expected value of $x = \tilde{x} - \Delta x$ (which is equal to $E[x] = \tilde{x} - E[\Delta x] = \tilde{x} - \Delta_s x$) belongs to the interval $\mathbf{E} = [\tilde{x} - \Delta_s, \tilde{x} + \Delta_s]$.

If we have this information for every x_i, then, in addition to the interval $\mathbf{y}$ of possible value of y, we would also like to know the interval of possible values of $E[y]$. This additional interval will hopefully provide us with the information on how repeated measurements can improve the accuracy of this indirect measurement. Thus, we arrive at the following problem:

Precise formulation of the problem. Given an algorithm computing a function $f(x_1, \ldots, x_n)$ from R^n to R, and values $\underline{x}_1, \bar{x}_1, \ldots, \underline{x}_n, \bar{x}_n, \underline{E}_1, \bar{E}_1, \ldots, \underline{E}_n, \bar{E}_n$, we want to find

$$\underline{E} \stackrel{\text{def}}{=} \min\{E[f(x_1, \ldots, x_n)] \mid \text{ all distributions of } (x_1, \ldots, x_n) \text{ for which}$$

$$x_1 \in [\underline{x}_1, \bar{x}_1], \ldots, x_n \in [\underline{x}_n, \bar{x}_n], E[x_1] \in [\underline{E}_1, \bar{E}_1], \ldots E[x_n] \in [\underline{E}_n, \bar{E}_n]\};$$

and $\bar{E}$ which is the maximum of $E[f(x_1, \ldots, x_n)]$ for all such distributions.

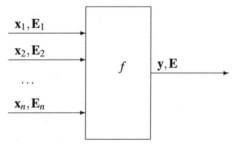

In addition to considering all possible distributions, we can also consider the case when all the variables x_i are independent.

How we solve this problem. The main idea behind straightforward interval computations can be applied here as well. Namely, first, we find out how to solve this problem for the case when $n = 2$ and $f(x_1, x_2)$ is one of the standard arithmetic operations. Then, once we have an arbitrary algorithm $f(x_1, \ldots, x_n)$, we parse it and replace each elementary operation on real numbers with the corresponding operation on quadruples $(\underline{x}, \underline{E}, \bar{E}, \bar{x})$.

To implement this idea, we must therefore know how to, solve the above problem for elementary operations.

For *addition*, the answer is simple. Since $E[x_1 + x_2] = E[x_1] + E[x_2]$, if $y = x_1 + x_2$, there is only one possible value for $E = E[y]$: the value $E = E_1 + E_2$. This value does not depend on whether we have correlation or nor, and whether we have any information about the correlation. Thus, $\mathbf{E} = \mathbf{E}_1 + \mathbf{E}_2$.

Similarly, the answer is simple for *subtraction:* if $y = x_1 - x_2$, there is only one possible value for $E = E[y]$: the value $E = E_1 - E_2$. Thus, $\mathbf{E} = \mathbf{E}_1 - \mathbf{E}_2$.

For *multiplication*, if the variables x_1 and x_2 are independent, then $E[x_1 \cdot x_2] = E[x_1] \cdot E[x_2]$. Hence, if $y = x_1 \cdot x_2$ and x_1 and x_2 are independent, there is only one possible value for $E = E[y]$: the value $E = E_1 \cdot E_2$; hence $\mathbf{E} = \mathbf{E}_1 \cdot \mathbf{E}_2$.

The first non-trivial case is the case of multiplication in the presence of possible correlation. When we know the exact values of E_1 and E_2, the solution to the above problem is as follows:

Theorem 1. *For multiplication $y = x_1 \cdot x_2$, when we have no information about the correlation,*

$$\underline{E} = \max(p_1 + p_2 - 1, 0) \cdot \bar{x}_1 \cdot \bar{x}_2 + \min(p_1, 1 - p_2) \cdot \bar{x}_1 \cdot \underline{x}_2 +$$

$$\min(1 - p_1, p_2) \cdot \underline{x}_1 \cdot \bar{x}_2 + \max(1 - p_1 - p_2, 0) \cdot \underline{x}_1 \cdot \underline{x}_2;$$

and

$$\overline{E} = \min(p_1, p_2) \cdot \bar{x}_1 \cdot \bar{x}_2 + \max(p_1 - p_2, 0) \cdot \bar{x}_1 \cdot \underline{x}_2 +$$

$$\max(p_2 - p_1, 0) \cdot \underline{x}_1 \cdot \bar{x}_2 + \min(1 - p_1, 1 - p_2) \cdot \underline{x}_1 \cdot \underline{x}_2,$$

where $p_i \stackrel{\text{def}}{=} (E_i - \underline{x}_i)/(\bar{x}_i - \underline{x}_i)$.

Theorem 2. *For multiplication under no information about dependence, to find $\underline{E}$, it is sufficient to consider the following combinations of p_1 and p_2:*

- $p_1 = \underline{p}_1$ and $p_2 = \underline{p}_2$; $p_1 = \underline{p}_1$ and $p_2 = \overline{p}_2$; $p_1 = \overline{p}_1$ and $p_2 = \underline{p}_2$; $p_1 = \overline{p}_1$ and $p_2 = \overline{p}_2$;
- $p_1 = \max(\underline{p}_1, 1 - \overline{p}_2)$ and $p_2 = 1 - p_1$ *(if $1 \in \mathbf{p}_1 + \mathbf{p}_2$); and*
- $p_1 = \min(\overline{p}_1, 1 - \underline{p}_2)$ and $p_2 = 1 - p_1$ *(if $1 \in \mathbf{p}_1 + \mathbf{p}_2$).*

The smallest value of $\underline{E}$ for all these cases is the desired lower bound $\underline{E}$.

Theorem 3. *For multiplication under no information about dependence, to find $\overline{E}$, it is sufficient to consider the following combinations of p_1 and p_2:*

- $p_1 = \underline{p}_1$ and $p_2 = \underline{p}_2$; $p_1 = \underline{p}_1$ and $p_2 = \overline{p}_2$; $p_1 = \overline{p}_1$ and $p_2 = \underline{p}_2$; $p_1 = \overline{p}_1$ and $p_2 = \overline{p}_2$;
- $p_1 = p_2 = \max(\underline{p}_1, \underline{p}_2)$ *(if $\mathbf{p}_1 \cap \mathbf{p}_2 \neq \emptyset$); and*
- $p_1 = p_2 = \min(\overline{p}_1, \overline{p}_2)$ *(if $\mathbf{p}_1 \cap \mathbf{p}_2 \neq \emptyset$).*

The largest value of $\overline{E}$ for all these cases is the desired upper bound $\overline{E}$.

For the *inverse* $y = 1/x_1$, the finite range is possible only when $0 \notin \mathbf{x}_1$. Without losing generality, we can consider the case when $0 < \underline{x}_1$. In this case, we get the following bound:

Theorem 4. *For the inverse $y = 1/x_1$, the range of possible values of E is $\mathbf{E} = [1/E_1, p_1/\bar{x}_1 + (1 - p_1)/\underline{x}_1]$.*

(Here p_1 denotes the same value as in Theorem 1).

Theorem 5. *For minimum $y = \min(x_1, x_2)$, when x_1 and x_2 are independent, we have $\overline{E} = \min(E_1, E_2)$ and*

$$\underline{E} = p_1 \cdot p_2 \cdot \min(\overline{x}_1, \overline{x}_2) + p_1 \cdot (1 - p_2) \cdot \min(\overline{x}_1, \underline{x}_2) +$$

$$(1 - p_1) \cdot p_2 \cdot \min(\underline{x}_1, \overline{x}_2) + (1 - p_1) \cdot (1 - p_2) \cdot \min(\underline{x}_1, \underline{x}_2).$$

Theorem 6. *For maximum $y = \min(x_1, x_2)$, when x_1 and x_2 are independent, we have $\underline{E} = \max(E_1, E_2)$ and*

$$\overline{E} = p_1 \cdot p_2 \cdot \max(\overline{x}_1, \overline{x}_2) + p_1 \cdot (1 - p_2) \cdot \max(\overline{x}_1, \underline{x}_2) +$$

$$(1 - p_1) \cdot p_2 \cdot \max(\underline{x}_1, \overline{x}_2) + (1 - p_1) \cdot (1 - p_2) \cdot \max(\underline{x}_1, \underline{x}_2).$$

Theorem 7. *For minimum $y = \min(x_1, x_2)$, when we have no information about the correlation between x_1 and x_2, we have $\overline{E} = \min(E_1, E_2)$,*

$$\underline{E} = \max(p_1 + p_2 - 1, 0) \cdot \min(\overline{x}_1, \overline{x}_2) + \min(p_1, 1 - p_2) \cdot \min(\overline{x}_1, \underline{x}_2) +$$

$$\min(1 - p_1, p_2) \cdot \min(\underline{x}_1, \overline{x}_2) + \max(1 - p_1 - p_2, 0) \cdot \min(\underline{x}_1, \underline{x}_2).$$

Theorem 8. *For maximum $y = \max(x_1, x_2)$, when we have no information about the correlation between x_1 and x_2, we have $\underline{E} = \max(E_1, E_2)$ and*

$$\overline{E} = \min(p_1, p_2) \cdot \max(\overline{x}_1, \overline{x}_2) + \max(p_1 - p_2, 0) \cdot \max(\overline{x}_1, \underline{x}_2) +$$

$$\max(p_2 - p_1, 0) \cdot \max(\underline{x}_1, \overline{x}_2) + \min(1 - p_1, 1 - p_2) \cdot \max(\underline{x}_1, \underline{x}_2).$$

Similar formulas can be produced for the cases when there is a strong correlation between x_i: namely, when x_1 is (non-strictly) increasing or decreasing in x_2.

For products of several random variables, the corresponding problem is already NP-hard [24].

Challenges. What is, in addition to intervals and first moments, we also know second moments (this problem is important for design of computer chips):

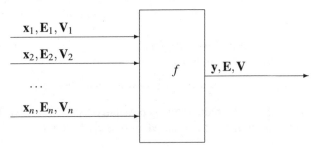

What if, in addition to moments, we also know p-boxes?

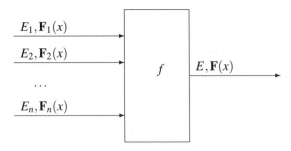

6 Additional Results and Challenges

Estimating bounds on statistical characteristics. The above techniques assume that we already know the moments etc. How can we compute them based on the measurement results – taking into account that these results represent the actual (unknown) values with measurement uncertainty.

For example, in the case of interval uncertainty, instead of the exact sample values, we have only interval ranges $[\underline{x}_i, \overline{x}_i]$ for the sample values $x_1, \ldots, x_n$. In this situation, we want to compute the ranges of possible values of the population mean $\mu = \frac{1}{n} \sum_{i=1}^n x_i$, population variance $V = \frac{1}{n} \sum_{i=1}^n (x_i - \mu)^2$, etc.

It turns out that most such problems are, in general, computationally difficult (to be more precise, NP-hard). Even computing the range $[\underline{V}, \overline{V}]$ of the population variance V is an NP-hard problem [16, 17]. In many practical situations, there exist feasible algorithms that compute the bounds of desirable statistical characteristics [4, 5, 11, 16, 17, 18, 22, 28, 40, 45, 49, 51]. For example, there exist efficient algorithms for computing $\underline{V}$ and efficient algorithms for computing $\overline{V}$ for several reasonable situations (e.g., when measurements are sufficiently accurate); efficient algorithms are also known for detecting outliers [12, 27].

An important issue is whether we can perform these computations on-line, updated the statistical characteristics as new measurements appear [29, 48].

When efficient algorithms are not known, we can use parallelization and quantum computing to speed up computation [26]. In many practical situations, there are still important open problems.

Computing amount of information. Another important problem is estimating amount of information, i.e., entropy. The traditional Shannon's definition described amount of information as the average number of "yes"-"no" questions that we need to ask to find the actual value (with given accuracy).

When we have finitely many alternatives, and we know the probabilities $p_1, \ldots, p_n$ of these alternatives, then this average number of "yes"-"no" questions is described by Shannon's entropy formula $S = -\sum_{i=1}^n p_i \cdot \log(p_i)$. In practice, we only have partial information about the probabilities, e.g., only intervals $[\underline{p}_i, \overline{p}_i]$ of possible values of p_i. Different values $p_i \in \mathbf{p}_i$ lead, in general, to different values S, so it is desirable to compute the range $[\underline{S}, \overline{S}]$ of possible values of S [20].

Since entropy S is a concave function, standard feasible algorithms for minimizing convex functions (= maximizing concave ones) enable us to compute $\overline{S}$; see, e.g., [20, 21]. Computing $\underline{S}$ is, in general, NP-hard [50], but for reasonable cases, feasible algorithms are possible [1, 2, 3, 31, 50].

Decision making. Computational aspects of decision making under interval and probabilistic uncertainty are discussed, e.g., in [52].

Acknowledgments. This work was supported in part by NASA under cooperative agreement NCC5-209, NSF grants EAR-0225670 and DMS-0532645, Star Award from the University of Texas System, and Texas Department of Transportation grant No. 0-5453.

References

[1] J. Abellan and S. Moral, Range of entropy for credal sets, In: M. López-Diaz et al. (eds.), *Soft Methodology and Random Information Systems*, Springer, Berlin and Heidelberg, 2004, pp. 157–164.

[2] J. Abellan and S. Moral, Difference of entropies as a nonspecificity function on credal sets, *Intern. J. of General Systems*, 2005, Vol. 34, No. 3, pp. 201–214.

[3] J. Abellan and S. Moral, An algorithm that attains the maximum of entropy for order-2 capacities, *Intern. J. of Uncertainty, Fuzziness, and Knowledge-Based Systems* (to appear).

[4] R. Aló, M. Beheshti, and G. Xiang, Computing Variance under Interval Uncertainty: A New Algorithm and Its Potential Application to Privacy in Statistical Databases, *Proceedings of the International Conference on Information Processing and Management of Uncertainty in Knowledge-Based Systems IPMU'06*, Paris, France, July 2–7, 2006 (to appear).

[5] J. B. Beck, V. Kreinovich, and B. Wu, Interval-Valued and Fuzzy-Valued Random Variables: From Computing Sample Variances to Computing Sample Covariances, In: M. Lopez, M. A. Gil, P. Grzegorzewski, O. Hrynewicz, and J. Lawry (eds.), *Soft Methodology and Random Information Systems*, Springer-Verlag, 2004, pp. 85–92.

[6] D. Berleant, M.-P. Cheong, C. Chu, Y. Guan, A. Kamal, G. Sheblé, S. Ferson, and J. F. Peters, Dependable handling of uncertainty, *Reliable Computing* 9(6) (2003), pp. 407–418.

[7] D. Berleant, L. Xie, and J. Zhang, Statool: a tool for Distribution Envelope Determination (DEnv), an interval-based algorithm for arithmetic on random variables, *Reliable Computing* 9(2) (2003), pp. 91–108.

[8] D. Berleant and J. Zhang, Using Pearson correlation to improve envelopes around the distributions of functions, *Reliable Computing*, 10(2) (2004), pp. 139–161.

[9] D. Berleant and J. Zhang, Representation and Problem Solving with the Distribution Envelope Determination (DEnv) Method, *Reliability Engineering and System Safety*, 85 (1–3) (July-Sept. 2004).

[10] D. Berleant and J. Zhang, Using Pearson correlation to improve envelopes around the distributions of functions, *Reliable Computing*, 10(2) (2004), pp. 139–161.

[11] E. Dantsin, V. Kreinovich, A. Wolpert, and G. Xiang, Population Variance under Interval Uncertainty: A New Algorithm, *Reliable Computing*, 2006, Vol. 12, No. 4, pp. 273–280.

[12] E. Dantsin, A. Wolpert, M. Ceberio, G. Xiang, and V. Kreinovich, Detecting Outliers under Interval Uncertainty: A New Algorithm Based on Constraint Satisfaction, *Proceedings of the International Conference on Information Processing and Management of Uncertainty in Knowledge-Based Systems IPMU'06*, Paris, France, July 2–7, 2006 (to appear).

[13] S. Ferson, L. Ginzburg, and R. Akcakaya, *Whereof One Cannot Speak: When Input Distributions Are Unknown*, Applied Biomathematics Report, 2001.

[14] S. Ferson, J. Hajagos, D. Berleant, J. Zhang, W. T. Tucker, L. Ginzburg, and W. Oberkampf, *Dependence in Dempster-Shafer Theory and Probability Bounds Analysis*, Technical Report SAND2004-3072, Sandia National Laboratory, 2004.

[15] S. Ferson, V. Kreinovich, L. Ginzburg, D. S. Myers, and K. Sentz, *Constructing Probability Boxes and Dempster-Shafer Structures*, Sandia National Laboratories, Report SAND2002-4015, January 2003.

[16] S. Ferson, L. Ginzburg, V. Kreinovich, L. Longpre, and M. Aviles, Exact Bounds on Finite Populations of Interval Data, *Reliable Computing*, 2005, Vol. 11, No. 3, pp. 207–233.

[17] S. Ferson, L. Ginzburg, V. Kreinovich, L. Longpré, and M. Aviles, Computing Variance for Interval Data is NP-Hard, *ACM SIGACT News*, 2002, Vol. 33, No. 2, pp. 108–118.

[18] L. Granvilliers, V. Kreinovich, and N. Mueller, Novel Approaches to Numerical Software with Result Verification, In: R. Alt, A. Frommer, R. B. Kearfott, and W. Luther (eds.), *Numerical Software with Result Verification*, International Dagstuhl Seminar, Dagstuhl Castle, Germany, January 19–24, 2003, Revised Papers, Springer Lectures Notes in Computer Science, 2004, Vol. 2991, pp. 274–305.

[19] L. Jaulin, M. Kieffer, O. Didrit, and E. Walter, *Applied interval analysis: with examples in parameter and state estimation, robust control and robotics*, Springer Verlag, London, 2001.

[20] G. J. Klir, *Uncertainty and Information: Foundations of Generalized Information Theory*, J. Wiley, Hoboken, New Jersey, 2005.

[21] V. Kreinovich, Maximum entropy and interval computations, *Reliable Computing*, 1996, Vol. 2, No. 1, pp. 63–79.

[22] V. Kreinovich, Probabilities, Intervals, What Next? Optimization Problems Related to Extension of Interval Computations to Situations with Partial Information about Probabilities, *Journal of Global Optimization*, 2004, Vol. 29, No. 3, pp. 265–280.

[23] V. Kreinovich and S. Ferson, Computing Best-Possible Bounds for the Distribution of a Sum of Several Variables is NP-Hard, *International Journal of Approximate Reasoning*, 2006, Vol. 41, pp. 331–342.

[24] V. Kreinovich, S. Ferson, and L. Ginzburg, Exact Upper Bound on the Mean of the Product of Many Random Variables With Known Expectations, *Reliable Computing* (to appear).

[25] V. Kreinovich, A. Lakeyev, J. Rohn, P. Kahl, *Computational complexity and feasibility of data processing and interval computations*, Kluwer, Dordrecht, 1997.

[26] V. Kreinovich and L. Longpre, Fast Quantum Algorithms for Handling Probabilistic and Interval Uncertainty, *Mathematical Logic Quarterly*, 2004, Vol. 50, No. 4/5, pp. 507–518.

[27] V. Kreinovich, L. Longpre, P. Patangay, S. Ferson, and L. Ginzburg, Outlier Detection Under Interval Uncertainty: Algorithmic Solvability and Computational Complexity, *Reliable Computing*, 2005, Vol. 11, No. 1, pp. 59–75.

[28] V. Kreinovich, L. Longpré, S. A. Starks, G. Xiang, J. Beck, R. Kandathi, A. Nayak, S. Ferson, and J. Hajagos, Interval Versions of Statistical Techniques, with Applications to Environmental Analysis, Bioinformatics, and Privacy in Statistical Databases", *Journal of Computational and Applied Mathematics* (to appear).

[29] V. Kreinovich, H. T. Nguyen, and B. Wu, On-Line Algorithms for Computing Mean and Variance of Interval Data, and Their Use in Intelligent Systems, *Information Sciences* (to appear).

[30] V. Kreinovich, G. N. Solopchenko, S. Ferson, L. Ginzburg, and R. Alo, Probabilities, intervals, what next? Extension of interval computations to situations with partial information about probabilities, *Proceedings of the 10th IMEKO TC7 International Symposium on Advances of Measurement Science*, St. Petersburg, Russia, June 30–July 2, 2004, Vol. 1, pp. 137–142.

[31] V. Kreinovich, G. Xiang, and S. Ferson, How the Concept of Information as Average Number of 'Yes-No' Questions (Bits) Can Be Extended to Intervals, P-Boxes, and more General Uncertainty, *Proceedings of the 24th International Conference of the North American Fuzzy Information Processing Society NAFIPS'2005*, Ann Arbor, Michigan, June 22–25, 2005, pp. 80–85.

[32] V. Kuznetsov, *Interval Statistical Models* (in Russian), Radio i Svyaz, Moscow, 1991.

[33] W. A. Lodwick and K. D. Jamison, Estimating and validating the cumulative distribution of a function of random variables: toward the development of distribution arithmetic, *Reliable Computing*, 2003, Vol. 9, No. 2, pp. 127–141.

[34] W. A. Lodwick, A. Neumaier, and F. Newman, Optimization under uncertainity: methods and applications in radiation therapy, *Proc. 10th IEEE Int. Conf. Fuzzy Systems*, December 2–5, 2001, Melbourne, Australia.

[35] C. Manski, *Partial Identification of Probability Distributions*, Springer-Verlag, New York, 2003.

[36] A. S. Moore, Interval risk analysis of real estate investment: a non-Monte-Carlo approach, *Freiburger Intervall-Berichte* 85/3, Inst. F. Angew. Math., Universitaet Freiburg I. Br., 23–49 (1985).

[37] R. E. Moore, *Automatic error analysis in digital computation*, Technical Report Space Div. Report LMSD84821, Lockheed Missiles and Space Co., 1959.

[38] R. E. Moore, Risk analysis without Monte Carlo methods, *Freiburger Intervall-Berichte* 84/1, Inst. F. Angew. Math., Universitaet Freiburg I. Br., 1–48 (1984).

[39] A. Neumaier, Fuzzy modeling in terms of surprise, *Fuzzy Sets and Systems* 135, 2003, 21–38.

[40] H. T. Nguyen, V. Kreinovich, and G. Xiang, Foundations of Statistical Processing of Set-Valued Data: Towards Efficient Algorithms, *Proceedings of the Fifth International Conference on Intelligent Technologies InTech'04*, Houston, Texas, December 2–4, 2004.

[41] M. Orshansky, W.-S. Wang, M. Ceberio, and G. Xiang, Interval-Based Robust Statistical Techniques for Non-Negative Convex Functions, with Application to Timing Analysis of Computer Chips, *Proceedings of the Symposium on Applied Computing SAC'06*, Dijon, France, April 23–27, 2006, pp. 1645–1649.

[42] M. Orshansky, W.-S. Wang, G. Xiang, and V. Kreinovich, Interval-Based Robust Statistical Techniques for Non-Negative Convex Functions, with Application to Timing Analysis of Computer Chips, *Proceedings of the Second International Workshop on Reliable Engineering Computing*, Savannah, Georgia, February 22–24, 2006, pp. 197–212.

[43] H. Regan, S. Ferson and D. Berleant, Equivalence of methods for uncertainty propagation of real-valued random variables, *International Journal of Approximate Reasoning*, in press.

[44] H.-P. Schröcker and J. Wallner, Geometric constructions with discretized random variables, *Reliable Computing*, 2006, Vol. 12, No. 3, pp. 203–223.

[45] S. A. Starks, V. Kreinovich, L. Longpre, M. Ceberio, G. Xiang, R. Araiza, J. Beck, R. Kandathi, A. Nayak, and R. Torres, Towards combining probabilistic and interval uncertainty in engineering calculations, *Proceedings of the Workshop on Reliable Engineering Computing*, Savannah, Georgia, September 15–17, 2004, pp. 193–213.

[46] W. T. Tucker and S. Ferson, *Probability Bounds Analysis in Environmental Risk Assessments*, Applied Biomathematics Report.

[47] P. Walley, *Statistical Reasoning with Imprecise Probabilities*, Chapman & Hall, N.Y., 1991.

[48] B. Wu, H. T. Nguyen, and V. Kreinovich, Real-Time Algorithms for Statistical Analysis of Interval Data, *Proceedings of the International Conference on Information Technology InTech'03*, Chiang Mai, Thailand, December 17–19, 2003, pp. 483–490.

[49] G. Xiang, Fast algorithm for computing the upper endpoint of sample variance for interval data: case of sufficiently accurate measurements, *Reliable Computing*, 2006, Vol. 12, No. 1, pp. 59–64.

[50] G. Xiang, O. Kosheleva, and G. J. Klir, Estimating Information Amount under Interval Uncertainty: Algorithmic Solvability and Computational Complex-

ity, *Proceedings of the International Conference on Information Processing and Management of Uncertainty in Knowledge-Based Systems IPMU'06*, Paris, France, July 2–7, 2006 (to appear).

[51] G. Xiang, S. A. Starks, V. Kreinovich, and L. Longpre, New Algorithms for Statistical Analysis of Interval Data, *Proceedings of the Workshop on State-of-the-Art in Scientific Computing PARA'04*, Lyngby, Denmark, June 20–23, 2004, Vol. 1, pp. 123–129.

[52] R. R. Yager and V. Kreinovich, Decision Making Under Interval Probabilities, *International Journal of Approximate Reasoning*, 1999, Vol. 22, No. 3, pp. 195–215.

[53] J. Zhang and D. Berleant, Envelopes around cumulative distribution functions from interval parameters of standard continuous distributions, *Proceedings, North American Fuzzy Information Processing Society (NAFIPS 2003)*, Chicago, pp. 407–412.

Part II

Soft Methods in Statistics and Random Information Systems

On Testing Fuzzy Independence

Olgierd Hryniewicz

Systems Research Institute, Newelska 6, 01-447 Warsaw, POLAND
hryniewi@ibspan.waw.pl

1 Introduction

Statistical analysis of dependencies existing in data sets is now one of the most important applications of statistics. It is also a core part of data mining - a rapidly developing in recent years part of information technology. Statistical methods that have been proposed for the analysis of dependencies in data sets can be roughly divided into two groups: tests of statistical independence and statistical measures of the strength of dependence.

Numerous statistical test of independence have been developed during the last one hundred (or even more) years. They have been developed for many parametric (like the test of independence for normally distributed data, based on the Pearson coefficient of correlation ρ) and non-parametric (like the test of independence based on the Spearman rank correlation statistic ρ_S) models. The relative ease of developing such tests stems from the fact that statistical independence is a very peculiar feature of data sets. In the case of independence, probability distributions that describe multivariate statistical data depend exclusively on the marginal probability distributions of separate components of vectors of random variables. This feature can exist unconditionally (as it is usually assumed in statistical analysis) or conditionally (when a value of a certain latent variable that influences the random variables of interest can be regarded as fixed for the analyzed data set). Despite the fact that independence can be rather frequently observed in carefully performed statistical experiments we are of the opinion that in case of real large data sets a perfect statistical independence exists rather seldom. On the other hand, however, the acceptance of the assumption of independence is sometimes necessary for, e.g., computational reasons. Therefore, there is often a practical need to soften the independence requirements by defining the state of "near-independence". The question arises then, how to evaluate this state using statistical data.

The concept of "near-independence" is definitely a vague one. In contrast to the case of independence, that is very precisely defined in terms of the theory of probability, it seems to be fundamentally impossible to define one measure of the strength

O. Hryniewicz: *On Testing Fuzzy Independence*, Advances in Soft Computing **6**, 29–36 (2006)
www.springerlink.com

dependence that could be used for the unique measurement of the deviation from independence. Therefore, there exist different measures of the strength of dependence which - depending on the context - may be used for the evaluation of the state of "near - independence". We claim that these measures might be used for the analysis of dependence when the state of independence is defined, using Zadeh's terminology, "to a degree". Suppose that there exist a certain measure of the strength of dependence α which in the case of independence adopts the value α_0. For this particular value the independence is definitely to a degree one. However, if we know that $0 < |\alpha - \alpha_0| \leq \varepsilon, \varepsilon > 0$ we can talk about the independence to a degree μ_ε depending on the value of ε and a given practical context. For example, the value of the Pearson correlation coefficient equal to 0,05 may indicate "near - independence" to a degree of 0,5, but this value equal to 0,1 may already indicate dependence ("near - independence to a degree of 0). This example shows that the concept of "near - independence" can be formally described using fuzzy sets. In Section 2 we present a general framework for dealing with this problem. We propose to use the concept of a statistical test of a fuzzy hypothesis for testing "near - independence".

In order to obtain useful statistical procedures for testing fuzzy "near - independence" we need statistical methods for the evaluation of statistical confidence intervals for the measures of dependence. These methods are not so frequently presented in statistical textbooks. Therefore in Section 3 we present some interesting results that have been published recently for the case when dependence structure is captured by some well known Archimedean copulas. In Section 4 we propose an alternative approach for testing "near - independence" using the Kendall τ statistic. The conclusions are presented in Section 5 of the paper.

2 Statistical Tests of Fuzzy Independence – General Approach

In classical statistics tests of independence are usually formulated as tests of a null hypothesis $H_0 : \alpha = \alpha_0$ against the alternative $H_1 : \alpha \neq \alpha_0$, where α is a parameter of the test statistic that for the case of independence adopts the value of α_0. For example, in the most frequently used model for the analysis of dependent statistical data it is assumed that data are modeled by a multivariate normal distribution characterized by the correlation coefficient ρ. In case of independence ρ is equal to zero, so the parametric test of independence is in this case equivalent to the test of the null hypothesis $H_0 : \rho = 0$ against the alternative $H_1 : \rho \neq 0$.

The construction of classical tests for independence is very simple if for a given value of the confidence level β there exist closed formulae for the confidence interval $(\hat{\alpha}_L(\beta), \hat{\alpha}_U(\beta))$ of the test statistic α that measures the strength of dependence. On the significance level $1 - \beta$ we reject the null hypothesis of independence if $\alpha_0 \notin (\hat{\alpha}_L(\beta), \hat{\alpha}_U(\beta))$.

Now, let us consider the case when we are not interested in the precisely defined independence (characterized by the precise value α_0 of the dependence parameter α, but in "near-independence" characterized by its fuzzy equivalent $\tilde{\alpha}_0$ described by the membership function $\mu(\alpha)$. The value of $\mu(\alpha)$ tells us to what degree we consider our data as independent.

Statistical tests of fuzzy hypotheses of a general form $H : \theta = \widetilde{\theta}_0$ have been considered by many authors (see , for example, the papers by Arnold [1], Römer and Kandel [11]). In this paper we apply the definition of a statistical test of a fuzzy hypothesis that was proposed by Grzegorzewski and Hryniewicz [6].

Suppose that we consider a null hypothesis $H : \theta \in \Theta_H$ against an alternative hypothesis $K : \theta \in \Theta_K = \neg\Theta_H$, where Θ_H and Θ_K are fuzzy subsets of the parameter space Θ, with membership functions $\mu_H : \Theta \to [0,1]$ and $\mu_K(x) = 1 - \mu_H(x)$, respectively. We'll transform such fuzzily formulated problem into a family of the crisp testing problems. If $\mathrm{supp}\mu_H$ is bounded, as in the case of testing fuzzy independence, then the problem of testing $H : \theta \in \Theta_H$ against $K : \theta \in \Theta_K = \neg\Theta_H$ can be transformed to a following family of the crisp testing problems

$$\left\{ H_{\theta_0} : \theta = \theta_0 \quad \text{vs.} \quad K_{\theta_0} : \theta \neq \theta_0, \text{where } \theta_0 \in \mathrm{supp}\mu_H \right\}; \tag{1}$$

Now let $\left\{ \varphi_{\theta_0} : \mathscr{X} \to \{0,1\}, \text{where } \theta_0 \in \mathrm{supp}\mu_H \right\}$ denotes a family of classical statistical tests on significance level δ for verifying crisp hypotheses (1). Then we get a following definition [6]:

Definition 1. *A function* $\psi : \mathscr{X} \to \mathscr{F}(\{0,1\})$ *such that*

$$\mu_\psi(0) = \begin{cases} \sup\limits_{\theta_0 \in supp\mu_H \,:\, \varphi_{\theta_0}(x)=0} \mu_H(\theta_0) & \text{if } \left\{ \theta_0 \in supp\mu_H : \atop \varphi_{\theta_0}(x) = 0 \right\} \neq \emptyset \\[2em] 0 & \text{if } \left\{ \theta_0 \in supp\mu_H : \atop \varphi_{\theta_0}(x) = 0 \right\} = \emptyset \end{cases} \tag{2}$$

$$\mu_\psi(1) = 1 - \mu_\psi(0) \tag{3}$$

is called a fuzzy test for verifying fuzzy hypotheses $H : \theta \in \Theta_H$ *against* $K : \theta \in \Theta_K = \neg\Theta_H$ *on significance level* δ.

Therefore our fuzzy test for fuzzy hypotheses does not always lead to binary decisions – to accept or to reject the null hypothesis – but to a fuzzy decision: $\psi = \xi/0 + (1-\xi)/1$, where $\xi \in (0,1)$, which can be interpreted as a degree of conviction that we should accept (ξ) or reject $(1-\xi)$ the hypothesis H. Hence, in the considered case of testing fuzzy "near-independence" it is sufficient to know confidence intervals for a measure of dependence α, and then to use Definition 1 in order to build an appropriate test for testing this hypothesis.

3 Parametric Methods for Measuring the Strength of Dependence in Copulas

Mathematical models used for the description of dependent random variables are well known for many years. Multivariate normal distribution, mentioned in the previous section, is the most popular one. However, it cannot be applied when the marginal distributions are not normal, as it often is the case, especially in the area of reliability and survival models. Moreover, in certain cases multivariate data are not described by the multivariate normal distribution despite the fact that all marginals are still normal. Therefore, statisticians have tried to find general methods for the construction of multivariate probability distributions using the information about marginals and the structure of dependence arriving at the notion of a copula.

Let $(X_1, \ldots, X_p)$ be a p-dimensional vector of random variables having continuous marginals $F_i(x_i), i = 1, \ldots, p$. The joint probability distribution of $(X_1, \ldots, X_p)$ is uniquely defined by its associated copula, defined for all $(u_1, \ldots, u_p) \in [0, 1]^p$ by

$$C(u_1, \ldots, u_p) = Pr(F_1(X_1) \le u_1, \ldots, F_p(X_p) \le u_p). \tag{4}$$

It can be shown that many well known multivariate probability distributions can be generated by parametric families C_α of copulas, where real- or vector-valued parameter α describes the strength of dependence between the components of the random vector. Pioneering works in this area can be found, for example, in papers by Gumbel [7], Clayton [3], Genest and McKay [4], and Marshall and Olkin [8]. The number of papers devoted to the theory and applications of copulas in multivariate statistics that have been published during last twenty years is huge, as copulas have found many applications in survival analysis, analysis of financial risks, and many other areas. For more recent results the reader should consult the book by Nelsen [9].

One of the most interesting, and often used in practice, classes of copulas is a class of symmetric copulas, named the Archimedean copulas. They are generated using a class Φ of functions $\varphi : [0, 1] \rightarrow [0, \infty]$, named generators, that have two continuous derivatives on $(0, 1)$ and fulfill the following conditions: $\varphi(1) = 1$, $\varphi'(t) < 0$, and $\varphi''(t) > 0$ for all $0 < t < 1$ (these conditions guarantee that φ has an inverse φ^{-1} that also has two derivatives). Every member of this class generates the following multivariate distribution function for the random vector $(X_1, \ldots, X_p)$:

$$C(u_1, \ldots, u_p) = Pr(F_1(X_1) \le u_1, \ldots, F_p(X_p) \le u_p) = \varphi^{-1}[\varphi(u_1) + \cdots + \varphi(u_p)] \tag{5}$$

The two-dimensional Archimedean copulas that are most frequently used in practice are defined by the following formulae (copulas and their respective generators):

- Clayton's

$$C(u, v) = \max\left(\left[u^{-\alpha} + v^{-\alpha} - 1\right]^{-1/\alpha}, 0\right), \alpha \in [-1, \infty) \setminus 0 \tag{6}$$

$$\varphi(t) = (t^{-\alpha} - 1)/\alpha, \alpha \in [-1, \infty) \setminus 0 \tag{7}$$

- Frank's

$$C(u, v) = -\frac{1}{\alpha} \ln\left(1 + \frac{(e^{-\alpha u} - 1)(e^{-\alpha v} - 1)}{e^{-\alpha} - 1}\right), \alpha \in (-\infty, \infty) \setminus 0 \tag{8}$$

$$\varphi(t) = \ln\left(\frac{1 - e^{-\alpha}}{1 - e^{-\alpha t}}\right), \alpha \in (-\infty, \infty) \setminus 0 \tag{9}$$

- Gumbel's

$$C(u,v) = \exp\left(-\left[(-\ln u)^{1+\alpha} + (-\ln v)^{1+\alpha}\right]^{\frac{1}{1+\alpha}}\right), \alpha \in (0, \infty) \qquad (10)$$

$$\varphi(t) = (-\ln(t))^{\alpha+1}, \alpha \in (0, \infty) \qquad (11)$$

In case of independence the dependence parameter α_{ind} adopts the value of 0 (in Clayton's and Frank's copulas as an appropriate limit). The copulas mentioned above are sometimes presented using different parametrization, and in such cases independence is equivalent to other values of α.

In order to test fuzzy independence defined in Section 2 it is necessary to know interval estimates of α, i.e. to know their confidence intervals. Unfortunately, we do not have any simple formulae that can be useful for this purpose for all sample sizes. We can only use asymptotic distributions of the estimators of α that depend upon the method of estimation, and whose properties for smaller sample sizes can be evaluated only by Monte Carlo methods.

In case of the known functional form of the marginal distributions $F_i(x_i), i = 1, \ldots, p$ for the estimation of α we can use the general methodology of maximal likelihood. If the sample size is sufficiently large all estimated parameters of the copula have a joint multivariate normal distribution with a covariance matrix obtained by the estimation of Fisher's information matrix. In case of general bivariate copulas the appropriate formulae are given by Shih and Louis [12]. Significant simplification of required computations can be achieved by using a two-stage parametric method. In this method at the first stage we assume independence and using the maximum likelihood method separately estimate the parameters of the marginal distribution. Then, we substitute the parameters of the marginals with their respective estimated values, and solve the maximum likelihood equation for the dependence parameter. The general formulae for the bivariate case are also given in Shih and Louis [12].

The parametric methodology has, according to some authors, some serious limitations. First, as we have already mentioned, it requires the knowledge of the class of marginal probability distributions. Second, the estimator of α depends on the estimators of other parameters, and this - according to these authors - may distort the information about the dependence structure. In order to avoid these problems a semi-parametric maximum likelihood method has been proposed by Shih and Louis [12], who proposed a semi-parametric two-stage procedure. At the first stage marginal distributions are estimated using the Kaplan-Meyer estimator, and at the second stage the dependence parameter is estimated using the maximum likelihood method. The formulae for the ML equations, and the asymptotic variances that are sufficient for the construction of the confidence interval of α are given in [12].

4 Testing of Fuzzy Independence Using Kendall τ Statistic

A fundamental for the analysis of copulas result was obtained by Genest and Rivest [5] who proved the equivalence between an Archimedean copula $C(X,Y)$

and the random variable V distributed as $K(v) = v - \varphi(v)/\varphi'(v), v \in (0,1)$. The similar equivalence between $K(v)$ and the probability distribution of a multivariate Archimedean copula was proved by Barbe et al. [2]. Genest and Rivest [5] noticed that for the Archimedean copulas the following relation holds for the population version of a measure of association known as Kendall's τ

$$\tau = 4E(V) - 1 = 4 \int_0^1 \frac{\varphi(v)}{\varphi'(v)} dv \tag{12}$$

From this equation we can find the following formulae for the expression of Kendall's τ in terms of the dependence parameter α [5]:

- Clayton's copula

$$\tau = \frac{\alpha}{\alpha + 2} \tag{13}$$

- Frank's copula

$$\tau = 1 + 4 \left(\frac{1}{\alpha} \int_0^\alpha \frac{t}{e^t - 1} dt - 1 \right) / \alpha \tag{14}$$

- Gumbel's copula

$$\tau = \frac{\alpha}{\alpha + 1} \tag{15}$$

In order to estimate Kendall's τ from bivariate data Genest and Rivest [5] introduced the following random variables

$$V_i = card\{(X_j, Y_j) : X_j < X_i, Y_j < Y_i\}/(n-1), i = 1, \ldots, n \tag{16}$$

Then, they showed that the sample version of Kendall's τ can be calculated from a very simple formula

$$\tau_n = 4\bar{V} - 1 \tag{17}$$

Similarly, if

$$W_i = card\{(X_j, Y_j) : X_j > X_i, Y_j > Y_i\}/(n-1), i = 1, \ldots, n, \tag{18}$$

then $\tau_n = 4\bar{W} - 1$. Moreover, Genest and Rivest [5] proved that if

$$S^2 = \sum_{i=1}^n (V_i + W_i - 2\bar{V})^2/(n-1), \tag{19}$$

then the large-sample distribution of $\sqrt{n}(\tau_n - \tau)/4S$ is standard normal. Thus, we can easily construct the confidence interval for the observed value of Kendall's τ.

Now, let us consider the main technical problem of this paper: verification of the hypothesis that the dependence parameter adopts values close (in a fuzzy sense) to zero. As it has been pointed out in Section 2 this can be achieved by the construction of the confidence interval for that parameter or for another parameter that is explicitly related to the dependence parameter. In the case of Gumbel's and Frank's copulas the solution is straightforward. We can easily express Kendall's τ as a function of the dependence parameter α, and thus to reformulate the fuzzy requirement on α as an appropriate fuzzy requirement on τ. It is worthwhile to note that in the case of nearly-independence, i.e. when α is close to zero, we have very simple relations: $\tau \approx \alpha$ for Gumbel's copula, and $\tau \approx \alpha/2$ for Clayton's copula. The relationship between Kendall's τ and the dependence parameter α in the case of Frank's copula requires individual analysis that is presented in the next paragraph.

The crucial point in the investigation of the relationship between the dependence parameter α and Kendall's τ in the case of Frank's copula is the evaluation of the integral in (14). Unfortunately, this integral cannot be expressed in a closed form. Prudnikov et al. [10] propose the following expansion of this integral for small values of the parameter α ($\alpha < \ln 2$):

$$\int_0^\alpha \frac{x}{e^x - 1} dx = \sum_{k=1}^\infty \frac{1}{k} (1 - e^\alpha)^k [\psi(1) - \psi(1+k)], \alpha < \ln 2 \qquad (20)$$

where $\psi(x)$ is the Euler's psi function that can be expressed by the following formula

$$\psi(x) = -0,577215.. - \sum_{k=0}^\infty \left(\frac{1}{k+x} - \frac{1}{k+1} \right) \qquad (21)$$

In the interesting us case of small values of α we can expand (20) in the Taylor series around zero arriving at a very simple relationship

$$\tau \approx \frac{\alpha}{9} \qquad (22)$$

Numerical investigations have shown that this approximate formula is very accurate (error smaller than 1%) for $\alpha < 0.9$ (i.e. for $\tau < 0.1$), and sufficiently good (error smaller than 5%) for $\alpha < 1.8$ (i.e. for $\tau < 0.2$). For example, for $\alpha = 0.9$ the exact value of τ is equal to 0.0992, and the approximate value is equal to 0.1. If $\alpha = 1.8$, then the exact value of τ is equal to 0.1939, and the approximate value is equal to 0.2.

The presented above analysis of the properties of Clayton's, Frank's, and Gumbel's copulas shows that in the case of testing fuzzy near-independence we can always use a very simple statistic such as Kendall's τ. Thus, the statistical test of fuzzy independence can be formulated as Kendall's test of the fuzzy hypothesis that τ adopts the value of fuzzy zero.

5 Conclusions

In the paper we have considered the case when we are interested in the statistical test of the hypothesis that statistical data are independent to certain degree. This requirement can be formally represented as the statistical test of a fuzzy hypothesis that the dependence parameter α is equal to a given fuzzy value $\tilde{\alpha}_0$. We have shown that in the case of the popular models of dependence such as certain Archimedean copulas (Clayton's, Frank's, and Gumbel's) a well known Kendall's τ statistic may be used for the construction of simple fuzzy tests of fuzzy independence.

References

[1] Arnold B. F. (1998), Testing fuzzy hypotheses with crisp data, *Fuzzy Sets and Systems* **94**, 323–333.

[2] Barbe P., Genest C., Ghoudi K., Rémillard B. (1996), On Kendall's Process. *Journal of Multivariate Analysis* **58**, 197–229.

[3] Clayton G.G. (1978), A Model for Association in Bivariate Life Tables and its Applications in Epidemiological Studies of Familial Tendency in Chronic Disease Incidence. *Biometrika* **65**, 141–151.

[4] Genest C., McKay R.J. (1986), Copules archimédiennes et familles de loi bidimensionnelles dont les marges sont données. *The Canadian Journal of Statistics* **14**, 145–159.

[5] Genest C., Rivest L-P. (1993), Statistical Inference Procedures for Bivariate Archimedean Copulas. *Journal of the American Statistical Association* **88**, 1034–1043.

[6] Grzegorzewski P., Hryniewicz O. (2001), Soft Methods in Hypotheses Testing. In: D. Ruan, J. Kacprzyk and M. Fedrizzi (Eds.): Soft computing for risk evaluation and management. Physica Verlag, Heidelberg and New York 2001, 55–72.

[7] Gumbel E.J. (1960), Distribution des valeurs extremês en plusieurs dimensions. *Publications de l'Institut de statistique de l'Universitè de Paris* **9**, 171–173.

[8] Marshall A.W., Olkin I. (1988), Families of Multivariate Distributions. *Journal of the American Statistical Association* **83**, 834–841.

[9] Nelsen R.B. (1999), An Introduction to Copulas. Springer-Verlag, New York.

[10] Prudnikov A.P., Bychkow Y.A., Marichev O.I. (1981), Integrals and Series (in Russian). Nauka, Moscow.

[11] Römer Ch., Kandel A., Statistical tests for fuzzy data, *Fuzzy Sets and Systems*, vol.72 (1995), 1–26.

[12] Shih J.H., Louis T.A. (1995), Inferences on the Association Parameter in Copula Models for Bivariate Survival Data. *Biometrics* **51**, 1384–1399.

Variance Decomposition of Fuzzy Random Variables

Andreas Wünsche[1] and Wolfgang Näther[2]

[1] TU Bergakademie Freiberg, Institut für Stochastik , 09596 Freiberg, Prüferstr. 9
   wuensche@math.tu-freiberg.de)
[2] TU Bergakademie Freiberg, Institut für Stochastik , 09596 Freiberg, Prüferstr. 9
   naether@math.tu-freiberg.de)

Summary. The conditional variance of random variables plays an important role for well-known variance decomposition formulas. In this paper, the conditional variance for fuzzy random variables and some properties of it are considered. Moreover possible applications of the variance decomposition formula are presented.

Key words: Fuzzy random variable, conditional variance, variance decomposition

1 Introduction

Conditional expectation and conditional variance play an important role in probability theory. Let be X a random variable on the probability space $(\Omega, \mathfrak{F}, P)$ and $\mathfrak{A} \subseteq \mathfrak{F}$ a sub-σ-algebra of $\mathfrak{F}$. Then the conditional expectation $\mathbf{E}(X|\mathfrak{A})$, for example, is the best mean squared approximation (best prediction) of X by a more rough, i.e. only $\mathfrak{A}$-measurable function.

Conditioning is one of the principles of variance reduction, i.e. the "more rough" random variable $\mathbf{E}(X|\mathfrak{A})$ has a smaller variance than X,

$$\mathbf{Var}(\mathbf{E}(X|\mathfrak{A})) \leq \mathbf{Var}X.$$

The difference $\mathbf{Var}X - \mathbf{Var}(\mathbf{E}(X|\mathfrak{A}))$ can be expressed mainly by the conditional variance of X which is defined by

$$\mathbf{Var}(X|\mathfrak{A}) = \mathbf{E}((X - \mathbf{E}(X|\mathfrak{A}))^2|\mathfrak{A}) \tag{1}$$

and which leads to the well known formula of variance decomposition

$$\mathbf{Var}X = \mathbf{E}(\mathbf{Var}(X|\mathfrak{A})) + \mathbf{Var}(\mathbf{E}(X|\mathfrak{A})). \tag{2}$$

This formula plays an important role in applications (see section 4).

Very often we meet the situation where the random variables X has only fuzzy outcomes. E.g. if an insurance company is interested in the claim sum X of the next

A. Wünsche and W. Näther: *Variance Decomposition of Fuzzy Random Variables*, Advances in Soft Computing **6**, 37–44 (2006)
www.springerlink.com

year, it would be wise to assume fuzzy claims. Or if we interested in the state X of health in a country the society of which is stratified e.g. wrt age or wrt to social groups then it seems to be a violation to restrict X on numbers. Linguistic expressions, however, lead more or less straightforward to a fuzzy valued X.

In section 2 we introduce necessary tools like frv's and their expectation and variance. In section 3 the conditional variance of a frv and some of its properties are investigated and in section 4 we discusses possible applications of the variance decomposition formula.

2 Preliminaries

A fuzzy subset $\widetilde{A}$ of $\mathbb{R}^n$ is characterized by its membership function $\mu_{\widetilde{A}} : \mathbb{R}^n \to [0,1]$ where $\mu_{\widetilde{A}}(x)$ is interpreted as the degree to which $x \in \mathbb{R}^n$ belongs to $\widetilde{A}$. The α-cuts of $\widetilde{A}$ for $0 < \alpha \leq 1$ are crisp sets and given by $\widetilde{A}_\alpha := \{x \in \mathbb{R}^n : \mu_{\widetilde{A}}(x) \geq \alpha\}$. Additionally, we call $\widetilde{A}_0 := \mathrm{cl}\{x \in \mathbb{R}^n : \mu_{\widetilde{A}}(x) > 0\}$, the support of $\widetilde{A}$.

Let $\mathcal{K}_c(\mathbb{R}^n)$ be the space of nonempty compact convex subsets of $\mathbb{R}^n$ and $\mathcal{F}_c(\mathbb{R}^n)$ the space of all fuzzy sets $\widetilde{A}$ of $\mathbb{R}^n$ with $\widetilde{A}_\alpha \in \mathcal{K}_c(\mathbb{R}^n)$ for all $\alpha \in (0,1]$. Using Zadeh's extension principle, addition between fuzzy sets from $\mathcal{F}_c(\mathbb{R}^n)$ and scalar multiplication (with $\lambda \in \mathbb{R}$) is defined as

$$\mu_{\widetilde{A} \oplus \widetilde{B}}(z) = \sup_{x+y=z} \min(\mu_{\widetilde{A}}(x), \mu_{\widetilde{B}}(y)) \quad ; \quad \mu_{\lambda\widetilde{A}}(x) = \mu_{\widetilde{A}}\left(\frac{x}{\lambda}\right), \lambda \neq 0.$$

Note that with Minkowski addition $\oplus$ between sets from $\mathcal{K}_c(\mathbb{R}^n)$ it holds

$$(\widetilde{A} \oplus \widetilde{B})_\alpha = \widetilde{A}_\alpha \oplus \widetilde{B}_\alpha \quad \text{and} \quad (\lambda\widetilde{A})_\alpha = \lambda\widetilde{A}_\alpha.$$

For $A \in \mathcal{K}_c(\mathbb{R}^n)$ the support function s_A is defined as

$$s_A(u) := \sup_{a \in A} a^T u \quad , \quad u \in \mathbb{S}^{n-1},$$

where $a^T u$ is the standard scalar product of a and u and $\mathbb{S}^{n-1} = \{t \in \mathbb{R}^n : ||t|| = 1\}$ the $(n-1)$-dimensional unit sphere in the Euclidean space $\mathbb{R}^n$. An natural extension of the support function of a fuzzy set $\widetilde{A} \in \mathcal{F}_c(\mathbb{R}^n)$ is:

$$s_{\widetilde{A}}(u,\alpha) = \begin{cases} s_{\widetilde{A}_\alpha}(u) & : \alpha > 0 \\ 0 & : \alpha = 0 \end{cases}, \quad u \in \mathbb{S}^{n-1}, \alpha \in [0,1].$$

Each fuzzy set $\widetilde{A} \in \mathcal{F}_c(\mathbb{R}^n)$ corresponds uniquely to its support function, i.e. different fuzzy subsets from $\mathcal{F}_c(\mathbb{R}^n)$ induce different support functions and for $\widetilde{A}, \widetilde{B} \in \mathcal{F}_c(\mathbb{R}^n)$ and $\lambda \in \mathbb{R}^+$ it holds

$$s_{\widetilde{A} \oplus \widetilde{B}} = s_{\widetilde{A}} + s_{\widetilde{B}} \tag{3}$$

$$s_{\lambda\widetilde{A}} = \lambda s_{\widetilde{A}}. \tag{4}$$

So we can consider $\widetilde{A}, \widetilde{B} \in \mathscr{F}_c(\mathbb{R}^n)$ in $L^2(\mathbb{S}^{n-1} \times [0,1])$ via its support function and we define

$$\delta_2(\widetilde{A}, \widetilde{B}) := \left(n \int_0^1 \int_{\mathbb{S}^{n-1}} |s_{\widetilde{A}}(u, \alpha) - s_{\widetilde{B}}(u, \alpha)|^2 v(du) d\alpha \right)^{\frac{1}{2}},$$

$$\left\langle \widetilde{A}, \widetilde{B} \right\rangle := \langle s_{\widetilde{A}}, s_{\widetilde{B}} \rangle = n \int_0^1 \int_{\mathbb{S}^{n-1}} s_{\widetilde{A}}(u, \alpha) s_{\widetilde{B}}(u, \alpha) v(du) d\alpha,$$

$$||\widetilde{A}||_2 := ||s_{\widetilde{A}}||_2 = \left(n \int_0^1 \int_{\mathbb{S}^{n-1}} s_{\widetilde{A}}(u, \alpha)^2 v(du) d\alpha \right)^{\frac{1}{2}}.$$

With $\delta_2(\widetilde{A}, \widetilde{B}) = ||s_{\widetilde{A}} - s_{\widetilde{B}}||_2$, $\mathscr{F}_c(\mathbb{R}^n)$ can be embedded isometrically and isomorph as closed convex cone in $L^2(\mathbb{S}^{n-1} \times [0,1])$.

Now, a fuzzy random variable (frv) can be defined as a Borel measurable function

$$\widetilde{X} : \Omega \to \mathscr{F}_c(\mathbb{R}^n)$$

from $(\Omega, \mathfrak{F}, P)$ to $(\mathscr{F}_c(\mathbb{R}^n), \mathfrak{B}_2)$ where $\mathfrak{B}_2$ is the σ-algebra induced by δ_2.

Then all α–cuts are compact convex random set (see Puri, Ralescu [11], too). There are further definitions of fuzzy random variables, which are equivalent under some constraints. For details see Krätschmer [6] [7]. The (Aumann-) expectation $\mathbf{E}\xi$ of a compact convex random set ξ is defined by the collection of all "pointwise" expectations $\mathbf{E}X$, the so called Bochner-integrals, with $X \in \xi$ almost surely and, i.e (see Aumann [1], too)

$$\mathbf{E}\xi = \{\mathbf{E}X : X : \Omega \to \mathbb{R}^n, X - \text{Bochner-integrable}, X(\omega) \in \xi(\omega) \text{ P-a.s.}\}.$$

Krätschmer shows in [8], that $\mathbf{E}\xi \in \mathscr{K}_c(\mathbb{R}^n)$ if and only if ξ is integrably bounded, i.e. if $\delta_2(\xi, \{0\})$ is integrable. A frv $\widetilde{X}$ is called integrably bounded if all α-cuts are integrably bounded. Then there exists a unique fuzzy set $\mathbf{E}\widetilde{X} \in \mathscr{F}_c(\mathbb{R}^n)$, called the Aumann expectation of $\widetilde{X}$, such that

$$(\mathbf{E}\widetilde{X})_\alpha = \mathbf{E}(\widetilde{X}_\alpha) \quad ; \quad 0 < \alpha \leq 1. \tag{5}$$

This expectation of a frv $\widetilde{X}$ was introduced by Puri/Ralescu [11]. Further we can define

$$\int_A \widetilde{X} dP := \mathbf{E}\left(\mathbf{I}_A \widetilde{X} \right),$$

where $\mathbf{I}_A$ denotes the indicator function of $A \in \mathfrak{F}$.

For an integrable bounded frv the measurable function

$$s_{\widetilde{X}(.)}(u, \alpha) : \quad \Omega \to \mathbb{R}, \ \omega \mapsto s_{\widetilde{X}(\omega)}(u, \alpha)$$

is integrable and the support function of the expectation is equal, the expectation of the support function (Vitale [14]):

$$s_{\mathbf{E}\widetilde{X}}(u, \alpha) = \mathbf{E}s_{\widetilde{X}}(u, \alpha), \quad u \in \mathbb{S}^{n-1}, \alpha \in (0,1]. \tag{6}$$

Following Körner [5] the variance of a frv $\widetilde{X}$ with $\mathbf{E}||\widetilde{X}||_2^2 < \infty$ is defined by

$$\mathbf{Var}\widetilde{X} = \mathbf{E}\delta_2^2(\widetilde{X}, \mathbf{E}\widetilde{X}) \tag{7}$$
$$= \mathbf{E}\langle s_{\widetilde{X}} - s_{\mathbf{E}\widetilde{X}}, s_{\widetilde{X}} - s_{\mathbf{E}\widetilde{X}}\rangle$$
$$= \mathbf{E}||\widetilde{X}||_2^2 - ||\mathbf{E}\widetilde{X}||_2^2.$$

Using (5) and (6) this can be written as

$$\mathbf{Var}\widetilde{X} = n\int_0^1 \int_{\mathbb{S}^{n-1}} \mathbf{Var}s_{\widetilde{X}}(u, \alpha)v(du)d\alpha.$$

For more details on the expectation and variance of frv's see e.g. Näther [9].

3 Conditional Variance

In this section, we present the definition of the conditional variance of a frv and prove same properties of it. As a corollary, we obtain a variance decomposition formula analogously to (2). We start with the definition of the conditional expectation of a frv.

Assumption 1
Let $(\Omega, \mathfrak{F}, P)$ be a probability space, $\mathfrak{A}$ a sub-σ-algebra of $\mathfrak{F}$ and $\widetilde{X}$ a frv with $\mathbf{E}(||\widetilde{X}||_2^2) < \infty$ (i.e. $\mathbf{Var}\widetilde{X} < \infty$).

Definition 1 (Conditional Expectation).
Under assumption 1, the *conditional expectation of a frv* $\widetilde{X}$ *with respect to* $\mathfrak{A}$ is the frv $\mathbf{E}(\widetilde{X}|\mathfrak{A})$ which:

(a) $\mathbf{E}(\widetilde{X}|\mathfrak{A})$ is $\mathfrak{A}$-measurable,

(b) $\int_A \mathbf{E}(\widetilde{X}|\mathfrak{A})dP = \int_A \widetilde{X}dP \quad \forall A \in \mathfrak{A}.$

Analogously to (5) it holds (see Puri and Ralescu [12])

$$(\mathbf{E}(\widetilde{X}|\mathfrak{A}))_\alpha = \mathbf{E}(\widetilde{X}_\alpha|\mathfrak{A}) \quad ; \quad 0 < \alpha \leq 1. \tag{8}$$

Moreover, similar to (6) it can be proven that

$$s_{\mathbf{E}(\widetilde{X}|\mathfrak{A})}(u, \alpha) = \mathbf{E}(s_{\widetilde{X}}(u, \alpha)|\mathfrak{A}), \quad u \in \mathbb{S}^{n-1}, \alpha \in (0, 1], \tag{9}$$

see M. Stojakovic\Z. Stojakovic [13] and Wünsche\Näther [15] and Hiai\Umegaki [4]. With the equations (6) and (9) further it can be proven that

$$\mathbf{E}(\mathbf{E}(\widetilde{X}|\mathfrak{A})) = \mathbf{E}(\widetilde{X}). \tag{10}$$

Definition 2 (Conditional Variance).
Under assumption 1, the *conditional variance of* $\widetilde{X}$ *wrt* $\mathfrak{A}$ is the real random variable

$$\mathbf{Var}(\widetilde{X}|\mathfrak{A}) := \mathbf{E}(\delta_2^2(\widetilde{X}, \mathbf{E}(\widetilde{X}|\mathfrak{A}))|\mathfrak{A}) \tag{11}$$

In the following we present some properties of the conditional variance.

Assumption 2
Let X be a non-negative almost surely bounded random variable on $(\Omega, \mathfrak{F}, P)$ which is conditional independent (for conditional independence see, for instance Chow/Teicher [3])) of $\widetilde{X}$ wrt $\mathfrak{A} \subset \mathfrak{F}$.

Theorem 1. Under assumption 1 and assumption 2 it holds

$$\mathbf{Var}(X\widetilde{X}|\mathfrak{A}) = \mathbf{E}(X^2|\mathfrak{A})\mathbf{E}(||\widetilde{X}||_2^2|\mathfrak{A}) - \mathbf{E}(X|\mathfrak{A})^2||\mathbf{E}(\widetilde{X}|\mathfrak{A})||_2^2. \tag{12}$$

For the proofs of the theorem and the following corollary see
Näther\Wünsche[10].

As a direct conclusions of theorem 1 we obtain the following rules for the conditional variance. Note that for $\mathfrak{A} = \{\emptyset, \Omega\}$ the conditional variance is the variance of the frv. Take the assumptions of theorem 1 and let be $\widetilde{A} \in \mathscr{F}_c(\mathbb{R}^n)$ and $\lambda \in \mathbb{R}$. Then it holds

$$\mathbf{Var}(\widetilde{X}|\mathfrak{A}) = \mathbf{E}(||\widetilde{X}||_2^2|\mathfrak{A}) - ||\mathbf{E}(\widetilde{X}|\mathfrak{A})||_2^2 \tag{13}$$

$$= n \int_0^1 \int_{\mathbb{S}^{n-1}} \mathbf{Var}(s_{\widetilde{X}}(u,\alpha)|\mathfrak{A}) v(du) d\alpha \tag{14}$$

$$\mathbf{Var}(\lambda\widetilde{X}|\mathfrak{A}) = \lambda^2 \mathbf{Var}(\widetilde{X}|\mathfrak{A}) \tag{15}$$

$$\mathbf{Var}(X\widetilde{A}|\mathfrak{A}) = ||\widetilde{A}||_2^2 \mathbf{Var}(X|\mathfrak{A}) \tag{16}$$

$$\mathbf{Var}\left(X\widetilde{X}|\mathfrak{A}\right) = \mathbf{E}\left(X^2|\mathfrak{A}\right)\mathbf{Var}(\widetilde{X}|\mathfrak{A}) + \mathbf{Var}(X|\mathfrak{A})||\mathbf{E}(\widetilde{X}|\mathfrak{A})||_2^2 \tag{17}$$

$$\mathbf{Var}\left(X\widetilde{X}|\mathfrak{A}\right) = \mathbf{E}\left(X|\mathfrak{A}\right)^2\mathbf{Var}(\widetilde{X}|\mathfrak{A}) + \mathbf{Var}(X|\mathfrak{A})\mathbf{E}\left(||\widetilde{X}||_2^2|\mathfrak{A}\right) \tag{18}$$

Now, we easily can obtain an analogon of variance decomposition formula (2).

Corollary 1. Under assumption 1 it holds

$$\mathbf{Var}\widetilde{X} = \mathbf{E}\left(\mathbf{Var}\left(\widetilde{X}|\mathfrak{A}\right)\right) + \mathbf{Var}\left(\mathbf{E}\left(\widetilde{X}|\mathfrak{A}\right)\right). \tag{19}$$

4 Applications of the Variance Decomposition Formula

4.1 Wald's Identity

Consider, for example, an insurance company with a random claim number N per year and N individual claims $C_1, .., C_N$ which, for simplicity, are assumed to be iid.

like C. Obviously, the company is interested in the variance of the claim sum $S :=$ $\sum_{i=1}^{N} C_i$ which easily can be computed by use of variance decomposition formula (2) and which leads to the well known Wald's identity

$$\mathbf{Var}S = \mathbf{E}N\mathbf{Var}C + \mathbf{Var}N(\mathbf{E}C)^2. \tag{20}$$

Now, let us discuss Wald's formula for a random number N of iid. fuzzy claims $\widetilde{C}_i; i = 1,..,N;$ distributed like the prototype claim $\widetilde{C}$. The claim sum

$$\widetilde{S} := \sum_{i=1}^{N} \widetilde{C}_i$$

is a frv, too. Applying (19) it holds

$$\mathbf{Var}\widetilde{S} = \mathbf{E}(\mathbf{Var}(\widetilde{S}|N)) + \mathbf{Var}(\mathbf{E}(\widetilde{S}|N)). \tag{21}$$

Obviously, we obtain

$$\mathbf{E}(\widetilde{S}|N) = \mathbf{E}(\sum_{i=1}^{N} \widetilde{C}_i|N)$$
$$= N\mathbf{E}\widetilde{C}.$$

Using (16) (with $\mathfrak{A} = \{\emptyset, \Omega\}$), we have

$$\mathbf{Var}(\mathbf{E}(\widetilde{S}|N)) = ||\mathbf{E}\widetilde{C}||_2^2 \mathbf{Var}N. \tag{22}$$

Since the $\widetilde{C}_i$ are iid the variance of the sum of the $\widetilde{C}_i$ is equal the sum of the variances i.e. it holds

$$\mathbf{Var}(\widetilde{S}|N) = N\mathbf{Var}\widetilde{C}$$
$$\implies \mathbf{E}(\mathbf{Var}(\widetilde{S}|N)) = \mathbf{E}N\mathbf{Var}\widetilde{C}.$$

Hence, (21) can be written as

$$\mathbf{Var}\widetilde{S} = \mathbf{E}N\mathbf{Var}\widetilde{C} + \mathbf{Var}N||\mathbf{E}\widetilde{C}||_2^2 \tag{23}$$

which is the direct analogon of Wald's identity (20).

4.2 Stratified Sampling

Consider a random characteristic X with $\mathbf{E}X = \mu$ on a stratified sample space Ω with the strata (decomposition) $\Omega_1,..,\Omega_k$. Let μ_i and σ_i^2 be expectation and variance of X in stratum Ω_i and $p_i = P(\Omega_i); i = 1,..,k$. Then, a consequence of (2) is

$$\mathbf{Var}X = \sum_{i=1}^{k} p_i \sigma_i^2 + \sum_{i=1}^{k} p_i (\mu_i - \mu)^2 \tag{24}$$

which is a well known formula in sampling theory (see e.g. Chaudhuri, Stenger [2]).

Consider a frv $\widetilde{X}$ on a probability space $(\Omega, \mathfrak{F}, P)$ which is stratified into strata $\Omega_i \in \mathfrak{F}; i = 1, .., k;$ with $\bigcup_{i=1}^{k} \Omega_i = \Omega$, $\Omega_i \cup \Omega_j = \emptyset$ for $i \neq j$ and $P(\Omega_i) =: p_i$. Let $\mathfrak{A} = \sigma(\Omega_1, .., \Omega_k)$ be the σ-algebra generated by the strata Ω_i. Obviously, it is $\mathfrak{A} \subseteq \mathfrak{F}$. Following (7) and having in mind $\mathbf{E}(\mathbf{E}(\widetilde{X}|\mathfrak{A})) = \mathbf{E}\widetilde{X}$ we obtain

$$\mathbf{Var}(\mathbf{E}(\widetilde{X}|\mathfrak{A})) = \mathbf{E}\delta_2^2(\mathbf{E}(\widetilde{X}|\mathfrak{A}), \mathbf{E}\widetilde{X})$$
$$= \sum_{i=1}^{k} p_i \delta_2^2(\mathbf{E}(\widetilde{X}|\Omega_i), \mathbf{E}\widetilde{X}).$$

On the other hand it holds

$$\mathbf{E}(\mathbf{Var}(\widetilde{X}|\mathfrak{A})) = \sum_{i=1}^{k} p_i \mathbf{Var}(\widetilde{X}|\Omega_i).$$

Using the abbreviations $\widetilde{\mu} := \mathbf{E}\widetilde{X}$, $\widetilde{\mu}_i := \mathbf{E}(\widetilde{X}|\Omega_i)$, $\sigma_i^2 := \mathbf{Var}(\widetilde{X}|\Omega_i)$; $i = 1, .., k$; formula (19)

$$\mathbf{Var}\widetilde{X} = \mathbf{E}\left(\mathbf{Var}\left(\widetilde{X}|\mathfrak{A}\right)\right) + \mathbf{Var}\left(\mathbf{E}\left(\widetilde{X}|\mathfrak{A}\right)\right)$$

can be specified as

$$\mathbf{Var}\widetilde{X} = \sum_{i=1}^{k} p_i \sigma_i^2 + \sum_{i=1}^{k} p_i \delta_2^2(\widetilde{\mu}_i, \widetilde{\mu})$$

which is a direct generalization of formula (24).

For more details and proofs see Näther\Wünsche[10].

References

[1] Aumann, R.J. (1965) Integrals of Set-Valued Functions. Journal of Mathematical Analysis and Applications, 12:1-12.

[2] Chaudhuri, A., Stenger, H. (1992) Survey Sampling, Theory and Methods. (Marcel Dekker, New York 1992).

[3] Chow, Y.S. and Teicher, H. (1997) Probability Theory. (Springer, New York).

[4] Hiai, F. and Umegaki, H. (1977) Integrals, Conditional Expectations and Martingales of Multivalued Functions. Journal of Multivariate Analysis, 7: 149-182.

[5] Körner, R. (1997) On the variance of fuzzy random variables. Fuzzy Sets and Systems, 92(1): 83-93.

[6] Krätschmer, V. (2001) A unified approach to fuzzy random variables. Fuzzy Sets and Systems, 123(1): 1-9.

[7] Krätschmer, V. (2004) Probability theory in fuzzy sample space. Metrika, 60(2): 167-189.

[8] Krätschmer, V. (2006) Integrals of random fuzzy sets, accepted for publication in Test.

[9] Näther, W. (2000) On random fuzzy variables of second order and their application to linear statistical inference with fuzzy data. Metrika, 51(3): 201-221.

[10] Näther, W., Wünsche, A. (2006) On the Conditional Variance of Fuzzy Random Variables, accepted for publication in Metrika.

[11] Puri, M.L., Ralescu, D.A. (1986) Fuzzy random Variables. Journal of Mathematical Analysis and Applications, 114: 409-422.

[12] Puri, M.L., Ralescu, D.A.(1991) Convergence Theorem for fuzzy martingale. Journal of Mathematical Analysis and Applications, 160(1): 107-122.

[13] Stojakovic, M. and Stojakovic, Z. (1996) Support functions for fuzzy sets. Proceedings of the Royal Society of London. Series A, 452(1946): 421-438.

[14] Vitale, R.V. (1988) An alternate formulation of mean value for random geometric figures. Journal of Microscopy, 151(3): 197-204.

[15] Wünsche, A., Näther, W. (2002) Least-squares fuzzy regression with fuzzy random variables. Fuzzy Sets and Systems, 130: 43-50.

Fuzzy Histograms and Density Estimation

Kevin Loquin[1] and Olivier Strauss[2]

LIRMM - 161 rue Ada - 34392 Montpellier cedex 5 - France
[1]Kevin.Loquin@lirmm.fr
[2]Olivier.Strauss@lirmm.fr

The *probability density function* is a fundamental concept in statistics. Specifying the density function f of a random variable X on Ω gives a natural description of the distribution of X on the universe Ω. When it cannot be specified, an estimate of this density may be performed by using a sample of n observations independent and identically distributed $(X_1, ..., X_n)$ of X.

Histogram is the oldest and most widely used density estimator for presentation and exploration of observed univariate data. The construction of a histogram consists in partitioning a given reference interval Ω into p bins A_k and in counting the number Acc_k of observations belonging to each cell A_k. If all the A_k have the same width h, the histogram is said to be uniform or regular. Let $\mathbb{1}_{A_k}$ be the characteristic function of A_k, we have

$$Acc_k = \sum_{i=1}^{n} \mathbb{1}_{A_k}(X_i).\tag{1}$$

By hypothesizing the density of the data observed in each cell to be uniform, an estimate $\hat{f}_{hist}(x)$ of the underlying probability density function $f(x)$ at any point x of A_k can be computed by:

$$\hat{f}_{hist}(x) = \frac{Acc_k}{nh}.\tag{2}$$

The popularity of the histogram technique is not only due to its simplicity (no particular skills are needed to manipulate this tool) but also to the fact that the piece of information provided by a histogram is more than a rough representation of the density underlying the data. In fact, a histogram displays the number of data (or observations) of a finite data set that belong to a given class i.e. in complete agreement with the concept summarized by the label associated with each bin of the partition thanks to the quantity Acc_k.

However, the histogram density estimator has some weaknesses. The approximation given by expression (2) is a discontinuous function. The choice of both reference interval and number of cells (i.e. bin width) have quite an effect on the estimated density. The apriorism needed to set those values makes it a tool whose robustness and reliability are too low to be used for statistical estimation.

K. Loquin and O. Strauss: *Fuzzy Histograms and Density Estimation*, Advances in Soft Computing **6**, 45–52 (2006)
www.springerlink.com   © Springer-Verlag Berlin Heidelberg 2006

In the last five years, it has been suggested by some authors that replacing the binary partition by a fuzzy partition will reduce the effect of arbitrariness of partitioning. This solution has been studied as a practical tool for Chi-squared tests [Run04], estimation of conditional probabilities in a learning context [VDB01], or estimation of percentiles [SCA00] and modes [SC02]. Fuzzy partitioning has received considerable attention in the literature especially in the field of control and decision theory. Recently, some authors have proposed to explore the universal approximation properties of fuzzy systems to solve system of equations [Per06, Per04, Wan98, HKAS03, Lee02].

In a first part, we will formally present the fuzzy partition as proposed in [Per06]. In section 2, a histogram based upon this previous notion will be defined, that will be called a fuzzy histogram. In a last section, some estimators of probability density functions will be shown, before concluding.

1 Fuzzy Partitions

1.1 Preliminary

In histogram technique, the accumulation process (see expression (1)) is linked to the ability to decide whether the element x belongs to a subset A_k of Ω, the universe, or not. This decision is tantamount to the question whether it is true that $x \in A_k$ or not (this is a binary question). However, in many practical cases, this question cannot be precisely answered : there exists a vagueness in the "frontiers" of A_k. A reasonable solution consists in using a scale whose elements would express various *degrees of truth* of $x \in A_k$, and A_k becomes a fuzzy subset of Ω. Let L be this scale of truth values. We usually put $L = [0, 1]$.

1.2 Strong Uniform Fuzzy Partition of the Universe

Here we will take an interval $\Omega = [a, b]$ (real) as the universe. Then,

Definition 1. *Let $m_1 < m_2 < ... < m_p$ be p fixed nodes of the universe, such that $m_1 = a$ and $m_p = b$, and $p \geq 3$. We say that the set of the p fuzzy subsets $A_1, A_2, ..., A_p$, identified with their membership functions $\mu_{A_1}(x), \mu_{A_2}(x), ..., \mu_{A_p}(x)$ defined on the universe, form a* strong uniform fuzzy partition of the universe, *if they fulfil the following conditions :*
   for $k = 1, ..., p$

1. *$\mu_{A_k}(m_k) = 1$ (m_k belongs to what is called the* core *of A_k),*
2. *if $x \notin [m_{k-1}, m_{k+1}]$, $\mu_{A_k}(x) = 0$ (because of the notation we should add : $m_0 = m_1 = a$ and $m_p = m_{p+1} = b$),*
3. *$\mu_{A_k}(x)$ is continuous,*
4. *$\mu_{A_k}(x)$ monotonically increases on $[m_{k-1}, m_k]$ and $\mu_{A_k}(x)$ monotonically decreases on $[m_k, m_{k+1}]$,*

5. $\forall x \in \Omega$, $\exists k$, such that $\mu_{A_k}(x) > 0$ (every element of the universe is treated in this partition).

6. for all $x \in \Omega$, $\sum_{k=1}^{p} \mu_{A_k}(x) = 1$

7. for $k \neq p$, $h_k = m_{k+1} - m_k = h = $ constant, so, $m_k = a + (k-1)h$,

8. for $k \neq 1$ and $k \neq p$, $\forall x \in [0, h]$ $\mu_{A_k}(m_k - x) = \mu_{A_k}(m_k + x)$ (μ_{A_k} is symmetric around m_k),

9. for $k \neq 1$ and $k \neq p$, $\forall x \in [m_k, m_{k+1}]$, $\mu_{A_k}(x) = \mu_{A_{k-1}}(x - h)$ and $\mu_{A_{k+1}}(x) = \mu_{A_k}(x - h)$ (all the μ_{A_k}, for $k = 2, ..., p-1$ have the same shape, with a translation of h. And as for μ_{A_1} and μ_{A_p}, they have the same shape, but truncated, with supports twice smaller than the other ones).

Condition 6 is known as the *strength condition*, which ensures a *normal weight* of 1, to each element x of the universe in a strong fuzzy partition. In the same way, conditions 7, 8 and 9 are the conditions for the uniformity of a fuzzy partition.

Proposition 1. *Let* $(A_k)_{k=1,...,p}$ *be a strong uniform fuzzy partition of the universe, then*

$$\exists K_A : [-1, 1] \longrightarrow [0, 1] \text{ pair, such that, } \mu_{A_k}(x) = K_A\left(\frac{x - m_k}{h}\right) \mathbb{1}_{[m_{k-1}, m_{k+1}]} \text{ and } \int K_A(u)du = 1.$$

Proof. We can take $K_A(u) = \mu_{A_k}(hu + m_k)$, $\forall k$. The support of K_A comes from the ones of the μ_{A_k}, and the parity is deduced from a translation of the symmetry of the μ_{A_k}. And, to end this proof, $\int_{-1}^{1} K_A(u)du = \int_{-1}^{1} \mu_{A_k}(hu + m_k)du = \int_{m_{k-1}}^{m_{k+1}} \frac{1}{h}\mu_{A_k}(x)dx = 1$.

Table 1. Strong uniform fuzzy partition examples

	Crisp	Triangular	Cosine
$\mu_{A_1}(x) =$	$\mathbb{1}_{[m_1, m_1 + \frac{h}{2}]}(x)$	$\frac{(m_2 - x)}{h} \mathbb{1}_{[m_1, m_2]}(x)$	$\frac{1}{2}(cos(\frac{\pi(x - m_1)}{h}) + 1)\mathbb{1}_{[m_1, m_2]}(x)$
$\mu_{A_k}(x) =$	$\mathbb{1}_{[m_k - \frac{h}{2}, m_k + \frac{h}{2}]}(x)$	$\frac{(x - m_{k-1})}{h} \mathbb{1}_{[m_{k-1}, m_k]}(x)$ $+$ $\frac{(m_{k+1} - x)}{h} \mathbb{1}_{[m_k, m_{k+1}]}(x)$	$\frac{1}{2}(cos(\frac{\pi(x - m_k)}{h}) + 1)\mathbb{1}_{[m_{k-1}, m_{k+1}]}(x)$
$\mu_{A_p}(x) =$	$\mathbb{1}_{[m_p - \frac{h}{2}, m_p]}(x)$	$\frac{(x - m_{p-1})}{h} \mathbb{1}_{[m_{p-1}, m_p]}(x)$	$\frac{1}{2}(cos(\frac{\pi(x - m_p)}{h}) + 1)\mathbb{1}_{[m_{p-1}, m_p]}$
$K_A(x) =$	$\mathbb{1}_{[-\frac{1}{2}, \frac{1}{2}]}(x)$	$(1 - |x|)\mathbb{1}_{[-1, 1]}(x)$	$0.5(cos(\pi x) + 1)\mathbb{1}_{[-1, 1]}(x)$

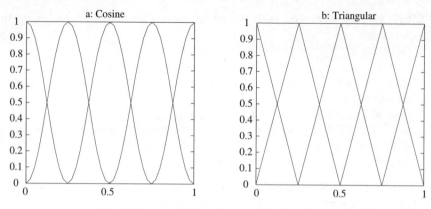

Fig. 1. Fuzzy partitions with $\Omega = [0,1]$ and $p = 5$

2 A Fuzzy-Partition Based Histogram

The accumulated value Acc_k is the key feature of the histogram technique. It is the number of observations in complete agreement with the label represented by the restriction of the real line to the interval (or bin) A_k. Due to the important arbitrariness of the partition, the histogram technique is known as being very sensitive to the choice of both reference interval and number of cells (or bin width). As mentioned before, the effect of this arbitrariness can be reduced by replacing the crisp partition by a fuzzy partition of the real line.

Let $(A_k)_{k=1,\ldots,p}$ be a strong uniform fuzzy partition of Ω, the natural extension of the expression (1) induces a distributed vote. The value of the accumulator Acc_k associated to the fuzzy subset A_k is given by:

$$Acc_k = \sum_{i=1}^{n} \mu_{A_k}(X_i). \tag{3}$$

Then, those accumulators still represent a "real" (generally not an integer) number of observations in accordance with the label represented by the fuzzy subset A_k. Moreover, the *strength* (Condition 6 of Definition 1) of the fuzzy partition $(A_k)_{k=1,\ldots,p}$ implies that the sum of the Acc_k equals to n,[1] the number of observations. Note that the classical crisp-partition based histogram is a particular case of the fuzzy-partition based histogram, when $(A_k)_{k=1,\ldots,p}$ is the crisp partition.

We propose to illustrate the softening property of the fuzzy histogram over the crisp histogram. Figure 2.(a) displays a crisp histogram of 35 observations drawn from a Gaussian process with mean $\mu = 0.3$ and variance $\sigma^2 = 1$. Figure 2.(b) displays a fuzzy triangular partition based histogram of the same observations with the same reference interval position. We have translated both crisp and fuzzy partitions by an amount of 30% of the bin width. As it can be seen on Figure 2.(c),

---

[1] indeed, $\sum_{k=1}^{p} Acc_k = \sum_{k=1}^{p} \sum_{i=1}^{n} \mu_{A_k}(X_i) = \sum_{i=1}^{n} \sum_{k=1}^{p} \mu_{A_k}(X_i) \sum_{i=1}^{n} 1 = n$

this translation has quite an effect on the crisp-partition based histogram, while the fuzzy-partition based histogram plotted on Figure 2.(d) still has the same general shape. The number of observations is too small, regarding the number of fuzzy subsets ($p = 8$) of the partition, to ensure that the convergence conditions are fulfilled (see theorem 1).

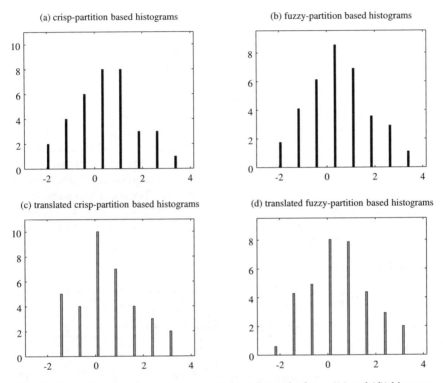

Fig. 2. Effect of the translation on a crisp ((a) and (b)) and a fuzzy ((c) and (d)) histogram

3 Fuzzy Histogram Density Estimators

Expression (2) can be used for both crisp and fuzzy histograms to estimate the density underlying a set of observations. However, since A_k is a fuzzy subset, this expression no longer holds for any $x \in A_k$, but normalized accumulators $\frac{Acc_k}{nh}$ now have *degrees of truth* inherited from the fuzzy nature of A_k (see the preliminary of the section 1). The value $\frac{Acc_k}{nh}$ is then more *true* at m_k than at any other point of Ω. Our proposal is to assign this value $\frac{Acc_k}{nh}$ to the estimated density at each node m_k of the partition. Therefore, the estimated density can be obtained, at any point $x \neq m_k$, by interpolation.

In this paper, we propose to use, once again, the concept of strong uniform fuzzy partition of p fuzzy subsets to provide an interpolation of those p points.

Proposition 2. *An interpolant of a fuzzy histogram (of the points* $(m_k, \frac{Acc_k}{nh})$*) is given by*

$$\hat{f}_{FH}(x) = \frac{1}{nh} \sum_{k=1}^{p} Acc_k K_B \left(\frac{x - m_k}{h} \right) \tag{4}$$

where K_B is defined as in proposition 1 for the strong uniform fuzzy partition $(B_k)_{k=1,\dots,p}$.

Proof. Conditions 1 and 6 of the definition 1 imply that $\mu_{B_k}(m_l) = \delta_{kl}$ for $k, l \in \{1, \dots, p\}$, where δ_{kl} is the Kronecker symbol, and, $K_B(\frac{m_l - m_k}{h}) = \mu_{B_k}(m_l)$. Then, $\hat{f}_{FH}(m_l) = \frac{Acc_l}{nh}$, for all $l \in \{1, \dots, p\}$, which means that $\hat{f}_{FH}$ goes though the p points $(m_k, \frac{Acc_k}{nh})$.

Therefore, this interpolant (which is a density estimator) has the continuity properties of the membership functions of the fuzzy partition $(B_k)_{k=1,\dots,p}$, except at the nodes m_k, where the smoothness is not guaranteed. We can now add a convergence property of the estimators given by expression (4). So, let the error between the underlying density $f(x)$ and the estimate $\hat{f}_{FH}(x)$ be measured by the mean squared error : $MSE(x) \triangleq E_f[\hat{f}_{FH}(x) - f(x)]^2$. We have proved Theorem 1 in a paper to be published [LS], which is in some sense, the technical part of this paper. This proof is inspired from the demonstrations of the consistency theorems of the kernel density estimator, that are in [Tsy04].

Theorem 1. *Let us suppose*

1. *$f : \Omega \to [0,1]$ is a density function such that f is bounded ($\forall x \in \Omega$, $f(x) \leq f_{max} < +\infty$) and f', its derivative, is bounded ($\forall x \in \Omega$, $|f'(x)| \leq f'_{max} < +\infty$),*
2. *K_A, as defined in proposition 1, verifies $\int_{-1}^{1} K_A^2(u)du < +\infty$.*

Then, for all $x \in \Omega$,

$$h \to 0 \ and \ nh \to +\infty \Rightarrow MSE(x) \to 0 \tag{5}$$

This theorem gives a mathematical evidence that the fuzzy histogram is a proper representation of the distribution of data, because a simple interpolation of a normalized histogram converges (in MSE) to the underlying density. It converges under classical conditions, which are, the reduction of the support of the membership functions, or the growth of the number of fuzzy subsets of the partition ($h \to 0$ or $p \to +\infty$), and the growth of the mean number of data in each accumulator ($nh \to +\infty$).

However, the use of the membership functions of a fuzzy partition as interpolation functions is not compulsory. Thus, well-known interpolation functions could be used, e.g. the polynomial interpolation (with the Lagrange or the Newton form), or the spline interpolation, which improves the smoothness management at the nodes.

Figure 3 shows four estimations of a bimodal gaussian distribution with parameters $(m_1 = 4, \sigma_1^2 = 4)$ and $(m_2 = -1, \sigma_2^2 = \frac{1}{4})$, based upon a fuzzy triangular histogram. The circles are the interpolation points $(m_k, \frac{Acc_k}{nh})$. The dashed line is the crisp interpolation (see expression (2)). The solid line is the estimator obtained by

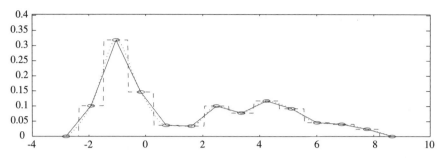

Fig. 3. Density estimation by interpolation

fuzzy triangular interpolation (see expression (4)). The dotted line is a spline interpolation of the points $\left(m_k, \frac{Acc_k}{nh}\right)$.

The estimations are obtained with $n = 100$ observations and $p = 14$ fuzzy subsets of the partition, which means that we are no longer in convergence conditions. Table 2 gives the empirical L_1 errors of interpolation, i.e. $\int_\Omega |\hat{f}_{FH}(x) - f(x)| dx$, obtained by repeating the experiment 100 times. This error is noted $m_{Error} \pm 3 * \sigma_{Error}$, where m_{Error} is the mean of the L_1 error over the 100 experiments and σ_{Error} its standard deviation.

Note that, whatever the interpolation scheme, compared to crisp histogram density estimators, the fuzzy histogram density estimators seem to be more stable (which can be measured by means of the standard deviation) and closer (in L_1 distance) to the underlying density (in that particular case).

Another important remark, deduced from Table 2, is that the fuzzy interpolants appear to be a good choice, because their error magnitudes are equivalent to those of the spline interpolant, which is known as being an optimal tool.

Table 2. L_1 errors of interpolation

	crisp accumulators	fuzzy accumulators
crisp interpolation	0.028307 ± 0.014664	0.024957 ± 0.01166
triangular interpolation	0.021579 ± 0.014963	0.020807 ± 0.013539
cosine interpolation	0.022524 ± 0.014832	0.021103 ± 0.01327
spline interpolation	0.021813 ± 0.01577	0.020226 ± 0.0139
Lagrange interpolation	3.7349 ± 7.893	2.2537 ± 4.3761

4 Conclusion

In this paper, we have presented density estimators based upon a fuzzy histogram. This latter being nothing else but a generalization of the popular crisp histogram, when replacing the crisp partition by a fuzzy partition. Those proposed density estimators consist in interpolations of the nodes' values of the density obtained in the usual way : $\frac{Acc_k}{nh}$.

References

[HKAS03] R. Hassine, F. Karray, A. M. Alimi, and M. Selmi. Approximation properties of fuzzy systems for smooth functions and their first order derivative. *IEEE Trans. Systems Man Cybernet.*, 33:160–168, 2003.

[Lee02] T. Leephakpreeda. Novel determination of differential-equation solutions : universal approximation method. *Journal of computational and applied mathematics*, 146:443–457, 2002.

[LS] K. Loquin and O. Strauss. Histogram density estimators based upon a fuzzy partition. *Laboratoire d'Informatique, de Robotique et de Microélectronique de Montpellier, Université Montpellier II*.

[Per04] I. Perfilieva. Fuzzy function as an approximate solution to a system of fuzzy relation equations. *Fuzzy sets and systems*, 147:363–383, 2004.

[Per06] I. Perfilieva. Fuzzy transforms: Theory and applications. *Fuzzy sets and systems*, 157:993–1023, 2006.

[Run04] T. A. Runkler. Fuzzy histograms and fuzzy chi-squared tests for independence. *IEEE international conference on fuzzy systems*, 3:1361–1366, 2004.

[SC02] O. Strauss and F. Comby. Estimation modale par histogramme quasi-continu. *LFA'02 Rencontres Francophones sur la Logique Floue et ses Applications, Montpellier*, pages 35–42, 2002.

[SCA00] O. Strauss, F. Comby, and M.J. Aldon. Rough histograms for robust statistics. *ICPR'2000 15th International Conference on Pattern Recognition, Barcelona, Catalonia, Spain, 3-8 September*, 2000.

[Tsy04] A. B. Tsybakov. *Introduction à l'estimation non-paramétrique*. Springer-Verlag, 2004.

[VDB01] J. Van Den Berg. Probabilistic and statistical fuzzy set foundations of competitive exception learning. *IEEE International Fuzzy Systems Conference*, pages 1035–1038, 2001.

[Wan98] L. X. Wang. Universal approximation by hierarchical fuzzy systems. *Fuzzy sets and systems*, 93:223–230, 1998.

Graded Stochastic Dominance as a Tool for Ranking the Elements of a Poset

Karel De Loof[1], Hans De Meyer[1], and Bernard De Baets[2]

[1] Department of Applied Mathematics and Computer Science, Ghent University, Krijgslaan 281 (S9), B-9000 Gent, Belgium
[2] Department of Applied Mathematics, Biometrics and Process Control, Ghent University, Coupure links 653, B-9000 Gent, Belgium

Summary. Three methods are outlined aiming at the (partial) ranking of the elements of a poset respecting the poset order. First, we obtain a partial ranking by applying first degree stochastic dominance to the rank probability distributions of the poset elements. Then we use the minimum, product or Łukasiewicz copula as an artefact for pairwisely coupling rank probability distributions into bivariate distributions. They serve as a basis for generating a probabilistic relation which constitutes a graded version of stochastic dominance. The transitivity of the probabilistic relations being characterizable in the framework of cycle-transitivity, a cutting level is specified that provides us with a strict partial order. Finally, we apply the graded stochastic dominance principle directly to the mutual rank probabilities. Based on exhaustive experiments, a conjecture on the transitivity of the associated probabilistic relation is made, and a (partial) ranking of the poset elements is extracted.

Keywords: Copula, Cycle-transitivity, Linear extension, Mutual ranking, Poset, Probabilistic relation, Ranking probability, Stochastic dominance, Stochastic transitivity.

1 Introduction

In a poset the elements are partially ordered: any two elements are either comparable or incomparable. The question arises whether the partial order can be consistently extended to a total order. Any such total order is called a linear extension of the poset, but the decision maker who wants to rank the elements, typically needs to select a single linear extension out of a manifold of linear extensions. His choice should, however, not be made arbitrarily. If, for example, a particular poset element is of low rank in most of the linear extensions, linear extensions in which that element occasionally appears at higher rank should be discarded from the option list. Ideally, his final choice must take into account the rank frequency distributions of all the poset elements.

A possible way to achieve this goal is to consider for each element of P its cumulative rank probability distribution (c.r.d.f.) and to establish a pairwise comparison of elements through the comparison of their distributions, relying, for instance, on the

K. De Loof et al.: *Graded Stochastic Dominance as a Tool for Ranking the Elements of a Poset*, Advances in Soft Computing **6**, 53–60 (2006)
www.springerlink.com

principle of stochastic dominance. This technique amounts in a partial order that extends the partial order on P. The iterative application of this technique can eventually generate a total order, though this is not generally the case.

A weakness of the stochastic dominance principle is that it only relies upon the information contained in the marginal c.r.d.f.'s, and therefore ignores any possible correlation between elements with respect to the positions they mutually occupy in the linear extensions. One can remedy this by considering the bivariate rank probability distributions and setting up mechanisms for extracting from these joint distributions degrees of precedence between every pair of poset elements. In this paper, two such methods are established. They are conform with a graded version of stochastic dominance.

First, we build artificial bivariate c.r.d.f.'s by coupling the marginal c.r.d.f.'s with a fixed copula. Three representative copulas are considered: the minimum copula T_M which describes comonotonic coupling, the product copula T_P which models the independent case, and the Łukasiewicz copula T_L which describes countermonotonic coupling. In all cases, we obtain a so-called probabilistic relation. Cutting this relation at some appropriate cutting level, yields a cycle-free crisp relation. Its transitive closure is a strict partial order that, in the case of the coupling with either T_M or T_L extends the given partial order. However, to find such appropriate cutting levels, one needs to know the type of transitivity of the probabilistic relation. For the three couplings mentioned, this typing has been realized in the framework of cycle-transitivity, previously established by some of the present authors.

Alternatively, we apply the graded stochastic dominance principle directly to the true bivariate c.r.d.f.'s. The latter are not explicitly required as only mutual rank probabilities need to be computed. On the other hand, we have not yet been able to characterize the transitivity of the generated probabilistic relation. Based on experimental results obtained on all posets containing at most 9 elements and on a random selection of posets containing up to 20 elements, we put forward a conjecture concerning this type of transitivity and derive a cutting level that is expected to provide us with a partial order that again extends the given partial order.

The outline of the paper is as follows. In section 2 we introduce concepts related to posets, linear extensions, rank probabilities and mutual rank probabilities. In Section 3, a partial ranking is obtained by applying the stochastic dominance principle. Section 4 is concerned with artificial couplings and graded stochastic dominance. Section 5 deals with graded stochastic dominance related to the mutual rank probabilities.

2 Preliminaries

2.1 Posets and Linear Extensions

A *poset* is denoted by $(P, \geq_P)$, where P is a set and $\geq_P$ an order relation on P, *i.e.* a reflexive, antisymmetric and transitive relation whose elements are written as $a \geq_P b$. Two elements x and y of P are called *comparable* if $x \geq_P y$ or $x \leq_P y$; otherwise they are called *incomparable*, and we write $x\|y$. A *complete order* is a poset in which every two elements are comparable. We say that x *covers* y, denoted as $x \succ_P y$, if $x >_P y$ and there exists no $z \in P$ such that $x >_P z >_P y$. By definition $x >_P y$ if $x \geq_P y$ and not $x \leq_P y$. A *chain* of a poset P is a subset of P in which every two elements are comparable. Dually, an *antichain* of a poset P is a subset of P in which every two elements are incomparable. A poset can be conveniently represented by a so-called *Hasse diagram* where $x >_P y$ if and only if there is a sequence of connected lines upwards from y to x. An example of a Hasse diagram is shown on the left of Figure 1.

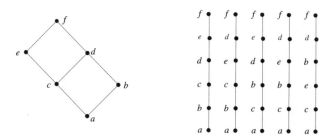

Fig. 1. left: Hasse diagram of a poset P; right: all linear extensions of P

An *ideal* of a poset $(P, \leq_P)$ is a subset $D \subseteq P$ such that $x \in D$, $y \in P$ and $y \leq_P x$ implies $y \in D$. The set of ideals forms a distributive lattice, called the *lattice of ideals* of P.

A poset $(Q, \leq_Q)$ is called an *extension* of a poset $(P, \leq_P)$ if for all $x, y \in Q$, $x \leq_P y$ imply $x \leq_Q y$. A *linear extension* or complete extension is an extension that is a chain. In Figure 1, also the five linear extensions of the poset P are given.

2.2 Rank Probabilities

The *rank probability* $p_x(i)$ of an element $x \in P$ is defined as the fraction of linear extensions of P where x occurs at position i. Letting i run over all positions from 1 to $|P|$, the function p_x is called the *rank probability distribution* of the element x. Note that we have identified a space of random events with the set of all linear extensions of a given poset P considered equally likely. To each element x of P is associated a discrete random variable X taking values in the set $\{1, 2, \ldots, |P|\}$. In Table 1 the rank probabilities are given for the poset of Figure 1.

Counting the number of linear extensions of a poset P with a specific element $x \in P$ on position i ($i = 1, 2, \ldots, |P|$) amounts to the problem of counting the number of paths in the ideal lattice representation containing the edges labelled x at height i, a problem that is easily solved by careful counting. Based on this equivalence, we have developed in [2] an algorithm that efficiently computes the rank probability distributions of all the elements in the poset in essentially two passes of the lattice of ideals. Its run time is proportional to the number of edges in the ideal lattice.

2.3 Mutual Rank Probabilities

The *mutual rank probability* $\text{Prob}(X > Y)$ of two elements $(x,y) \in P^2$ is defined as the fraction of linear extensions of P where element x is ranked higher than element y. The rank of an element of P in a linear extension is its position number in the chain. The mutual rank probability $\text{Prob}(X > Y)$ can be immediately computed from the *bivariate probability distribution* $p_{x,y}$ of (x,y), $p_{x,y}(i,j)$ denoting the joint probability that x occurs at position i and y at position j in a linear extension of P. In [2] we have proposed an efficient algorithm that gradually builds up a two-dimensional table initialized with zeroes, and which, after execution of the algorithm, contains all mutual rank probabilities $\text{Prob}(x > y)$. The main idea is to traverse the ideal lattice recursively in a depth-first manner.

In Table 2 the mutual rank probabilities of the poset of Figure 1 are presented in matrix form. Note that this matrix has zeroes everywhere on the diagonal and is reciprocal, *i.e.* any matrix element is the 1-complement of the diagonally opposite element. These properties clearly reflect that two elements of a poset cannot share the same position in a linear extension.

Table 1. Rank probabilities.

$X \backslash i$	1	2	3	4	5	6
a	1	0	0	0	0	0
b	0	$\frac{2}{5}$	$\frac{2}{5}$	$\frac{1}{5}$	0	0
c	0	$\frac{3}{5}$	$\frac{2}{5}$	0	0	0
d	0	0	0	$\frac{2}{5}$	$\frac{3}{5}$	0
e	0	0	$\frac{1}{5}$	$\frac{2}{5}$	$\frac{2}{5}$	0
f	0	0	0	0	0	1

Table 2. Mutual rank probabilities.

$X \backslash Y$	a	b	c	d	e	f
a	0	0	0	0	0	0
b	1	0	$\frac{3}{5}$	0	$\frac{1}{5}$	0
c	1	$\frac{2}{5}$	0	0	0	0
d	1	1	1	0	$\frac{3}{5}$	0
e	1	$\frac{4}{5}$	1	$\frac{2}{5}$	0	0
f	1	1	1	1	1	0

3 Stochastic Dominance

Let us consider the c.r.d.f. F_X of $x \in P$ defined by $F_X(s) = \sum_{i \le s} p_x(i)$. A well-known concept for comparing two random variables is that of *stochastic dominance* [5], which is particularly popular in financial mathematics. A random variable X with c.d.f. F_X *stochastically dominates in first degree* a random variable Y with c.d.f. F_Y, denoted as $X \ge_{\text{FSD}} Y$, if for any real t it holds that $F_X(t) \le F_Y(t)$. The condition for first degree stochastic dominance is rather severe, as it requires that the graph of the function F_X lies entirely beneath the graph of the function F_Y. The need to relax this condition has led to other types of stochastic dominance, such as second degree and third degree stochastic dominance.

Clearly, the relation $\ge_{\text{FSD}}$ yields a partial order on P. If in the given poset P two elements x and y are comparable, say $x >_P y$, then in all linear extensions of P, x is ranked higher than y and F_X lies entirely beneath F_Y, whence $x >_{\text{FSD}} y$. The converse, however, is not generally true as it can happen for incomparable elements x and y that $x >_{\text{FSD}} y$, in other words, elements that are incomparable at the level of the poset can become comparable at the level of the partial order induced by first degree stochastic dominance. Hence, we obtain a partial order that extends $>_P$. This procedure can be repeated until convergence. Full ranking of the elements can be achieved if the limiting partial order is a linear order (chain). However, this is not generally true [7].

4 Graded Stochastic Dominance

The principle of stochastic dominance relies upon the information contained in the marginal rank probability distributions only. More information concerning the rank of the elements in the linear extensions of P is contained in the bivariate c.r.d.f. but, as far as we know, the stochastic dominance principle has not been extended to deal with this additional information.

A first method for generating a graded stochastic dominance relation is based on the construction of artificial bivariate c.r.d.f.'s from the marginal c.r.d.f.'s. To that aim, we fix *copula* that is representative for a coupling with positive, negative or zero-correlation.

4.1 Copulas

Copulas were introduced in statistics to express the coupling of two univariate c.d.f.'s into a single bivariate c.d.f. [6]. Today, there is a renewed interest in copulas, also out of the scope of probability theory.

A binary operator $C : [0,1]^2 \rightarrow [0,1]$ is a *copula* if it has neutral element 1, absorbing element 0 and the so-called property of *moderate growth*, *i.e.* for all $x_1 \leq x_2$ and all $y_1 \leq y_2$, it holds that

$$C(x_1,y_1) + C(x_2,y_2) \geq C(x_1,y_2) + C(x_2,y_1).$$

Any copula is (pointwisely) situated between the Łukasiewicz copula T_L defined by $T_L(u,v) = \max(u+v-1,0)$ and the minimum copula T_M defined by $T_M(u,v) = \min(u,v)$, *i.e.* for any copula C it holds that $T_L \leq C \leq T_M$. Another important copula is the product copula T_P defined by $T_P(u,v) = uv$.

Sklar's theorem says that for any random vector (X,Y) there exists a copula C_{XY} s.t. $F_{X,Y}(x,y) = C_{XY}(F_X(x),F_Y(y))$. This shows that a copula captures the dependence structure irrespective of the marginals. If two random variables are coupled by means of T_M (resp. T_L) they are called *comonotonic* (resp. *countermonotonic*). With the copula T_P, the bivariate c.d.f. is the product of the univariate marginal c.d.f.'s, implying that the random variables are independent. Considering the three bivariate c.d.f.'s of three random variables (X,Y,Z), the compatibility problem states that not all combinations of copulas are possible. In particular, C_{XY}, C_{YZ} and C_{XZ} can be all equal to T_M or all equal to T_P, but they cannot be all equal to T_L.

4.2 Cycle-Transitivity

Let Q be a probabilistic relation on A, and a,b,c any three elements of A. We define

$$\alpha_{abc} = \min\{Q(a,b), Q(b,c), Q(c,a)\},$$
$$\beta_{abc} = \text{median}\{Q(a,b), Q(b,c), Q(c,a)\},$$
$$\gamma_{abc} = \max\{Q(a,b), Q(b,c), Q(c,a)\}.$$

The relation Q is called *cycle-transitive* w.r.t. an *upper bound function* (u.b.f.) U if for any $a, b, c \in A$ it holds that

$$L(\alpha_{abc}, \beta_{abc}, \gamma_{abc}) \le \alpha_{abc} + \beta_{abc} + \gamma_{abc} - 1 \le U(\alpha_{abc}, \beta_{abc}, \gamma_{ab})$$

with the *lower bound function* function (l.b.f.) L defined by $L(\alpha, \beta, \gamma) = 1 - U(1 - \gamma, 1 - \beta, 1 - \alpha)$.

Various types of transitivity of probabilistic relations fit into this framework[1]; we consider only a few here. A probabilistic relation Q that is cycle-transitive w.r.t. the u.b.f. $U_{12}(\alpha, \beta, \gamma) = \alpha + \beta - \alpha\beta$ is called *product-transitive*. It is equivalent with stating that for any $a, b, c \in A^3$ it holds that $Q(a,c) \ge Q(a,b)Q(b,c)$. Cycle-transitivity w.r.t. the u.b.f. $U_{23}(\alpha, \beta, \gamma) = \beta + \gamma - \beta\gamma$ is also known as *dice-transitivity* [3]. We will also consider cycle-transitivity w.r.t. the u.b.f. $U_{13}(\alpha, \beta, \gamma) = \alpha + \gamma - \alpha\gamma$. Cycle-transitivity w.r.t. the u.b.f. $U_M(\alpha, \beta, \gamma) = \beta$ is equivalent to min-transitivity. At the other end, there is cycle-transitivity w.r.t. the u.b.f. $U_L(\alpha, \beta, \gamma) = 1$, which is equivalent to T_L-transitivity, in many contexts, the weakest possible type of transitivity, requiring $Q(a,b) + Q(b,c) + Q(c,a) \le 2$ for any $a, b, conA$. Finally, we mention cycle-transitivity w.r.t. the u.b.f. $U_{ps}(\alpha, \beta, \gamma) = \gamma$, which is also known as *partial stochastic transitivity* and is equivalent to requiring that for any $a, b, c \in A$ for which $Q(a,b) > 1/2$ and $Q(b,c) > 1/2$, it holds that $Q(a,c) \ge \min(Q(a,b), Q(b,c))$.

4.3 The Graded Stochastic Dominance Relation

In [3] we have established a method for comparing pairwisely random variables in a pairwise manner. We defined a degree of preference between two random variables. This method can be used in the present context as with each $x \in P$ we associate a random variable X.

If for each pair of random variables (X, Y) is given a function $F_{X,Y}$ that has all the properties of a binary c.d.f., then we can construct a relation Q by $Q(x,y) = \text{Prob}(X > Y)$ wich clearly depends on these functions $F_{X,Y}$. The relation Q is a probabilistic relation as it satisfies $Q(x,y) + Q(y,x) = 1$. $Q(x,y)$ can also be interpreted as the degree by which x stochastically dominates y. If $x >_P y$, then $Q(x,y) = 1$, in agreement with $X >_{FSD} Y$. This explains why the probabilistic relation can be called a *graded stochastic dominance* relation.

4.4 Artificial Couplings

We construct functions $F_{X,Y}$ by artificially coupling all pairs of random variables (X, Y) by means of a same copula $C \in \{T_M, T_P, T_L\}$, i.e. $F_{X,Y}(u, v) = C(F_X(u), F_Y(v))$. Clearly, because C is a copula, these functions can be regarded as bivariate c.d.f.'s and we can compute the probabilistic relation Q. We will provide Q with a superscript referring to this copula. In [3], we have proven that

- if $C = T_M$, then Q^M is cycle-transitive w.r.t. the u.b.f. U_L (T_L-transitive);
- if $C = T_P$, then Q^P is cycle-transitive w.r.t. the u.b.f. U_{23} (dice-transitive);
- if $C = T_L$, then Q^L is cycle-transitive w.r.t. the u.b.f. U_{ps} (partial stochastic transitive).

Knowing the type of transitivity, we are able to determine for each case the minimal cutting level δ, such that the graph of the crisp relation obtained from Q by setting its elements strictly smaller than δ equal to 0 and its other elements equal to 1, is free from cycles. In particular, for the three couplings of interest, the minimum value of the cutting level δ is:

- if $C = T_M$ then $\delta^M = \dfrac{|P|-1}{|P|}$;
- if $C = T_P$ then $\delta^P = 1 - \dfrac{1}{4\cos^2(\pi/(|P|+2))}$;
- if $C = T_L$ then $\delta^L = 1/2 + \varepsilon$,   (arbitrarily small $\varepsilon > 0$).

Since $x >_P y$ implies $Q^M(x,y) = 1$, cutting Q^M at the level 1 yields the crisp relation $>_P$, whereas cutting it at the level δ^M yields a cycle-free relation whose transitive closure is a strict partial order that extends $>_P$. Similarly, $x >_P y$ implies $Q^L(x,y) > 1/2$ so that cutting Q^L at the level δ^M yields a cycle-free relation whose transitive closure is a strict partial order that extends $>_P$. Finally, as $x >_P y$ implies $Q^P(x,y) > 1/2$, the transitive closure of the cycle-free relation obtained by cutting Q^P at the level δ^P is not necessarily an extension of $>_P$. For our actual purposes, the artificial coupling with T_P is therefore not recommendable.

5 Ranking Based on the Mutual Ranking Probabilities

Instead of defining artificial bivariate c.r.d.f.'s for the computation of Q, we can identify Q directly with the relation built from the mutual ranking probabilities, i.e. $Q(x,y) = \text{Prob}(X > Y)$ for all $(x,y) \in P^2$. However, as the characterizing type of transitivity of the probabilistic relation Q is not known, we cannot directly advance an appropriate cutting level δ.

It can be easily proven that Q is at least T_L-transitive. Moreover, in [4] it has been pointed out that Q is *proportional-transitive* in the sense that there exists a threshold $\lambda < 1$ such that for all $a,b,c \in A$ it holds that $Q(a,b) \geq \lambda \wedge Q(b,c) \geq \lambda$ implies $Q(a,c) \geq \lambda$. The sharpest value of λ presently known is approximately 0.78 (see [8]).

We have analyzed all posets with at most 9 elements and a random selection of posets containing up to 20 elements within the framework of cycle-transitivity and have checked their transitivity w.r.t. a variety of u.b.f.'s. These experimental results allow us to conjecture the following.

Conjecture: The probabilistic relation Q is cycle-transitive w.r.t. the u.b.f. U_{23} (dice-transitivity).

In fact, all our experiments indicate that Q might have an even stronger type of transitivity, namely cycle-transitivity w.r.t. the u.b.f. U_{13}. However, under the assumption that the weaker conjecture formulated above is true, we can use the above-mentioned cutting level δ^P to generate first a cycle-free crisp relation and then a strict partial order that extends $>_P$.

References

[1] B. De Baets and H. De Meyer, *Cycle-transitivity versus FG-transitivity*, Fuzzy Sets and Systems **152**, 2005, 249–270.

[2] K. De Loof, B. De Baets and H. De Meyer, *Exploiting the lattice of ideals representation of a poset*, Fundamenta Informaticae, 2006, to appear.

[3] B. De Schuymer, H. De Meyer and B. De Baets, *Cycle-transitive comparison of independent random variables*, Journal of Multivariate Analysis **96**, 2005, 352–373.

[3] B. De Schuymer, H. De Meyer and B. De Baets, *On the transitivity of comonotonic and countermonotonic comparison of random variables*, Journal of Multivariate Analysis, to appear.

[4] P. Fishburn, *Proportional transitivity in linear extensions of ordered sets*, J. Comb. Theory **B41**, 1986, 48–60.

[5] H. Levy, *Stochastic Dominance*, Kluwer Academic Publishers, Norwell, MA, 1998.

[6] R. Nelson, *An Introduction to Copulas*, Lecture Notes in Statistics **139**, Springer, New York, 1998.

[7] G. Patil and C. Taillie, *Multiple indicators, partially ordered sets, and linear extensions: Multi-criterion ranking and priorization*, Environmental and Ecological Statistics **11**, 2004, 199–228.

[8] Y. Yu, *On proportional transitivity of ordered sets*, Order **15**, 1998, 87–95.

On Neyman-Pearson Lemma for Crisp, Random and Fuzzy Hypotheses

Adel Mohammadpour[1,2] and Ali Mohammad-Djafari[2]

[1] Dept. of Stat., Faculty of Math. and Computer Sciences, Amirkabir Univ. of Thec. (Tehran Polytechnic) and Statistical Research Center
adel@aut.ac.ir
[2] Laboratoire des Signaux et Systèmes, L2S, CNRS-Supélec-Univ. Paris 11
{mohammadpour,djafari}@lss.supelec.fr

Summary. We show that the best test for fuzzy hypotheses in the Bayesian framework is equivalent to Neyman-Pearson lemma in the classical statistics.

Keywords: Neyman-Pearson lemma, random and fuzzy hypotheses

1 Introduction

In this work we try to clarify the relations between simple crisp hypothesis testing in the Bayesian and fuzzy framework. To do this in a rigorous and simple way, we start first by presenting these approaches using the same notations, then we give a simple example and finally we show the relation which exists among them in the next sections.

Let X be a random variable with probability density function (pdf) $f(x|\theta)$, denoted by $X \sim f(x|\theta)$, where θ is an unknown parameter. A hypothesis H is called simple if distribution (pdf) of X does not depend on unknown parameter under H, [22]. For example, the following hypotheses are simple

$$\begin{cases} H_0 : \theta = \theta_0 \\ H_1 : \theta = \theta_1 \end{cases},$$

(1)

where θ_0 and θ_1 are known fixed numbers.

Now assume that $X \sim f(x)$ and

$$f(x) = \int_{\mathbb{R}} f(x|\theta)\, \pi(\theta)\, \mathrm{d}\theta,$$

(2)

where $\pi(\theta)$ is a prior pdf for the random parameter θ. The hypotheses

A. Mohammadpour and A. Mohammad-Djafari: *On Neyman-Pearson Lemma for Crisp, Random and Fuzzy Hypotheses*,
Advances in Soft Computing **6**, 61–69 (2006)
www.springerlink.com

$$\begin{cases} H_0 : \theta \sim \pi_0(\theta) \\ H_1 : \theta \sim \pi_1(\theta) \end{cases}, \tag{3}$$

where $\pi_0(\theta)$ and $\pi_1(\theta)$ are known pdfs, called simple random hypotheses.

There is also another case of hypotheses. Similar to the above two cases the observations are crisp (ordinary), but we may have some fuzzy knowledge about the unknown parameter under the hypotheses. These fuzzy hypotheses can be expressed by membership functions. Let X be a random variable with pdf $f(x|\theta)$. Then the hypotheses

$$\begin{cases} H_0 : \theta \text{ is approximately } m_0(\theta) \\ H_1 : \theta \text{ is approximately } m_1(\theta) \end{cases} \text{ or } \begin{cases} H_0 : \theta \simeq m_0(\theta) \\ H_1 : \theta \simeq m_1(\theta) \end{cases}, \tag{4}$$

where m_0 and m_1 are known membership functions, called fuzzy hypotheses.

The best test can be found for simple versus simple hypothesis by the Neyman-Pearson lemma. A few authors had tried to find the best test for testing fuzzy hypotheses with crisp data, Arnold [1, 2] and Taheri and Behboodian [26]. In this paper we show that the best test function for testing random and fuzzy hypotheses in (3) and (4), based on crisp data, are equal if $\pi_0 = m_0$ and $\pi_1 = m_1$. On the other hand, we show that, the best test function for testing crisp and random hypotheses in (1) and (3) are equal if the distribution of X under each hypothesis H_i, $i = 0, 1$ for crisp and random hypotheses are equal.

We recall that, some authors work on testing fuzzy hypotheses and hypothesis testing with fuzzy observations. We classify some of previous works in hypothesis testing in Table 1.

Table 1. Trends on hypothesis testing for fixed sample size.

	Crisp Observations	Fuzzy Observations
Crisp Hypotheses in Classical Framework	... Fisher (1925) [16] Neyman & Pearson (1933) [23] ...	Casals et al. (1986,1989) [11, 9] Gil & Casals (1988) [18] Bandemer & Näther (1992) [4] Filzmoser & Viertl (2003) [15]
Fuzzy Hypotheses in Classical Framework	Watanabe & Imaizumi (1993) [29] Arnold (1995,1996,1998) [1, 3, 2] Taheri & Behboodian (1999) [26] Arnold & Gerke (2003) [3]	Saade & Schwarzlander [25] Saade (1994) [24] Kang et al. (2001) [21] Grzegorzewski (2002) [19]
Crisp Hypotheses in Bayesian Framework	... Box & Tiao (1973) [7] Cox & Hinkley (1974) [13] ...	Casals et al. (1986) [12] Frühwirth-Schnatter (1993) [17]
Fuzzy Hypotheses in Bayesian Framework	Delgado et al. (1985) [14] Taheri & Behboodian (2001) [27]	Casals & Gil (1990) [10] Casals (1993) [8] Taheri & Behboodian (2002) [28]

To make our point clear we propose to consider the following artificial example. Suppose we are interested in evaluating the diameters of washers produced by a factory and we know that the distribution of the difference of such diameters from a norm diameter, is normal, $N(\theta, \sigma^2)$, where σ^2 is known. For example, we test the hypotheses

$$
\begin{cases} H_0 : \theta = 0 \\ H_1 : \theta = 0.01 \end{cases} \quad \text{or} \quad \begin{cases} H_0 : f(x) = \frac{1}{\sqrt{2\pi\sigma^2}} e^{\frac{-1}{2\sigma^2} x^2} \\ H_1 : f(x) = \frac{1}{\sqrt{2\pi\sigma^2}} e^{\frac{-1}{2\sigma^2} (x-0.01)^2} \end{cases}, \tag{5}
$$

if we are interested in finding a positive shift. Now, consider testing prior distribution for random parameter θ based on an observation of X with pdf (2), where $X|\theta$ has a normal distribution with known variance, $N(\theta, \tau^2)$, and θ has normal distribution with pdf $\pi(\theta) = \frac{1}{\sqrt{2\pi}} e^{\frac{-1}{2}(\theta-\mu)^2}$ and μ is an unknown fixed parameter. We want to test

$$
\begin{cases} H_0 : \theta \sim \frac{1}{\sqrt{2\pi}} e^{\frac{-1}{2}\theta^2} \\ H_1 : \theta \sim \frac{1}{\sqrt{2\pi}} e^{\frac{-1}{2}(\theta-0.01)^2} \end{cases} \quad \text{or} \tag{6}
$$

$$
\begin{cases} H_0 : f(x) = \int_{\mathbb{R}} \frac{1}{\sqrt{2\pi\tau^2}} e^{\frac{-1}{2\tau^2}(x-\theta)^2} \frac{1}{\sqrt{2\pi}} e^{\frac{-1}{2}\theta^2} d\theta = \frac{1}{\sqrt{2\pi(\tau^2+1)}} e^{\frac{-x^2}{2(\tau^2+1)}} \\ H_1 : f(x) = \int_{\mathbb{R}} \frac{1}{\sqrt{2\pi\tau^2}} e^{\frac{-1}{2\tau^2}(x-\theta)^2} \frac{1}{\sqrt{2\pi}} e^{\frac{-1}{2}(\theta-0.01)^2} d\theta = \frac{1}{\sqrt{2\pi(\tau^2+1)}} e^{\frac{-(x-0.01)^2}{2(\tau^2+1)}} \end{cases}
$$

Note that, the hypotheses (5) and (6) are equal if $\sigma^2 = \tau^2 + 1$.

Finally, we want to test the following fuzzy hypotheses

$$
\begin{cases} H_0 : \text{There is no shift} \\ H_1 : \text{There is a positive shift} \end{cases} \quad \text{or} \quad \begin{cases} H_0 : \theta \text{ is approximately } 0 \\ H_1 : \theta \text{ is approximately } 0.01 \end{cases},
$$

which may be interpreted by

$$
\begin{cases} H_0 : \theta \simeq m_0(\theta) = 100((0.01+\theta)I_{[-0.01,0)} + (0.01-\theta)I_{[0,0.01]}(\theta)) \\ H_1 : \theta \simeq m_1(\theta) = 100(\theta I_{[0,0.01)}(\theta) + (0.02-\theta)I_{[0.01,0.02]}(\theta)) \end{cases} \tag{7}
$$

In Figure 1, we plot three types of hypotheses in this example. We can see that, crisp hypotheses can be considered as a special case of random or fuzzy hypotheses; and a random hypothesis is the same as fuzzy hypothesis if their corresponding prior and membership function is equal.

2 The Best Test

Lemma 1. *(Neyman-Pearson) Let X be a random vector with unkown pdf $f(x)$, $(X \sim f(x))$. For testing*

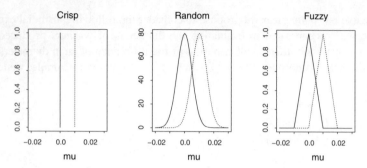

Fig. 1. The graphs of crisp, random and fuzzy hypotheses for H_0 and H_1 (left and right graphs in each figure respectively)

$$\begin{cases} H_0 : f(x) = f_0(x) \\ H_1 : f(x) = f_1(x) \end{cases},$$ (8)

where f_0 and f_1 are known pdfs. Any test with test function

$$\varphi(x) = \begin{cases} 1 & \text{if } f_1(x) > k f_0(x) \\ 0 & \text{if } f_1(x) < k f_0(x) \end{cases}$$ (9)

for some $k \geq 0$ is most powerful of its size.

See e.g., [22], for a complete version of Neyman-Pearson lemma. We recall that a MP test $\varphi(x)$, for a given α (probability of type *I* error) maximizes the power under H_1, where power is equal to $1 - \beta$ and β is the probability type *II* error. For simplicity sake (without loss of generality) we assume that all functions (except $\varphi(x)$) in this section are continuous.

2.1 Crisp Hypotheses

If $X \sim f(x|\theta)$, where θ is an unknown real parameter, then the following hypotheses are the same

$$\begin{cases} H_0 : f(x|\theta) = f(x|\theta_0) \\ H_1 : f(x|\theta) = f(x|\theta_1) \end{cases} \iff \begin{cases} H_0 : \theta = \theta_0 \\ H_1 : \theta = \theta_1 \end{cases},$$ (10)

where θ_0 and θ_1 are fixed known numbers. Note that (10) is a particular case of (8) when $f_0(x) = f(x|\theta_0)$ and $f_1(x) = f(x|\theta_1)$. We can define the probability of type *I* error as follow:

$$\alpha = \int_{\mathbb{R}^n} \varphi(x) f(x|\theta_0) \, \mathrm{d}x = \mathbb{E}[\varphi(X)|\theta = \theta_0],$$ (11)

and the probability of type *II* error, β, by the same way:

$$\beta = \int_{\mathbb{R}^n} (1 - \varphi(x)) f(x|\theta_1) \, \mathrm{d}x = \mathbb{E}[(1 - \varphi(X))|\theta = \theta_1].$$ (12)

2.2 Random Hypotheses

If $X \sim f(x) = \int_{\mathbb{R}} f(x|\theta) \, \pi(\theta) \, d\theta$, where θ is a random parameter (in the Bayesian framework) with unknown pdf π, then we call the hypotheses

$$\begin{cases} H_0 : \theta \sim \pi_0(\theta) \\ H_1 : \theta \sim \pi_1(\theta) \end{cases} \tag{13}$$

simple versus simple random hypothesis (π_0 and π_1 are two known prior pdfs). In this case we can simplify the problem of testing by calculating marginal pdf of $f(x)$ under H_0 and H_1 (call $f_0(x)$ and $f_1(x)$ respectively) as follows

$$f_0(x) = \int_{\mathbb{R}} f(x|\theta) \, \pi_0(\theta) \, d\theta \ \text{ and } \ f_1(x) = \int_{\mathbb{R}} f(x|\theta) \, \pi_1(\theta) \, d\theta. \tag{14}$$

Theorem 1. *By* (14), *the hypotheses* (13) *and* (8) *are the same, and the best test for testing* (13) *is given by the Neyman-Pearson lemma,* (9).

The probability of type *I* (or *II*) error, in (11), can be calculated by

$$\begin{aligned} \alpha &= \int_{\mathbb{R}^n} \varphi(x) \, f_0(x) \, dx \\ &= \int_{\mathbb{R}^n} \varphi(x) \int_{\mathbb{R}} f(x|\theta) \, \pi_0(\theta) \, d\theta \, dx \\ &= \int_{\mathbb{R}} \pi_0(\theta) \int_{\mathbb{R}^n} \varphi(x) \, f(x|\theta) \, dx \, d\theta \\ &= \int_{\mathbb{R}} \pi_0(\theta) \, \mathbb{E}(\varphi(X)|\theta) \, d\theta. \end{aligned} \tag{15}$$

Let π_0 and π_1 be two *unnormalized* priors, Gelman et al. (2004) page 62, such that

$$\int_{\mathbb{R}} \pi_0(\theta) \, d\theta = M < \infty \ \text{ and } \ \int_{\mathbb{R}} \pi_1(\theta) \, d\theta = N < \infty. \tag{16}$$

Then, such as (15), the probability of type *I* and *II* errors, in (11) and (12), are equal to

$$\alpha = \frac{1}{M} \int_{\mathbb{R}} \pi_0(\theta) \, \mathbb{E}[\varphi(X)|\theta] \, d\theta, \tag{17}$$

$$\beta = \frac{1}{N} \int_{\mathbb{R}} \pi_1(\theta) \, \mathbb{E}[(1 - \varphi(X))|\theta] \, d\theta. \tag{18}$$

2.3 Fuzzy Hypotheses

Some of statisticians, specially Bayesians, claim that the hypotheses such as (10) are not realistic, e.g.

> "It is rare and perhaps impossible to have a null hypothesis that can be exactly modeled as $\theta = \theta_0$", Berger and Delampady [5].

They suggest the following hypotheses instead of (10):

$$\begin{cases} H_0 : |\theta - \theta_0| \leq \varepsilon \\ H_1 : |\theta - \theta_1| > \varepsilon \end{cases}, \quad \text{where } \varepsilon \text{ is small and crisp.}$$

Even if some of the statisticians (e.g. Zellner [30]) do not agree with this extension, the people of fuzzy community are agree with this point of view and they extend this idea and propose the following hypotheses

$$\begin{cases} H_0 : |\theta - \theta_0| \text{ is } small \\ H_1 : |\theta - \theta_1| \text{ is } large \end{cases},$$

which consider to be closer to the real world problems, [26]. In this proposition *small* and *large* are expressed by two membership functions. That is, instead of testing (10) they test the fuzzy hypotheses denoted by:

$$\begin{cases} H_0 : \theta \simeq m_0(\theta) \\ H_1 : \theta \simeq m_1(\theta) \end{cases}, \quad \text{where } m_0 \text{ and } m_1 \text{ are known membership functions.} \quad (19)$$

We recall the following definitions of Taheri and Behboodian [26], which had proved the Neyman-Pearson lemma for testing (19), based on them.

Definition 1. *Any hypothesis of the form "$H : \theta$ is $m(\theta)$" (or $H : \theta \simeq m(\theta)$) is called a* fuzzy hypothesis*; where $m(\theta)$ is a membership function, a function from $\mathbb{R}$ to $[0,1]$.*

Definition 2. *Let $\varphi(X)$ be a test function, and let*

$$\int_{\mathbb{R}} m_0(\theta) \, d\theta = M < \infty \quad and \quad \int_{\mathbb{R}} m_1(\theta) \, d\theta = N < \infty.$$

The probability of type I error of $\varphi(X)$ is

$$\alpha_\varphi = \frac{1}{M} \int_{\mathbb{R}} m_0(\theta) \, \mathbb{E}[\varphi(X)|\theta] \, d\theta,$$

and the probability of type II error of $\varphi(X)$ is

$$\beta_\varphi = \frac{1}{N} \int_{\mathbb{R}} m_1(\theta) \, \mathbb{E}[(1 - \varphi(X))|\theta] \, d\theta.$$

Definition 3. *A test φ is said to be a test of* (significance) *level α if $\alpha_\varphi \leq \alpha$, where $\alpha \in [0,1]$, and we call α_φ the size of φ.*

Definition 4. *A test φ of level α is said to be a* best test *of level α , if $\beta_\varphi \leq \beta_{\varphi^*}$ for all tests φ^* of level α.*

Theorem 2. *Let X be a random vector, with pdf $f(x|\theta)$, where θ is an unknown parameter. For testing (19), any test with test function*

$$\varphi(x) = \begin{cases} 1 & \text{if } \int_{\mathbb{R}} f(x|\theta) \, m_1(\theta) \, d\theta > k \int_{\mathbb{R}} f(x|\theta) \, m_0(\theta) \, d\theta \\ 0 & \text{if } \int_{\mathbb{R}} f(x|\theta) \, m_1(\theta) \, d\theta < k \int_{\mathbb{R}} f(x|\theta) \, m_0(\theta) \, d\theta \end{cases} \quad (20)$$

for some $k \geq 0$, is the best test of its size.

Let π_0 and π_1 be unnormalized priors with range $[0, 1]$. Then the probability of type I and II errors in Definition 2 and $\{(16), (17), (18)\}$ are exactly the same.

Theorem 3. *Let m_0 and m_1 be unnormalized priors with range $[0, 1]$. Then the best test function for testing random hypotheses (13) and fuzzy hypotheses (19) are equal to (20).*

Therefore, we can prove the Neyman-Pearson lemma for testing fuzzy hypotheses (19), Theorem 2, by Theorem 1.

3 Conclusion

We showed that there is no difference between *simple versus simple* and *simple random versus simple random* hypotheses testing and then we showed that the best test function for *random* and *fuzzy* hypotheses are equal. That is, the best test for fuzzy hypotheses in the Bayesian framework is the Neyman-Pearson lemma in the classical statistics. But they may have different meanings.

Acknowledgments

The authors would like to thank reviewers for their helpful comments. The first author also grateful to LSS (CNRS-Supélec-Univ. Paris 11) and Statistical Research Center (Tehran) for the support of this work.

References

[1] B.F. Arnold: An approach to fuzzy hypothesis testing. Metrika **44** 119–126 (1996)
[2] B.F. Arnold: Testing fuzzy hypotheses with crisp data. Fuzzy Sets and Systems **94** 323–333 (1998)
[3] B.F. Arnold, O. Gerke: Testing fuzzy linear hypotheses in linear regression models. Metrika **57** 81–95 (2003)
[4] H. Bandemer, W. Näther: Fuzzy data analysis. Kluwer Academic Publishers, Dordrecht, (1992)
[5] J. Berger, M. Delampady: Testing precise hypotheses. Statistical Science **2** 317–352 (1987)
[6] A. Gelman, J.B. Carlin, H.S. Stern, D.B. Rubin: Bayesian Data Analysis. 2nd ed. Chapman & Hall/CRC, New York (2004)
[7] G.E. Box, G.C. Tiao: Bayesian inference in statistical analysis, Addison-Wesley, Reading, Mass (1973)
[8] M.R. Casals: Bayesian testing of fuzzy parametric hypotheses from fuzzy information, RAIRO, Operations Research **27** 189–199 (1993)

[9] M.R. Casals, M.A. Gil: A note on the operativness of Neyman-Pearson tests with fuzzy information. Fuzzy Sets and Systems **30** 215–220 (1989)

[10] M.R. Casals, P. Gil: Bayes testing of fuzzy hypotheses based on fuzzy information. (Spanish) Rev. Real Acad. Cienc. Exact. Fís. Natur. Madrid **84** 441–451 (1990)

[11] M.R. Casals, M.A. Gil, P. Gil: On the use of Zadeh's probabilistic definition for testing statistical hypotheses from fuzzy information. Fuzzy Sets and Systems **20** 175–190 (1986)

[12] M.R. Casals, M.A. Gil, P. Gil: The fuzzy decision problem: an approach to the problem of testing statistical hypotheses with fuzzy information, Eur. J. Oper. Res. **27** 371–382 (1986)

[13] D.R. Cox, D.V. Hinkley: Theoretical statistics. Wiley, New York, (1974)

[14] M. Delgado, J.L. Verdegay, M.A. Vila: Testing fuzzy hypotheses, a Bayesian approach. In M.M. Gupta, editor, Approximate reasoning in expert systems, North-Holland, Amsterdam, 307–316 (1985)

[15] P. Filzmoser, R. Viertl: Testing hypotheses with fuzzy data: the fuzzy p-value. Metrika, **59** 21–29 (2004)

[16] R.A. Fisher: Statistical methods for research workers: Oliver and Boyd, London (1925)

[17] S. Frühwirth-Schnatter: On fuzzy Bayesian inference, Fuzzy Sets and Systems **60** 41–58 (1993)

[18] M.A. Gil, M.R. Casals, An operative extension of the likelihood ratio test from fuzzy data. Statist. Hefte **29** 191–203 (1988)

[19] P. Grzegorzewski, Testing fuzzy hypotheses with vague data. Statistical modeling, analysis and management of fuzzy data, Stud. Fuzziness Soft Comput., Physica, Heidelberg **87** 213–225 (2002)

[20] L. Haekwan, H. Tanaka: Fuzzy approximations with non-symmetric fuzzy parameters in fuzzy regression analysis. J. Oper. Res. Soc. Japan **42** 98–112 (1999)

[21] M.-K. Kang, S.-W Jang, O.-J. Kwon: Testing statistical hypotheses for fuzzy data with agreement index, Far East J. Theor. Stat. **5** 343–352 (2001)

[22] E.L. Lehmann: *Testing statistical hypotheses.* Wiley, New York (1986)

[23] J. Neyman, E.S. Pearson: On the problem of the most efficient tests of statistical hypotheses. Philosophical Transactions of the Royal Society (A), **231** 289–337 (1933)

[24] J.J. Saade, Extension of fuzzy hypothesis testing with hybrid data: Fuzzy Sets and Systems **63** 57–71 (1994)

[25] J.J. Saade, H. Schwarzlander: Fuzzy hypothesis testing with hybrid data. Fuzzy Sets and Systems **35** 197–212 (1990)

[26] S.M. Taheri, J. Behboodian: Neyman-Pearson lemma for fuzzy hypotheses testing. Metrika **49,** 3–17 (1999)

[27] S.M. Taheri, J. Behboodian, A Bayesian approach to fuzzy hypotheses testing, Fuzzy Sets and Systems **123** 39–48 (2001)

[28] S.M. Taheri, J. Behboodian: Fuzzy hypotheses testing with fuzzy data: a Bayesian approach. In R.P. Nikhil and M. Sugeno, editors, Advances in Soft Computing, AFSS 2002, Springer, Berlin 527–533 (2002)

[29] N. Watanabe, T. Imaizumi: A fuzzy statistical test of fuzzy hypotheses, Fuzzy Sets and Systems **53** 167–178 (1993)

[30] A. Zellner: Comment on "Testing precise hypotheses" by Berger and Delampady. Statistical Science **2,** 339–341 (1987)

Fuzzy Probability Distributions Induced by Fuzzy Random Vectors

Wolfgang Trutschnig

Institute of Statistics and Probability Theory, Vienna University of Technology, Wiedner Hauptstraße 8 / 107, A-1040 Vienna
trutschnig@statistik.tuwien.ac.at

1 Introduction

As a matter of fact in many real situations uncertainty is not only present in form of randomness (stochastic uncertainty) but also in form of fuzziness (imprecision), for instance due to the inexactness of measurements of continuous quantities. From the probabilistic point of view the unavoidable fuzziness of measurements has (amongst others) the following far-reaching consequence: According to the classical Strong Law of Large Numbers (SLLN), the probability of an event B can be regarded as the limit of the relative frequencies of B induced by a sequence of identically distributed, independent, integrable random variables $(X_n)_{n \in \mathbb{N}}$ (with probability one).

Incorporating into considerations the fact that a realistic sample of a d-dimensional continuous quantity consists of d-dimensional fuzzy vectors, it is first of all necessary to generalize relative frequencies to the case of fuzzy samples, which yields so-called *fuzzy relative frequencies* and furthermore mandatory to consider and analyze *fuzzy-valued 'probabilities'* as generalization of classical probabilities.

In the sequel the definition of fuzzy relative frequencies and the most important properties of fuzzy relative frequencies regarded as fuzzy-valued set functions are stated. After that it is shown that similar to fuzzy relative frequencies every fuzzy random vector $X^\star$ naturally induces a fuzzy-valued 'probability', which will be called *fuzzy probability distribution induced by $X^\star$*.

Finally a SLLN for these fuzzy probability distributions will be stated.

Remark: All symbols used throughout the paper without explicit definition are explained at the end of the paper in the **List of Symbols**.

2 Fuzzy Relative Frequencies

Given a sample $x_1^\star, \cdots, x_n^\star \in \mathscr{F}_c^d$ of d-dimensional fuzzy vectors and a partition $K_1, K_2, \cdots, K_j$ of $\mathbb{R}^d$ the first important thing concerning frequencies and histograms

W. Trutschnig: *Fuzzy Probability Distributions Induced by Fuzzy Random Vectors*, Advances in Soft Computing **6**, 71–78 (2006)
www.springerlink.com

that has to be considered is, that it may happen, that an element $x_i^\star$ is not contained in a single class but partially lies in different classes.

In order to get a grip on this classification problem one can proceed as follows:

For every $\alpha \in (0,1]$ and every set $K \subseteq \mathbb{R}^d$ define the *lower relative frequency of level* α, denoted by $\underline{h}_{n,\alpha}(K)$, and the *upper relative frequency of level* α, denoted by $\overline{h}_{n,\alpha}(K)$, by:

$$\overline{h}_{n,\alpha}(K) := \frac{\#\{i : [x_i^\star]_\alpha \cap K \neq \emptyset\}}{n}$$
$$\underline{h}_{n,\alpha}(K) := \frac{\#\{i : [x_i^\star]_\alpha \subseteq K\}}{n} \tag{1}$$

Thus the lower relative frequency of level α counts all $i \in \{1,2,\cdots,n\}$ for which the α-cut of $x_i^\star$ is *contained* in the set K and divides by n, whereas the upper relative frequency of level α counts all $i \in \{1,2,\cdots,n\}$ for which the α-cut of $x_i^\star$ has *non-empty intersection* with the set K and divides by n.

Since obviously $\underline{h}_{n,\alpha}(K) \leq \overline{h}_{n,\alpha}(K)$ holds for every $n \in \mathbb{N}$, for every $\alpha \in (0,1]$ and every $K \subseteq \mathbb{R}^d$, it follows immediately that $\left([\underline{h}_{n,\alpha}(K), \overline{h}_{n,\alpha}(K)]\right)_{\alpha \in (0,1]}$ is a family of compact non-empty intervals (for n and K fixed) in α.

Furthermore (again for n and K fixed) it follows immediately from the definition that $\underline{h}_{n,\alpha}(K)$ increases if α increases and that $\overline{h}_{n,\alpha}(K)$ decreases if α increases. Consequently $\left([\underline{h}_{n,\alpha}(K), \overline{h}_{n,\alpha}(K)]\right)_{\alpha \in (0,1]}$ is a family of compact non-empty intervals that decreases if α increases, i.e.

$$\left[\underline{h}_{n,\alpha}(K), \overline{h}_{n,\alpha}(K)\right] \supseteq \left[\underline{h}_{n,\beta}(K), \overline{h}_{n,\beta}(K)\right]$$

holds for $\alpha \leq \beta$ and $\alpha, \beta \in (0,1]$.

It can be seen easily that unfortunately in general $\left([\underline{h}_{n,\alpha}(K), \overline{h}_{n,\alpha}(K)]\right)_{\alpha \in (0,1]}$ is not a family of α-cuts of a fuzzy number, but at least there exists a fuzzy number denoted by $h_n^\star(K) \in \mathscr{F}_c^1$ (in fact the *convex hull* as in equation (8)) and a finite set $N \subseteq (0,1]$, such that

$$[h_n^\star(K)]_\alpha = \left[\underline{h}_{n,\alpha}(K), \overline{h}_{n,\alpha}(K)\right] \tag{2}$$

holds for all $\alpha \in (0,1] \setminus N$.

In other words, building the convex hull of $\left([\underline{h}_{n,\alpha}(K), \overline{h}_{n,\alpha}(K)]\right)_{\alpha \in (0,1]}$ does not change the intervals $[\underline{h}_{n,\alpha}(K), \overline{h}_{n,\alpha}(K)]$ for every $\alpha \in (0,1] \setminus N$ (compare for instance [5]).

For every $K \subseteq \mathbb{R}^d$ the fuzzy number $h_n^\star(K)$ will be called the *fuzzy relative frequency of the set* K induced by the sample $x_1^\star, \cdots, x_n^\star \in \mathscr{F}_c^d$.

Regarding the fuzzy relative frequency $h_n^\star(\cdot)$ as fuzzy-valued set function on $\mathfrak{p}(\mathbb{R}^d)$ it has the following properties, which are easy to prove (notation as explained in the List of Symbols):

Lemma 1. *Suppose that* $x_1^\star, x_2^\star, \ldots, x_n^\star \in \mathscr{F}_c^d$ *is a fuzzy sample of size* n *and that* B, C *are arbitrary subsets of* $\mathbb{R}^d$, *then:*

1. $supp\big(h_n^\star(B)\big) \subseteq [0,1]$
2. $h_n^\star(\mathbb{R}^d) = \mathbf{1}_{\{1\}}, \quad h_n^\star(\emptyset) = \mathbf{1}_{\{0\}}$ *(Crisp extremal events)*
3. $B \subseteq C \subseteq \mathbb{R} \Rightarrow h_n^\star(B) \preceq h_n^\star(C)$ *(Monotonicity)*
4. $B \cap C = \emptyset \Rightarrow h_n^\star(B \cup C) \subseteq h_n^\star(B) \oplus h_n^\star(C)$ *(Sub/Superadditivity)*
5. $h_n^\star(B^c) = \mathbf{1}_{\{1\}} \ominus h_n^\star(B)$

Having in mind both the Strong Law of Large Numbers (compare [1]) and the fact that $h_n^\star(B)$ is a fuzzy number for every set $B \subseteq \mathbb{R}^d$ it is inevitable to consider fuzzy-valued 'probabilities' as generalizations of classical probabilities. Of course the question immediately arises, which properties a fuzzy-valued mapping $\mathbb{P}^\star : \mathscr{A} \to \mathscr{F}$ on a σ-algebra $\mathscr{A}$ should fulfill in order to be called fuzzy-valued 'probability', however Lemma 1 suggests what properties a meaningful notion at least must satisfy.

In the sequel such a notion will be called *fuzzy probability distribution* and it will be demonstrated how an arbitrary fuzzy random vector $X^\star$ (similar to the relative frequencies) induces a fuzzy probability distribution on the Borel σ-field $\mathscr{B}(\mathbb{R}^d)$ of $\mathbb{R}^d$. (For so-called *fuzzy probability distributions induced by fuzzy probability densities* compare [9].)

3 Fuzzy Random Vectors

If $(\Omega, \mathscr{A}, \mathscr{P})$ is a probability space and $X^\star : \Omega \to \mathscr{F}_c^d$ $(d \geq 1)$ is a fuzzy-vector-valued function on Ω, then the following abbreviation will be used for every $\omega \in \Omega$ and every $\alpha \in (0,1]$:

$$X_\alpha(\omega) := [X^\star(\omega)]_\alpha = \big\{x \in \mathbb{R}^d : \big(X^\star(\omega)\big)(x) \geq \alpha\big\} \tag{3}$$

Furthermore the graph $\Gamma(X_\alpha)$ is for every $\alpha \in (0,1]$ defined by

$$\Gamma(X_\alpha) := \big\{(\omega,x) : \omega \in \Omega \text{ and } x \in X_\alpha(\omega)\big\} \subseteq \Omega \times \mathbb{R}^d. \tag{4}$$

Definition 1. *Let* $(\Omega, \mathscr{A}, \mathscr{P})$ *be a probability space and* $\mathscr{B}(\mathbb{R}^d)$ *denote the Borel subsets of* $\mathbb{R}^d$. *Then a function* $X^\star : \Omega \to \mathscr{F}_c^d$ *is called* (d-dimensional) *fuzzy random vector if*

$$\{\omega \in \Omega : X_\alpha(\omega) \cap B \neq \emptyset\} \in \mathscr{A} \tag{5}$$

holds for every $B \in \mathscr{B}(\mathbb{R}^d)$ *and every* $\alpha \in (0,1]$.

Remark: If $X^\star$ satisfies Definition 1 and $d = 1$ then $X^\star$ is called fuzzy random variable.

It is well known from the theory of multifunctions (compare [3]) and the theory of random sets (compare [6]) that there exists a multitude of measurability conditions equivalent to (5) in Definition 1 if the probability space $(\Omega, \mathscr{A}, \mathscr{P})$ is complete. So for instance the following result holds (compare [6]):

Lemma 2. *Let* $(\Omega, \mathscr{A}, \mathscr{P})$ *be a complete probability space. Then for a fuzzy vector-valued mapping* $X^\star : \Omega \to \mathscr{F}_c^d$ *the following conditions are equivalent:*

1. $X^\star$ is a d-dimensional fuzzy random vector.
2. X_α is Effros-measurable measurable for every $\alpha \in (0,1]$, i.e. for G open
   $\{\omega \in \Omega : X_\alpha(\omega) \cap G \neq \emptyset\} \in \mathscr{A}$ holds for every $\alpha \in (0,1]$.
3. $\Gamma(X_\alpha)$ defined according to (4) is measurable for every $\alpha \in (0,1]$, i.e. $\Gamma(X_\alpha) \in \mathscr{A} \otimes \mathscr{B}(\mathbb{R}^d)$ holds for every $\alpha \in (0,1]$.
4. For every $\alpha \in (0,1]$ the mapping $X_\alpha : \Omega \to \mathscr{K}_c^d$ is measurable with respect to $\mathscr{B}((\mathscr{K}_c^d, \delta_H))$.

The following definition of independence will be used:

Definition 2 (Independence of fuzzy random vectors).
Suppose that $(\Omega, \mathscr{A}, \mathscr{P})$ is a complete probability space and that $X^\star : \Omega \to \mathscr{F}_c^d$ and $Y^\star : \Omega \to \mathscr{F}_c^d$ are d-dimensional fuzzy random vectors. Then $X^\star$ and $Y^\star$ are said to be independent if for arbitrary Borel sets $B_1, B_2 \in \mathscr{B}(\mathbb{R}^d)$ the following equality holds for every $\alpha \in (0,1]$:

$$\mathscr{P}(X_\alpha(\omega) \subseteq B_1, Y_\alpha(\omega) \subseteq B_2) = \mathscr{P}(X_\alpha(\omega) \subseteq B_1) \cdot \mathscr{P}(Y_\alpha(\omega) \subseteq B_2) \qquad (6)$$

4 Fuzzy Probability Distributions Induced by Fuzzy Random Variables and Fuzzy Random Vectors

Every d-dimensional fuzzy random vector $X^\star : \Omega \to \mathscr{F}_c^d$ induces families $(\underline{\pi}_\alpha)_{\alpha \in (0,1]}$ and $(\overline{\pi}_\alpha)_{\alpha \in (0,1]}$ of real-valued functions on $\mathscr{B}(\mathbb{R}^d)$ in the following way: For every $\alpha \in (0,1]$ and every $B \in \mathscr{B}(\mathbb{R}^d)$ define

$$\begin{aligned}\overline{\pi}_\alpha(B) &:= \mathscr{P}(\{\omega \in \Omega : X_\alpha(\omega) \cap B \neq \emptyset\}) \\ \underline{\pi}_\alpha(B) &:= \mathscr{P}(\{\omega \in \Omega : X_\alpha(\omega) \subseteq B\}).\end{aligned} \qquad (7)$$

Obviously for every $B \in \mathscr{B}(\mathbb{R}^d)$ and $\alpha \in (0,1]$ $\underline{\pi}_\alpha(B) \leq \overline{\pi}_\alpha(B)$ holds.

Using the fact that $\mathscr{P}$ is a probability measure, this shows that $[\underline{\pi}_\alpha(B), \overline{\pi}_\alpha(B)]$ is a nonempty, compact subinterval of $[0,1]$ for every $\alpha \in (0,1]$ and every $B \in \mathscr{B}(\mathbb{R}^d)$.

Suppose for the moment that $B \in \mathscr{B}(\mathbb{R}^d)$ is fixed and that $\alpha, \beta \in (0,1], \alpha \leq \beta$ holds, then it follows that $X_\alpha(\omega) \supseteq X_\beta(\omega)$ for every $\omega \in \Omega$, and that

$$\{\omega \in \Omega : X_\alpha(\omega) \subseteq B\} \subseteq \{\omega \in \Omega : X_\beta(\omega) \subseteq B\},$$

which shows that $\underline{\pi}_\alpha(B) \leq \underline{\pi}_\beta(B)$. Moreover

$$\{\omega \in \Omega : X_\beta(\omega) \cap B \neq \emptyset\} \subseteq \{\omega \in \Omega : X_\alpha(\omega) \cap B \neq \emptyset\},$$

which gives that $\overline{\pi}_\beta(B) \leq \overline{\pi}_\alpha(B)$.

This proves that for fixed $B \in \mathscr{B}(\mathbb{R}^d)$, $([\underline{\pi}_\alpha(B), \overline{\pi}_\alpha(B)])_{\alpha \in (0,1]}$ is a nested, monotonically decreasing family of non-empty compact intervals in α.

Unfortunately in general $([\underline{\pi}_\alpha(B), \overline{\pi}_\alpha(B)])_{\alpha \in (0,1]}$ is not a family of α-cuts of a fuzzy number as the following simple counter-example shows easily:

Example 1. Suppose that $(\Omega, \mathscr{A}, \mathscr{P})$ is an arbitrary probability space and let $\eta^{\star} \in \mathscr{F}_c^1$ be the triangular fuzzy number with α-cuts $[\eta^{\star}]_{\alpha} = [\alpha - 1, 1 - \alpha]$ for every $\alpha \in (0,1]$. Define $X^{\star} : \Omega \to \mathscr{F}_c^1$ by simply setting $X^{\star}(\omega) = \eta^{\star}$ for every $\omega \in \Omega$. Obviously $X^{\star}$ is a fuzzy random variable (the measurability condition obviously is fulfilled). Choosing $B = [-\frac{1}{2}, \frac{1}{2}] \in \mathscr{B}(\mathbb{R})$ gives

$$[\underline{\pi}_{\alpha}(B), \overline{\pi}_{\alpha}(B)] = \begin{cases} [0,1] & \text{if } \alpha < \frac{1}{2} \\ \{1\} & \text{if } \alpha \geq \frac{1}{2}. \end{cases}$$

If $\left([\underline{\pi}_{\alpha}(B), \overline{\pi}_{\alpha}(B)] \right)_{\alpha \in (0,1]}$ was a family of α-cuts of a fuzzy number, then

$$[\underline{\pi}_{\beta}(B), \overline{\pi}_{\beta}(B)] = \bigcap_{\alpha < \beta} [\underline{\pi}_{\alpha}(B), \overline{\pi}_{\alpha}(B)]$$

would hold for every $\beta \in (0,1]$. Choosing $\beta = \frac{1}{2}$ therefore would give

$$[\underline{\pi}_{\frac{1}{2}}(B), \overline{\pi}_{\frac{1}{2}}(B)] = \bigcap_{\alpha < \frac{1}{2}} [\underline{\pi}_{\alpha}(B), \overline{\pi}_{\alpha}(B)] = [0,1],$$

which is a contradiction to $[\underline{\pi}_{\frac{1}{2}}(B), \overline{\pi}_{\frac{1}{2}}(B)] = \{1\}$.

Nevertheless one can easily construct a fuzzy number denoted by $\mathbb{P}^{\star}(B) \in \mathscr{F}_c^1$ for every $B \in \mathscr{B}(\mathbb{R}^d)$ by again simply building the convex hull, i.e. for $x \in \mathbb{R}$ define

$$(\mathbb{P}^{\star}(B))(x) := \begin{cases} 0 & \text{if } x \notin [\underline{\pi}_{\alpha}(B), \overline{\pi}_{\alpha}(B)] \quad \forall \alpha \in (0,1] \\ \sup\{\alpha \in (0,1] : x \in ([\underline{\pi}_{\alpha}(B), \overline{\pi}_{\alpha}(B)]\} & \text{otherwise}. \end{cases} \tag{8}$$

It is well known (compare [5]) that the α-cuts

$$[\mathbb{P}^{\star}(B)]_{\alpha} =: [\underline{p}_{\alpha}(B), \overline{p}_{\alpha}(B)] \tag{9}$$

of $\mathbb{P}^{\star}(B)$ coincide with $[\underline{\pi}_{\alpha}(B), \overline{\pi}_{\alpha}(B)]$ for λ-almost every $\alpha \in (0,1]$.

Building the convex hull for every $B \in \mathscr{B}(\mathbb{R}^d)$ defines a fuzzy-valued mapping $\mathbb{P}^{\star} : \mathscr{B}(\mathbb{R}^d) \to \mathscr{F}_c^1$ that satisfies all properties stated in Lemma 1 on $\mathscr{B}(\mathbb{R}^d)$, which is now going to be proved in three steps:

Lemma 3. *Suppose that $(\Omega, \mathscr{A}, \mathscr{P})$ is an arbitrary probability space and that $X^{\star} : \Omega \to \mathscr{F}_c^d$ is a fuzzy random vector. Then the families $(\underline{\pi}_{\alpha})_{\alpha \in (0,1]}$ and $(\overline{\pi}_{\alpha})_{\alpha \in (0,1]}$, defined according to (7), fulfill the following assertions.*

1. *$\underline{\pi}_{\alpha}(\mathbb{R}^d) = \overline{\pi}_{\alpha}(\mathbb{R}^d) = 1, \quad \underline{\pi}_{\alpha}(\emptyset) = \overline{\pi}_{\alpha}(\emptyset) = 0 \quad \forall \alpha \in (0,1]$*
2. *If $A, B \in \mathscr{B}(\mathbb{R}^d), A \subseteq B$, then $\underline{\pi}_{\alpha}(A) \leq \underline{\pi}_{\alpha}(B)$ and $\overline{\pi}_{\alpha}(A) \leq \overline{\pi}_{\alpha}(B)$ holds for all $\alpha \in (0,1]$.*
3. *For every $\alpha \in (0,1]$ $\underline{\pi}_{\alpha}$ is superadditive and $\overline{\pi}_{\alpha}$ is subadditive, i.e. $\underline{\pi}_{\alpha}(A \cup B) \geq \underline{\pi}_{\alpha}(A) + \underline{\pi}_{\alpha}(B)$ and $\overline{\pi}_{\alpha}(A \cup B) \leq \overline{\pi}_{\alpha}(A) + \overline{\pi}_{\alpha}(B)$ holds if $A, B \in \mathscr{B}(\mathbb{R}^d)$ and $A \cap B = \emptyset$.*

4. *For every $A \in \mathscr{B}(\mathbb{R}^d)$ and every $\alpha \in (0,1]$ the identities $\underline{\pi}_\alpha(A^c) = 1 - \overline{\pi}_\alpha(A)$ and $\overline{\pi}_\alpha(A^c) = 1 - \underline{\pi}_\alpha(A)$ hold.*

Proof: The last assertion for example is an immediate consequence of the following identities:

$$
\begin{aligned}
\underline{\pi}_\alpha(A^c) &= \mathscr{P}(\{\omega \in \Omega : X_\alpha(\omega) \subseteq A^c\}) = 1 - \mathscr{P}(\{\omega \in \Omega : X_\alpha(\omega) \not\subseteq A^c\}) \\
&= 1 - \mathscr{P}(\{\omega \in \Omega : X_\alpha(\omega) \cap A \neq \emptyset\}) = 1 - \overline{\pi}_\alpha(A)
\end{aligned}
$$

$$
\begin{aligned}
\overline{\pi}_\alpha(A^c) &= \mathscr{P}(\{\omega \in \Omega : X_\alpha(\omega) \cap A \neq \emptyset\}) = 1 - \mathscr{P}(\{\omega \in \Omega : X_\alpha(\omega) \cap A^c = \emptyset\}) \\
&= 1 - \mathscr{P}(\{\omega \in \Omega : X_\alpha(\omega) \subseteq A\}) = 1 - \underline{\pi}_\alpha(A)
\end{aligned}
$$

The other assertions can be proved similarly. ■

Remark: In fact more properties of $\underline{\pi}_\alpha$ and $\overline{\pi}_\alpha$ can be proved.

The properties of $\underline{\pi}_\alpha$ and $\overline{\pi}_\alpha$ can be transfered to $\underline{p}_\alpha$ and $\overline{p}_\alpha$ respectively by using the following simple lemma.

Lemma 4. *Let α be an arbitrary but fixed real number in $(0,1]$ and suppose that $(\alpha_n)_{n\in\mathbb{N}}$ is a strictly increasing sequence in $(0,1)$ that converges to α. Then the following equality holds for every $B \in \mathscr{B}(\mathbb{R}^d)$:*

$$
\begin{aligned}
[\underline{p}_\alpha(B), \overline{p}_\alpha(B)] &= \bigcap_{\beta<\alpha} [\underline{\pi}_\alpha(B), \overline{\pi}_\alpha(B)] = \bigcap_{n=1}^{\infty} [\underline{\pi}_{\alpha_n}(B), \overline{\pi}_{\alpha_n}(B)] \\
&= \left[\lim_{n\to\infty} \underline{\pi}_{\alpha_n}(B), \lim_{n\to\infty} \overline{\pi}_{\alpha_n}(B) \right]
\end{aligned}
\tag{10}
$$

Proof: The first equality is an immediate consequence of the definition of the convex hull, the second equality follows directly from the properties of the sequence $(\alpha_n)_{n\in\mathbb{N}}$ and the third equality can be proved easily. ■

Theorem 1. *Suppose that $(\Omega, \mathscr{A}, \mathscr{P})$ is an arbitrary probability space and that $X^\star : \Omega \to \mathscr{F}_c^d$ is a fuzzy random vector. For every $B \in \mathscr{B}(\mathbb{R}^d)$ and every $\alpha \in (0,1]$ let $\underline{\pi}_\alpha(B)$ and $\overline{\pi}_\alpha(B)$ be defined according to (7), $\mathbb{P}^\star(B)$ defined according to (8) and $\underline{p}_\alpha(B)$ and $\overline{p}_\alpha(B)$ defined according to (9).*
Then $\mathbb{P}^\star : \mathscr{B}(\mathbb{R}^d) \to \mathscr{F}_c^1$ fulfills all the properties of $h_n^\star$ stated in Lemma 1.

Proof: For example if $B, C \in \mathscr{B}(\mathbb{R}^d)$ with $B \subseteq C$ and $\alpha \in (0,1]$ arbitrary, then applying Lemma 3 and equation (10) immediately shows that

$$
\overline{p}_\alpha(B) = \lim_{n\to\infty} \overline{\pi}_{\alpha_n}(B) \leq \lim_{n\to\infty} \overline{\pi}_{\alpha_n}(C) = \overline{p}_\alpha(C) \qquad \text{and}
$$

$$
\underline{p}_\alpha(B) = \lim_{n\to\infty} \underline{\pi}_{\alpha_n}(B) \leq \lim_{n\to\infty} \underline{\pi}_{\alpha_n}(C) = \underline{p}_\alpha(C)
$$

which proves monotonicity.

The other assertions can be proved in a similar manner. ■

Definition 3. *Under the above assumptions* $\mathbb{P}^\star$ *will be called* fuzzy probability distribution induced by $X^\star$.

Before stating a SLLN for fuzzy probability distributions induced by fuzzy random vectors and fuzzy relative frequencies define the metric $\delta_{H,p}^\star(\cdot,\cdot)$ by

$$\delta_{H,p}^\star(A^\star,B^\star) := \left(\int_{(0,1]} \left(\delta_H([A^\star]_\alpha, [B^\star]_\alpha) \right)^p d\lambda(\alpha) \right)^{1/p} \tag{11}$$

for every pair $A^\star, B^\star \in \mathscr{F}_c^d$ for which the integral exists (compare [5] and [8]).

Using this metric the following result can be proved:

Theorem 2 (SLLN for fuzzy relative frequencies).
Suppose that $(\Omega, \mathscr{A}, \mathscr{P})$ is a complete probability space, that $X^\star, X_1^\star, X_2^\star, \cdots$ are pairwise independent, identically distributed d-dimensional fuzzy random vectors, and that $B \in \mathscr{B}(\mathbb{R}^d)$ is an arbitrary Borel set.

Then there exists a set $N \in \mathscr{A}$, fulfilling $\mathscr{P}(N) = 0$, such that for every $\omega \in N^c$ the following identity holds ($p \in [1, \infty)$ arbitrary):

$$\lim_{n \to \infty} \delta_{H,p}^\star \left(h_n^\star(B, \omega), \mathbb{P}^\star(B) \right) = 0 \tag{12}$$

Proof: The paper with the corresponding proof is on the verge of submission.

Acknowledgement

This work was partly supported by the Jubiläumsfondsprojekt No. 11460 of the National Bank of Austria.

5 List of symbols

$\oplus$	Common sum of fuzzy numbers (and fuzzy vectors), defined via the Minkowski sum of the α-cuts
$\ominus$	Common difference of fuzzy numbers (and fuzzy vectors), defined via the Minkowski difference of the α-cuts
$[\cdot]_\alpha$	α-cut of a fuzzy number or of a fuzzy vector
#	Cardinality of a set
$\mathbf{1}_A$	Indicator function of the set A
$\mathscr{A}_1 \otimes \mathscr{A}_2$	Product σ-algebra
$[a,c] \preceq [b,d]$	Semiordering of intervals, defined by $[a,c] \preceq [b,c] :\Longleftrightarrow a \leq b$ and $c \leq d$
$\mathscr{B}(\mathbb{R}^d)$	Borel σ-algebra on $\mathbb{R}^d$
$\mathscr{B}\left((\mathscr{K}_c^d, \delta_H) \right)$	Borel σ-algebra generated by the Hausdorff metric δ_H

on $\mathcal{K}_c^d$

δ_H .... Hausdorff metric

$\mathcal{F}_c^1$ .... Set of all fuzzy numbers $x^\star$ having non-empty compact intervals as α-cuts for every $\alpha \in (0, 1]$

$\mathcal{F}_c^d$ .... Set of all d-dimensional fuzzy vectors $x^\star$ having non-empty compact convex sets α-cuts for every $\alpha \in (0, 1]$

$\mathcal{K}^d$ .... Family of all non-empty compact subsets of $\mathbb{R}^d$

$\mathcal{K}_c^d$ .... Family of all non-empty compact convex subsets of $\mathbb{R}^d$

$\mathfrak{p}(\mathbb{R}^d)$ .... Power set of $\mathbb{R}^d$

$\mathrm{supp}(\cdot)$ .... Support of a fuzzy number or of a fuzzy vector

$\xi^\star \preceq \eta^\star$ .... Semiordering of fuzzy numbers, defined by
$\xi^\star \preceq \eta^\star :\Longleftrightarrow [\xi^\star]_\alpha \preceq [\eta^\star]_\alpha$ for every $\alpha \in (0, 1]$

$\xi^\star \subseteq \eta^\star$ .... Inclusion of fuzzy numbers, defined by
$\xi^\star \subseteq \eta^\star :\Longleftrightarrow [\xi^\star]_\alpha \subseteq [\eta^\star]_\alpha$ for every $\alpha \in (0, 1]$

References

[1] H. Bauer: *Wahrscheinlichkeitstheorie*, W. de Gruyter Verlag, Berlin New York, 2002

[2] J.J. Buckley: *Fuzzy Probabilities*, Physica, Heidelberg New York, 2003

[3] C. Castaing, M. Valadier: *Convex Analysis and Measurable Multifunctions*, Lecture Notes in Mathematics, Springer, Berlin Heidelberg New York, 1977

[4] D. Hareter, R. Viertl: Fuzzy Information and Bayesian Statistics, in M. Lopez-Diaz, M.A. Gil, P. Grzegorzewski, O. Hryniewicz, J. Lawry (Eds.): *Soft Methodology and Random Information Systems*, Springer-Verlag, Heidelberg, pp. 392-398 (2004)

[5] V. Krätschmer: Some complete metrics on spaces of fuzzy subsets, *Fuzzy sets and systems* **130**, 357-365 (2002)

[6] I. Molchanov: *Theory of Random Sets*, Springer, London, 2005

[7] S. Niculescu, R. Viertl: Bernoulli's Law of Large Numbers for Vague Data, *Fuzzy Sets and Systems* **50**, 167-173 (1992)

[8] M.L. Puri, D.A. Ralescu: Fuzzy random variables, *J. Math. Anal. Appl.* **114**, 409-422 (1986)

[9] W. Trutschnig, D. Hareter: Fuzzy Probability Distributions, in M. Lopez-Diaz, M.A. Gil, P. Grzegorzewski, O. Hryniewicz, J. Lawry (Eds.): *Soft Methodology and Random Information Systems*, Springer-Verlag, Heidelberg, pp. 399-406 (2004)

On the Identifiability of TSK Additive Fuzzy Rule-Based Models

José Luis Aznarte M. and José Manuel Benítez

Department of Computer Science and Artificial Intelligence, University of Granada
{jlaznarte|j.m.benitez}@decsai.ugr.es

1 Introduction

Fuzzy Set Theory has been developed during the second half of last century, with a starting point in L.A. Zadeh seminal paper [14]. From that moment on, there has been a harsh debate between scientifics supporting it and others believing that it was an unnecesary mathematical construct, generally opposing it to Probability Theory. Those researchers usually complained about the alleged lack of mathematical soundness of Fuzzy Logic and its applications. For a succint review on this debate, see Section 1 of [2].

The truth is that, at least in the framework of time series analysis, the applications of Fuzzy Logic concepts have been characteristically naive in terms of their mathematical and statistical foundations. Fuzzy-based models tailored (or just applied) to forecast time series have bloomed in the last decades, and many of them have at least one thing in common: their almost complete ignorance of the proposals, tools and ideas of the Statistical time series modelling approach.

Although there have been signs of some advantages in the Soft-Computing framework for time series over the traditional one, this should not at all mean a complete disregard of the latter. In fact, hybridizing concepts and technologies is one of the core ideas of Soft-Computing, so a deep look into the Statistical time series approach should be a priority for any Soft-Computing researcher trying to face the problem of forecasting future values of a time series.

In this paper, we prove the global identifiability of a fuzzy rule-based model for time series modelling. This is a first step in a new look into time series modelling via Soft-Computing. The final aim is to develop a sound statistical framework for fuzzy models to gather the benefits of the traditional approach and of the new developments.

J.L. Aznarte M. and J.M. Benítez: *On the Identifiability of TSK Additive Fuzzy Rule-Based Models*, Advances in Soft Computing **6**, 79–86 (2006)
www.springerlink.com

2 Fuzzy Rule-Based Models. Notation

Let us start by defining the notation used throughout this paper. Henceforth we will refer to y_t as the value at time t of a time series $\{y_t\}$. $\tilde{\mathbf{x}}_t \in \mathbb{R}^D$ will refer to a $D \times 1$ vector of lagged values of y_t and/or exogenous variables. In order to further ease the reading, $\mathbf{x}_t \in \mathbb{R}^{D+1}$ is defined as $\mathbf{x}_t = [1, \tilde{\mathbf{x}}_t']'$, where the first element is sometimes called *intercept*.

The general nonlinear stochastic model is expressed as

$$y_t = \Psi(\tilde{\mathbf{x}}_t; \psi) + \varepsilon_t, \tag{1}$$

where $\Psi(\tilde{\mathbf{x}}_t; \psi)$ is a nonlinear function of the variables $\tilde{\mathbf{x}}_t$ with parameter vector ψ. As usual, $\{\varepsilon_t\}$ is a sequence of independently normally distributed random variables with zero mean and variance ς^2.

Fuzzy rule-based models (FRBM) are usually divided into two main types: the Mamdani model [5] and the Takagi-Sugeno-Kang (TSK) model [11]. These two types differ in the shape of the fuzzy rules employed, more precisely, in the consequents of those rules. The functional (usually linear) consequents of TSK-type models are considered to be a crucial factour determining their better approximation capabilities, and hence they have been traditionally preferred to Mamdani models for function approximation and time series forecasting. In this work, we will center out attention in TSK FRBM.

A fuzzy rule of type TSK has the following shape:

$$R^k : \quad \text{IF } x_1 \text{ IS } A_1^k \ \wedge \ x_2 \text{ IS } A_2^k \ \wedge \ \dots \ \wedge \ x_D \text{ IS } A_D^k \text{ THEN } y = f^k(\mathbf{x}). \tag{2}$$

This rule is premised on the vector $\mathbf{x} \in \mathbb{R}^D$. A_d^k is a fuzzy set in the input variable x_d in the kth rule domain, and $\wedge$ is a fuzzy conjunction operator (for more information on Fuzzy Logic-related concepts, see for example [4]).

The consequent $f^k(\mathbf{x})$ of these rules, which can be seen as a varying singleton, is a function describing the global input-output relathionship in a *localized* input-output space, that is, describes a part of the input-output map. This consequent is usually of a linear form, i.e.,

$$f^k(\mathbf{x}) = \mathbf{b}^{k'}\mathbf{x} = b_0^k + b_1^k x_1 + b_2^k x_2 + \dots + b_D^k x_D. \tag{3}$$

The so-called *firing strength* of the kth rule is obtained by taking the fuzzy conjunction of the membership functions of a rule's IF-part, that is,

$$\mu^k(\mathbf{x}) = \mu_1^k(x_1) \wedge \mu_2^k(x_2) \wedge \dots \wedge \mu_D^k(x_D), \tag{4}$$

where $\mu_d^k(x_d)$ is the membership degree of x_d to the fuzzy set A_d^k.

The inference procedure associated to such rules uses the fuzzy implication operator, $\rightarrow$, to obtain the fuzzy set represented by each rule, and the fuzzy disjunction operator, $\vee$, to join the mapped regions for all K rules in the output space. To obtain the final output of the model, a weighted average gravity method is usually used.

Depending on the choices made for the conjunction, implication and disjunction operators, different classes of the fuzzy model can be derived. In this work, we shall center our attention in the class using *multiplicative* conjunction, *multiplicative* implication and *additive* disjunction. The final defuzzified output of the system is hence

$$y^o = \sum_{k=1}^{K} \tilde{\mu}^k(\mathbf{x}) f^k(\mathbf{x}),$$ (5)

where $\tilde{\mu}^k(\mathbf{x}) = \frac{\mu^k(\mathbf{x})}{\sum_{k'=1}^{K} \mu^{k'}(\mathbf{x})}$.

The membership function, μ_d^k, can be chosen amongst several types, attending to diverse criteria. Traditionally, the most common have been triangular, trapezoidal, sigmoid, Gaussian... The latter is given by

$$\mu_d^k(x_d; c_d^k, \sigma_d^k) = \frac{1}{\sqrt{2\pi\sigma_d^k}} \exp\left(-\frac{(x_d - c_d^k)^2}{2\sigma_d^k}\right).$$ (6)

2.1 TSK FRBM for Univariate Time Series Analysis

If used to model a univariate time series $\{y_t\}$, we shall write the output of the FRBM as

$$y_t = \sum_{k=1}^{K} \tilde{\mu}^k(\mathbf{x}_t) f^k(\mathbf{x}_t) + \varepsilon_t.$$ (7)

Note that the presence of the error term ε_t is not common in the fuzzy literature, being usual in the probabilistic approach to time series. Notwithstanding, in this context we should consider the FRBM as a nonlinear stochastic model (1), and hence we must admit that not all the variability of the series can be explained by it. The unexplained remainder is the series $\{\varepsilon_t\}$. The parameter vector of this nonlinear stochastic model is

$$\psi = [\psi'_\mu, \psi'_f]' = [\psi'_{\mu^1}, \ldots, \psi'_{\mu^K}, \psi'_{f^1}, \ldots, \psi'_{f^K}],$$ (8)

where $\psi_{\mu^k} = [c_1^k, \sigma_1^k, \ldots, c_D^k, \sigma_D^k]$ are known as *nonlinear parameters* and $\psi_{f^k} = \mathbf{b}^k = [b_1^k, \ldots, b_D^k]$ as *linear parameters*.

We can now rewrite the model as

$$y_t = \sum_{k=1}^{K} \tilde{\mu}(\mathbf{x}_t; \psi_{\mu^k}) f(\mathbf{x}_t; \psi_{f^k}) + \varepsilon_t.$$ (9)

As it was noted [1], when applied in the time series framework, the consequent $f^k(\mathbf{x}_t)$ of a TSK fuzzy rule is equivalent to a linear AR model of order D, and the whole model can be seen as a generalisation of the smooth transition autoregressive (STAR) model [12] or as equivalent to the NCSTAR [7]. We will now use the results in [1] to derive the necessary and sufficient conditions for identifiability of FRBM.

3 Identifiability of FRBMs

If we consider the use of FRBM as Statistical modelling, we can see it as a procedure to specify the probability of the observations by a family of distributions, indexed by parameters. This procedure includes the statistical inference, the simulation and the prediction. All these depend on identifiable models, so it is important to study the identifiability conditions for FRBM.

For example, we must explicitly specify the sources of uniqueness of the model in order to guarantee convergence of the mean squared error (MSE) estimator function. This issue has been deeply studied in the nonlinear statistical models framework, including the feedforward neural network [6] and some derived models [9, 8].

Here we will adapt those results for the FRBM model, stating under which conditions identifiability is guaranteed. In order to do so, we will first discuss the concepts of minimality [10] or "nonredundancy" [3] and the concept of model reducibility.

Definition 1. *An FRBM model is* minimal *(or nonredundant) if its input-output map cannot be obtained from another FRBM with fewer rules.*

One of the sources of unidentifiability in an FRBM is the presence of irrelevant rules, that can be removed without affecting its modelling capabilities. Obviously, the minimality condition holds only for irreducible models.

Definition 2. *An FRBM model is* reducible *if one of the following conditions hold:*

i. *Some of the consequents of the rules vanish (*$\mathbf{b}^k \to 0$ *for some k).*
ii. *Some of the membership functions vanish (*$\sigma_d^k \to 0$ *for some k,d).*

Furthermore, we can define the property of identifiability as

Definition 3. *An FRBM is* identifiable *if there are no two sets of parameters such that the corresponding input-output maps are identical.*

There are two properties of FRBM that cause unidentifiability:

(P. 1) The *interchangeability* of the rules. The order in which rules are considered is totally irrelevant for the computations of the model but affects the search in the parameter space (giving place to multiple local maxima for the log-likelihood function).

(P. 2) The presence of *irrelevant* rules, i.e., if there is at least one rule with zero consequent ($\mathbf{b}^k = 0$ for some k) or if the conjunction of the membership functions is zero for at least one rule ($\sigma_d^k = 0$ for some k,d).

If we ensure that the model is irreducible, then we know that the only way to change the input-output map is through property (P. 1). This can be achieved in the style of [8] by applying a "specific-to-general" model building strategy based on statistical inference through Lagrange Multiplier (LM) linearity tests.

As it was proved in [3, 10], an irreducible model is minimal. This equivalence implies that there are no means, apart from the conditions stated in Definition 2 of reducibility, to further reduce the number of rules of a FRBM without changing the functional input-output map.

The problem of interchangeability of rules (P. 1) can be prevented by establishing a unique order among them. This might be ensured by defining (and forcing) a lexicographical order, $\prec$, among the rule antecendent parts. We first establish an order among every variable's membership functions, which is induced by the order of their location parameters c_d^k (not caring about their width σ_d^k). This order is usually given by their linguistic definition. Then, to compare (and sort) the rules, we apply the lexicographical order, which would result in the following Restriction:

$$\mu(\mathbf{x}_t; \psi_{\mu^k}) \prec \mu(\mathbf{x}_t; \psi_{\mu^{k+1}}), \quad k = 1, \dots, K \qquad (\text{R. 1})$$

This restriction defines a complete ordering for rules, which would allow us to write $R_i \prec R_{i+1}$.

By imposing (R. 1), we prevent the interchangeability of rules. We can thus guarantee that, if irrelevant rules do not exist, the model is identifiable and minimal.

In order to formally state the sufficient conditions under which the FRBM model is globally identifiable, and following [9], we need the following assumptions.

Assumption 1 *The linear parameters $\mathbf{b}^k$ do not vanish for any k. Furthermore, $\sigma_d^k > 0 \ \forall d, \forall k$.*

Assumption 2 *The covariate vector $\mathbf{x}_t$ has an invariant distribution that has a density everywhere positive in an open ball.*

Assumption 1 prevents from the effects of property (P. 2) and Assumption 2 avoids problems related to multicollinearity.

We will also make use of the following

Lemma 1. *The family of n-dimensional Gaussian cumulative distribution functions is linearly independent.*

Proof. Trivial in light of Proposition 2 and Theorem of [13].

Theorem 1. *Under restriction (R. 1) and Assumptions 1 and 2, the TSK additive FRBM is globally identifiable.*

Proof. Let us suppose two vector of parameters, $\psi = [\psi'_\mu, \psi'_f]'$ and $\overline{\psi} = [\overline{\psi'_\mu}, \overline{\psi'_f}]'$ such that

$$\sum_{i=1}^{K} \tilde{\mu}(\mathbf{x}_t; \psi_{\mu^i}) f(\mathbf{x}_t; \psi_{f^i}) = \sum_{j=1}^{K} \tilde{\mu}(\mathbf{x}_t; \overline{\psi_{\mu^j}}) f(\mathbf{x}_t; \overline{\psi_{f^j}}). \qquad (10)$$

To prove global identifiability of the FRBM we need to show that, under restriction (R. 1) and the assumptions, (10) is satisfied if and only if $\psi_{\mu^k} = \overline{\psi_{\mu^k}}$ and $\psi_{f^k} = \overline{\psi_{f^k}}$ for $k = 1, \dots, K$.

Assumption 1 clearly excludes the possibility of (10) being true when both sides of the equality are zero, so we shall study the other possibilities.

To ease the notation, we will note $\mu_i(\mathbf{x}_t) = \tilde{\mu}(\mathbf{x}_t; \psi_{\mu^i})$, $\overline{\mu_j}(\mathbf{x}_t) = \tilde{\mu}(\mathbf{x}_t; \overline{\psi_{\mu^j}})$, $\mathbf{b}_i\mathbf{x}_t = f(\mathbf{x}_t; \psi_{f^i})$ and $\overline{\mathbf{b}_i\mathbf{x}_t} = f(\mathbf{x}_t; \overline{\psi_{f^j}})$ henceforth in this proof. We can thus rewrite (10) as:

$$\sum_{i=1}^{K} \mu_i(\mathbf{x}_t)\mathbf{b}_i\mathbf{x}_t - \sum_{j=1}^{K} \overline{\mu_j}(\mathbf{x}_t)\overline{\mathbf{b}_j}\mathbf{x}_t = 0 \tag{11}$$

This equality can be true under two different situations:

i) If every $\mu_i(\mathbf{x}_t)$ is different from every $\overline{\mu_j}(\mathbf{x}_t)$, then we know that $\mathbf{b}_i\mathbf{x}_t = \overline{\mathbf{b}_j}\mathbf{x}_t = 0$, by Lemma 1.
Obviously, this would contradict Assumption 1.

ii) There exist i_1, j_1 such that $\mu_{i_1}(\mathbf{x}_t) = \overline{\mu_{j_1}}(\mathbf{x}_t)$.
We know that $\mu_l(\mathbf{x}_t) \neq \mu_m(\mathbf{x}_t)$ for $l \neq m$ and $\overline{\mu_l}(\mathbf{x}_t) \neq \overline{\mu_m}(\mathbf{x}_t)$ for $l \neq m$. Hence, we could write (11) as

$$\left(\mathbf{b}_{i_1} - \overline{\mathbf{b}_{j_1}}\right) \mathbf{x}_t \mu_{i_1}(\mathbf{x}_t) + \sum_{\substack{i=1 \\ i \neq i_1}}^{K} \mathbf{b}_i\mathbf{x}_t \mu_i(\mathbf{x}_t) - \sum_{\substack{j=1 \\ j \neq j_1}}^{K} \overline{\mathbf{b}_j}\mathbf{x}_t \overline{\mu_j}(\mathbf{x}_t) = 0 \tag{12}$$

This equation is similar to (11) in that it would be true under the same two situations. Hence, following the same rationale, we could further write it as

$$\left(\mathbf{b}_{i_1} - \overline{\mathbf{b}_{j_1}}\right) \mathbf{x}_t \mu_{i_1}(\mathbf{x}_t) + \left(\mathbf{b}_{i_2} - \overline{\mathbf{b}_{j_2}}\right) \mathbf{x}_t \mu_{i_2}(\mathbf{x}_t) +$$
$$\sum_{\substack{i=1 \\ i \neq i_1 \\ i \neq i_2}}^{K} \mathbf{b}_i\mathbf{x}_t \mu_i(\mathbf{x}_t) - \sum_{\substack{j=1 \\ j \neq j_1 \\ j \neq j_2}}^{K} \overline{\mathbf{a}_j}\mathbf{x}_t \overline{\mu_j}(\mathbf{x}_t) = 0 \tag{13}$$

Hence, we can proceed inductively (in k steps) up to

$$\left(\mathbf{b}_{i_1} - \overline{\mathbf{b}_{j_1}}\right) \mathbf{x}_t \mu_{i_1}(\mathbf{x}_t) + \ldots + \left(\mathbf{b}_{i_K} - \overline{\mathbf{b}_{j_K}}\right) \mathbf{x}_t \mu_{i_K}(\mathbf{x}_t) = 0, \tag{14}$$

which, as all the $\mu_{i_k}(\mathbf{x}_t)$ are distinct and hence linearly independent, forces $\mathbf{b}_{i_k} = \mathbf{b}_{j_k}$ for every k, resulting in $\psi = \overline{\psi}$. It also remarkable that, in 14, actually $i_k = j_k$ for $k = 1, \ldots, K$ because restriction (R. 1) holds, q.e.d.

We can also provide an alternative proof as follows: [2] stated the functional equivalence between fuzzy rule-based systems and Gaussian mixtures. In particular, TSK rule-based systems were proven to be equivalent to Gaussian mixtures of *equal priors*. Using this result, and knowing that Proposition 2 in [13] guarantees the identifiability of Gaussian mixtures, restriction (R. 1) gives as a result identifiable fuzzy rule-based systems. □

Theorem 1 applies to all FRBMs of the type mentioned in section 2, i.e., those that use TSK type rules, Gaussian membership functions and multiplicative conjunction and implication toghether with additive disjunction. This is the most common configuration for time series modelling.

The extention to other types of membership functions is straightforward as long as we can derive a result similar to Lemma 1. For example, if we were to prove identifiability for a FRBM whose membership function were the Cauchy density function,

$$\mu(x_d; \alpha, u) = \frac{\alpha}{\pi(\alpha^2 + (x - u)^2)},\tag{15}$$

where $\psi_{\mu^k} = [\alpha, u]$, we would follow the same schema except that we would have a Lemma equivalent to Lemma 1 by relying on Proposition 4 of [13].

Another example would be an FRBM which used as membership functions the difference of two sigmoids, in the spirit of the regime switching model of [9]. This membership function is defined in the multidimensional case as

$$\mu(\mathbf{x}_t; \psi_{\mu^k}) = -\left(\frac{1}{1 + \exp\left(\gamma^k(\mathbf{d}^k \cdot \mathbf{x}_t - \beta_1^k)\right)} - \frac{1}{1 + \exp\left(\gamma^k(\mathbf{d}^k \cdot \mathbf{x}_t - \beta_2^k)\right)} \right)\tag{16}$$

where $\psi_{\mu^k} = [\gamma^k, d_1^k, \ldots, d_D^k, \beta_1^k, \beta_2^k]$. In light of Lemma A.1 of [9], such a FRBM would be globally identifiable as well.

Another question worth to study is the effect of the restrictions posed by Theorem 1 concerning the input space fuzzy partitions allowed. One might think that the rule ordering restriction can somehow limit the validity of the result to just some cases of TSK FRBM.

As the reader might know, there are two main ways to partition the input space in the fuzzy subspaces which are covered by each rule. One alternative, called *grid partition* consists in setting a number of one-dimensional membership functions on each dimension and use as many rules as combinations of different membership functions there are. This results in every part of the input space covered by at least one rule. The other alternative is called *patched partition* and places multidimensional membership functions only in relevant parts of the space.

Actually, the ordering restriction allows for two rules to share at most all the one-dimensional membership functions but one. This is the usual situation when we have a grid type input space partition. Of course, it also allows for no multidimensional membership functions being shared amongst two rules, which is the case for patched type partition. Hence both main input space fuzzy partitioning schemes are covered by Theorem 1.

4 Conclusions

Due to their good performance, many models originated from the Soft-Computing area are increasingly being applied to time series modelling. They provide new perspectives and estimation procedures. However, authors in this area have not usally paid too much attention to important and desirable Statistical properties, like identifiability.

In our continuous work for a better and profitable merge of knowledge from both areas (Soft-Computing and Statistics), we address the identifiability of a wide class of Fuzzy Rule-Based Models. We have provided a formal proof of their global identifiability. This result will be completed with future research towards formally stating the properties of Soft-Computing approaches for time series modelling and forecast.

References

[1] J.L. Aznarte M., J.M. Benítez, and J.L. Castro. Equivalence relationships between fuzzy additive systems for time series analysis and smooth transition models. *IFSA 2005 World Conference*, Beijing, China, 2005.

[2] M-T. Gan, M. Hanmandlu, and A.H. Tan. From a gaussian mixture model to additive fuzzy systems. *IEEE Transactions on Fuzzy Systems*, 13(3):303–316, 2005.

[3] J.T.G. Hwang and A.A. Ding. Prediction intervals for artificial neural networks. *Journal of the American Statistical Association*, 92:109–125, 1997.

[4] G.J. Klir and B. Yuan Fuzzy sets and fuzzy logic. Prentice-Hall, 1995.

[5] E.H. Mamdani. Application of fuzzy logic to approximate reasoning using linguistic synthesis. *IEEE Transactions on Computers*, 26(12):1182–1191, 1977.

[6] M.C. Medeiros, T. Teräsvirta, and G. Rech. Building neural network models for time series: A statistical approach. *Journal of Forecasting*, 25(1):49–75, 2006.

[7] M.C. Medeiros and A. Veiga. A hybrid linear-neural model for time series forecasting. *IEEE Transactions on Neural Networks*, 11(6):1402–1412, 2000.

[8] M.C. Medeiros and A. Veiga. A flexible coefficient smooth transition time series model. *IEEE Transactions on Neural Networks*, 16(1):97–113, January 2005.

[9] M. Suarez-Fariñas, C.E. Pedreira, and M.C. Medeiros. Local global neural networks: a new approach for nonlinear time series modelling. *Journal of the American Statistical Association*, 2004.

[10] H.J. Sussman. Uniqueness of the weights for minimal feedforward nets with a given input-output map. *Neural Networks*, 5:589–593, 1992.

[11] T. Takagi and M. Sugeno. Fuzzy identification of systems and its applications to modeling and control. *IEEE Transactions on Systems, Man and Cybernetics*, 15:116–132, 1985.

[12] T. Teräsvirta. Specification, estimation and evaluation of smooth transition autoregresive models. *Journal of the American Statistical Association*, 89:208–218, 1994.

[13] S.J. Yakowitz and J.D. Spragins. On the identifiability of finite mixtures. *The Annals of Mathematical Statistics*, 38(1):209–214, 1968.

[14] L.A. Zadeh. Fuzzy sets. *Information and control*, 3(8):338–353, 1965.

An Asymptotic Test for Symmetry of Random Variables Based on Fuzzy Tools

González-Rodríguez. G.[1], Colubi, A.[1], D'Urso P.[2] and Giordani, P.[3]

[1] Dpto. de Estadística e I.O. Universidad de Oviedo. 33007 Spain
    {gil,colubi}@uniovi.es
[2] Dpto. di Scienze Economiche, Gestionali e Sociali. Università degli Studi del Molise.
    86100. Italy
    durso@unimol.it
[3] Dpto. di Statistica, Probabilità e Statistiche Applicate, Università degli Studi di Roma, "La
    Sapienza". 00185 Italy
    paolo.giordani@uniroma1.it

A new measure of skewness for real random variables is proposed in this paper. The measure is based on a fuzzy representation of real-valued random variables which can be used to characterize the distribution of the original variable through the *expected value* of the 'fuzzified' random variable. Inferential studies concerning the expected value of fuzzy random variables provide us with a tool to analyze the asymmetry degree from random samples. As a first step, we propose an asymptotic test of symmetry. We present some examples and simulations to illustrate the behaviour of the proposed test.

1 Introduction

In González-Rodríguez *et al.* [6] a family of useful fuzzy representations of real-valued random variables was introduced. Fuzzy representations map each real value into a fuzzy one such that the fuzzification of a real-valued random element leads to a fuzzy random variable (in Puri and Ralescu's sense [15]). The (fuzzy) expected values of some of these fuzzifications (which will be referred to as *characterizing fuzzy representations*) capture the whole information about the distribution of the original random variable. In this way, the distance between the distributions of two random variables can be quantified through the distance between the expected values of their characterizing fuzzy representations.

Since a random variable X has a symmetric distribution about $\theta \in \mathbb{R}$ if, and only if, $X - \theta$ and $\theta - X$ are identically distributed, we introduce a measure of the skewness based on a distance between the 'characterizing fuzzy expected values' of $X - \theta$ and $\theta - X$. This measure will be illustrated by means of some examples in Section 3.

G. González-Rodríguez et al.: *An Asymptotic Test for Symmetry of Random Variables Based on Fuzzy Tools*, Advances in Soft Computing **6**, 87–94 (2006)
www.springerlink.com

In the last years some inferential studies concerning the expected value of fuzzy random variables have been developed (see, for instance, [4], [5], [7], [8], [9], [10], [11], [12], [13] and [14]). Given that the proposed skewness measure is based on certain (fuzzy) expected values, some of these studies could be useful to analyze its inferential properties. As a first approach, we introduce an asymptotic test of symmetry. In order to illustrate the empirical behaviour of this test some simulation results are shown.

2 Preliminaries

Let $\mathcal{K}_c(\mathbb{R})$ be the class of the nonempty compact intervals of $\mathbb{R}$, and let $\mathcal{F}_c(\mathbb{R})$ be the class of the normal upper semicontinuous fuzzy sets of $\mathbb{R}$ with bounded closure of the support, that is,

$$\mathcal{F}_c(\mathbb{R}) = \left\{ U : \mathbb{R} \to [0,1] \mid U_\alpha \in \mathcal{K}_c(\mathbb{R}) \text{ for all } \alpha \in [0,1] \right\}$$

where U_α is the α-level of U (i.e. $U_\alpha = \{x \in \mathbb{R} \mid U(x) \geq \alpha\}$) for all $\alpha \in (0,1]$, and $U_0 = \mathrm{cl}\{x \in \mathbb{R} \mid U(x) > 0\}$. Zadeh's extension principle [16] allows us to define on $\mathcal{F}_c(\mathbb{R})$ a sum and a product by a scalar compatible with the usual arithmetic in $\mathcal{K}_c(\mathbb{R})$, namely, $(U+V)_\alpha = U_\alpha + V_\alpha$ and $(\lambda U)_\alpha = \lambda U_\alpha$ for all $U,V \in \mathcal{F}_c(\mathbb{R})$, $\lambda \in \mathbb{R}$ and $\alpha \in [0,1]$.

The support function of a fuzzy set $U \in \mathcal{F}_c(\mathbb{R})$ is $s_U(u,\alpha) = \sup_{w \in U_\alpha} \langle u, w \rangle$ for all $u \in \{-1,1\}$ and $\alpha \in [0,1]$, where $\langle \cdot, \cdot \rangle$ denotes the inner product. The support function allows to embed $\mathcal{F}_c(\mathbb{R})$ onto a cone of continuous and Lebesgue integrable functions $\mathcal{L}(\{-1,1\} \times [0,1])$ by means of the mapping $s : \mathcal{F}_c(\mathbb{R}) \to \mathcal{L}(\{-1,1\} \times [0,1])$ where $s(U) = s_U$ (see Diamond and Kloeden, [3]).

The (φ, W)-distance was introduced by Bertoluzza et al. [1] and it is defined by

$$D_W^\varphi(U,V) = \sqrt{\int_{[0,1]} \int_{[0,1]} [f_U(\alpha,\lambda) - f_V(\alpha,\lambda)]^2 \, dW(\lambda) d\varphi(\alpha)}$$

for all $U,V \in \mathcal{F}_c(\mathbb{R})$, with $f_U(\alpha,\lambda) = \lambda \sup U_\alpha + (1-\lambda) \inf U_\alpha$. The weight measures W and φ can be formalized as probability measures on $([0,1], \mathcal{B}_{[0,1]})$ ($\mathcal{B}_{[0,1]}$ being the Borel σ-field on $[0,1]$), W is assumed to be associated with a non-degenerate distribution and φ is assumed to correspond to a strictly increasing distribution function on $[0,1]$.

Let $(\Omega, \mathscr{A}, P)$ be a probability space. A mapping $\mathscr{X} : \Omega \to \mathcal{F}_c(\mathbb{R})$ is a *fuzzy random variable* (FRV) in Puri and Ralescu's sense (see Puri and Ralescu [15]) if for each $\alpha \in [0,1]$ the α-level mappings $\mathscr{X}_\alpha : \Omega \to \mathcal{K}_c(\mathbb{R})$, defined so that $\mathscr{X}_\alpha(\omega) = (\mathscr{X}(\omega))_\alpha$ for all $\omega \in \Omega$, are random sets (that is, Borel-measurable mappings w.r.t. the Borel σ-field generated by the topology associated with the well-known Hausdorff metric d_H on $\mathscr{K}(\mathbb{R})$). Alternatively, an FRV is an $\mathcal{F}_c(\mathbb{R})$-valued random element (i.e. a Borel-measurable mapping) when the Skorohod metric is considered on $\mathcal{F}_c(\mathbb{R})$ (see Colubi et al. [2]).

The *expected value (or mean)* of an integrably bounded FRV $\mathscr{X}$ (that is, $\mathscr{X}$ verifying that $\max\{\inf \mathscr{X}_0, \sup \mathscr{X}_0\} \in L^1(\Omega, \mathscr{A}, P)$), is the unique $\widetilde{E}(\mathscr{X}) \in \mathscr{F}_c(\mathbb{R})$ such that $\left(\widetilde{E}(\mathscr{X})\right)_\alpha =$ Aumman's integral of the random set $\mathscr{X}_\alpha$ for all $\alpha \in [0,1]$ (see Puri and Ralescu [15]), that is,

$$\left(\widetilde{E}(\mathscr{X})\right)_\alpha = \left\{E(f) \mid f : \Omega \to \mathbb{R}, f \in L^1, f \in \mathscr{X}_\alpha \text{ a.s. } [P]\right\}.$$

Consider the mapping $\gamma^{\mathcal{C}} : \mathbb{R} \to \mathscr{F}_c(\mathbb{R})$ which transforms each value $x \in \mathbb{R}$ into the fuzzy number whose α-level sets are

$$\left(\gamma^{\mathcal{C}}(x)\right)_\alpha = \left[f_L(x) - (1-\alpha)^{1/h_L(x)}, f_R(x) + (1-\alpha)^{1/h_R(x)}\right]$$

for all $\alpha \in [0,1]$, where $f_L : \mathbb{R} \to \mathbb{R}$, $f_R : \mathbb{R} \to \mathbb{R}$, $f_L(x) \le f_R(x)$ for all $x \in \mathbb{R}$, and $h_L : \mathbb{R} \to (0, +\infty)$, $h_R : \mathbb{R} \to (0, +\infty)$ are continuous and bijective. In González-Rodríguez *et al.* [6] it is proved that if $X : \Omega \to \mathbb{R}$ is a random variable and $f_L(X), f_R(X) \in L^1(\Omega, \mathscr{A}, P)$, then $\widetilde{E}(\gamma^{\mathcal{C}} \circ X)$ characterizes the distribution of X.

3 A Measure of the Skewness of a Random Variable

A random variable $X : \Omega \to \mathbb{R}$ is symmetric around $\theta \in R$ if, and only if, $X - \theta$ and $\theta - X$ are identically distributed. Consequently, if $f_L(X), f_R(X) \in L^1(\Omega, \mathscr{A}, P)$, we have that X is symmetric around θ if, and only if, $\widetilde{E}(\gamma^{\mathcal{C}} \circ (X - \theta)) = \widetilde{E}(\gamma^{\mathcal{C}} \circ (\theta - X))$ and hence $D_w^\varphi(\widetilde{E}(\gamma^{\mathcal{C}} \circ (X - \theta)), \widetilde{E}(\gamma^{\mathcal{C}} \circ (\theta - X)) = 0$. Intuitively, the greater this distance the lower the symmetry of X. Thus, in order to quantify the degree of skewness of X we define the $\gamma^{\mathcal{C}}$-*skewness measure* as

$$k_{\gamma^{\mathcal{C}}} = D_w^\varphi(\widetilde{E}(\gamma^{\mathcal{C}} \circ (X - \theta)), \widetilde{E}(\gamma^{\mathcal{C}} \circ (\theta - X)).$$

Example: In this context, a useful choice of characterizing fuzzy representation (which will be denoted by γ^0) is the one determined by $f_L(x) = f_R(x) = 0$, and

$$h_L(x) = \begin{cases} \dfrac{1}{1+x} & \text{if } x \ge 0 \\ 1-x & \text{if } x < 0 \end{cases}$$

and

$$h_R(x) = \frac{1}{h_L(x)} \quad \text{for all } x \in \mathbb{R}$$

In order to compare graphically and numerically the skewness through this fuzzification, we have considered three distributions. As an example of symmetric distribution, we have assumed X_1 to be normally distributed as an $\mathscr{N}(4, 1)$. To represent a skewed distribution, we have assumed X_2 behaving as a χ_4^2 random variable, and to consider an intermediate situation, we have supposed X_3 to be a mixture of the preceding distributions with mixing proportion $p = .5$.

The expected value of the three variables is 4, thus we will focus on the symmetry around $\theta = 4$.

In Figures 1, 2 and 3 we represent (on the right side) the characterizing expected values of $X_i - 4$ in comparison with $4 - X_i$ through γ^0, as well as the distance between them, for $i = 1, 2, 3$. In addition, as a reference we have also represented (on the left) the density functions of these variables.

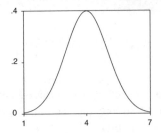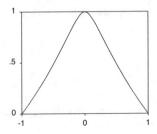

Fig. 1. $\mathcal{N}(4,1)$ distribution: density function (left) and characterizing expected values (right) of $X_1 - 4$ and $4 - X_1$ ($k_{\gamma^0} = 0$)

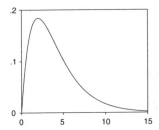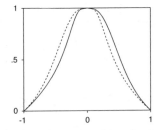

Fig. 2. χ_4^2 distribution: density function (left) and characterizing expected values (right) of $X_2 - 4$ and $4 - X_2$ ($k_{\gamma^0} = .0754$)

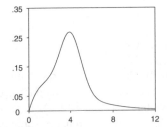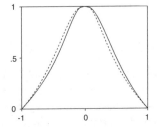

Fig. 3. .5-mixture of an $\mathcal{N}(4,1)$ and a χ_4^2 distribution: density function (left) and characterizing expected values (right) of $X_3 - 4$ and $4 - X_3$ ($k_{\gamma^0} = .03858$)

It should be noted that one of the main contributions of the applied fuzzification is to express the skewness by means of the distance between two *expected values*, which will make easier the inferential treatment. We observe clearly the effect of the skewness on the difference between the characterizing expected values corresponding to each distribution.

4 An Asymptotic Test of Symmetry of a Random Variable

In Section 3 we have indicated that a random variable X is symmetric around θ if, and only if, $\widetilde{E}\big(\gamma^{\mathcal{C}} \circ (X - \theta)\big) = \widetilde{E}\big(\gamma^{\mathcal{C}} \circ (\theta - X)\big)$ provided that $f_L(X), f_R(X) \in L^1(\Omega, \mathscr{A}, P)$. Thus, if we choose a fuzzification such that $f_L(X), f_R(X) \in L^1(\Omega, \mathscr{A}, P)$, it is equivalent testing $H_0 : X$ is symmetric than testing that fuzzy random variables $\gamma^{\mathcal{C}} \circ (X - \theta)$ and $\gamma^{\mathcal{C}} \circ (\theta - X)$ have the same expected value, that is,

$$H_0 : X \text{ is symmetric} \quad vs \quad H_1 : X \text{ is not symmetric}$$

$$\Leftrightarrow \quad H_0 : \widetilde{E}\big(\gamma^{\mathcal{C}} \circ (X - \theta)\big) = \widetilde{E}\big(\gamma^{\mathcal{C}} \circ (\theta - X)\big)$$

$$vs \quad H_1 : \widetilde{E}\big(\gamma^{\mathcal{C}} \circ (X - \theta)\big) \neq \widetilde{E}\big(\gamma^{\mathcal{C}} \circ (\theta - X)\big).$$

As a result we can employ the developments in [7] concerning the test for the equality of the expected values of two dependent fuzzy random variables to obtain the following testing procedure for symmetry.

Theorem 1. *Let $(\Omega, \mathscr{A}, P)$ be a probability space, $X : \Omega \to \mathbb{R}$ a random variable, $\gamma^{\mathcal{C}}$ a fuzzy representation so that $f_L(X), f_R(X) \in L^1(\Omega, \mathscr{A}, P)$ and $(X_1, \ldots, X_n)$ be a sequence of independent random elements distributed as X. If $D_W^\varphi(\gamma^{\mathcal{C}} \circ (X - \theta), \gamma^{\mathcal{C}} \circ (X - \theta)) \in L^1(\Omega, \mathscr{A}, P)$ is nondegenerate, to test at the nominal significance level $\alpha \in [0, 1]$*

$$H_0 : X \text{ is symmetric} \quad vs \quad H_1 : X \text{ is not symmetric}$$

the null hypothesis H_0 should be rejected if

$$\sqrt{n}\,\widehat{k}_{\gamma^{\mathcal{C}}}^{\,n} = \sqrt{n}\,D_W^\varphi\big(\gamma^{\mathcal{C}} \circ (\overline{X}_n - \theta), \gamma^{\mathcal{C}} \circ (\theta - \overline{X}_n)\big) > z_\alpha,$$

where $\overline{X}_n$ denotes the sample mean of X, and z_α is the $100(1 - \alpha)$ fractile of the distribution of a Gaussian variable on $\mathscr{L}(\{-1, 1\} \times [0, 1])$ with mean 0 and covariance function $C(u, \beta_1, v, \beta_x) = \mathrm{Cov}((s_{\gamma^{\mathcal{C}} \circ (X - \theta)} - s_{\gamma^{\mathcal{C}} \circ (\theta - X)})(u, \beta_1), (s_{\gamma^{\mathcal{C}} \circ (X - \theta)} - s_{\gamma^{\mathcal{C}} \circ (\theta - X)})(v, \beta_2))$ for all $(u, \beta_1), (v, \beta_2) \in \{-1, 1\} \times [0, 1]$.

In [7] several methods to apply in practice the testing procedure are analyzed, as well as an algorithm and some simulations that show a quite good behaviour for moderate/large samples irrespectively of the dependence degree between the involved fuzzy random variables. In this setting we can express the testing algorithm as follows:

Asymptotic testing procedure

Step 1: Compute the value of the statistic $T = \sqrt{n}\, D_w^{\varphi}\big(\widehat{G}_n(X-\theta), \widehat{G}_n(\theta-X)\big)$

Step 2: Obtain a random sample $(X_1^*, \ldots, X_m^*)$ from $(X_1, \ldots, X_n)$ and compute the
value $T = \sqrt{n}\, D_w^{\varphi}\big(\gamma^{\mathcal{L}} \circ (\overline{X}_m^* - \theta) + \gamma^{\mathcal{L}} \circ (\theta - \overline{X}_n), \gamma^{\mathcal{L}} \circ (\overline{X}_n - \theta) + \gamma^{\mathcal{L}} \circ (\theta - \overline{X}_m^*)\big).$

Step 3: Repeat step 2 a large number b of times and approximate the p–value as the
proportion of values in $\{T_1^*, \ldots, T_b^*\}$ greater than T.

5 Simulation Studies

In order to show the empirical behaviour of the proposed test, we have simulated
samples from the populations considered in Example 1. Specifically, ???

Each simulation corresponds to 10,000 iterations of the test at a nominal significance level .05 for different sample sizes n. Both measures W and φ involved in the
the distance have been chosen to be the Lebesgue one on $[0,1]$. In order to approximate the Gaussian process, we have considered $m = 10,000$. In Table 1 we see that,
moderate/large samples are required to order to obtain suitable results, as usual for
the asymptotic test in [7].

Table 1. Empirical percentage of rejections under H_0.

	$n = 30$	$n = 100$	$n = 300$
$p = .0$	6.39	5.78	4.96

6 Concluding Remarks

This paper means an introductory work for analyzing the skewness of a real distribution by means of certain fuzzy representations. In order to illustrate the effect of
the fuzzy representations, we have shown a simple one, but they can be chosen to
reflect the asymmetry depending on different parameters. It seems to be useful to
take advantage of the versatility of the family of fuzzy representations in connection
with this topic both from a descriptive and an inferential point of view. Further investigations in this respect are being carried out. In addition, although the behaviour
of the $\gamma^{\mathcal{L}}$-skewness measure in the cases analyzed in this paper seems suitable, it is
essential to make comparison with other approaches in the literature.

Acknowledgement

The research in this paper has been partially supported by the Spanish Ministry of
Education and Science Grant MTM2005-00045. Its financial support is gratefully
acknowledged.

References

[1] C. Bertoluzza, N. Corral, and A. Salas. On a new class of distances between fuzzy numbers. *Mathware Soft Comput.*, 2:71–84, 1995.

[2] A. Colubi, J. S. Domínguez-Menchero, M. López-Díaz, and D. A. Ralescu. A $d_e[0,1]$-representation of random upper semicontinuous functions. *Proc. Amer. Math. Soc.*, 130:3237–3242, 2002.

[3] P. Diamond and P. Kloeden. *Metric Spaces of Fuzzy Sets: Theory and Applications*. World Scientific, Singapore, 1994.

[4] D. García, M. A. Lubiano, and M. C. Alonso. Estimating the expected value of fuzzy random variables in the stratified random sampling from finite populations. *Inform. Sci.*, 138:165–184, 2001.

[5] M.A. Gil, M. Montenegro, G. González-Rodríguez, A. Colubi, and M.R. Casals. Bootstrap approach to the multi-sample test of means with imprecise data. *Comput. Statist. Data Anal.*, 2006. (accepted, in press).

[6] G. González-Rodríguez, A. Colubi, and M.A. Gil. A fuzzy representation of random variables: an operational oool in exploratory analysis and hypothesis testing. *Comput. Statist. Data Anal.*, 2006. (accepted, in press).

[7] G. González-Rodríguez, A. Colubi, M.A. Gil, and P. D'Urso. An asymptotic two dependent samples test of equality of means of fuzzy random variables. In *COMPSTAT*, 2006. (in press).

[8] G. González-Rodríguez, M. Montenegro, A. Colubi, and M.A. Gil. Bootstrap techniques and fuzzy random variables: Synergy in hypothesis testing with fuzzy data. *Fuzzy Sets and Systems*, 2006. (accepted, in press).

[9] R. Körner. An asymptotic α-test for the expectation of random fuzzy variables. *J. Statist. Plann. Inference*, 83:331–346, 2000.

[10] M. A. Lubiano and M. A. Gil. Estimating the expected value of fuzzy random variables in random samplings from finite populations. *Statist. Papers*, 40:277–295, 1999.

[11] M. A. Lubiano, M. A. Gil, and M. López-Díaz. On the rao-blackwell theorem for fuzzy random variables. *Kybernetika*, 35:167–175, 1999.

[12] M. Montenegro, M. R. Casals, M. A. Lubiano, and M. A. Gil. Two-sample hypothesis tests of means of a fuzzy random variable. *Inform. Sci.*, 133:89–100, 2001.

[13] M. Montenegro, A. Colubi, M. R. Casals, and M. A. Gil. Asymptotic and bootstrap techniques for testing the expected value of a fuzzy random variable. *Metrika*, 59:31–49, 2004.

[14] M. Montenegro, G. González-Rodríguez, M. A. Gil, A. Colubi, and M. R. Casals. Introduction to ANOVA with fuzzy random variables. In M. López-Díaz, M. A. Gil, P. Grzegorzewski, O. Hryniewicz, and J. Lawry, editors, *Soft Methodology and Random Information Systems*, pages 487–494. Springer-Verlag, Berlin, 2004.

[15] M. L. Puri and D. A. Ralescu. Fuzzy random variables. *J. Math. Anal. Appl.*, 114:409–422, 1986.

[16] L.A. Zadeh. The concept of a linguistic variable and its application to approximate reasoning, II. *Inform. Sci.*, 8:301–353, 1975.

Exploratory Analysis of Random Variables Based on Fuzzifications

Colubi, A., González-Rodríguez. G., Lubiano, M.A. and Montenegro, M.[1]

Dpto. de Estadística e I.O. Universidad de Oviedo. 33007 Spain
{gil,colubi,lubiano,mmontenegro}@uniovi.es

In this paper we propose a new way of representing the distribution of a real random variable by means of the expected value of certain kinds of fuzzifications of the original variable. We will analyze the usefulness of this representation from a descriptive point of view. We will show that the graphical representation of the fuzzy expected value displays in a visible way relevant features of the original distribution, like the central tendency, the dispersion and the symmetry. The fuzzy representation is valuable for representing continuous or discrete distributions, thus, it can be employed both for representing population distributions and for exploratory data analysis.

1 Introduction

A family of fuzzy representations of real random variables has been proposed in [2]. Some of them were used to characterize the real distributions with inferential purposes. Some other ones capture visual information about the distributions by focusing mainly on the mean value and the variance, although these fuzzifications do not characterize the distribution and loose valuable information in descriptive analysis.

Actually, it seems quite complex to find a fuzzification in this family allowing to visualize properly any kind of distribution. However, on the basis on the same intuitive ideas, we find another family of fuzzification with valuable graphical properties. The fuzzy representation of a real random variable allows us to associate the distribution with the expected value of a fuzzy random variable. The one obtained in this paper will be referred to as *exploratory fuzzy expected value*.

The exploratory fuzzy expected value will allow to represent both continuous and discrete distribution, which leads to a double use. On one hand, population distributions will be graphically represented by displaying important features (mean value, variability, skewness, "density"). In this sense, it can be interpreted as a kind of "parametrical" density or distribution function.

On the other hand, it can be used with exploratory purposes. The aim of the exploratory and descriptive analysis is to gain understanding of data, which is one of the most important targets of the statistical analysis. Data visualization associated

A. Colubi et al.: *Exploratory Analysis of Random Variables Based on Fuzzifications*, Advances in Soft Computing **6**, 95–102 (2006)
www.springerlink.com

with the exploratory fuzzy expected value will allow to capture information about important features of the data, which will allow to formulate reasonable hypotheses that can later be checked using some of the inferential methods above-mentioned.

2 Preliminaries

Let $\mathscr{K}_c(\mathbb{R})$ be the class of the nonempty compact intervals of $\mathbb{R}$ and let $\mathscr{F}_c(\mathbb{R})$ be the class of the fuzzy subsets U of $\mathbb{R}$ such that the α-level sets $U_\alpha \in \mathscr{K}_c(\mathbb{R})$ for all $\alpha \in (0,1]$, where $U_\alpha = \{x \in \mathbb{R} \,|\, U(x) \geq \alpha\}$, and $U_0 = \mathrm{cl}\{x \in \mathbb{R} \,|\, U(x) > 0\}$. In this context, the *sendograph* of $U \in \mathscr{F}_c(\mathbb{R})$ is the region enclosed by U and the x-axis on U_0, and $\mathbf{A}(U)$ will denote the corresponding area.

The space $\mathscr{F}_c(\mathbb{R})$ can be endowed with a semilinear structure, induced by a sum and the product by a scalar, both based upon Zadeh's extension principle [4], in accordance with which the following properties can be derived $(U+V)_\alpha = U_\alpha + V_\alpha$ and $(\lambda U)_\alpha = \lambda U_\alpha$ for all $U, V \in \mathscr{F}_c(\mathbb{R})$, $\lambda \in \mathbb{R}$ and $\alpha \in [0,1]$.

Given a probability space $(\Omega, \mathscr{A}, P)$, a *fuzzy random variable* (FRV) associated with $(\Omega, \mathscr{A})$ is intended to be, in accordance with Puri and Ralescu [3], a mapping $\mathscr{X} : \Omega \to \mathscr{F}_c(\mathbb{R})$ such that for each $\alpha \in [0,1]$ the α-level mapping $\mathscr{X}_\alpha : \Omega \to \mathscr{K}_c(\mathbb{R})$, defined so that $\mathscr{X}_\alpha(\omega) = (\mathscr{X}(\omega))_\alpha$ for all $\omega \in \Omega$, is a random set (that is, a Borel-measurable mapping w.r.t. the Borel σ-field generated by the topology associated with the well-known Hausdorff metric d_H on $\mathscr{K}(\mathbb{R})$). Alternatively, an FRV is an $\mathscr{F}_c(\mathbb{R})$-valued random element (i.e. a Borel-measurable mapping) when the Skorohod metric is considered on $\mathscr{F}_c(\mathbb{R})$ (see Colubi *et al.* [1]).

A fuzzy random variable $\mathscr{X} : \Omega \to \mathscr{F}_c(\mathbb{R})$ is said to be *integrably bounded* if and only if, $\max\{|\inf X_0|, |\sup X_0|\} \in L^1(\Omega, \mathscr{A}, P)$. If $\mathscr{X}$ is an integrably bounded fuzzy random variable, the *expected value (or mean)* of $\mathscr{X}$ is the unique $\widetilde{E}(\mathscr{X}) \in \mathscr{F}_c(\mathbb{R})$ such that $(\widetilde{E}(\mathscr{X}))_\alpha = $ Aumman's integral of the random set $\mathscr{X}_\alpha$ for all $\alpha \in [0,1]$ (see Puri and Ralescu [3]), that is,

$$\left(\widetilde{E}(\mathscr{X})\right)_\alpha = \left\{E(f) \,\middle|\, f : \Omega \to \mathbb{R}, f \in L^1, f \in \mathscr{X}_\alpha \text{ a.s. } [P]\right\}.$$

3 The Exploratory Fuzzy Representation

A *fuzzy representation of a random variable* transforms crisp data (variable values) into fuzzy sets (the associated FRV values). The representations in [2] are mappings $\gamma^C : \mathbb{R} \to \mathscr{F}_c(\mathbb{R})$ which transforms each value $x \in \mathbb{R}$ into the fuzzy number whose α-level sets are

$$\left(\gamma^C(x)\right)_\alpha = \left[f_L(x) - (1-\alpha)^{1/h_L(x)}, f_R(x) + (1-\alpha)^{1/h_R(x)}\right]$$

for all $\alpha \in [0,1]$, where $f_L : \mathbb{R} \to \mathbb{R}$, $f_R : \mathbb{R} \to \mathbb{R}$, $f_L(x) \leq f_R(x)$ for all $x \in \mathbb{R}$, and $h_L : \mathbb{R} \to (0,+\infty)$, $h_R : \mathbb{R} \to (0,+\infty)$ are continuous and bijective. By varying functions f_L, f_R, h_L and h_R it is possible to get representing fuzzy random variables whose expected value capture visual information about different parameters of the distribution, however it seems complex to show jointly the most important ones.

In order to overcome this inconveniency, we will consider a new family of fuzzifications based on the same idea, that is, in such a way that the fuzzy expected value of the transformed random element capture important information about the original distribution.

Let $f : [0, \infty) \to [0, 1]$ be an injective function. We define the auxiliar functional $\gamma_f : \mathbb{R} \to \mathscr{F}_c(\mathbb{R})$ so that,

$$[\gamma_f(x)]_\alpha = \begin{cases} \left[0, x^2 + x^2 \left(\dfrac{1 - f(x)}{f(x)}\right)\left(\dfrac{f(x) - \alpha}{f(x)}\right)\right] & \text{if } 0 \leq \alpha \leq f(x) \\[2em] \left[0, x^2 \left(\dfrac{1 - \alpha}{1 - f(x)}\right)\right] & \text{if } f(x) < \alpha \leq 1 \end{cases} \quad (1)$$

for all $\alpha \in [0, 1]$ and $x \in [0, \infty)$. Term $x^2(1 - f(x))/f(x)$ has been defined to guarantee that the area of sendograph of $\gamma_f(x)$ is equal to x^2 (see Figure 1). This functional depends on the square values to make the variance visible in the exploratory fuzzy expected value.

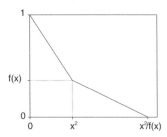

Fig. 1. Representation of the fuzzy set $\gamma_f(x)$

The family of *exploratory fuzzy representation* depends on a triple θ in a class

$$\Theta = \{(x_0, a, f) \mid x_0 \in \mathbb{R}, a \in \mathbb{R}^+, f : [0, \infty) \to [0, 1] \text{ injective}\}$$

where x_0 will be a kind of 'symmetry' point, a a scale parameter and f the function above defined. Thus, if $\text{sig}(x)$ is the sign of x, $\gamma^\theta : \mathbb{R} \to \mathscr{F}_c(\mathbb{R})$ is defined for $\theta = (x_0, a, f)$ so that

$$\gamma^\theta(x) = \mathbf{1}_{\{x\}} + \text{sig}(x - x_0)\gamma_f\left(\left|\dfrac{x - x_0}{a}\right|\right)$$

for all $x \in \mathbb{R}$.

If $X : \Omega \to \mathbb{R}$ is a real-valued random variable so that $EX^2 < \infty$ and $f : [0, \infty) \to [0, 1]$ is an injective function so that $(f(X))^{-1} \in L^1(\Omega, \mathscr{A}, P)$, then the *exploratory fuzzy expected value* is $\widetilde{E}(\gamma^\theta \circ X)$. It should be noted that condition $(f(X))^{-1} \in$

$L^1(\Omega, \mathscr{A}, P)$ is not restrictive, because functions like $f_p^\delta(x) = (p^x + \delta)/(1 + \delta)$ with $p \in (0, 1)$ and $\delta > 0$ for all $x \in \mathbb{R}$ verifies it irrespectively of X.

In this paper, we have considered $\theta_s = (EX, 1, f_{.6}^{.001}) \in \Theta$, which is a very simple and useful choice. Thus, the γ^{θ_s}-fuzzy representation of a random variable allows us to easily visualize features like the central tendency, variability, skewness, type of variable (discrete/continuous), and the existence of extreme values. More precisely, we can state that

If X is a random variable and

$$\gamma^{\theta_s} = 1_{\{x\}} + sig(x - EX)\gamma_{f_{.6}^{.001}}(|x - EX|)$$

for all $x \in \mathbb{R}$, where γ_f is defined as in (1) and

$$f_{.6}^{.001}(x) = \frac{.6^x + .001}{1.001},$$

then

i) $(\widetilde{E}(\gamma^{\theta_s} \circ X))_1 = \{EX\}$ (that is, the 1-level set shows a **mean value** of X).
ii) $A(\widetilde{E}(\gamma^{\theta_s} \circ X)) = Var(X)$ (that is, the area of the sendograph shows the **variance** of X).
iii) The symmetry of $\widetilde{E}(\gamma^{\theta_s} \circ X)$ is connected with the symmetry of X around its mean value. The more skewness of X the more asymmetry of $\widetilde{E}(\gamma^{\theta_s} \circ X)$. Thus, the asymmetry of the exploratory fuzzy expected value shows the **skewness** of X.
iv) If X is a continuous variable, then $\widetilde{E}(\gamma^{\theta_s} \circ X)$ will be "smooth" (excepting at EX), whereas if it is discrete, the exploratory fuzzy expected value will show non-smooth changes of slope in each of the values X takes on (that is, the "smoothness" allows us to distinguish the **discrete** and **continuous** distributions).
v)   Large values of X will be associated with large-spread 0-level sets (that is, thus the spread of the lower α-level sets can be useful to determine the presence of **extreme values**).

In the following sections we will illustrate this properties by representing the exploratory fuzzy expected value of some relevant population/sample distributions.

4 Exploratory Analysis of Random Variables Through the Fuzzy Representation

In this Section the graphical representation of the exploratory fuzzification of different parametric distributions will be shown. Concretely, we will focus on the binomial, the poisson, the exponential, the normal and the χ^2 distribution. They have been chosen in order to show the different features of the exploratory fuzzy expected value that we have indicated in the preceding section. The distributions were approximated by Monte Carlo method on the basis of 100000 simulations.

In Figure 2 we show the exploratory fuzzy expected value of two random variables with binomial distributions. In both cases $n = 5$, but $p = .5$ at the left graphic

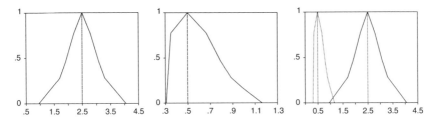

Fig. 2. Exploratory fuzzy expected value associated with $\mathscr{B}(5, .5)$ (left) and $\mathscr{B}(5, .1)$ (center) distributions. Comparison (right)

and $p = .1$ at the center one. We can see the respective mean values at the 1-level sets. The symmetry of the $\mathscr{B}(5, .5)$ and the skewness of the $\mathscr{B}(5, .1)$ is evident. It should be noted that the area of the sendograph shows the variance, although to make comparisons we have to take into account the range of the supports. The graphic on the right shows both fuzzy representations in the same scale. The difference in the areas, associated to the variabilities, is clear. If the aim were to compare the two distributions irrespectively of the variance, we could make use of the scale parameter a. The discrete character of the binomial distribution is connected with the lack of smoothness of the fuzzy sets and the right spread of the 0-level of the binomial $\mathscr{B}(5, .1)$ shows the presence of values far away from the mean.

In Figure 3, random variables with Poisson and exponential distributions, both with expected value equal to 4, are represented. The most remarkable difference is the large left-spreads with respect to the mean value of the exponential distribution, which indicates that in the exponential distribution the values lower than the mean have a greater density than in the Poisson distribution. In this case, since the Poisson is discrete but not finite, the lack of smoothness is less evident than for the binomial. We can also observe than the exponential distribution is considerably more asymmetric and variable than the Poisson.

The χ^2 distributed random variables were chosen with 1 and 2 degrees of freedom (see Figure 4). The left-spreads with respect to the mean values are more homogeneous than those for the Poisson and the exponential distributions, which indicates that the low values w.r.t. the corresponding expected value are relatively less fre-

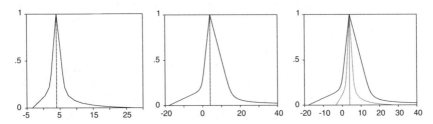

Fig. 3. Exploratory fuzzy expected value associated with $\mathscr{P}(4)$ (left) and Exp$(.25)$ (center) distributions. Comparison (right)

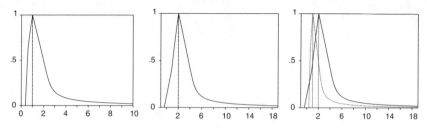

Fig. 4. Exploratory fuzzy expected value associated with χ_1^2 (left) and χ_2^2 (center) distributions. Comparison (right)

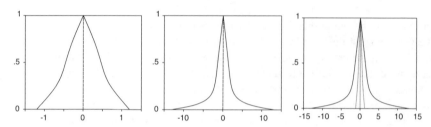

Fig. 5. Exploratory fuzzy expected value associated with $\mathcal{N}(0,1)$ (left) and $\mathcal{N}(0,2)$ (center) distributions. Comparison (right)

quent, mainly for the χ_1^2. The asymmetry of both distributions is evidenced and the greater variability of the χ_2^2 is easily noticed.

Finally, the exploratory fuzzy expected values corresponding to centered normal distributions with variances 1 and 4 are shown in Figure 5. The difference with the preceding distributions is obvious. As expected, the most similar shape to the standard normal distribution is the $\mathcal{B}(5,0.5)$, although we can see the difference in the smoothness of the curve. We observe the greater variability, the greater area and, in this case, the greater spreads for the 0-level.

5 Exploratory Data Analysis Through the Fuzzy Representation

When only data are available and the aim is to gain understanding of them, we can also make use of the graphical representation of the fuzzy mean. To illustrate it, we have simulated 4 samples with different sample sizes.

The exploratory fuzzy expected value associated with the first simulated samples are presented in Figure 6. We can observe the same features that we have commented in the preceding section. The distribution of sample 1 seems to be more skewed than the one in sample 2. The right spread of the 0-level in sample 1 seems to point out the presence of values quite greater than the mean. On the contrary, sample 2 seems to be quite symmetric around its mean. The sample sizes are quite low, although the clear lack of smoothness points out that there are repeated values, which indicates that they could come from discrete population distributions.

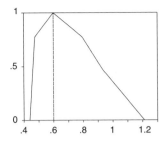
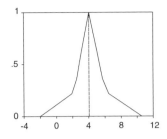

Fig. 6. Exploratory fuzzy expected value associated with sample 1 $n = 10$ (left) and sample 2 $n = 20$ (right)

In Figure 7 we present the graphical representation corresponding to the other simulated samples. We observe that the sample 4 is strongly asymmetric, with extreme values much greater than the sample mean, while sample 3 seems to be slightly asymmetric. The range of the supports suggests that the sample 3 is quite less variable than the sample 4. In this case the sample sizes are larger than in the preceding case, and the curves seems to be quite smooth, which suggests that the population distributions could be continuous.

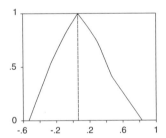
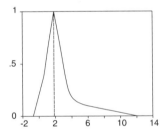

Fig. 7. Exploratory fuzzy expected value associated with sample 3 $n = 30$ (left) and sample 4 $n = 50$ (right)

Actually, sample 1 have been simulated form a $\mathscr{B}(5, .1)$, sample 2 from a $\mathscr{P}(4)$, sample 3 from a $\mathscr{N}(0, 1)$ and sample 4 from a χ_2^2. If we compare the population distributions with the sample ones, we can note the similarities.

Acknowledgement

The research in this paper has been partially supported by the Spanish Ministry of Education and Science Grant MTM2005-00045. Its financial support is gratefully acknowledged.

References

[1] A. Colubi, J. S. Domínguez-Menchero, M. López-Díaz, and D. A. Ralescu. A $d_e[0,1]$-representation of random upper semicontinuous functions. *Proc. Amer. Math. Soc.*, 130:3237–3242, 2002.

[2] G. González-Rodríguez, A. Colubi, and M.A. Gil. A fuzzy representation of random variables: an operational tool in exploratory analysis and hypothesis testing. *Comput. Statist. Data Anal.*, 2006. (accepted, in press).

[3] M. L. Puri and D. A. Ralescu. Fuzzy random variables. *J. Math. Anal. Appl.*, 114:409–422, 1986.

[4] L.A. Zadeh. The concept of a linguistic variable and its application to approximate reasoning, II. *Inform. Sci.*, 8:301–353, 1975.

A Method to Simulate Fuzzy Random Variables

González-Rodríguez. G.[1], Colubi, A.[1], Gil, M.A.[1] and Coppi, R.[2]

[1] Dpto. de Estadística e I.O. Universidad de Oviedo. 33007 Spain
{gil,colubi,magil}@uniovi.es
[2] Dpto. di Statistica, Probabilità e Statistiche Applicate, Università degli Studi di Roma, "La Sapienza"
renato.coppi@uniroma1.it

In this paper a method is introduced to simulate fuzzy random variables by using the support function. On the basis of the support function, the class of values of a fuzzy random variable can be 'identified' with a closed convex cone of a Hilbert space, and we now suggest to simulate Hilbert space-valued random elements and to project later into such a cone. To make easier the projection above we will consider isotonic regression. The procedure will be illustrated by means of several examples.

1 Introduction

In the literature on fuzzy-valued random variables, there are only a few references to modeling the distribution of these random elements. These models (for instance, see [12]) are theoretically well stated, but they are not soundly supported by empirical evidence, since they correspond to quite restrictive random mechanisms and hence they are not realistic in practice (see [4]).

Nevertheless, many probabilistic and statistical studies on fuzzy random variables would be better developed if simulation studies could be carried out (cf. [8], [9], [11]).

A similar situation arises in connection with functional data, to which a lot of attention is being paid in the last years, especially in which concerns random elements taking on values in Hilbert spaces (see, for instance, [14], [15]). The assumption of the Hilbert space structure is very helpful for simulation purposes (see [7], [1] or [16]).

The key idea in the methodology to be presented is first based on passing from the space of fuzzy random variable values into the Hilbert space of the corresponding integrable functions through the support function; then, one can generate Hilbert space-valued random elements and project them into the convex cone of the image of the space of fuzzy values. The projection theorem in Hilbert spaces validates the way to proceed and theoretically it would be possible to simulate all possible distributions on the space.

G. González-Rodríguez et al.: *A Method to Simulate Fuzzy Random Variables*, Advances in Soft Computing **6**, 103–110 (2006)
www.springerlink.com   © Springer-Verlag Berlin Heidelberg 2006

This idea can be easily implemented from a theoretical viewpoint. In practice, when fuzzy values to be dealt with are fuzzy sets of the one-dimensional Euclidean space the implementation does not entail important difficulties, since the support function of a fuzzy value is characterized by two real-valued functions on the unit interval, namely, the one associated with the infima and that associated with the suprema. These two functions are in the cone of the monotonic functions, and they are subject to the constraint of the infimum being lower than the supremum for each level. They have been analyzed in connection with some probabilistic problems (see [2]). However, for fuzzy sets of multi-dimensional Euclidean spaces, the practical developments become much more complex, although some alternatives to simplify them will be commented along the paper.

In this paper a procedure to simulate fuzzy random variables for which the shape of fuzzy values is not constrained will be introduced. In case there are some preferences on the shape of the considered fuzzy values the procedure could also adapted.

2 Preliminaries

Let $\mathcal{K}_c(\mathbb{R}^p)$ be the class of the nonempty compact convex subsets of $\mathbb{R}^p$ endowed with the Minkowski sum and the product by a scalar, that is, $A + B = \{a + b \,|\, a \in A, b \in B\}$ and $\lambda A = \{\lambda a \,|\, a \in A\}$ for all $A, B \in \mathcal{K}_c(\mathbb{R}^p)$ and $\lambda \in \mathbb{R}$. We will consider the *class of fuzzy sets*

$$\mathcal{F}_c(\mathbb{R}^p) = \left\{ U : \mathbb{R}^p \to [0, 1] \,\middle|\, U_\alpha \in \mathcal{K}_c(\mathbb{R}^p) \text{ for all } \alpha \in [0, 1] \right\}$$

where U_α is the α-level of U (i.e. $U_\alpha = \{x \in \mathbb{R}^p \,|\, U(x) \geq \alpha\}$) for all $\alpha \in (0, 1]$, and U_0 is the closure of the support of U. The space $\mathcal{F}_c(\mathbb{R}^p)$ can be endowed with the sum and the product by a scalar based on Zadeh's extension principle [17], which satisfies that $(U + V)_\alpha = U_\alpha + V_\alpha$ and $(\lambda U)_\alpha = \lambda U_\alpha$ for all $U, V \in \mathcal{F}_c(\mathbb{R}^p)$, $\lambda \in \mathbb{R}$ and $\alpha \in [0, 1]$.

The support function of a fuzzy set $U \in \mathcal{F}_c(\mathbb{R}^p)$ is $s_U(u, \alpha) = \sup_{w \in U_\alpha} \langle u, w \rangle$ for any $u \in \mathbb{S}^{p-1}$ and $\alpha \in [0, 1]$, where $\mathbb{S}^{p-1}$ is the unit sphere in $\mathbb{R}^p$ and $\langle \cdot, \cdot \rangle$ denotes the inner product. The support function allows us to embed $\mathcal{F}_c(\mathbb{R}^p)$ onto a cone of the continuous and Lebesgue integrable functions $\mathcal{L}(\mathbb{S}^{p-1})$ by means of the mapping $s : \mathcal{F}_c(\mathbb{R}^p) \to \mathcal{L}(\mathbb{S}^{p-1} \times [0, 1])$ where $s(U) = s_U$ (see [5]).

We will consider the *generalized metric* by Körner and Näther [10] D_K, which is defined so that

$$[D_K(U, V)]^2 =$$

$$\int_{(\mathbb{S}^{p-1})^2 \times [0,1]^2} \left(s_U(u, \alpha) - s_V(u, \alpha) \right) \left(s_U(v, \beta) - s_V(v, \beta) \right) dK(u, \alpha, v, \beta),$$

for all $U, V \in \mathcal{F}_c(\mathbb{R}^p)$, where K is a positive definite and symmetric kernel; thus, D_K coincides with a generic L_2 distance $\|\cdot\|_2$ on the Hilbert space $\mathcal{L}(\mathbb{S}^{p-1} \times [0, 1])$.

Let $(\Omega, \mathcal{A}, P)$ be a probability space. A *fuzzy random variable* (FRV) in Puri & Ralescu's sense [13] is a mapping $\mathcal{X} : \Omega \to \mathcal{F}_c(\mathbb{R}^p)$ so that the α-level mappings

$\mathscr{X}_{\alpha} : \Omega \rightarrow \mathscr{K}_c(\mathbb{R}^p)$, defined so that $\mathscr{X}_{\alpha}(\omega) = (\mathscr{X}(\omega))_{\alpha}$ for all $\omega \in \Omega$, are random sets (that is, Borel-measurable mappings with the Borel σ-field generated by the topology associated with the well-known Hausdorff metric d_H on $\mathscr{K}(\mathbb{R}^p)$). Alternatively, an FRV is an $\mathscr{F}_c(\mathbb{R}^p)$-valued random element (i.e. a Borel-measurable mapping) when the Skorokhod metric is considered on $\mathscr{F}_c(\mathbb{R}^p)$ (see [3]).

If $\mathscr{X} : \Omega \rightarrow \mathscr{F}_c(\mathbb{R}^p)$ is a fuzzy random variable such that $d_H(\{0\}, \mathscr{X}_0) \in L^1(\Omega, \mathscr{A}, P)$, then the *expected value (or mean)* of $\mathscr{X}$ is the unique $E(\mathscr{X}) \in \mathscr{F}_c(\mathbb{R}^p)$ such that $(E(\mathscr{X}))_{\alpha} =$ Aumman's integral of the random set $\mathscr{X}_{\alpha}$ for all $\alpha \in [0,1]$, that is,

$$(E(\mathscr{X}))_{\alpha} = \{E(X|P) \mid X : \Omega \rightarrow \mathbb{R}^p, X \in L^1(\Omega, \mathscr{A}, P), X \in \mathscr{X}_{\alpha} \ a.s. \ [P]\}.$$

3 Simulation of Fuzzy Random Variables Through Functional Random Variables

The space of fuzzy values $\mathscr{F}_c(\mathbb{R}^p)$ is a closed convex cone of the Hilbert space $\mathscr{L}(\mathbb{S}^{p-1} \times [0,1])$, and hence there exists a unique projection. As a consequence, given an arbitrary $f \in \mathscr{L}(\mathbb{S}^{p-1} \times [0,1])$ there is a unique fuzzy set $P(f) = A_f$ which corresponds to the anti-image of the support function of the projection of f onto the cone $s(\mathscr{F}_c(\mathbb{R}^p))$. We will denote by $P : \mathscr{L}(\mathbb{S}^{p-1} \times [0,1]) \rightarrow s(\mathscr{F}_c(\mathbb{R}^p))$ the projection function.

For any random element X taking on values in $\mathscr{L}(\mathbb{S}^{p-1} \times [0,1])$, the mapping $s^{-1} \circ P \circ X$ is a fuzzy random variable. In this way, if random elements of $\mathscr{L}(\mathbb{S}^{p-1} \times [0,1])$ are generated, random elements of $s(\mathscr{F}_c(\mathbb{R}^p))$ could be obtained through the projection P. Due to the fact that $s(\mathscr{F}_c(\mathbb{R}^p)) \subset \mathscr{L}(\mathbb{S}^{p-1} \times [0,1])$, we can guarantee that this method involves all the possible distributions on $s(\mathscr{F}_c(\mathbb{R}^p))$ and, since s is an isometry, by applying s^{-1} we would get all the possible distributions on $\mathscr{F}_c(\mathbb{R}^p)$.

The *theoretical method to generate $\mathscr{F}_c(\mathbb{R}^p)$-valued fuzzy random variables* consists in

Step 1

Simulating random elements on $\mathscr{L}(\mathbb{S}^{p-1} \times [0,1])$ by following the current directions in Functional Data Analysis (i.e., by considering bases either from a given function plus a noise term, or from discretized brownian motions, and so on).

Step 2

Projecting the simulated elements into the isometric cone of $\mathscr{F}_c(\mathbb{R}^p)$.

Step 3

Identifying the fuzzy set associated with the generated support function.

This theoretical method seems to be complex to implement in practice, although it would be feasible in some particular cases. Thus, in case $p = 1$, the unit sphere $\mathbb{S}^{p-1}$ reduces to the set $\{-1, 1\}$ whence the fuzzy set $A \in \mathscr{F}_c(\mathbb{R})$ can be characterized by means of two monotonic functions $s_A(-1, \cdot)$ and $s_A(1, \cdot)$ (see [2]) which satisfy certain constraints (since the infimum should always be lower than the supremum).

To make the problem easy to handle, fuzzy values can be reparameterized in terms of the left and right spreads with respect to the center of the 1-level. Once fuzzy values are reparameterized in such a way, arbitrary functions can be generated to construct later the function of the left spreads (for the infima) and the function of the right spreads (for the suprema). Since these two functions are monotonic and nonnegative, we can apply an algorithm of the isotonic regression restricted to positive values (see [6]). Later, the mid point of the 1-level would be generated at random and, along with the spreads simulated before, the infimum and supremum functions defining the fuzzy value would be obtained.

The *'practical' method to generate $\mathscr{F}_c(\mathbb{R})$-valued fuzzy random variables* we suggest in this paper can be summarized as follows:

Step $\mathscr{F}_c(\mathbb{R})$-1

To generate at random the mid-point of the 1-level, x_0, as well as two random functions on the Hilbert space $\mathscr{L}([0,1])$, $f_l, f_r : [0,1] \to R$ (there is no need for these functions to be generated independently).

Step $\mathscr{F}_c(\mathbb{R})$-2

To find the antitonic regressions of f_l^* and f_r^* to get the left and right spreads $s_l, s_r : [0,1] \to [0, \infty)$, respectively.

Step $\mathscr{F}_c(\mathbb{R})$-3

The α-levels of the fuzzy value A generated through *Steps $\mathscr{F}_c(\mathbb{R})$-1* and *$\mathscr{F}_c(\mathbb{R})$-2* would be given by $A_\alpha = [x_0 - s_l(\alpha), x_0 + s_r(\alpha)]$ (which is well-defined).

As we have commented before, the procedure above does not involve constraints on the shape of fuzzy values to be generated, although this type of constraint (like, for instance, to assume that x_0 is deterministic, functions f_i are linear functions, etc.) could be incorporated if required.

4 Some Illustrative Examples

We now illustrate the ideas in Section 3 by means of two examples. Since *Steps $\mathscr{F}_c(\mathbb{R})$-2* and *$\mathscr{F}_c(\mathbb{R})$-3* do not involve any random process, the differences in applying the algorithm are restricted to Step *$\mathscr{F}_c(\mathbb{R})$-1*. There are many ways of simulating random functions in the Hilbert space $\mathscr{L}([0,1])$. Some of them, as those based on a function plus a noise term or considering a class depending on real random parameters, can be easily imitated in $\mathscr{F}_c(\mathbb{R})$. However, the Hilbert spaces present some distinguishing characteristics, such as the generating basis, that can be taken into account to simulate random elements in a wider context.

In this section two ways of simulating from generating bases the functions f_1 and f_2 in *Step $\mathscr{F}_c(\mathbb{R})$-1* of the above-described procedure are detailed.

Consider a referential triangular fuzzy set Tri$(-1, 0, 1)$, which is equivalent to consider the spread functions $f_1(\alpha) = f_2(\alpha) = 1 - \alpha$ for all $\alpha \in [0,1]$. Since these spread functions correspond to linear functions, the trigonometric basis will be suitable to represent them. This basis is given by

$$\varphi_j(x) = \begin{cases} 1 & \text{if } j = 0 \\ \sqrt{2}\cos(\pi jx) & \text{if } j = 1, 2, \ldots \end{cases}$$

Coefficients of the spread functions in this basis are given by

$$\theta_j = \begin{cases} .5 & \text{if } j = 0 \\ 0 & \text{if } j \text{ is an even number} \\ \dfrac{2\sqrt{2}}{\pi^2 j^2} & \text{if } j \text{ is an odd number} \end{cases}$$

For practical purposes we will consider the approximation of the function corresponding to the first 21 terms of the linear combination (i.e., $j = 0, \ldots, 20$). Coefficients are distorted in a random way so that all the generated random functions follow the expression

$$\sum_{j=0}^{20} (\theta_j + \varepsilon_j)\varphi_j$$

where $(\varepsilon_0, \ldots, \varepsilon_{20})$ is a random vector.

The way of distorting the coefficients is crucial, since small perturbations can produce shapes completely different from the original one. It should be recalled that, in order to get well-defined fuzzy sets, we will need to apply an antitonic regression algorithm after the simulation of the functions in $\mathscr{L}([0,1])$. Thus, if the simulated functions are highly variable (in the sense of showing many monotonicity changes), the antitonic regression corresponding to the spreads will have many constant parts, and hence the obtained fuzzy set will present a lot of discontinuities. In order to illustrate this behaviour, we will firstly consider the following:

Case A. For the left spread a sequence of independent realizations, $\varepsilon_0^l, \ldots, \varepsilon_{20}^l$, are simulated from the normal distribution $\mathscr{N}(0, .01)$, and for the right spread a sequence of independent realizations, $\varepsilon_0^r, \ldots, \varepsilon_{20}^r$, are simulated from the normal distribution $\mathscr{N}(0, .1)$. Thus, we get two random functions

$$f_l = \sum_{j=0}^{20} (\theta_j + \varepsilon_j^l)\varphi_j \text{ and } f_r = \sum_{j=0}^{20} (\theta_j + \varepsilon_j^r)\varphi_j.$$

The mid-point of the 1-level is chosen at random from a normal distribution $\mathscr{N}(2, 1)$. By applying *Steps $\mathscr{F}_c(\mathbb{R})$-2* and *$\mathscr{F}_c(\mathbb{R})$-3*, a random fuzzy set is obtained.

In order to compare some particular realizations of the simulated fuzzy random variable with the expected value of such an element, we have made 10,000 simulations and we have approximated the (fuzzy) mean value by Monte Carlo method. In Figures 1 and 2 three simulated values and the corresponding mean value are shown. We can see that, although the perturbations were chosen to follow distributions with

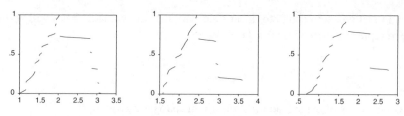

Fig. 1. Simulated values in **Case A**

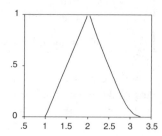

Fig. 2. Approximated mean value in the simulation in **Case A**

a relative small variability, the simulated fuzzy sets are quite different from the referential triangular fuzzy number and have many discontinuities. Nonetheless, the shape of the expected value is quite smooth and more similar to the referential fuzzy number. The difference between this mean value and the original triangular one is mainly due to the application of the antitonic regression algorithm (the expected value of the antitonic regression can be different from the antitonic regression of the expected value).

In order to obtain smoother shapes, we can simulate the perturbations in the coefficients with a decreasing weight as follows.

Case B. For the left spread a sequence of independent realizations $U_0^l, \ldots, U_{20}^l$ from the uniform distribution $\mathcal{U}_{(0,1)}$ are simulated, and the perturbations are considered so that $\varepsilon_0^l = U_0^l$, $\varepsilon_j^l = U_j^l \cdot \varepsilon_{j-1}^l$. For the right spread the same process is followed but using the beta distribution $\beta(5,3)$ instead of the uniform one. Again, the mid-point is chosen at random from a normal distribution $\mathcal{N}(2,1)$ and Steps $\mathcal{F}_c(\mathbb{R})$-2 and $\mathcal{F}_c(\mathbb{R})$-3 are followed to get the random fuzzy set. In Figures 3 and 4 three simulated values and the corresponding mean value (approximated by 10,000 realizations of the process) are shown. As expected, we can see smoother shapes than those in Case A, although they are also quite different and the greater the magnitude of right perturbations the greater the probability of discontinuities.

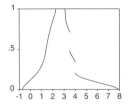 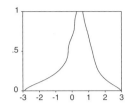 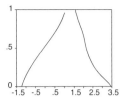

Fig. 3. Simulated values in **Case B**

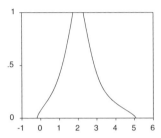

Fig. 4. Approximated mean value in the simulation in **Case B**

Acknowledgement

The research in this paper has been partially supported by the Spanish Ministry of Education and Science Grant MTM2005-00045. Its financial support is gratefully acknowledged.

References

[1] H. Cardot. Nonparametric regression for functional responses with application to conditional functional principal components analysis. Technical Report LSP-2005-01, Universite Paul Sabatier, 2005.

[2] A. Colubi, J. S. Domínguez-Menchero, M. López-Díaz, and R. Körner. A method to derive strong laws of large numbers for random upper semicontinuous functions. *Statist. Probab. Lett.*, 53:269–275, 2001.

[3] A. Colubi, J. S. Domínguez-Menchero, M. López-Díaz, and D. A. Ralescu. A $d_e[0,1]$-representation of random upper semicontinuous functions. *Proc. Amer. Math. Soc.*, 130:3237–3242, 2002.

[4] A. Colubi, C. Fernández-García, M. López-Díaz, , and M. A. Gil. Simulation of random fuzzy variables: an empirical approach to statistical/probabilistic studies with fuzzy experimental data. *IEEE Trans. Fuzzy Syst.*, 10:384–390, 2002.

[5] P. Diamond and P. Kloeden. *Metric Spaces of Fuzzy Sets: Theory and Applications*. World Scientific, Singapore, 1994.

[6] J.S. Domínguez-Menchero and G. González-Rodríguez. Analyzing an extension of the isotonic regression problem. 2006. (submitted).

[7] R. Fraiman and G. Muñiz. Trimmed means for functional data. *Test*, 10:419–440, 2001.

[8] M.A. Gil, M. Montenegro, G. González-Rodríguez, A. Colubi, and M.R. Casals. Bootstrap approach to the multi-sample test of means with imprecise data. *Comput. Statist. Data Anal.*, 2006. (accepted, in press).

[9] G. González-Rodríguez, A. Colubi, and M.A. Gil. A fuzzy representation of random variables: an operational tool in exploratory analysis and hypothesis testing. *Comput. Statist. Data Anal.*, 2006. (accepted, in press).

[10] R. Körner and W. Näther. On the variance of random fuzzy variables. In C. Bertoluzza, M.A. Gil, and D.A. Ralescu, editors, *Statistical Modeling, Analysis and Management of Fuzzy Data*, pages 22–39. Physica-Verlag, Heidelberg, 2002.

[11] M. Montenegro, A. Colubi, M. R. Casals, and M. A. Gil. Asymptotic and bootstrap techniques for testing the expected value of a fuzzy random variable. *Metrika*, 59:31–49, 2004.

[12] M. L. Puri and D. A. Ralescu. The concept of normality for fuzzy random variables. *Ann. Probab.*, 11:1373–1379, 1985.

[13] M. L. Puri and D. A. Ralescu. Fuzzy random variables. *J. Math. Anal. Appl.*, 114:409–422, 1986.

[14] J. O. Ramsay and B. W. Silverman. *Functional data analysis*. Springer Series in Statistics. Springer–Verlag, New York, 1997.

[15] J. O. Ramsay and B. W. Silverman. *Applied functional data analysis. Methods and case studies*. Springer Series in Statistics. Springer–Verlag, New York, 2002.

[16] F. Yao, H.-G. Müller, and J.-L. Wang. Functional data analysis for sparse longitudinal data. *J. Amer. Statist. Assoc.*, 100:577–590, 2005.

[17] L.A. Zadeh. The concept of a linguistic variable and its application to approximate reasoning, II. *Inform. Sci.*, 8:301–353, 1975.

Friedman's Test for Ambiguous and Missing Data

Edyta Mrówka and Przemysław Grzegorzewski

Systems Research Institute Polish Academy of Sciences, Newelska 6, 01-447 Warsaw, Poland
{mrowka, pgrzeg}@ibspan.waw.pl

Summary. Friedman's test is traditionally applied for testing independence between k orderings ($k > 2$). In the paper we show how to generalize Friedman's test for situations with missing information or non-comparable outputs. This contribution is a corrected version of our previous paper [8].

Keywords: Friedman's test, IF-sets, missing data, ranks, testing independence.

1 Introduction

Suppose we observe k sets ($k > 2$) of rankings of n subjects and we want do decide whether there is any association or dependence between this rankings. A typical example is a situation with k judges or experts, called observers, each of whom is presented with the same set of n objects to be ranked. Then Friedman's test will provide a nonparametric test of independence of their rankings.

Classical Friedman's test have been constructed for unambiguous rankings (i.e. for k linear orderings). However, in a real life we often meet ambiguous answers or even situations when one or more observers cannot rank all the objects (i.e. we have partial orderings only). In the paper we have proposed how to generalize Friedman's test to situations when not all elements could be univocally ordered.

The paper is organized as follows: In Sec. 2 we recall the classical Friedman's test. Then we show how to apply IF-sets for modelling ambiguous orderings (Sec. 3). And finally, in Sec. 4, we suggest how to generalize Friedman's test for such orderings.

2 Friedman's Test – A Classical Approach

Let $X = \{x_1, \ldots, x_n\}$ denote a finite universe of discourse. Suppose that elements (objects) $x_1, \ldots, x_n$ are ordered according to preferences of k observers $A_1, \ldots, A_k$.

E. Mrówka and P. Grzegorzewski: *Friedman's Test for Ambiguous and Missing Data*, Advances in Soft Computing **6**, 111–118 (2006)

Then our data could be presented in the form of a two-way layout (or matrix) M with k rows and n columns. Let R_{ij}, $i = 1, 2, \ldots, k$, $j = 1, 2, \ldots, n$, denote the rank given by ith observer to the jth object. Then $R_{i1}, R_{i2}, \ldots, R_{in}$ is a permutation of the first n integers while $R_{1j}, R_{2j}, \ldots, R_{kj}$ is a set of rankings given to object number j by successive observers.

Suppose we are interested in testing the null hypothesis that there is no association between rankings given by k observers. If the jth object x_j has the same preference relative to all other objects $x_1, \ldots, x_{j-1}, x_{j+1}, \ldots x_n$ in the opinion of each of the k observers, then all ranks in the jth column will be identical. Therefore, the ranks in each column are indicative for the agreement among observers. Let $R = (R_1, \ldots, R_n)$ denote observed column totals, i.e.

$$R_j = \sum_{i=1}^{k} R_{ij}, \quad j = 1, \ldots, n, \tag{1}$$

and let $\bar{R}$ denote the average column total. It can be shown that $\bar{R}$ equals $\bar{R}^* = \frac{k(n+1)}{2}$ for perfect agreement between rankings. Then the sum of squares of deviations between actually observed column total and average column total for perfect agreement is given by:

$$S(R) = \sum_{j=1}^{n} \left[R_j - \frac{k(n+1)}{2} \right]^2. \tag{2}$$

It can be shown that for any sets of k rankings S ranges between zero and $k^2 n (n^2 - 1)/12$, with the maximum value attained where there is a perfect agreement and the minimum value attained when each observer's rankings are assigned completely at random. Therefore, S maybe used to test the null hypothesis H that the rankings are independent (see, e.g., [3]). In practise we use a linear function of statistic (2) defined as follows

$$Q = \frac{12S}{kn(n+1)}. \tag{3}$$

If the null hypothesis holds then statistic (3) approaches the chi-square distribution with $n-1$ degrees of freedom as k increases. Numerical comparisons have shown this to be a good approximation as long as $k > 7$ (see [3]). Therefore, we reject the null hypothesis if $Q \geq \chi^2_{n-1,\alpha}$, where $\chi^2_{n-1,\alpha}$ is a crital value of the chi-square distribution with $n-1$ degrees of freedom corresponding to assumed significance level α. A test based on Q is called *Friedman's test*.

Friedman's test could be used provided all objects are univocally classified by all observers. However, it may happen that one or more observers cannot rank all the objects under study (e.g. he is not familiar with all elements) or they have problems with specifying their preferences. A way-out is then to remove all objects which are not ordered by all of the observers and not to include them into considerations but this approach involves always a loss of information. Moreover, if the number of ill-classified objects is large, eliminating them would be preclusive of applying Friedman's test. Below we show how to generalize the classical Friedman's test to make it possible to infer about possible association between rankings with missing information or non-comparable outputs.

3 IF-Sets in Modelling Rankings

In this section we show how to apply IF-sets in modelling orderings or rankings. A method given below seems to be useful especially if not all elements under consideration could be ranked (see [5–7]). Let us recall (see [1]) that an IF-set C in a universe of discourse X is given by a set of ordered triples

$$C = \{\langle x, \mu_C(x), \nu_C(x)\rangle : x \in X\}, \tag{4}$$

where $\mu_C, \nu_C : X \to [0,1]$ are functions such that

$$0 \leq \mu_C(x) + \nu_C(x) \leq 1 \qquad \forall x \in X. \tag{5}$$

The numbers $\mu_C(x)$ and $\nu_C(x)$ represent the degree of membership and degree of nonmembership of the element $x \in X$ to C, respectively. For each element $x \in X$ we can compute the, so called, IF-index of x in C defined by

$$\pi_C(x) = 1 - \mu_C(x) - \nu_C(x), \tag{6}$$

which quantifies the amount of indeterminacy associated with x_i in C. It is seen immediately that $\pi_C(x) \in [0,1] \ \forall x \in X$.

In our approach we will attribute an IF-set to the ordering corresponding to each observer. For simplicity of notation we will further on identify orderings expressed by the observers with the corresponding IF-sets $A_1, \ldots, A_k$. Thus, for each $i = 1, \ldots, k$ let

$$A_i = \{\langle x_j, \mu_{A_i}(x_j), \nu_{A_i}(x_j)\rangle : x_j \in X\} \tag{7}$$

denote an intuitionistic fuzzy subset of the universe of discourse $X = \{x_1, \ldots, x_n\}$, where membership function $\mu_{A_i}(x_j)$ indicates the degree to which x_j is the most preferred element by ith observer, while nonmembership function $\nu_{A_i}(x_j)$ shows the degree to which x_j is the less preferred element by ith observer. The main problem now is to determine these membership and nonmembership functions when the only available information are orderings that admit ties and elements that cannot be ranked.

However, for each observer one can always specify two functions $w_{A_i}, b_{A_i} : X \to \{0, 1, \ldots, n-1\}$ defined as follows: for each given $x_j \in X$ let $w_{A_i}(x_j)$ denote the number of elements $x_1, \ldots, x_{j-1}, x_{j+1}, \ldots, x_n$ surely worse than x_j, while $b_{A_i}(x_j)$ let be equal to the number of elements surely better than x_j in the ordering corresponding to the preferences expressed by observer A_i. Then, using functions $w_{A_i}(x_j)$ and $b_{A_i}(x_j)$, we may determine the requested membership and nonmembership functions as $\mu_{A_i}(x_j) = \frac{w_{A_i}(x_j)}{n-1}$ and $\nu_{A_i}(x_j) = \frac{b_{A_i}(x_j)}{n-1}$.

Our IF-sets $A_1, \ldots, A_k$ have some interesting properties. For example, $\pi_{A_i}(x_j) = 0$ for each $x_j \in X$ if and only if all elements are ranked by ith observer and there are no ties. Conversely, if there is an element $x_j \in X$ such that $\pi_{A_i}(x_j) > 0$ then there are ties or non-comparable elements in the ordering made by ith observer. Moreover, more ties or elements that are not comparable with the others, then bigger values of the IF-index are observed. One may also notice that $\pi_{A_i}(x_j) = 1$ if and only if element $x_j \in X$ is either non-comparable with all other elements or all elements $x_1, \ldots, x_n$ have obtained the same rank in the ordering made by ith observer. Hence, it seems that IF-sets might be used as a natural and useful tool for modelling nonlinear orderings.

4 Generalization of Friedman's Test

According to (2) the test statistic for testing independence might be expressed in a following way

$$S(R) = d(R, \overline{R}^*), \tag{8}$$

where $d(R, \overline{R}^*)$ denotes a distance between the observed column totals $R = (R_1, \ldots, R_n)$ and the average column totals $\overline{R}^*$ obtained for perfect agreement between rankings. Now to construct a straightforward generalization of Friedman's test for orderings containing elements that cannot be ranked by all observers, we have to find counterparts of R and $\overline{R}^*$ and a suitable measure of distance between these two objects (see [7]). As we have suggested in the previous section, we would consider appropriate IF-sets $A_1, \ldots, A_k$ for modelling ill-defined rankings. Thus instead of R will also consider an IF-set A, defined as follows

$$A = \{\langle x_j, \mu_A(x_j), \nu_A(x_j) \rangle : x_j \in X\}, \tag{9}$$

where the membership and nonmembership functions μ_A and ν_A are given by $\mu_A(x_j) = \frac{1}{k} \sum_{i=1}^{k} \mu_{A_i}(x_j)$, and $\nu_A(x_j) = \frac{1}{k} \sum_{i=1}^{k} \nu_{A_i}(x_j)$, respectively.

If there is a perfect agreement within the group of observers and all objects are ranked without ties, then the resulting IF-set is of a form

$$A^* = \{\langle x_j, \mu_{A^*}(x_j), \nu_{A^*}(x_j) \rangle : x_j \in X\}, \tag{10}$$

such that the membership function is given by $\mu_{A^*}(x_{j_1}) = \frac{n-1}{n-1} = 1$, $\mu_{A^*}(x_{j_2}) = \frac{n-2}{n-1}$, $\mu_{A^*}(x_{j_3}) = \frac{n-3}{n-1}, \ldots, \mu_{A^*}(x_{j_{n-1}}) = \frac{1}{n-1}$, $\mu_{A^*}(x_{j_n}) = 0$, where $x_{j_1}, \ldots, x_{j_n}$ is a permutation of elements $x_1, \ldots, x_n$ and the nonmembership function is $\nu_{A^*}(x_j) = 1 - \mu_{A^*}(x_j)$ for each $j = 1, \ldots, n$. Therefore, for perfect agreement between rankings, instead of the average column totals $\overline{R}^*$ we obtain an IF-set

$$\overline{A}^* = \{\langle x_j, \mu_{\overline{A}^*}(x_j), \nu_{\overline{A}^*}(x_j) \rangle : x_j \in X\} \tag{11}$$

such that $\mu_{\overline{A}^*}(x_1) = \ldots = \mu_{\overline{A}^*}(x_n) = \frac{1}{2}$ and $\nu_{\overline{A}^*}(x_1) = \ldots = \nu_{\overline{A}^*}(x_n) = \frac{1}{2}$. Now, after substituting R and $\overline{R}^*$ by IF-sets A and $\overline{A}^*$, respectively, we have to choose a suitable distance between these two IF-sets. Several measures of distance between IF-sets were considered in the literature (see, e.g. [4]). In this paper we will apply a distance proposed by Atanassov [2], i.e. such function $d : IFS(X) \times IFS(X) \rightarrow$

$\mathbb{R}^+ \cup \{0\}$ which for any two IF-subsets $B = \{\langle x_j, \mu_B(x_j), \nu_B(x_j)\rangle : x_j \in X\}$ and $C = \{\langle x_j, \mu_C(x_j), \nu_C(x_j)\rangle : x_j \in X\}$ of the universe of discourse $X = \{x_1, \ldots, x_n\}$ is defined as

$$d(B,C) = \sum_{j=1}^{n} \left[(\mu_B(x_j) - \mu_C(x_j))^2 + (\nu_B(x_j) - \nu_C(x_j))^2 \right]. \tag{12}$$

For actual observed rankings, modelled by IF-sets $A_1, \ldots, A_k$, test statistic is a distance (12) between IF-set A obtained from (9) and $\overline{A}^*$ obtained from (11) and is given by

$$\widetilde{S}(A) = d(A, \overline{A}^*) = \sum_{j=1}^{n} \left[\left(\mu_A(x_j) - \frac{1}{2} \right)^2 + \left(\nu_A(x_j) - \frac{1}{2} \right)^2 \right]. \tag{13}$$

It can be shown that a following lemma holds:

Lemma 1. *For any $A \in IFS$ there exist two intuitionistic fuzzy sets $A^\sharp$ and $A^\flat$ such that:*

$$\min \left\{ \widetilde{S}(A^\flat), \widetilde{S}(A^\sharp) \right\} \leq \widetilde{S}(A) \leq \max \left\{ \widetilde{S}(A^\flat), \widetilde{S}(A^\sharp) \right\} \tag{14}$$

where

$$A^\flat = \{\langle x_j, \mu_A(x_j), \nu_A(x_j) + \pi_A(x_j)\rangle : x_j \in X\}, \tag{15}$$
$$A^\sharp = \{\langle x_j, \mu_A(x_j) + \pi_A(x_j), \nu_A(x_j)\rangle : x_j \in X\}. \tag{16}$$

Hence our test statistic $\widetilde{S}(A)$ based on ill-defined data is bounded by two other statistics $\widetilde{S}(A^\flat)$ and $\widetilde{S}(A^\sharp)$ corresponding to situations with perfect rankings. Indeed, for each $x_j \in X$

$$\mu_{A^\flat}(x_j) = 1 - \nu_{A^\flat}(x_j) \Rightarrow \pi_{A^\flat}(x_j) = 0 \tag{17}$$
$$\mu_{A^\sharp}(x_j) = 1 - \nu_{A^\sharp}(x_j) \Rightarrow \pi_{A^\sharp}(x_j) = 0 \tag{18}$$

which means that $A^\flat$ and $A^\sharp$ describe situations when all elements are univocally classified. Therefore, there exist two systems of rankings (in a classical sense) $R^\flat$ and $R^\sharp$ and one-to-one mapping transforming $A^\flat$ and $A^\sharp$ onto $R^\flat$ and $R^\sharp$, respectively. Thus both statistics

$$T_1(A) = \frac{6k(n-1)^2}{n(n+1)} \widetilde{S}\left(A^\flat\right), \tag{19}$$

$$T_2(A) = \frac{6k(n-1)^2}{n(n+1)} \widetilde{S}\left(A^\sharp\right), \tag{20}$$

are chi-square distributed with $n-1$ degrees of freedom. Unfortunately, T_1 and T_2 are not independent. According to (14) we get a following inequality

$$T_{\min} \leq T(A) \leq T_{\max} \tag{21}$$

$\alpha=0.05$	k					
	5	6	7	8	9	10
m	16.46	16.07	18.68	14.97	15.64	15.64
0	15.37	14.89	14.66	14.66	14.38	14.26
1	16.16	15.78	14.22	14.43	14.82	12.48
2	14.59	13.75	14.03	14.82	17.56	14.26
3	14.32	14.71	15.09	15.35	15.59	15.22
4	16.64	14.84	15.69	16.64	15.08	14.85
5	16.72	15.62	16.34	18.41	15.62	18.14
6	18.12	18.94	16.99	17.85	16.59	16.43
7	21.04	18.82	18.43	17.32	17.17	17.37
8	21.83	20.35	19.68	19.68	18.83	18.34
9	23.84	22.25	22.70	21.86	20.39	20.15
10	26.72	23.38	23.70	22.54	22.48	21.36
11	28.09	26.45	24.69	24.68	23.23	21.59
12	31.23	28.95	27.47	27.00	25.36	23.01
13	33.89	31.29	29.06	27.78	26.62	25.66
14	36.45	33.91	31.36	30.64	29.74	27.77
15	39.28	36.60	34.59	32.49	32.14	29.19
16	42.99	39.36	36.51	35.14	33.36	31.85
17	45.22	41.89	40.15	36.63	35.52	33.90
18	48.58	45.02	42.15	40.77	37.61	35.61
19	52.20	47.80	45.23	42.63	40.20	38.55
20	54.99	51.18	48.54	45.71	43.18	40.71

Tab. 1. Critical values of T_{min} (number of objects n=10)

$\alpha=0.05$	k					
	5	6	7	8	9	10
m	16.46	16.07	18.68	14.97	15.64	15.64
0	16.42	16.78	16.37	16.30	16.56	16.18
1	18.21	17.47	16.65	16.47	17.10	14.73
2	17.99	17.15	16.31	18.27	20.25	17.14
3	18.91	17.51	18.74	19.67	18.48	19.01
4	18.34	20.16	19.92	19.69	17.32	18.71
5	20.65	21.55	21.73	19.96	19.33	19.59
6	21.92	22.89	21.30	21.64	19.93	22.32
7	24.14	21.96	22.04	20.97	23.21	20.99
8	25.76	24.00	25.22	23.52	22.75	23.05
9	27.46	26.71	26.78	24.91	25.32	22.69
10	30.43	27.64	28.53	25.68	26.67	26.40
11	31.39	31.27	30.96	29.66	27.84	25.52
12	34.44	33.42	31.99	31.19	29.86	28.58
13	36.97	35.43	33.58	31.98	31.69	30.27
14	39.72	36.53	34.51	34.98	33.19	32.89
15	42.47	40.00	37.96	37.15	36.32	34.27
16	46.40	42.97	40.84	38.74	36.59	36.49
17	47.88	46.04	43.43	40.65	41.29	37.72
18	51.24	47.16	44.86	43.30	43.25	40.17
19	54.25	50.45	48.48	46.30	44.54	42.59
20	57.09	53.29	52.65	48.76	45.84	46.70

Tab. 2. Critical values of T_{max} (number of objects n=10)

where

$$T_{\min} = \min\{T_1(A), T_2(A)\}, \tag{22}$$

$$T_{\max} = \max\{T_1(A), T_2(A)\}, \tag{23}$$

$$T(A) = \frac{6k(n-1)^2}{n(n+1)}\widetilde{S}(A). \tag{24}$$

Traditionally, in hypothesis testing we reject the null hypothesis if test statistic belongs to critical region or accept it otherwise. In our problem with missing data a final decision would be based on these two statistics T_1 and T_2. Let us denote by $q_\alpha^{\min}$ and $q_\alpha^{\max}$ the critical values of $T_{\min}$ and $T_{\max}$, respectively, i.e. we find them as the solutions of the equations $P(T_{\min} > q_\alpha^{\min}) = \alpha$ and $P(T_{\max} > q_\alpha^{\max}) = \alpha$, respectively, when the null hypothesis holds.. The results were obtained via Monte Carlo simulations. Some critical values are shown in Table 1 and Table 2.

Numerical comparisons have shown that $T_{\min}$ and $T_{\max}$ approaches the gamma distribution as k increases. Additionally those approximations work well if the number of missing or ill defined data does not exceed the number of correctly classified data. Moreover, we can find regression formulae showing how the parameters of shape $\lambda_{\min}$ and $\lambda_{\max}$ corresponding to approximate distributions of $T_{\min}$ and $T_{\max}$ depend on the number of observers k, number of objects n and number of ill defined data m, e.g.:

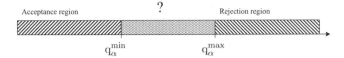

$$q_\alpha^{min} \qquad\qquad q_\alpha^{max}$$

Fig. 1. Decision region

$$\lambda_{min} = 362.684 - 9.7702n - 5.7933k + 2.8427m, \qquad (25)$$
$$\lambda_{max} = 378.743 - 9.7426n - 6.1272k + 2.8512m. \qquad (26)$$

It is worth noting that for both distributions the parameter of scale β is constants and is equal to $\beta_{min} = 1.3848$ and $\beta_{max} = 1.6076$, respectively.

Going back to our hypothesis testing problem, we should reject H on the significance level α if $T(A) \geq q_\alpha^{max}$ while there are no reasons for rejecting H (i.e. we accept H) if $T(A) < q_\alpha^{min}$ These two situations are quite obvious. However, it may happen that $q_\alpha^{min} \leq T(A) < q_\alpha^{max}$. In such a case we are not completely convinced neither to reject nor to accept H (see Fig. 1).

Thus instead of a binary decision we could indicate a degree of conviction that one should accept or reject H. The measure describing degree of necessity for rejecting H is given by following formula:

$$Ness(\text{reject } H) = \begin{cases} 1 & \text{if} \quad q_{min} \geq T(A) \\ \frac{q_{max} - T(A)}{q_{max} - q_{min}} & \text{if} \quad q_{min} < T(A) \leq q_{max} \\ 0 & \text{if} \quad q_{max} < T(A) \end{cases} \qquad (27)$$

Simultaneously we get another measure

$$Poss(\text{accept } H) = 1 - Ness(\text{reject } H) \qquad (28)$$

describing the degree of possibility for accepting H.

Thus, $Ness(\text{reject } H) = 1$ means that the null hypothesis should be rejected and hence there is no possibility for accepting H while $Ness(\text{reject } H) = \xi \in (0,1)$ shows how strong our data are for or against $(1 - \xi)$ the null hypothesis H. Filially $Ness(\text{reject } H) = 0$ means that there is no reason for rejecting null hypothesis H.

5 Conclusion

In the paper we have proposed how to generalize the well-known Friedman's test to situations in which not all elements could be ordered. We have discussed Friedman's test as a nonparametric tool for testing independence of k variates. However, this very test could be also applied as nonparametric two-way analysis of variance for the balanced complete block design. In this case kn subjects are grouped into k blocks each containing n subjects and within each block n treatments are assigned randomly to the matched subjects. In order to determine whether the treatment effects are all the same, Friedman's test could be used. It should be stressed that our generalized version of Friedman's test also works for two-way analysis of variance by ranks with missing data.

References

[1] K. Atanassov, "Intuitionistic fuzzy sets", Fuzzy Sets and Systems, vol. 20 (1986), 87-96.

[2] K. Atanassov, "Intuitionistic Fuzzy Sets: Theory and Applications", Physica-Verlag, 1999.

[3] J.D. Gibbons, S. Chakraborti, "Nonparametric Statistical Inference", Marcel Dekker, 2003.

[4] P. Grzegorzewski, "Distances between intuitionistic fuzzy sets and/or interval-valued fuzzy sets based on the Hausdorff metric", Fuzzy Sets and Systems, vol. 148 (2003), 319-328.

[5] P. Grzegorzewski, "The generalized Spearman's rank correlation coefficient", In: Proceedings of the Tenth International Conference on Information Processing and Management of Uncertainty in Knowledge-Based Systems, Perugia, Italy, July 4-9, 2004, pp. 1413-1418.

[6] P. Grzegorzewski, "On measuring association betwwen preference systems", In: Proceedings of the 2004 IEEE International Conference on Fuzzy Systems, Budapest, Hungary, July 25-29, 2004, pp. 133-137.

[7] P. Grzegorzewski, "The generalized coefficient of concordance", In: Current Issues in Data nad Knowledge Engineering, B. De Baets at all. (Eds.) Warszawa 2004, pp. 241-251.

[8] E. Mrówka, P. Grzegorzewski, "Friedman's test with missing observations, In: Proceedings of the 4th Conference of the European Society for Fuzzy Logic and Technology - Eusflat 2005, Barcelona, September 7-9, 2005, pp. 627-632.

Part III

Probability of Imprecisely-Valued Random Elements with Applications

Measure-Free Martingales with Application to Classical Martingales

S.F. Cullender, W.-C. Kuo, C.C.A. Labuschagne and B.A. Watson

School of Mathematics, University of the Witwatersrand, Private Bag 3, WITS 2050, South Africa
scullender@maths.wits.ac.za, wpkuo@global.co.za,
cola@maths.wits.ac.za, bwatson@maths.wits.ac.za

Summary. The aim of this work is to give a summary of some of the known properties of sets of measure-free martingales in vector lattices and Banach spaces. In particular, we consider the relationship between such sets of martingales and the ranges of the underlying filtration of conditional expectation operators.

1 Introduction

There are examples in the literature where certain aspects of martingale theory are considered in a suitable framework which avoids the use of an underlying measure space (cf. [2, 3, 4, 5, 6, 7, 8, 9, 15, 16, 17, 18]). The aim of this work is to give a summary of some of the known properties of sets of measure-free martingales. In particular, we consider the relationship between such sets of martingales and the ranges of the underlying filtration of conditional expectation operators.

We assume that the reader is familiar with the terminology and notation of vector lattices (i.e. Riesz spaces) and Banach lattices, as can be found in [12, 14, 19].

Some general notation and terminology on martingales are in order at this stage, so as to avoid unnecessary repetition later.

Let E be a vector space. A sequence (T_i) of linear projections defined on E for which $T_i T_m = T_m T_i = T_i$ for each $m \geq i$ is called a *filtration* of linear projections on E. If $\mathcal{R}(T_i)$ denotes the range of T_i, then a filtration of linear projections (T_i) is a commuting family of linear projections with increasing ranges, i.e. $\mathcal{R}(T_i) \uparrow_i$. A sequence $(f_i, T_i)_{i \in \mathbb{N}}$, where (T_i) a filtration of linear projections on E and $f_i \in \mathcal{R}(T_i)$ for each $i \in \mathbb{N}$, is called a *martingale* if $f_i = T_i f_m$, for each $m \geq i$.

Let E be a vector space and (T_i) a filtration of linear projections on E and $M(E, T_i) := \{(f_i, T_i) : (f_i, T_i) \text{ is a martingale on } E\}$. Then $M(E, T_i)$ is a vector space if we define addition and scalar multiplication by

$$(f_i, T_i) + (g_i, T_i) = (f_i + g_i, T_i) \text{ and } \lambda(f_i, T_i) = (\lambda f_i, T_i) \text{ for each } \lambda \in \mathbb{R}.$$

If E is an ordered vector space and (T_i) a filtration of positive (i.e. $T_i x \geq 0$ for all $x \geq 0$) linear projections on E, define $\leq$ on $M(E, T_i)$ by $(f_i, T_i) \leq (g_i, T_i) \iff f_i \geq$

S.F. Cullender et al.: *Measure-Free Martingales with Application to Classical Martingales*, Advances in Soft Computing
6, 121–128 (2006)
www.springerlink.com

g_i for all $i \in \mathbb{N}$, and let $M_+(E,T_i) := \{(f_i,T_i) : f_i \geq 0 \text{ for all } i \in \mathbb{N}\}$. Then $M(E,T_i)$ is an ordered vector space with positive cone $M_+(E,T_i)$.

If E is a vector lattice, then $e \in E_+$ is called a *weak order unit* for E if $x \in E_+$ implies that $x \wedge ne \uparrow x$. If (Ω, Σ, μ) is a probability space, then $\mathbf{1}$, defined by $\mathbf{1}(s) = 1$ for all $s \in \Omega$, is a weak order unit for $L^p(\mu)$ for all $1 \leq p < \infty$.

2 Martingales in Vector Lattices

In the setting of vector lattices with weak order units, the following definition is taken from [5], where a motivation is also given:

Definition 1. *Let E be a vector lattice with weak order unit e. A positive order continuous projection $T : E \rightarrow E$ for which $T(w)$ is a weak order unit in E for each weak order unit $w \in E_+$ and $\mathscr{R}(T)$ is a Dedekind complete Riesz subspace of E, is called a conditional expectation on E.*

A proof is given in [7] that the statement "$T(w)$ is a weak order unit in E for each weak order unit $w \in E_+$" in the preceding definition is equivalent to the statement "$T(e) = e$".

Let E be a vector lattice and (T_i) a filtration of positive linear projections on E. Let

$$M_{oc}(E,T_i) := \{(f_i,T_i) \in M(E,T_i) : (f_i) \text{ is order convergent in } E\},$$

$$M_{ob}(E,T_i) := \{(f_i,T_i) \in M(E,T_i) : (f_i) \text{ is an order bounded in } E\}.$$

It is easy to show that the above defined sets of martingales are ordered vector subspaces of $M(E,T_i)$. Moreover, since order convergent sequences are order bounded, $M_{oc}(E,T_i) \subseteq M_{ob}(E,T_i)$. However, [8, Corollary 5.2] shows that equality holds in the setting of vector lattices with weak order units:

Theorem 1. *Let E be a Dedekind complete vector lattice with weak order unit e, and let (T_i) be a filtration of conditional expectations on E. Then $M_{oc}(E,T_i) = M_{ob}(E,T_i)$.*

If E is a vector lattice and $T : E \rightarrow E$ is a positive linear map, then T is said to be *strictly positive* if $\{x \in E : T(|x|) = 0\} = \{0\}$.

There is a connection between $M_{oc}(E,T_i)$ and $\overline{\bigcup_{i=1}^{\infty} \mathscr{R}(T_i)}$, where the latter denotes the order closure of $\bigcup_{i=1}^{\infty} \mathscr{R}(T_i)$:

Theorem 2. *[8, Theorem 5.8] Let E be a Dedekind complete vector lattice with weak order unit e and let (T_i) be a filtration of conditional expectations on E with T_1 strictly positive. Then $M_{oc}(E,T_i)$ is a Dedekind complete vector lattice and $L: M_{oc}(E,T_i) \rightarrow \overline{\bigcup_{i=1}^{\infty} \mathscr{R}(T_i)}$, defined by $L((f_i,T_i)) = \lim_i f_i$ (order), is an order continuous surjective Riesz isomorphism.*

Let E be a vector lattice and (T_i) a filtration of positive linear projections on E. Let $M_r(E,T_i)$ denote the set of all *regular martingales* on E; i.e., those martingales (f_i,T_i) on E for which there exist $(g_i,T_i),(h_i,T_i) \in M_+(E,T_i)$ such that $f_i = g_i - h_i$.

It is readily verified that $M_r(E,T_i)$ is an ordered vector subspace of $M(E,T_i)$ and $M_r(E,T_i) = M_+(E,T_i) - M_+(E,T_i)$.

The simple proof given in [11] for the following result, is based on the ideas in [8] and the main idea in the proof of [18, Theorem 7]:

Theorem 3. *If E is a Dedekind [σ-Dedekind] complete vector lattice and (T_n) a filtration of order [σ-order] continuous positive linear projections on E, then $M_r(E,T_i)$ is a Dedekind [σ-Dedekind] complete vector lattice.*

3 Martingales in Banach Spaces and Banach Lattices

Let (Ω,Σ,μ) denote a probability space. Then, for $1 \leq p < \infty$ and X a Banach space, let $L^p(\mu,X)$ denote the space of (classes of a.e. equal) Bochner p-integrable functions $f \colon \Omega \to X$ and denote the Bochner norm on $L^p(\mu,X)$ by Δ_p, i.e. $\Delta_p(f) = \left(\int_\Omega \|f\|_X^p d\mu\right)^{1/p}$.

If one wants to apply a measure-free approach to martingales on $L^p(\mu,X)$-spaces, a measure-free approach to Banach spaces has to be considered. In [2, 3], such an approach is followed:

Let X be a Banach space and (T_i) a filtration of contractive linear projections on X. Define $\|\cdot\|$ on $M(X,T_i)$ by $\|(f_i,T_i)\| = \sup_i \|f_i\|$ and let $\mathcal{M}(X,T_i)$ denote the space of *norm bounded martingales* on X; i.e., $\mathcal{M}(X,T_i) = \{(f_i,T_i) \in M(X,T_i) : \|(f_i,T_i)\| < \infty\}$. Then $\mathcal{M}(X,T_i)$ is a Banach space with respect to $\|\cdot\|$.

Let $\mathcal{M}_{nc}(X,T_i)$ denote the space of *norm convergent martingales* on X; i.e., $\mathcal{M}_{nc}(X,T_i) = \{(f_i,T_i) \in \mathcal{M}(X,T_i) : (f_i) \text{ is norm convergent in } X\}$.

To describe $\mathcal{M}_{nc}(X,T_i)$, the following results are used in [2]:

Proposition 1. *Let X be a Banach space and let (T_i) be a filtration of contractive linear projections on X. Then $f \in \overline{\bigcup_{i=1}^{\infty} \mathcal{R}(T_i)}$, the latter denoting the norm closure of $\bigcup_{i=1}^{\infty} \mathcal{R}(T_i)$, if and only if $\|T_i f - f\| \to 0$.*

Corollary 1. *Let X be a Banach space and let (f_i,T_i) be a martingale in X, where (T_i) is a filtration of contractive linear projections on X. Then (f_i,T_i) converges to f if and only if $f \in \overline{\bigcup_{i=1}^{\infty} \mathcal{R}(T_i)}$ and $f_i = T_i f$ for all $i \in \mathbb{N}$.*

An application in [2] of Proposition 1 and Corollary 1 yields:

Proposition 2. *Let X be a Banach space and (T_i) a filtration of contractive linear projections on X. Then $L \colon \mathcal{M}_{nc}(X,T_i) \to \overline{\bigcup_{i=1}^{\infty} \mathcal{R}(T_i)}$, defined by $L((f_i,T_i)) = \lim_i f_i$ (norm), is a surjective isometry.*

Another application in [3] of Proposition 1 and Corollary 1 provides a proof, via martingale techniques, for the following well known result:

Proposition 3. *Let X be a Banach space and (x_i) a basic sequence in X. Then (x_i) is an unconditional basic sequence if and only if the closure of the span of (x_i), denoted $[x_i]$, can be renormed so that it is an order continuous Banach lattice with positive cone*

$$C_+^{(x_i)} := \left\{ \sum_{i=1}^{\infty} \alpha_i x_i \in [x_i] : \alpha_i \geq 0 \text{ for each } i \in \mathbb{N} \right\}.$$

Motivated by [18], Proposition 2 is specialized in [2] to Banach lattices to obtain:

Proposition 4. *Let E be a Banach lattice and (T_i) a filtration of positive contractive linear projections on E for which $\overline{\bigcup_{i=1}^{\infty} \mathscr{R}(T_i)}$ is a closed Riesz subspace of E. If $L: \mathscr{M}_{nc}(E, T_i) \to \overline{\bigcup_{i=1}^{\infty} \mathscr{R}(T_i)}$ is defined by $L((f_i, T_i)) = \lim_i f_i$, then $\mathscr{M}_{nc}(E, T_i)$ is a Banach lattice and $L: \mathscr{M}_{nc}(E, T_i) \to \overline{\bigcup_{i=1}^{\infty} \mathscr{R}(T_i)}$ is a surjective Riesz isometry.*

By Corollary 1 we have

$$\mathscr{M}_{nc}(E, T_i) = \{(f_i, T_i) \in \mathscr{M}_{nb}(E, T_i) : \exists f \in E \text{ such that } f_i = T_i f \to f\}.$$

Corollary 2. *Let E be a Banach lattice and (T_i) a filtration of positive contractive linear projections on E for which $\overline{\bigcup_{i=1}^{\infty} \mathscr{R}(T_i)}$ is a closed Riesz subspace of E. Then $\mathscr{M}_{nc}(E, T_i)$ is a Banach lattice in which the following formulas hold:*

$$\left(\lim_{m\to\infty} T_n f_m, T_n \right)^+ = \left(\lim_{m\to\infty} T_n f_m^+, T_n \right);$$

$$\left(\lim_{m\to\infty} T_n f_m, T_n \right)^- = \left(\lim_{m\to\infty} T_n f_m^-, T_n \right);$$

$$\left(\lim_{m\to\infty} T_n f_m, T_n \right) \vee \left(\lim_{m\to\infty} T_n g_m, T_n \right) = \left(\lim_{m\to\infty} T_n (f_m \vee g_m), T_n \right); \tag{1}$$

$$\left(\lim_{m\to\infty} T_n f_m, T_n \right) \wedge \left(\lim_{m\to\infty} T_n g_m, T_n \right) = \left(\lim_{m\to\infty} T_n (f_m \wedge g_m), T_n \right);$$

$$\left| \left(\lim_{m\to\infty} T_n f_m, T_n \right) \right| = \left(\lim_{m\to\infty} T_n |f_m|, T_n \right).$$

Proof. By Proposition 4, we have that $\mathscr{M}_{nc}(E, T_i)$ is a Banach lattice.

The formulas are easy to prove. Since L is a bijective Riesz homomorphism, it follows from $L(T_n|f|, T_n) = |f| = |L(T_n f, T_n)|$ that $(T_n|f|, T_n) = L^{-1}(L(T_n|f|, T_n)) = L^{-1}(|L(T_n f, T_n)|) = |(L^{-1}(L(T_n f, T_n))| = |(T_n f, T_n)|$. The other formulas follow in a similar manner. □

Let (T_i) be a filtration of positive contractive linear projections on a Banach lattice E. As in [18], we now consider the space

$$\mathscr{M}_r(E, T_i) = \{(f_i, T_i) \in \mathscr{M}(E, T_i) : \exists (g_i, T_i) \in M_+(E, T_i), f_i \leq g_i \, \forall \, i \in \mathbb{N}\},$$

the elements of which are called *regular norm bounded* martingales.

Troitski proves in [18] that the formulas in (1) also hold in $\mathscr{M}_r(E, T_i)$ and in $\mathscr{M}(E, T_i)$. He uses less stringent assumptions on (T_i) than in Corollary 2, but he makes additional assumptions on E:

Theorem 4. ([18, Theorems 7 and 13]) *Let E be a Banach lattice and* (T_i) *a filtration of positive contractive linear projections on a Banach lattice E.*

(a) *If E is an order continuous Banach lattice, then* $\mathscr{M}_r(E,T_i)$ *is a Dedekind complete Banach lattice with lattice operations given by* (1) *and martingale norm given by* $\|(f_n,T_n)\| = \sup_n \|f_n\|$.

(b) *If E is a KB-space, then* $\mathscr{M}(E,T_i)$ *is a Banach lattice with lattice operations given by* (1) *and martingale norm* $\|(f_n,T_n)\| = \sup_n \|f_n\|$.

It follows easily that if $\overline{\bigcup_{i=1}^{\infty} \mathscr{R}(T_i)}$ is a closed Riesz subspace of E, then

$$\mathscr{M}_{oc}(E,T_i) \subseteq \mathscr{M}_r(E,T_i) \subseteq \mathscr{M}(E,T_i). \tag{2}$$

One can say more about the inclusions in (2) under additional assumptions on E (see [18, Proposition 16]):

Corollary 3. *Let E be a Banach lattice with order continuous norm and* (T_i) *a filtration of positive contractive linear projections on a Banach lattice E for which* $\overline{\bigcup_{i=1}^{\infty} \mathscr{R}(T_i)}$ *is a closed Riesz subspace of E.*

(a) *If E is an order continuous Banach lattice, then* $\mathscr{M}_{nc}(E,T_i)$ *is an ideal in* $\mathscr{M}_r(E,T_i)$.

(b) *If E is a KB-space, then* $\mathscr{M}_r(E,T_i) = \mathscr{M}(E,T_i)$ *and* $\mathscr{M}_{nc}(E,T_i)$ *is a projection band in* $\mathscr{M}(E,T_i)$.

4 Martingales in $L^p(\mu,X)$

Chaney and Schaefer extended the Bochner norm to the tensor product of a Banach lattice and a Banach space (see [1] and [14]). If E is a Banach lattice and Y is a Banach space, then the l-norm of $u = \sum_{i=1}^{n} x_i \otimes y_i \in E \otimes Y$ is given by $\|u\|_l = \inf\{ \| \sum_{i=1}^{n} \|y_i\| |x_i| \| : u = \sum_{i=1}^{n} x_i \otimes y_i \}$.

Furthermore, if $E = L^p(\mu)$ where (Ω, Σ, μ) is a σ-finite measure space, then we have that $E \widetilde{\otimes}_l Y$ is isometric to $L^p(\mu,Y)$.

Let E and F be Banach lattices. We denote the *projective cone* of $E \otimes F$ by $E_+ \otimes F_+ := \{ \sum_{i=1}^{n} x_i \otimes y_i : (x_i, y_i) \in E_+ \times F_+ \}$. It was shown by Chaney and Schaefer that $E \widetilde{\otimes}_l F$ is a Banach lattice with positive cone the l-closure of $E_+ \otimes F_+$.

Let E_1 and E_2 be Banach lattices and Y_1 and Y_2 Banach spaces. If $S: E_1 \to E_2$ is a positive linear operator and $T: Y_1 \to Y_2$ a bounded linear operator, then $\|(S \otimes T)u\|_l \le \|S\| \|T\| \|u\|_l$ for all $u \in E_1 \otimes Y_1$ (see [10]).

The following is proved in [2]:

Theorem 5. *Let E be a Banach lattice and Y be a Banach space [lattice]. If* (S_i) *is a filtration of positive contractive linear projections on E with each* $\mathscr{R}(S_i)$ *a closed Riesz subspace of E, and* (T_i) *is a filtration of [positive] contractive linear projections on Y [and each* $\mathscr{R}(T_i)$ *is a closed Riesz subspace of Y], then* $(S_i \otimes_l T_i)$ *is a filtration of [positive] contractive linear projections on* $E \widetilde{\otimes}_l Y$ *with each* $S(E_i) \widetilde{\otimes}_l T(Y_i)$ *a closed [Riesz] subspace of* $E \widetilde{\otimes}_l Y$.

To consider tensor product versions of some of the martingale results stated earlier, we need the following result noted by Popa, [13]:

Theorem 6. *Let E and F be Banach lattices.*

(a) If E and F are order continuous Banach lattices, then $E \widetilde{\otimes}_l F$, is an order continuous Banach lattice.
(b) If E and F are KB-spaces, then $E \widetilde{\otimes}_l F$ is a KB-space.

The following is an l-tensor product version of Corollary 3.

Theorem 7. *Let E and F be Banach lattices and let (S_i) and (T_i) be filtrations of positive contractive linear projections on E and F respectively with each $\mathscr{R}(S_i)$ and each $\mathscr{R}(T_i)$ a closed Riesz subspace of E and F respectively.*

(a) If E and F are order continuous Banach lattices, then $\mathscr{M}_r(E \widetilde{\otimes}_l F, T_i \otimes_l S_i)$ is a Banach lattice and $\mathscr{M}_{nc}(E \widetilde{\otimes}_l F, T_i \otimes_l S_i)$ is an ideal in $\mathscr{M}_r(E \widetilde{\otimes}_l F, T_i \otimes_l S_i)$.
(b) If E and F are KB-spaces, then $\mathscr{M}(E \widetilde{\otimes}_l F, T_i \otimes_l S_i)$ is a Banach lattice and $\mathscr{M}_{nc}(E \widetilde{\otimes}_l F, T_i \otimes_l S_i)$ is a projection band in $\mathscr{M}(E \widetilde{\otimes}_l F, T_i \otimes_l S_i)$.

Proof (a) Since E and F are order continuous Banach lattices, $E \widetilde{\otimes}_l F$ is an order continuous Banach lattice, by Popa's result. By Proposition 5, we get that $(S_i \otimes_l T_i)$ is a filtration of positive contractive linear projections on $E \widetilde{\otimes}_l F$ with $\overline{\bigcup_{i=1}^{\infty} \mathscr{R}(T_i \otimes S_i)}$ a closed Riesz subspace of $E \widetilde{\otimes}_l F$. But then $\mathscr{M}_{nc}(E \widetilde{\otimes}_\alpha F, T_i \otimes S_i)$ is an ideal in the Banach lattice $\mathscr{M}_r(E \widetilde{\otimes}_l F, T_i \otimes_l S_i)$, by Corollary 3 (a).

(b) Since E and F are KB-spaces, $E \widetilde{\otimes}_l F$ is a KB-space, by Popa's result. Similar reasoning as in (a), but by using Corollary 3 (b), shows that $\mathscr{M}_{nc}(E \widetilde{\otimes}_l F, T_i \otimes S_i)$ is a projection band in the Banach lattice $\mathscr{M}(E \widetilde{\otimes}_l F, T_i \otimes S_i) = \mathscr{M}_r(E \widetilde{\otimes}_l F, T_i \otimes S_i)$. □

In [2], we show that, if (S_i) is a filtration of positive contractive linear projections on the Banach lattice E such that each $\mathscr{R}(S_i)$ is a closed Riesz subspace of E and (T_i) is a filtration of contractive linear projections on the Banach space Y, then $\overline{\bigcup_{i=1}^{\infty} \mathscr{R}(S_i)} \widetilde{\otimes}_l \overline{\bigcup_{i=1}^{\infty} \mathscr{R}(T_i)} = \overline{\bigcup_{i=1}^{\infty} \mathscr{R}(S_i \otimes_l T_i)}$.

In [10], it is shown that, if E is a Banach lattice and Y a Banach space, then $u \in E \widetilde{\otimes}_l Y$ if and only if $u = \sum_{i=1}^{\infty} x_i \otimes y_i$, where $\left\| \sum_{i=1}^{\infty} |x_i| \right\|_E < \infty$ and $\lim_{i \to \infty} \|y_i\|_Y = 0$.

As a consequence, the following result is derived in [2].

Theorem 8. *Let (S_i) be a filtration of positive contractive linear projections on the Banach lattice E such that each $\mathscr{R}(S_i)$ is a closed Riesz subspace of E and (T_i) a filtration of contractive linear projections on the Banach space Y. Then, in order for $M = (f_n, S_n \otimes_l T_n)_{n=1}^{\infty}$ to be a convergent martingale in $E \widetilde{\otimes}_l Y$, it is necessary and sufficient that, for each $i \in \mathbb{N}$, there exist convergent martingales $\left(x_i^{(n)}, S_n \right)_{n=1}^{\infty}$ and $\left(y_i^{(n)}, T_n \right)_{n=1}^{\infty}$ in E and Y respectively such that, for each $n \in \mathbb{N}$, we have*

$$f_n = \sum_{i=1}^{\infty} x_i^{(n)} \otimes y_i^{(n)},$$

where

$$\left\| \sum_{i=1}^{\infty} \left| \lim_{n \to \infty} x_i^{(n)} \right| \right\| < \infty \ and \ \lim_{i \to \infty} \left\| \lim_{n \to \infty} y_i^{(n)} \right\| = 0.$$

As a simple consequence of Theorem 8, the following representation result is noted in [2]:

Theorem 9. *Let (Ω, Σ, μ) denote a probability space, $(\Sigma_n)_{n=1}^{\infty}$ a filtration, X a Banach space and $1 \le p < \infty$. Then, in order for $(f_n, \Sigma_n)_{n=1}^{\infty}$ to be a convergent martingale in $L^p(\mu, X)$, it is necessary and sufficient that, for each $i \in \mathbb{N}$, there exist a convergent martingale $\left(x_i^{(n)}, \Sigma_n \right)_{n=1}^{\infty}$ in $L^p(\mu)$ and $y_i \in X$ such that, for each $n \in \mathbb{N}$, we have*

$$f_n(s) = \sum_{i=1}^{\infty} x_i^{(n)}(s) y_i \ for \ all \ s \in \Omega,$$

where $\left\| \sum_{i=1}^{\infty} \left| \lim_{n \to \infty} x_i^{(n)} \right| \right\|_{L^p(\mu)} < \infty \ and \ \lim_{i \to \infty} \|y_i\| = 0.$

Acknowledgement

The third and fourth named authors were funded by the John Knopfmacher Centre for Applicable Analysis and Number Theory. The third named author also gratefully acknowledges funding by the Anderson Capelli Fund and thanks Professor Shoumei Li and her students – Li Guan and Xiang Li – for their warm hospitality during his stay at the Beijing University of Technology in March/April 2006.

References

[1] Chaney, J. (1972). Banach lattices of compact maps. *Math. Z.* **129**, 1-19.

[2] Cullender, S.F. and Labuschagne, C.C.A. (2005). A description of norm-convergent martingales on vector valued L^p-spaces. *J. Math. Anal. Appl.*, (to appear).

[3] Cullender, S.F. and Labuschagne, C.C.A. (2006). Unconditional martingale difference sequences in a Banach space. *J. Math. Anal. Appl.*, (to appear).

[4] DeMarr, R. (1966). A martingale convergence theorem in vector lattices. *Canad. J. Math.* **18**, 424-432.

[5] Kuo, W.-C., Labuschagne, C.C.A. and Watson, B.A. (2004). Discrete-time stochastic processes on Riesz spaces. *Indag. Math. (N.S.)* **15**, 435-451.

[6] Kuo, W.-C., Labuschagne, C.C.A. and Watson, B.A. (2004). Riesz space and fuzzy upcrossing theorems. In: Soft methodology and random information systems. *Advances in Soft Computing Springer*, Berlin, pp. 101-108.

[7] Kuo, W.-C., Labuschagne, C.C.A. and Watson, B.A. (2005). Conditional expectations on Riesz spaces. *J. Math. Anal. Appl.* **303**, 509-521.

[8] Kuo, W.-C., Labuschagne, C.C.A. and Watson, B.A. (2005). Convergence of Riesz space martingales. *Indag. Math. (N.S.)*, (to appear).

[9] Kuo, W.-C., Labuschagne, C.C.A. and Watson, B.A. (2006). Ergodic theory and the strong law of large numbers on Riesz spaces. *J. Math. Anal. Appl.*, (to appear).

[10] Labuschagne, C.C.A. (2004). Characterizing the one-sided tensor norms Δ_p and $^t\Delta_p$. *Quaest. Math.* **27**, 339-363.

[11] Labuschagne, C.C.A., Pinchuck, A.L. and van Alten, C.J. (2006). Rradström embeddings of vector lattice cones. (Preprint).

[12] Meyer-Nieberg, P. (1991). *Banach lattices*. Springer Verlag.

[13] Popa, N. (1979). Die Permanenzeigenschaften der Tensorprodukte von Banachverbänden. In: *Romanian-Finnish Seminar on Complex Analysis (Proc. Bucharest 1976)*, LNM Berlin-Heidelberg-New York, **743**, 627-647.

[14] Schaefer, H.H. (1974). *Banach Lattices and Positive Operators*. Springer Verlag.

[15] Stoica, G. (1990). Martingales in vector lattices. *Bull. Math. Soc. Sci. Math. Roumanie. (N.S.)* **34(82)**, 357-362.

[16] Stoica, G. (1991). Martingales in vector lattices II. *Bull. Math. Soc. Sci. Math. Roumanie. (N.S.)* **35(83)**, 155-157.

[17] Stoica, G. (1994). The structure of stochastic processes in normed vector lattices. *Stud. Cerc. Mat.* **46**, 477-486.

[18] Troitsky, V. (2005). Martingales in Banach lattices. *Positivity* **9**, 437-456.

[19] Zaanen, A.C. (1997). *Introduction to Operator Theory in Riesz Space*. Springer Verlag.

A Note on Random Upper Semicontinuous Functions

Hung T. Nguyen[1], Yukio Ogura[2], Santi Tasena[1] and Hien Tran[1]

[1] Department of Mathematical Sciences, New Mexico State University Las Cruces, NM
    88003-8001, USA
    hunguyen@nmsu.edu
[2] Department of Mathematics, Saga University, Saga 840-8502, JAPAN.
    ogura@cc.saga-u.ac.jp

Summary. This note aims at presenting the most general framework for a class $\mathcal{U}$ of random upper semicontinuous functions, namely random elements whose sample paths are upper semicontinuous (u.s.c.) functions, defined on some locally compact, Hausdorff and second countable base space, extending Matheron's framework for random closed sets. It is shown that while the natural embedding process does not provide compactness for $\mathcal{U}$, the *Lawson* topology does.

1 Introduction

Among many applications, random sets, i.e. random elements taking sets as values, are appropriate models for coarse data analysis (e.g. Nguyen, 2006). More generally, perception-based information can be viewed as realizations of random fuzzy sets, i.e. random elements whose values are fuzzy subsets (e.g. Nguyen, 2005). Unlike general stochastic processes, and similar to random vectors, the theory of random closed sets of a locally compact, Hausdorff and second countable topological space (LCHS), as developed by Matheron (1975), is tractable due to the Choquet theorem characterizing their distributions at a simpler level, via the concept of capacity functionals. As such, when extending random closed sets to fuzzy sets, it is desirable to consider fuzzy sets whose membership functions are generalizations of indicator functions of closed sets, i.e. upper semicontinuous functions. While general stochastic processes can have their sample paths as u.s.c. functions with values in the extended real line, we focus here our attention to the range of the unit interval $[0, 1]$ in view of the theory of fuzzy sets.

The efforts in defining random fuzzy sets of $\mathbb{R}^d$ as bona fide random elements taking values in function spaces were started since the mid 1980's (see Li et al, 2002) with no relation to Matheron's topology on the space of closed sets. While the space of closed sets of a LCHS space, equipped with Matheron's topology (the hit-or-miss topology), is compact and second countable (hence metrizable, and thus a separable metric space, idealistic for developing probability theory, see Dudley, 1989), the

H.T. Nguyen et al.: *A Note on Random Upper Semicontinuous Functions*, Advances in Soft Computing **6**, 129–135 (2006)

extension to fuzzy sets was not fully developed, but only up to the case of generalizations of random *compact* sets (see Li et al, 2002), i.e. the study of appropriate
topologies on the space of u.s.c. functions (with values in $[0,1]$) seems lacking. It
is conceivable that Matheron's hit-or-miss topology on the space of closed sets of a
LCHS space could be extended to the space of u.s.c. functions (via identification of
sets with their indicator functions) preserving all desirable properties, namely compactness and second countability. It is surprising that this has not been done satisfactorily, as far as we know! The purpose of our Note is to clarify the above situation by
indicating a concrete topology (from lattice theory) for the space of u.s.c. functions
leading to a satisfactory framework for random (*closed*) fuzzy sets.

2 A Survey of Literature

The problem we are investigating is somewhat similar to that of extending the topological space $C[0,1]$ (the space of continuous functions, defined on $[0,1]$, with the
sup norm, the space of sample paths of, say, Brownian motion) to the bigger space
$D[0,1]$ of functions which are right continuous and having left limits (sample paths
of processes with jumps) where the Skorohod topology is separable and metrizable
(see Billingsley, 1968).

Throughout, E denotes a locally compact, Hausdorff and second countable topological space (LCHS), and $(\Omega, \mathscr{A}, P)$ a probability space. The spaces of closed, open
and compact subsets of E are denoted as $\mathscr{F}, \mathscr{G}, \mathscr{K}$, respectively. Note that E is
metrizable. To define random elements with values in $\mathscr{F}$, one looks for some topol-
ogy τ for $\mathscr{F}$ and considers its associated borel σ-field $\mathscr{B}(\tau)$. For various topologies
on $\mathscr{F}$ for general topological spaces, see Beer (1993). Matheron (1975) considered
the hit-or-miss topology, denoted as τ, (or the Fell topology, see Molchanov, 2005)
which is suitable for establishing Choquet theorem. This topology is generated by
the base consisting of

$$\mathscr{F}^K_{G_1,...,G_n} = \mathscr{F}^K \cap \mathscr{F}_{G_1} \cap \cdots \cap \mathscr{F}_{G_n}, \qquad \text{for } n \in \mathbb{N}, K \in \mathscr{K}, G_i \in \mathscr{G},$$

where $\mathscr{F}^K = \{F \in \mathscr{F} : F \cap K = \varnothing\}$ and $\mathscr{F}_G = \{F \in \mathscr{F} : F \cap G \neq \varnothing\}$. For $n = 0$, the
above elements mean $\mathscr{F}^K$.

It is shown by Matheron that this topology makes $\mathscr{F}$ a compact and second
countable topological space (hence metrizable). For $E = \mathbb{R}^d$, for example, a metric compatible with Matheron's topology is the Hausdorff-Buseman metric (see
e.g. Molchanov, 2005) $d_{HB}(A,B) = \sup_{x \in E} e^{-\rho(0,x)} |\rho(x,A) - \rho(x,B)|$, where ρ is
some metric compatible with the topology of E and $\rho(x,A)$ denotes the distance
from x to the closed set A.

Now we wish to extend the concept of random closed sets to that of random
fuzzy sets, in fact to random "closed" fuzzy sets. As fuzzy sets are defined as generalizations of ordinary (crisp) sets, by generalizing indicator functions of ordinary
sets, i.e. for each subset A of E, its indicator function is

$$1_A : E \to \{0,1\}, \qquad 1_A(x) = 1 \text{ or } 0 \text{ according to } x \in A \text{ or } x \notin A,$$

we identify ordinary subsets with their indicator functions, and *define* a fuzzy subset of E as a function from E to the unit interval $[0,1]$. Since the indicator function f of a closed set is upper semicontinuous (u.s.c.) functions on E, i.e. for any $\alpha \geq 0$, the α-level set $A_\alpha(f) = \{x \in E : f(x) \geq \alpha\}$ is a closed, we call the space of u.s.c. functions, defined on E, with values in $[0,1]$ the space of closed fuzzy sets of E, and denote it from now on as $\mathcal{U}(E)$. By identification, $\mathcal{F} \subseteq \mathcal{U}$. It is desirable to define a topology η on $\mathcal{U}$ extending Matheron topology on $\mathcal{F}$ and making $\mathcal{U}$ a separable metric space. Moreover, if $\mathcal{B}(\eta)$ is the borel σ-field generated by such a topology, it is desirable that a map $X : \Omega \to \mathcal{U}$ is $\mathcal{A}/\mathcal{B}(\eta)$-measurable if and only if for each $\alpha \geq 0$, the set-valued map $X_\alpha(\omega) = \{x \in E : X(\omega)(x) \geq \alpha\}$ is a random closed set in the sense of Matheron.

In the study of random u.s.c. functions, e.g. Teran (2005), the base space E is in general a Banach space. But even for the case of $\mathbb{R}^d$, the separable metric is only obtained for some *subclasses* of $\mathcal{U}(E)$. Note that the recent work of Colubi et al (2002) is about u.s.c. functions defined on $[0,1]$ with values in E and not about random u.s.c. fuzzy sets although they did consider random *compact* fuzzy sets.

3 Topologies on $\mathcal{U}(E)$

In the following, $\mathcal{U}(E)$ denotes the space of u.s.c. functions on E with values either in the extended real line $\mathbb{R}^*$, or $[0,1]$. As in Ogura and Li (2004), it is first natural to search for a topology on $\mathcal{U}(E)$ by an embedding process. This is carried out as follows. It is well-known that any function $f : E \to [0,1]$ (or more generally to $\mathbb{R}$) is determined completely by its α-sets $A_\alpha(f)$, in fact, for $\alpha \in \mathbb{Q}_1 = \mathbb{Q} \cap [0,1]$, where $\mathbb{Q}$ denotes the rationals. In other words, the mapping $\Psi : \mathcal{U}(E) \to \prod_{\alpha \in \mathbb{Q}_1} \mathcal{F}_\alpha$, the countable product of identical copies $\mathcal{F}_\alpha$ of $\mathcal{F}$, sending f to $(A_\alpha(f), \alpha \in \mathbb{Q}_1)$, is an embedding. Now equip $\prod_{\alpha \in \mathbb{Q}_1} \mathcal{F}_\alpha$ with the product topology of the $(\mathcal{F}_\alpha, \tau)$'s. One can induce on $\mathcal{U}$ a topology γ via the relative topology of its image $\Psi(\mathcal{U})$ in $\prod_{\alpha \in \mathbb{Q}_1} \mathcal{F}_\alpha$. Note that $f_n \to f$ in $\mathcal{U}$ iff and only if $A_\alpha(f_n) \to A_\alpha(f)$ in $\mathcal{F}$ for all $\alpha \in \mathbb{Q}_1$.

Remark. Using epigraphs of u.s.c. functions, $\mathcal{U}$ can be also embedded into the space of closed sets of the product space $E \times \mathbb{R}$, Molchanov (2005).

A subbase for the induced topology on $\mathcal{U}$ consists of

$$\mathcal{U}^K(\alpha) = \{f : A_\alpha(f) \cap K = \varnothing\} \text{ and } \mathcal{U}_G(\alpha) = \{f : A_\alpha(f) \cap G \neq \varnothing\},$$
$$K \in \mathcal{K}, G \in \mathcal{G}, \alpha \in \mathbb{Q}_1.$$

This can be seen as follows. Let γ be the induced topology on $\mathcal{U}$ and γ' be the topology generated by the above subbase. The embedding Ψ is continuous under γ'. Thus, $\gamma' \subseteq \gamma$, but γ being the smallest topology making Ψ continuous, we have that $\gamma = \gamma'$. $\square$

Remark. The map $(f,g) \to f \vee g$ is continuous. Indeed, let $f_n \to f$ and $g_n \to g$. Then $A_\alpha(f_n) \to A_\alpha(f)$ and $A_\alpha(g_n) \to A_\alpha(g)$ for all $\alpha \in \mathbb{Q}$. Since the union of random closed sets is continuous under the hit-or-miss topology (Matheron, 1975), we have $A_\alpha(f_n \vee g_n) = A_\alpha(f_n) \cup A_\alpha(g_n) \rightarrow A_\alpha(f) \cup A_\alpha(g) = A_\alpha(f \vee g)$, i.e. $f_n \vee g_n \rightarrow f \vee g$. $\square$

While the countable product space $\prod_{\alpha \in \mathbb{Q}_1} \mathscr{F}_\alpha$ is compact and second countable (hence metrizable), the induced topology on $\mathscr{U}(E)$ does not make $\mathscr{U}(E)$ a compact space. Indeed :

Theorem 1. $\Psi(\mathscr{U})$ is *not closed* in the product topology of $\prod_{\alpha \in \mathbb{Q}_1} \mathscr{F}_\alpha$.

This negative result can be proved by a counter-example !

Let $f_n \in \Psi(\mathscr{U})$ such that $A_\alpha(f_n) \rightarrow M_\alpha$, for all $\alpha \in \mathbb{Q}_1$. Observe that if $f_n \rightarrow f \in \Psi(\mathscr{U})$, then, for all α, $M_\alpha = A_\alpha(f)$ and $M_\alpha = \cap_{\beta<\alpha} M_\beta$ for all $\alpha \in \mathbb{Q}_1$. Let x_1, x_2 be two distinct points in E, and $\{\alpha_n\} \subset \mathbb{Q}_1$ be such that $\alpha_1 < \alpha_2 < ... \rightarrow \alpha_o \in \mathbb{Q}_1$. Define the u.s.c. functions f_n by $f_n(x_1) = 1$, $f_n(x_2) = \alpha_n$ and $f_n(x) = 0$ for $x \neq x_1, x_2$, $n \geq 1$. Then obviously, $A_\alpha(f_n) = \{x_1\}, \{x_1, x_2\}, E$, according to $\alpha \in (\alpha_n, 1], \alpha \in (0, \alpha_n]$ or $\alpha = 0$, respectively. Then $(A_\alpha(f_n), \alpha \in \mathbb{Q}_1) \rightarrow (M_\alpha, \alpha \in \mathbb{Q}_1)$ where $M_\alpha = \{x_1\}, \{x_1, x_2\}, E$, for $\alpha \in [\alpha_o, 1] \cap \mathbb{Q}_1$, $(0, \alpha_o) \cap \mathbb{Q}_1$ or $\alpha = 0$, respectively. Clearly, $M_\alpha \neq \cap_{\beta<\alpha} M_\beta$ for $\alpha = \alpha_o$. $\square$

Let $\Psi_{[0,1]}$ be the embedding of $\mathscr{U}$ into the product space $\prod_{\alpha \in [0,1]} \mathscr{F}_\alpha$. One can then induce on $\mathscr{U}$ another topology γ_* via the relative topology of its image $\Psi_{[0,1]}(\mathscr{U})$ in $\prod_{\alpha \in [0,1]} \mathscr{F}_\alpha$. We thus have two topologies γ and γ_* in $\mathscr{U}$. $\square$

Theorem 2. The topology γ is strictly coarser than γ_*.

This can also be proved by the following counter-example.

Let $\Psi_{\mathbb{Q}_1}$ and $\Psi_{[0,1]}$ be the embeddings of $\mathscr{U}$ into the two product spaces $\prod_{\alpha \in \mathbb{Q}_1} \mathscr{F}_\alpha$ and $\prod_{\alpha \in [0,1]} \mathscr{F}_\alpha$, respectively. Let x_1, x_2 be two distinct points in E, and $\beta \in (0, 1) \backslash \mathbb{Q}_1$. Consider the u.s.c. function f_o defined by $f_o(x_1) = 1$, $f_o(x_2) = \beta$ and $f_o(x) = 0$ otherwise. Then $A_\alpha(f_o) = \{x_1\}, \{x_1, x_2\}, E$, according to $\alpha \in (\beta, 1], \alpha \in (0, \beta]$ or $\alpha = 0$. Now, let $G = \{x \in E : \rho(x_1, x) < \rho(x_1, x_2)/2\}$, which is an open set in E (where ρ denotes some compatible metric on E). Then $\mathscr{F}_G$ is a neighborhood of $A_\beta(f_o) = \{x_1, x_2\}$ in the Matheron topology so that the set $V = \{f \in \mathscr{U} : A_\beta(f) \in \mathscr{F}_G\}$ is a neighborhood of f_o in the topology γ_*. However, one can see that any neighborhood of f_o in the topology γ is not included in V. This means that the topology γ is strictly coarser than γ_*. $\square$

As in Ogura and Li (2004), one may want to take a general dense subset Q of $[0, 1]$ which includes 0 and 1 in place of $\mathbb{Q}_1$ or $[0, 1]$ in the above. In this case, one gets another topology γ_Q via the relative topology of the image $\Psi_Q(\mathscr{U})$ in $\prod_{\alpha \in Q} \mathscr{F}_\alpha$, where Ψ_Q is the embedding of $\mathscr{U}$ into the product space $\prod_{\alpha \in Q} \mathscr{F}_\alpha$. The argument in the counter-example above also proves

Corollary. If $Q_1 \setminus Q_2 \neq \emptyset$ and $Q_2 \setminus Q_1 \neq \emptyset$, then the topologies γ_{Q_1} and γ_{Q_2} are not comparable each other.

Theorem 3. The restriction of γ to $\mathscr{F}^* = \{1_F : F \in \mathscr{F}\}$ coincides with the hit-or-miss topology τ.

Proof. Clearly a subbase of the relative topology for $\mathscr{F}^*$ consists of sets of the forms

$$\{1_F : F \in \mathscr{F}^K\} \text{ and } \{1_F : F \in \mathscr{F}_G\} \text{ with } K \in \mathscr{K}, G \in \mathscr{G}.$$

Thus $\mathscr{F}^*$ is homeomorphic to $\mathscr{F}$ under the identification $F \to 1_F$.

Theorem 4. A map $X : \Omega \to \mathscr{U}$ is $\mathscr{A}/\mathscr{B}(\gamma)$ measurable if and only if all level sets map $A_\alpha(X) : \Omega \to \mathscr{F}$ are $\mathscr{A}/\mathscr{B}(\tau)$ measurable, where $A_\alpha(X)(\omega) = \{x \in E : X(\omega)(x) \geq \alpha\}$, $\alpha \in [0,1]$.

Proof. (i) Assuming that X is a random element. Then $\Psi_{\mathbb{Q}_1} \circ X$ is measurable. Now the projection $\pi_\alpha : \prod_{\alpha \in \mathbb{Q}_1} \mathscr{F}_\alpha \to \mathscr{F}_\alpha$ is measurable (continuous), so is $\pi_\alpha \circ \Psi_{\mathbb{Q}_1} \circ X = A_\alpha(X)$. For $\alpha \in [0,1] \backslash \mathbb{Q}_1$, $A_\alpha(X) = \cap_{\beta < \alpha, \beta \in \mathbb{Q}_1} A_\beta(X)$ and hence measurable.

(ii) Suppose $A_\alpha(X)$ is measurable for all α. Then the map $\Psi_{\mathbb{Q}_1} \circ X : \Omega \to (\prod_{\alpha \in \mathbb{Q}_1} \mathscr{F}_\alpha, \prod_{\alpha \in \mathbb{Q}_1} \mathscr{B}(\tau))$ is measurable as well. By second countability, we have $\mathscr{B}(\prod_{\alpha \in \mathbb{Q}_1} \mathscr{F}_\alpha) = \prod_{\alpha \in \mathbb{Q}_1} \mathscr{B}(\tau)$ and the results follows. $\square$

The above investigations lead to a separable, metrizable space $\mathscr{U}$ as a subspace of $\prod_{\alpha \in \mathbb{Q}_1} \mathscr{F}_\alpha$. However, as the Matheron's topology on $\mathscr{F}(E)$ is precisely the Lawson topology on the continuous lattice $\mathscr{F}(E)$ making it a Hausdorff, compact and second countable space, see Gierz et al (2003), it is interesting to find out whether the Lawson topology on the continuous lattice $\mathscr{U}(E)$ could provide the same properties, namely compactness and second countability for it. It turns out that the answer is yes ! While the following results are buried in Gierz et al (2003), we think they should be made known to the probability community in some explicit way (see however Noberg, 1992). For that purpose, we will here describe topologies on lattices and state clearly the Lawson topology for the continuous lattice $\mathscr{U}(E)$ for the LCHS space E.

Let $(L, \leq)$ be a partially ordered set (poset). When L is a lattice, the meet and join operations will be denoted as $\wedge$ and $\vee$, respectively. A lattice L is complete if every subset of L has a sup and an inf. Recall that the relation "way-below" is defined as follows. For x, y in a poset L, x is said to be way-below y, in symbol, $x << y$, if for all directed subsets D of L (i.e. for each $u, v \in D$, there is a $z \in D$ such that with $u, v \leq z$) for which sup$D$ exists, if $y \leq$ sup$D$, then there exists a $z \in D$ with $x \leq z$. A complete lattice L is called a *continuous lattice* if it satisfies the axiom of approximation, i.e. for all $x \in L$, the set $\{y \in L : y << x\}$ is directed and $x = \sup\{y \in L : y << x\}$. Note that, the lattice of open sets (or closed sets) $\mathscr{O}(E)$ (or $\mathscr{F}(E)$) is a continuous lattice. In the following E always denotes a LCHS topological space. The space $\mathscr{L}(E)$ (resp. $\mathscr{U}(E)$) of lower semicontinuous functions on E with values in the extended real line (resp. u.s.c. functions) is a continuous lattice.

To facilitate the reading from Gierz et al (2003), we will state results in terms of open sets and lower semicontinuous functions. The reader simply needs to translate in terms of closed sets and upper semicontinuous functions.

The Lawson topology on a continuous lattice L is the smallest topology containing the following sets

$$\{y \in L : x << y\} \text{ and } \{y \in L : x \not\le y\}, x \in L$$

The following facts can be found in Gierz et al (2003).

Theorem: If L is a continuous lattice, then the Lawson topology is the unique Hausdorff, compact topology on L making the meet operation $\wedge : L \times L \to L$ continuous.

In view of the bijection between the $\mathcal{O}(E)$ and $\mathcal{F}(E)$, the Lawson topology on $\mathcal{F}(E)$ is the hit-or-miss topology of Matheron. Consider $\mathcal{O}(E)$ as a sublattice of $\mathcal{L}(E)$ (via identification of indicator functions of sets), the Lawson topology on $\mathcal{O}(E)$ is the subspace of the Lawson topology on $\mathcal{L}(E)$. Furthermore, $\mathcal{O}(E)$ is a closed subspace of $\mathcal{L}(E)$.

Theorem 5. The Lawson topology on $\mathcal{U}(E)$ is Hausdorff, compact and second countable.

Final remark. It is clear that the purpose of this Note is to suggest the use of the Lawson topology (in lattice theory) to define random u.s.c. functions, and in particular, *random fuzzy closed sets*, in its most elegant framework. Exactly like random closed sets on LCHS spaces, this Lawson topology makes the space of u.s.c. functions a Hausdorff, compact and second countable topological space. It is also clear that, in immediate future research, details and measurability considerations with respect to Lawson topology need to be worked out. This should set up the right framework for studying Choquet theorem for u.s.c. random functions.

Acknowledgments

We thank J. Harding for mentioning to us the connections with continuous lattices.

References

[1] Beer, G. (1993), *Topologies on Closed and Closed Convex Sets*,Kluwer Academic, Dordrecht.

[2] Billingsley, P. (1968), *Convergence of Probability Measures*, J.Wiley, New York.

[3] Colubi, A., Dominguez-Menchero, Lopez-Diaz, M. and Ralescu, D. (2002), A D[0,1] representation of random upper semicontinuous functions, *Proc. Amer. Math. Soc.* 130(11), 3237-3242.

[4] Dudley, R. (1989), *Real Analysis and Probability*, Wadsworth and Brooks/Cole, Belmon.

[5] Gierz, G., Hofmann, K. H., Keimeil, K., Lawson, J. D., Mislove, M. W. and Scott, D. S. (2003), *Continuous Lattices and Domains*. Cambridge Univ. Press, Cambridge, UK.

[6] Li, S., Ogura, Y. and Kreinovich, V. (2002), *Limit Theorems and Applications of Set-Valued and Fuzzy Set-Valued Random Variables,* Kluwer Academic, Dordreccht.

[7] Matheron, G. (1975), *Random Sets and Integral Geometry,* J.Wiley, New York.

[8] Molchanov, I. (2005), *Theory of Random Sets,* Springer-Verlag.

[9] Nguyen, H. T. (2005), On modeling of perception-based information for intelligent technology and statistics, *J. of Taiwan Intelligent Technology and Applied Statistics* 3(2), 25-43.

[10] Nguyen, H. T. (2006), *An Introduction to Random Sets,* Chapman and Hall/CRC.

[11] Norberg, T. (1992), On the existence of ordered coupling of random sets-with applications, *Israel Journal of Mathematics* (77), 241-264.

[12] Ogura, Y. and Li, S. (2005), On limit theorems for random fuzzy sets including large deviation principles, In *Soft Methodology and Random Information Systems* (Lopez-Dias, M. et al, eds.), Springer-Verlag, 32-44.

[13] Teran, P. (2005), A large deviation principle for random upper semicontinuous functions, *Proc. Amer. Math. Soc.* 134(2), 571-580.

Optional Sampling Theorem and Representation of Set-Valued Amart

Shoumei Li[1]* and Li Guan[2]

[1] Department of Applied Mathematics, Beijing University of Technology, 100 Pingleyuan, Chaoyang District, Beijing, 100022, P.R.China
lisma@bjut.edu.cn
[2] guanli@mails.bjut.edu.cn

Summary. In this paper, we shall prove some properties of set-valued asymptotic martingale (amart for short) and provide an optional sampling theorem. We also prove a quasi Risez decomposition theorem for set-valued amarts. Then we shall discuss the existence of selections of set-valued amarts and give a representation theorem.

1 Introduction

It is well known that classical martingale theory plays an important role in probability theory and applications. By the development of stopping time techniques, it is allowed the generalization of martingale concepts. The outcome of this effort was the introduction and detailed study of vector-valued asymptotic martingale (amart for short) and uniform amart. Readers may mainly refer to papers of those of Bellow [4] and [5], Chacon and Sucheston [6], Edgar-Sucheston [7]-[9]. Especially we would like to mention that Edgar and Sucheston discussed the properties, almost sure convergence theorems and the Riesz decomposition theorems of vector-valued amarts in [7] and [9], which are relative to our this paper.

For the theory of set-valued martingale, many good results have been obtained. For examples, representation theorem of set-valued martingales was proved by Luu by means of martingale selections [16]; convergence theorems of set-valued martingales, submartingales and supermartingales under various settings were obtained by many authors, such as Hess [10], Hiai and Umegaki [11], Korvin and Kleyle [12], Li and Ogura [13] and [14], Papageorgiou [20] and [21], Wang and Xue [24]. The concept of a set-valued uniform amart was introduced by Luu [18], and he obtained the representation theorem of set-valued uniform amart. Luu also introduced the concept of set-valued L^1-amart and got many good results in [17] and [19]. Papageorgiou [22] discussed convergence of set-valued uniform amarts in the sense of Kuratowski-

---

[†] Research partial supported by NSFC, NSFB, Projects of Beijing Bai Qian Wan Ren Cai and Chuang Xin Tuan Dui, P.R. China

Mosco, and also obtained the weak convergence theorems of set-valued amarts. In [23], Zhang, Wang and Gao discussed equivalent definitions of a set-valued amart.

As we have known that in vector-valued case, there are an optional sampling theorem and Risez decomposition theorem (cf. [7]. Can we get the similar results for set-valued amarts? Does there exist an amart selection of a set-valued amart? Can we provide a sequence of amart selections of a set-valued amart so that the set-valued amart can be represented by this sequence of amart selections? These are what shall focus on in this paper.

This paper is organized as follows. In section 2, we shall briefly introduce some concepts and notations on set-valued random variables. In section 3, we shall prove some basic properties of set-valued amart, and then state optimal sampling theorem of set-valued amarts. We shall also provide a quasi Risez decomposition theorem for set-valued amarts. In section 4, we shall give the representation theorem of compact convex set-valued amarts. Since the page limitation, we omit the proofs of theorems here. If readers are interested in the whole paper, please contact us.

2 Preliminary on Set-Valued Random Variables

Throughout this paper, we assume that $(\Omega, \mathscr{A}, \mu)$ is a nonatomic complete probability space, $(\mathfrak{X}, \|\cdot\|)$ is a real separable Banach space with its dual space $\mathfrak{X}^*$, $\mathbf{K}(\mathfrak{X})$ is the family of all nonempty closed subsets of $\mathfrak{X}$, $\mathbf{K}_b(\mathfrak{X})$ is the family of all nonempty bounded closed subsets of $\mathfrak{X}$, $\mathbf{K}_c(\mathfrak{X})$ is the family of all nonempty closed convex subsets of $\mathfrak{X}$, and $\mathbf{K}_k(\mathfrak{X})$ ($\mathbf{K}_k(\mathfrak{X})$, resp.) is the family of all nonempty compact (compact convex, resp.) subsets of $\mathfrak{X}$.

Let A and B be two nonempty subsets of $\mathfrak{X}$ and let $\lambda \in \mathbb{R}$, the set of all real numbers. We define addition and scalar multiplication by

$$A + B = \{a + b : a \in A, b \in B\},$$

$$\lambda A = \{\lambda a : a \in A\}.$$

The Hausdorff metric on $\mathbf{K}_b(\mathfrak{X})$ is defined by

$$d_H(A, B) = \max\{\sup_{a \in A} \inf_{b \in B} \|a - b\|, \ \sup_{b \in B} \inf_{a \in A} \|a - b\|\} \tag{1}$$

for A, $B \in \mathbf{K}_b(\mathfrak{X})$. The equivalent definition of Hausdorff metric is

$$d_H(A, B) = \max\{\inf\{\lambda : B \subset A + \lambda\}, \inf\{\lambda : A \subset B + \lambda\}\}, \tag{2}$$

where $A + \lambda = \{x : d(x, A) \leq \lambda\}$.

The metric space $(\mathbf{K}_b(\mathfrak{X}), d_H)$ is complete, $\mathbf{K}_{bc}(\mathfrak{X}), \mathbf{K}_k(\mathfrak{X})$ and $\mathbf{K}_{kc}(\mathfrak{X})$ are closed subsets of $(\mathbf{K}_b(\mathfrak{X}), d_H)$, and $\mathbf{K}_k(\mathfrak{X})$ and $\mathbf{K}_{kc}(\mathfrak{X})$ are separable (cf. [15], Theorems 1.1.2 and 1.1.3). For an A in $\mathbf{K}_b(\mathfrak{X})$, let $\|A\|_{\mathbf{K}} = d_H(\{0\}, A)$. For more properties of the Hausdorff metric, readers could refer to [3].

For each $A \in \mathbf{K}_b(\mathfrak{X})$, define the support function by

$$s(x^*, A) = \sup_{a \in A} < x^*, a >, \quad x^* \in \mathfrak{X}^*,$$

where $\mathfrak{X}^*$ is the dual space of $\mathfrak{X}$.

A set-valued mapping $F : \Omega \to \mathbf{K}(\mathfrak{X})$ is called set-valued random variable (or measurable) if, for each open subset O of $\mathfrak{X}$, $F^{-1}(O) = \{\omega \in \Omega : F(\omega) \cap O \neq \emptyset\} \in \mathscr{A}$. For two set-valued random variables F, G, $F = G$ if and only if $F(\omega) = G(\omega)$ i.e.(μ).

A set-valued random variable F is called *integrably bounded* (cf. [11] or [15]) if $\int_\Omega \|F(\omega)\|_\mathbf{K} d\mu < \infty$. Let $L^1[\Omega; \mathbf{K}(\mathfrak{X})]$ denote the space of all integrably bounded random variables, and $L^1[\Omega, \mathscr{A}_0, \mu; \mathbf{K}_f(\mathfrak{X})]$ denote the space of all $\mathscr{A}_0$-measurable integrably bounded random variables taking values in $\mathbf{K}_f(\mathfrak{X})$, where "f" can be "c", "k", "kc" etc., and $\mathscr{A}_0$ is a sub-σ-filed of $\mathscr{A}$. If $\mathscr{A}_0 = \mathscr{A}$, we may write $L^1[\Omega; \mathbf{K}_f(\mathfrak{X})]$ for short.

For each set-valued random variable F, the expectation of F, denoted by $E[F]$, is defined by

$$E[F] = \left\{ \int_\Omega f d\mu : f \in S_F \right\},$$

where $\int_\Omega f d\mu$ is the usual Bochner integral in $L^1[\Omega, \mathfrak{X}]$, the family of integrable $\mathfrak{X}$-valued random variables, and $S_F = \{f \in L^1[\Omega; \mathfrak{X}] : f(\omega) \in F(\omega), a.e.(\mu)\}$. This kind of integral is called Aumann integral (cf. [1]) in literature. For more concepts such as conditional expectation of a set-valued random variable, set-valued martingale and more results, readers can refer to [11] and [15].

Assume that $\{\mathscr{A}_n : n \in \mathbb{N}\}$ is an increasing sequence of sub-σ-fields of $\mathscr{A}$ such that $\mathscr{A} = \sigma(\bigcup_{n \geq 1} \mathscr{A}_n)$.

A function $\tau : \Omega \to \mathbb{N} \bigcup \{+\infty\}$ is said to be a stopping time with respect to $\{\mathscr{A}_n : n \in \mathbb{N}\}$, if for each $n \geq 1, \{\tau = n\} =: \{\omega \in \Omega : \tau(\omega) = n\} \in \mathscr{A}_n$. The set of all stopping times is denoted by T^*. And we say that $\tau_1 \leq \tau_2$ if and only if $\tau_1(\omega) \leq \tau_2(\omega)$ for all $\omega \in \Omega$. Let T denote the set of all bounded stopping times, and $T(\sigma) = \{\tau : \tau \geq \sigma, \tau \in T\}$. Given $\tau \in T$, we define

$$\mathscr{A}_\tau = \{A \in \mathscr{A} : A \bigcap \{\tau = n\} \in \mathscr{F}_n, n \geq 1\}.$$

Then $\mathscr{A}_\tau$ is a sub-σ-field of $\mathscr{A}$. If $X_n \in L^1[\Omega; \mathbf{K}(\mathfrak{X})]$ for any $n \in \mathbb{N}$, we define $X_\tau(\omega) = X_{\tau(\omega)}(\omega)$ for all $\omega \in \Omega$. Then $X_\tau : \Omega \to \mathbf{K}(\mathfrak{X})$ is $\mathscr{A}_\tau$-measurable.

3 Properties, Optional Sampling Theorem and a Quasi Risez Decomposition of Set-Valued Amart

In this section, we shall first prove some basic properties of set-valued amart. We introduce the following definition of set-valued amarts.

Definition 3.1 An adapted set-valued sequence $\{F_n, \mathscr{A}_n : n \in \mathbb{N}\} \subset L^1[\Omega; \mathbf{K}(\mathfrak{X})]$ is called *a set-valued amart*, if the net $\{\int_\Omega F_\tau d\mu\}_{\tau \in T}$ convergent in the sense of d_H.

If for any $x^* \in \mathfrak{X}^*$, $\{s(x^*, F_n), \mathscr{A}_n : n \in \mathbb{N}\}$ is a real-valued amart, we call $\{F_n, \mathscr{A}_n : n \geq 1\}$ is a set-valued weak amart. The next following Theorem tell us the relationship between set-valued amart and set-valued weak amart.

Theorem 3.2 If $\{F_n, \mathscr{A}_n : n \in \mathbb{N}\}$ is a set-valued amart in $L^1[\Omega; \mathbf{K}_c(\mathfrak{X})]$, then it is a set-valued weak amart.

It is clear that a linear combination of set-valued amarts is a set-valued amart. We next prove that finite union and intersection of set-valued amarts are also amarts. We need to prepare two Lemmas for it.

Lemma 3.3 If $\{A, B, A_n, B_n : n \in \mathbb{N}\} \subset \mathbf{K}_b(\mathfrak{X})$, $\lim_{n \to \infty} d_H(A_n, A) = 0$ and $\lim_{n \to \infty} d_H(B_n, B) = 0$, then

(i) for any given nonempty set C, we have $\lim_{n \to \infty} d_H(A_n \cap C, A \cap C) = 0$,

(ii) $\lim_{n \to \infty} d_H(A_n \cap B_n, A_n \cap B) = 0$,

(iii) $\lim_{n \to \infty} d_H(A_n \cap B_n, A \cap B) = 0$.

Lemma 3.4 If $\{A, B, A_n, B_n : n \in \mathbb{N}\} \subset \mathbf{K}_b(\mathfrak{X})$, $\lim_{n \to \infty} d_H(A_n, A) = 0$, and $\lim_{n \to \infty} d_H(B_n, B) = 0$, then

(i) For any given set $C \in \mathbf{K}_b(\mathfrak{X})$, we have $\lim_{n \to \infty} d_H(A_n \cup C, A \cup C) = 0$,

(ii) $\lim_{n \to \infty} d_H(A_n \cup B_n, A_n \cup B) = 0$,

(iii) $\lim_{n \to \infty} d_H(A_n \cup B_n, A \cap B) = 0$.

Theorem 3.5 Let $\{F_n, \mathscr{A}_n : n \geq 1\}$ and $\{G_n, \mathscr{A}_n : n \geq 1\}$ be set-valued adapted L^1-bounded sequences.

(a) If $\{\int F_\tau d\mu\}_{\tau \in T}$ and $\{\int G_\tau d\mu\}_{\tau \in T}$ are bounded, then $\{\int F_\tau \cup G_\tau d\mu\}_{\tau \in T}$ and $\{\int F_\tau \cap G_\tau d\mu\}_{\tau \in T}$ are bounded.

(b) If $\{F_n, \mathscr{A}_n : n \geq 1\}$ and $\{G_n, \mathscr{A}_n : n \geq 1\}$ are amarts, then $\{F_n \cup G_n, \mathscr{A}_n : n \geq 1\}$ and $\{F_n \cap G_n, \mathscr{A}_n : n \geq 1\}$ are amarts.

Theorem 3.6 Let $\{F_n, \mathscr{A}_n : n \geq 1\}$ be a set-valued amart, then $\{\int_\Omega F_\tau d\mu : \tau \in T\}$ is bounded.

We next prove an "optional sampling theorem".

Theorem 3.7 Let $\{F_n, \mathscr{A}_n : n \geq 1\}$ be a set-valued amart and let $\{\tau_k : k \geq 1\}$ be a nondecreasing sequence of bounded stopping times. Define $G_k = F_{\tau_k}$, then $\{G_k, \mathscr{A}_{\tau_k} : k \in \mathbb{N}\}$ is a set-valued amart.

Corollary 3.8 (Optional stopping theorem) Let $\{F_n : n \geq 1\}$ be a set-valued amart, σ a stopping time (possibly infinite). Then $G_n = F_{n \wedge \sigma}$ is a set-valued amart.

By Theorem 3.2 of [7], we know that if $\{f_n, \mathscr{A}_n : n \geq 1\}$ is a vector-valued amart in $L^1[\Omega, \mathfrak{X}]$, then f_n has Riesz decomposition, i.e. $f_n = m_n + p_n$ with $\{m_n, \mathscr{A}_n : n \geq 1\}$ a martingale and $p_n \to 0$ in L^1, in addition, $\{p_n : n \geq 1\}$ is uniformly integral and $p_n \to 0$ a.e.. $\{p_n : n \geq 1\}$ is called a potential.

Since the space $\mathbf{K}(\mathfrak{X})$ is not a linear space with respect to the addition and multiplication, it is difficult to obtain the Riesz decomposition for a set-valued amart. Now we prove a quasi Risez decomposition theorem of a set-valued amart.

Theorem 3.9 If $\{F_n, \mathscr{A}_n : n \geq 1\} \subset L^1[\Omega; \mathbf{K}_{kc}(\mathfrak{X})]$ is a set-valued amart satisfies $F_n(\omega) \subset G(\omega)$, a.e. with $G \in L^1[\Omega; \mathbf{K}_{kc}(\mathfrak{X})]$, then there exists a set-valued martingale $\{M_n, \mathscr{A}_n : n \geq 1\}$ and $\{Z_n : n \geq 1\}$ such that

$$F_n(\omega) \subset M_n(\omega) + Z_n(\omega), \quad a.e.$$

$$\|Z_n\|_{\mathbf{K}} \to 0 \ (n \to \infty).$$

Corollary 3.10   Under the Assumption as Theorem 3.9, then there exists a set-valued martingale $\{M_n, \mathscr{A}_n : n \geq 1\} \subset L^1[\Omega; \mathbf{K}_{kc}(\mathfrak{X})]$ such that the sequence $\{\rho_n = d_H(F_n, M_n), \mathscr{A}_n : n \geq 1\}$ is a potential.

4 Representation Theorem for Closed Convex Set-Valued Amarts

Definition 4.1   A sequence $\{f_n, \mathscr{A}_n : n \geq 1\}$ is called an amart selection of $\{F_n, \mathscr{A}_n : n \geq 1\}$ if
(i) $f_n \in S_{F_n}(\mathscr{A}_n)$ for all $n \in \mathbb{N}$.
(ii) $\{f_n, \mathscr{A}_n : n \geq 1\}$ is an amart in $L^1[\Omega, \mathfrak{X}]$.

In this case we write $\{f_n\} \in AMS(\{F_n\})$, and $AMS(\{F_n\})$ denotes the set of all amart selections of $\{F_n, \mathscr{A}_n : n \in \mathbb{N}\}$.

Example 4.2   Let $\{f_n, \mathscr{A}_n : n \in \mathbb{N}\}$ be an $\mathfrak{X}$-valued amart and $\{r_n, \mathscr{A}_n : n \in \mathbb{N}\}$ be a real-valued amart. Take a bounded closed convex subset B of $\mathfrak{X}$. Define

$$F_n = f_n + r_n B,$$

then $\{F_n, \mathscr{A}_n : n \in \mathbb{N}\}$ is a set-valued amart.

It is easy to see that every sequence $\{g_n : n \in \mathbb{N}\}$, defined by $g_n = f_n + r_n x$ for some $x \in B$, is an amart selection of $\{F_n, \mathscr{A}_n : n \in \mathbb{N}\}$.

We shall have the natural question: does $\{F_n, \mathscr{A}_n : n \geq 1\}$ always have $\mathfrak{X}$-valued amart selections? The following theorem will answer this question by using Steiner method in finite dimensional space.

Theorem 4.3   Assume $\mathfrak{X}$ is a d-dimensional Banach space and $\{F_n, \mathscr{A}_n : n \geq 1\} \subset L^1[\Omega, \mathscr{A}, \mu; \mathbf{K}_{bc}(\mathfrak{X})]$ is a set-valued amart. Then it admits an amart selection.

To get further representation theorem, we need the following Lemmas and notations.

Lemma 4.4   Let $\{F_n, \mathscr{A}_n : n \geq 1\} \subset L^1[\Omega; \mathbf{K}_c(\mathfrak{X})]$ be a set-valued amart, then for any $A \in \mathscr{A}$, $\{I_A F_n, \mathscr{A}_n : n \geq 1\}$ is a set-valued amart.

Let $\{F_n, \mathscr{A}_n : n \geq 1\} \subset L^1[\Omega; \mathbf{K}_c(\mathfrak{X})]$ is a set-valued amart and $\{r_n, \mathscr{A}_n : n \geq 1\}$ a potential. Denote

$$AMS(F_n, r_n) = \{\{f_n\} \in AMS(F_n) : \|p_n(\omega)\| \leq r_n(\omega), \ a.e.,$$

$$p_n \text{ is the potential in the Riesz decomposition of } f_n\}.$$

Lemma 4.5(cf. [19]) If $\mathscr{A}_1 \subset \mathscr{A}_0$ are two sub-σ-fields of $\mathscr{A}$, $F \in L^1[\Omega, \mathbf{K}_c(\mathfrak{X})]$, $G \in L^1[\Omega, \mathscr{A}_0, \mathbf{K}_c(\mathfrak{X})]$ and $\theta : \Omega \to \mathbb{R}^+ \backslash \{0\}$ is a $\mathscr{A}_1$-measurable function, then for each $f \in S_F(\mathscr{A}_1)$, we can find $g \in S_G(\mathscr{A}_0)$ such that

$$\|f(\omega) - E[g(\omega)|\mathscr{A}_1]\| \leq d_H(F(\omega), E[G(\omega|\mathscr{A}_1)]) + \theta(\omega), \ a.e..$$

Consequently, if G is $\mathscr{A}_1$-measurable then there is some $g \in S_G(\mathscr{A}_1)$ such that

$$\|f(\omega) - g(\omega)\| \leq d_H(F(\omega), E[G(\omega|\mathscr{A}_1)]) + \theta(\omega), \ a.e..$$

Theorem 4.6 If $\{F_n, \mathscr{A}_n : n \geq 1\} \subset L^1[\Omega; \mathbf{K}_{kc}(\mathfrak{X})]$ is a set-valued amart satisfies $F_n(\omega) \subset G(\omega), a.e.$ with $G \in L^1[\Omega; \mathbf{K}_{kc}(\mathfrak{X})]$, then there is a positive potential $\{r_n, \mathscr{A}_n : n \geq 1\}$ such that for $k \geq 1$

$$S_{F_k}(\mathscr{A}_k) = \pi_k(AMS(F_n, r_n)),$$

where for every $\{f_n : n \geq 1\} \in AS(F_n, r_n)$, $\pi_k(\{f_n\}) = f_k$ (the usual projection to the kth element of the sequence $\{f_n : n \geq 1\}$).

Remark 4.7 From the proof of above theorem, we can see that there exist amart selections for a set-valued amart under the conditions of Theorem, even if $\mathfrak{X}$ is an infinite dimensional Banach space.

Now we are ready to state the following representation theorem of a set-valued amart.

Theorem 4.8 If $\{F_n, \mathscr{A}_n : n \geq 1\} \subset L^1[\Omega; \mathbf{K}_{kc}(\mathfrak{X})]$ is a set-valued amart satisfies $F_n(\omega) \subset G(\omega), a.e.$ with $G \in L^1[\Omega; \mathbf{K}_{kc}(\mathfrak{X})]$, then there exist a positive potential $\{r_n, \mathscr{A}_n : n \geq 1\}$ and a sequence $\{f_n^k : k \geq 1\} \subset AMS(F_n, r_n)$ such that for every $n \geq 1, F_n(\omega) = cl\{f_n^k(\omega) : k \geq 1\}$ for all $\omega \in \Omega$.

References

[1] R. Aumann (1965). Integrals of set valued functions, *J. Math. Anal. Appl.* **12**, 1-12.

[2] S. Bagchi (1985). On a.s. convergence of classes of multivalued asymptotic martingales, *Ann. Inst. H. Poincaré, Probabilités et Statistiques*, **21**, 313-321.

[3] G. Beer (1993). *Topologies on Closed and Closed Convex Sets*, Kluwer Academic Publishers.

[4] A. Bellow (1978). Some aspects of the theory of the vector-valued amarts, *"Vector Space Measure and Applications I"* (ed. R. Aron and S. Dineen), Lect. Notes Math.", **644**, 57-67.

[5] A. Bellow (1978). Uniform amarts: A class of asymptotic martingales for which strong almost sure convergence obtains. *Z. Wahr. verw. Gebiete*, **41**, 177-191.

[6] R.V. Chacon and L. Sucheston (1975). On convergence of vector-valued asymptotic martingales. *Z. Wahr. verw. Gebiete.* **33**, 55-59.

[7] G.A. Edgar and L. Suchistion (1976). Amarts: A class of asympotic martingales B. Discrete parameter. *J. Multiv. Anal.*, **6**, 193-221.

[8] G.A. Edgar and L. Suchistion (1976). Amarts: A class of asympotic martingales A. continuous parameter. *J. Multiv. Anal.*, **6**, 572-591.

[9] G.A. Edgar and L. Suchistion (1976). The Riesz decomposition for vector-valued amarts, *Z. Wahr. verw. Gebiete.* **36**, 85-92.

[10] C. Hess (1983). Measurability and integrability of the weak upper limit of a sequence of multifunctions, *J. Math. Anal. Appl.*, **153**, 226-249.

[11] F. Hiai and H. Umegaki (1977). Integrals, conditional expectations and martingales of multivalued functions, *Jour. Multiv. Anal.*, **7** 149-182.

[12] A.de Korvin and R. Kleyle (1985). A convergence theorem for convex set valued supermartingales. *Stoch. Anal. Appl.*, **3**, 433-445.

[13] S. Li and Y. Ogura (1998). Convergence of set valued sub-supermartingales in the Kuratowski-Mosco sense, *Ann. Probab.*, **26**, 1384-1402.

[14] S. Li and Y. Ogura (1999). Convergence of set valued and fuzzy valued martingals. *Fuzzy Sets and Syst.*, **101**, 453-461.

[15] S. Li, Y. Ogura and V. Kreinovich (2002). *Limit Theorems and Applications of Set-Valued and Fuzzy Sets-Valued Random Variables*, Kluwer Academic Publishers.

[16] D.Q. Luu (1981). Representations and regularity of multivalued martingales, *Acta Math. Vietn.*, **6**, 29-40.

[17] D.Q. Luu (1984). Applications of set-valued Randon-Nikodym theorms to convergence of multivalued L^1-amarts, *Math. Scand.*, **54**, 101-114.

[18] D.Q. Luu (1985). Quelques resultats des amarts uniform nultivoques dans les espaces de Banach, *CRAS, Paris*, **300**, 63-65.

[19] D.Q. Luu (1986). Representation theorems for multi-valued (regular) L^1-amarts, *Math. Scand.*, **58**, 5-22.

[20] N.S. Papageorgiou (1985). On the theorey of Banach space valued multifunctions. 1. integration and conditional wxpectation, *J. Multiv. Anal.*, **17**, 185-206.

[21] N.S. Papageorgiou (1987). A convergence theorem for set valued multifunctions. 2. set valued martingales and set valued measures, *J. Multiv. Anal.*, **17**, 207-227.

[22] N.S. Papageorgiou (1995). On the conditional expectation and convergence properties of random sets. *Trans. Amer. Math. Soc.*, **347**, 2495-2515.

[23] W.X. Zhang, Z.P. Wang and Y. Gao (1996). *Set-valued random processes*, Science Publisher. (in Chinese).

[24] Z.P. Wang and X. Xue (1994). On convergence and closedness of multivalued martingales, *Trans. Ameri. Math. Soc.*, **341**, 807-827.

On a Choquet Theorem for Random Upper Semicontinuous Functions

Yukio Ogura *

Department of Mathematics, Saga University, Saga 840-8502, Japan
ogura@cc.saga-u.ac.jp

Summary. We extend some topologies on the space of upper semicontinuous functions with compact support to those on that of general upper semicontinuous functions and see that graphical topology and modified L^p topology are the same. We then define random upper semicontinuous functions using their topological Borel field and finally give a Choquet theorem for random upper semicontinuous functions.

1 Introduction

In the study of random upper semicontinuous (u. s. c. in brief) functions, one sometimes restrict the random functions to take values in $[0,1]$. Those are called random fuzzy sets or fuzzy set valued random variable, and conceives some applications such as data analysis for imprecise or incomplete data. The mathematical treatment of such u. s. c. functions begun in late 70's, and since then various interesting topologies as well as the notions of random variables are proposed. Among them, the uniform Hausdorff topology (sometimes denoted by d_∞ or d_U) is the strongest and regarded as most important. However, it is too strong for some purposes. For example, it does not provide a separable space and its Borel field is so large that one might find difficulty to check the measurability. This is one of the reason that various topologies have been proposed. The L^p topology was already given in an early work by Puri and Ralescu [11], and the graphical and Skorohod topology in [9] and [2] respectively. For a systematic study of such topologies, one is referred to [10], although the concerned u. s. c. functions there are restricted to those with compact supports.

On the other hand, in the study random sets, the Choquet theorem was successfully used by Matheron[4], who exploited Fell topology (hit and error topology) on the space of closed sets (see also [5]). This suggests us to study u. s. c. functions without the compact support restriction, which also extends the area of applications.

The object of this paper is to inspect those topologies above on the space of general u. s. c. functions taking values in $[0,1]$ and obtain a Choquet theorem. For

---
* Partially supported by Grant-in-Aid for Scientific Research 17540123.

Y. Ogura: *On a Choquet Theorem for Random Upper Semicontinuous Functions*, Advances in Soft Computing **6**, 145–151 (2006)

this purpose, we first review the Fell topology on a LCHS (locally compact Hausdorff second countable) space in Section 2. In Section 3, we extend some topologies on the space of u. s. c. functions with compact support to those on that of general u. s. c. functions and see that graphical topology and the modified L^p topology are the same (see Section 3 below for the definition). In the last Section 4, we first study the topological Borel fields for the topologies given in Section 2. We then define random u. s. c. functions properly and finally obtain a Choquet theorem for them.

We close this section with noting that a forthcoming paper [7] will provide the proofs left in this paper. Also in [8], we refer to Lawson topology in continuous lattice theory, which conceives another approach to the Choquet theorem than in this article.

2 Topologies on the Space of Closed Sets

Let S be a LCHS space with a compatible metric ρ^S. Let also

$$\mathscr{F}(S) \quad \text{the space of all closed subsets of } S,$$
$$\mathscr{F}'(S) \quad \text{the space of all non-empty closed subsets of } S,$$
$$\mathscr{K}(S) \quad \text{the space of all compact subsets of } S,$$
$$\mathscr{G}(S) \quad \text{the space of all open subsets of } S.$$

For a family $\mathscr{H}$ of subsets of S, denote

$$\mathscr{H}^K = \{A \in \mathscr{H} : A \cap K = \emptyset\}, \qquad \mathscr{H}_G = \{A \in \mathscr{H} : A \cap G \neq \emptyset\}.$$

Definition 1. *The* Fell topology *on $\mathscr{F}(S)$ has a sub-base $\mathscr{F}(S)_G$ for all $G \in \mathscr{G}(S)$ and $\mathscr{F}(S)^K$ for all $K \in \mathscr{K}(S)$.*

The closed set space $\mathscr{F}(S)$ is a compact metric space but $\mathscr{F}'(S)$ is not closed in $\mathscr{F}(S)$. As a compatible metric, the *Huasdorff-Buseman metric*

$$d_{HB}^S(A, B) = \sup_{x \in S} e^{-\rho^S(0,x)} |\rho^S(x, A) - \rho^S(x, B)|,$$

where $\rho^S(x, A) = \inf_{y \in A} \rho^S(x, y)$, is introduced (see [5]). We understand that this is for the case where S is a Banach space (actually of finite dimensional in our case), because the origin 0 appears in the formula. In that case, the metric is compatible with the Fell topology in $\mathscr{F}'(S)$. However, we could not find any good convention for $\rho^S(x, \emptyset)$ with which the metric is compatible with the Fell topology in $\mathscr{F}(S)$. For a construction of a comaptible metric, see [12].

3 Topologies on the Space of Upper Semicontinuous Functions

Let E be a LCHS space with a compatible metric ρ^E and $I = [0, 1]$. We endow the product space $I \times E$ with the product topology. For the convenience of the following arguments, we use the maximum metric

$$\rho^{I\times E}((\alpha,x),(\beta,y)) = |\alpha - \beta| \vee \rho^E(x,y),^2$$

which is compatible with the product topology. Let $\mathcal{U} = \mathcal{U}(E)$ be the space of u. s. s. functions u on a closed set D_u in E taking values in I, which is refered to the space of fuzzy sets. We denote by $\mathcal{U}' = \mathcal{U}'(E)$ the space of all normal u. s. s. functions, that is all $u \in \mathcal{U}$ satisfying $u(x) = 1$ for some $x \in E$. Let $\mathcal{D} = \mathcal{D}(I;\mathcal{F}(E))$ be the space of all functions from I to $\mathcal{F}(E)$ which is decreasing in I (with respect to the set inclusion order in E), and left continuous in $(0,1]$ (note that u has right limit in $[0,1)$ automatically). We denote by $\mathfrak{G} = \mathfrak{G}(I \times E)$ the space of all closed graphs in $I \times E$, that is, the space of all closed sets G in $I \times E$ such that $G_\alpha = G \cap (\{\alpha\} \times E)$ is decreasing in $\alpha \in I$. Clearly $\mathcal{U}$ is identified with $\mathcal{D}$ through the map $L : \mathcal{U} \to \mathcal{D}$ defined by

$$L(u)(\alpha) = \{x \in D_u : u(x) \geq \alpha\}, \qquad \alpha \in I,$$

and with $\mathfrak{G}$ through the map $G : \mathcal{U} \to \mathfrak{G}$ given by

$$G(u) = \{(\alpha,x) \in I \times D_u : 0 \leq \alpha \leq u(x)\}.$$

We sometimes denote $u(\alpha) = L(u)(\alpha)$, where no confusion occurs. Thus $G(u)$ has another expression $G(u) = \cup_{\alpha \in I}(\{\alpha\} \times u(\alpha))$. Finally, for each $u \in \mathcal{U}$ and $K \in \mathcal{K}(E)$, the u. s. c. function $u \sqcap K$ is defined as the restriction of u to the domain $D_u \cap K$. Thus, $u \sqcap K(\alpha) = u(\alpha) \cap K$, $\alpha \in I$, and $G(u \sqcap K) = G(u) \cap (I \times K)$.

The topologies on $\mathcal{U}$ can be defined from those in $\mathcal{D}$ or $\mathfrak{G}$ through the identications. The strongest topology is that induced by the uniform norm of the Hausdorff-Fell distance, that is

$$d_U(u_1,u_2) = \|d^E(u_1(\cdot),u_2(\cdot))\|_I.^3$$

Then d_U is a metric on $\mathcal{U}$ and the metric space $(\mathcal{U},d_U)$ is complete but not separable in general. To define the Skorohod metric, denote by Λ the class of strictly increasing continuous mappings of I onto itself. Then the Skorohod metric d_S is defined by

$$d_S(u_1,u_2) = \inf_{\lambda \in \Lambda} \{\|\lambda - 1\|_I \vee \|d^E(u_1 \circ \lambda(\cdot),u_2(\cdot))\|_I\},$$

where 1 is the identical mapping. Following [1], we define the *Billingsley metric*

$$d_B(u_1,u_2) = \inf_{\lambda \in \Lambda} \{\|\lambda\|^0 \vee \|d^E(u_1 \circ \lambda(\cdot),u_2(\cdot))\|_I\},$$

where $\|\lambda\|^0 = \sup_{\alpha<\beta} \left| \log \dfrac{\lambda(\beta) - \lambda(\alpha)}{\beta - \alpha} \right|$.

The proof of the following assertions will be given in our forthcoming paper [7].

Lemma 1. *Let E be a metric space.*

(1) If E contains three distinct points, then the space $\mathcal{U}$ is not complete under d_S.

(2) If E has an accumulation point, then the space $\mathcal{U}$ is neither locally compact under d_S nor under d_B.

---

2 $a \vee b$ stands for the maximum of a and b, respectively.

3 For a map $x : I \to \mathbb{R}$. we denote $\|x\|_I = \sup_{\alpha \in I} |x(\alpha)|$.

We next introduce d_Q-metric and the graphical metric. Let $Q = \{\alpha_k : k \in \mathbb{N}\}$ be a countable dense subset of I including 0. Then the d_Q-metric is defined by

$$d_Q(u_1, u_2) := \sum_{k=1}^{\infty} 2^{-k} (d^E(u_1(\alpha_k), u_2(\alpha_k)) \wedge 1).$$

This is actually a metric on $\mathscr{U}$, since $d_Q(u_1, u_2) = 0$ implies $u_1 = u_2$. The topology induced by d_Q is same as that by the product space $\prod_{\alpha \in Q} \mathscr{F}(E)_\alpha$, where $\mathscr{F}(E)_\alpha$ is a copy of $\mathscr{F}(E)$, and is studied more in detail in [8] in the case where $Q = \mathbb{Q} \cap I$. In our our forthcoming paper [7], we will also show

Theorem 1. (1) *The space $(\mathscr{U}, d_Q)$ is a LCHS space.*
   (2) *The space $(\mathscr{U}', d_Q)$ is not a locally compact space.*

The graphical topology on $\mathfrak{G}$ (and so on $\mathscr{U}$ by the identification above) is defined as the relative topology of the Fell topology on $\mathscr{F}(I \times E)$. Since the Fell topology on $\mathscr{F}(I \times E)$ is metrizable, so is graphical topology on $\mathfrak{G}$. We denote a comaptible metric by d_G.

Theorem 2. (1) *For any countable dense subset Q of I including 0, the topology on $\mathscr{U}$ induced by d_Q is stronger than the graphical topology.*
   (2) *The topology on $\mathscr{U}$ induced by d_S is stronger than the graphical topology.*

We will next concern the d_p-metric for $p \geq 1$. Although the usual L^p metric d_p is given by

$$d_p(u_1, u_2) = \left(\int_I d_H^E(u_1(\alpha), u_2(\alpha))^p d\alpha \right)^{1/p},$$

we would like to introduce the modified L^p metric:

$$d_p^0(u_1, u_2) = \left(\int_I d^E(u_1(\alpha), u_2(\alpha))^p d\alpha + d^E(u_1(0), u_2(0))^p \right)^{1/p}.$$

Theorem 3. *The graphical topology and that induced by the modified L^p metric d_p^0 are the same.*

4 Choquet Theorem for Random Upper Semicontinuous Functions

We start with some definitions of various σ-fields.

Definition 2. *The cylindrical σ-field $\mathscr{C} = \mathscr{C}(\mathscr{U})$ is the σ-field generated by the family*
$$\{u \in \mathscr{U} : u(\alpha) \in A\}, \qquad \alpha \in I, \ A \in \mathscr{B}(\mathscr{F}(E)),$$
where $\mathscr{B}(\mathscr{F}(E))$ is the topological Borel field of $\mathscr{F}(E)$ with Fell topology.

Definition 3. (1) *For a metric d in $\mathscr{U}$, $\mathscr{B}_d^0$ is the σ-field generated d-open balls.*
(2) *For a metric d on $\mathscr{U}$, $\mathscr{B}_d$ is the σ-field generated d-open sets.*

We then have the following

Theorem 4. *For each countable dense subset Q of $[0,1)$ including 0, it holdsp*

$$\mathscr{C} = \mathscr{B}_{d_S}^0 = \mathscr{B}_{d_S} = \mathscr{B}_{d_B}^0 = \mathscr{B}_{d_B} = \mathscr{B}_{d_G}^0 = \mathscr{B}_{d_G}$$
$$= \mathscr{B}_{d_p^0}^0 = \mathscr{B}_{d_p^0} = \mathscr{B}_{d_Q}^0 = \mathscr{B}_{d_Q},$$

which we simply denote by $\mathscr{B}$.

We now define random u. s. c. functions.

Definition 4. *A map $X : \Omega \to \mathscr{U}$ is called a* random u. s. c. function *or* random fuzzy set (fuzzy set valued random variable) *on a probability space $(\Omega, \mathscr{A}, P)$ if it is $\mathscr{A}/\mathscr{B}$-measurable.*

For a LCHS space S, a set function $T : \mathscr{K}(S) \to [0,1]$ is called an *alternating Choquet capacity of infinite order on S* if it satisfies

(i) $T(\emptyset) = 0$,
(ii) $\lim_n T(K_n) = T(K)$ *for all decreasing sequence $\{K_n\} \subset \mathscr{K}(S)$ with $\bigcap_n K_n = K$,*

(iii) $T\left(\bigcap_{j=1}^n K_j\right) \le \sum_{\emptyset \ne I \subset \{1,2,,\ldots,n\}} (-1)^{|I|+1} T\left(\bigcup_{i \in I} K_i\right)$ *for all $n \ge 2$ and $K_1, K_2, \ldots, K_n \in \mathscr{K}(S)$.*

Then the celebrated Choquet theorem says

Theorem A. *([4, 5]) Let S be a LCHS space. Then, a functional $T : \mathscr{K}(S) \to [0,1]$ is a capacity functional of a unique random set $X : \Omega \to \mathscr{F}(S)$ on a probability space $(\Omega, \mathscr{A}, P)$ which satisfies*

$$T(K) = P(K \cap X \ne \emptyset), \qquad K \in \mathscr{K}(S),$$

if and only if T is an alternating Choquet capacity of infinite order on S.

For a $K \in \mathscr{F}(I \times E)$, we define the upper shadow K^+ of K by

$$K^+ = \bigcup_{(\alpha,x) \in K} [\alpha, 1] \times \{x\}$$

Now we have the following

Theorem 5. *Let E be a LCHS space. Then, a functional $T : \mathscr{K}(I \times E) \to [0,1]$ is a capacity functional of a unique random set $X : \Omega \to \mathscr{F}(I \times E)$ on a probability space $(\Omega, \mathscr{A}, P)$ which satisfies*

$$T(K) = P(K \cap X \neq \emptyset), \qquad K \in \mathscr{K}(I \times E),$$

*if and only if T is an alternating Choquet capacity of infinite order on I ×
E. Further, $G^{-1} \circ X$ is a random u. s. c. function, that is $G^{-1} \circ X$ is $\mathscr{A}/\mathscr{B}$-
measurable and satisfies $P(G^{-1} \circ X \in \mathscr{U}) = 1$, if and only if T satisfies*

$$T(K) = T(K^+), \qquad K \in \mathscr{K}(I \times E). \tag{1}$$

Proof. Since $I \times E$ is also a LCHS space, the first part of Theorem follows from
Theorem A. Further, the sufficiency in the second part is also clear. Hence, we have
only to show the necessity in the second part. We thus assume (1) and will show
that X is $\mathscr{A}/\mathscr{B}(\mathfrak{G})$-measurable and satisfies $P(G^{-1} \circ X \in \mathscr{U}) = 1$, where $\mathscr{B}(\mathfrak{G})$ is
the topological Borel field on $\mathfrak{G}$ with respect to the relative topology of the Fell
topology on $\mathscr{F}(I \times E)$. The measurability is clear from the definitions. To show the
latter assertion, take a countable base $\{V_i\}_{i\in\mathbb{N}}$ of the topology on E. Since E is locally
compact, we may assume that each V_i has compact closure $\bar{V}_i$. Similarly, since I is a
compact metric space, we can also take s countable base $\{U_i\}_{i\in\mathbb{N}}$ of the Euclidean
topology on I. Obviously closure $\bar{U}_i$ in I is compact. Let now $K_{ij} = \bar{U}_i \times \bar{V}_j$, $i, j \in \mathbb{N}$.
Then it is a countable base of $I \times E$. Hence, for each closed set $F \in \mathscr{F}(I \times E) \setminus \mathfrak{G}$,
we can find a $i, j \in \mathbb{N}$ such that

$$F \cap K_{ij} = \emptyset, \qquad F \cap K_{ij}^+ \neq \emptyset.$$

This leads us to

$$P(X \notin \mathfrak{G}) \leq \sum_{i,j\in\mathbb{N}} P(X \cap K_{ij} = \emptyset, X \cap K_{ij}^+ \neq \emptyset).$$

Since $P(X \cap K_{ij} = \emptyset, X \cap K_{ij}^+ \neq \emptyset) = T(K_{ij}^+) - T(K_{ij}) = 0$, we have $P(X \notin \mathfrak{G}) = 0$,
which implies $P(P(G^{-1} \circ X \in \mathscr{U}) = P(X \in \mathfrak{G}) = 1$. $\square$

Acknowledgments

We thank H. Nguyen for his drawing our attention to this subject and informing us
of the paper [12].

References

[1] P. Billingsley, *Convergence of Probability Measures*, Second Edition, John
Wiley & Sons, Inc., 1999.

[2] A. Colubi, J. S. Domínguez Menchero, M. López-Díaz and D. Ralescu,
A $D_E[0, 1]$ representation of random upper semicontinuous functions, *Proc.
Amer. Math. Soc.*, **130**(2002), 3237-3242.

[3] S. Li, Y. Ogura and V. Kreinovich, *Limit Theorems and Applications of Set-valued and Fuzzy-valued Random Variables*, Kluwer Academic Publishers, 2002.

[4] G. Matheron, *Random Sets and Integral Geometry*, John Wiley and Sons, 1975.

[5] I. Molchanov, *Theory of Random Sets*, Springer, 2005.

[6] H. T. Nguyen, *An Introduction to Random Sets*, CRC press, 2006.

[7] H. T. Nguyen and Y. Ogura, On random upper semicontinuous functions, *in preparation*.

[8] H. T. Nguyen, Y. Ogura, S. Tasena and H. Tran, A note on random upper semicontinuous functions, in *this proceedings*.

[9] Y. Ogura and S. Li, Separability for graph convergence of sequences of fuzzy valued random variables, *Fuzzy Sets and Syst.*, **123**(2001), 19-27.

[10] Y. Ogura and S. Li, On limit theorems for random fuzzy sets including large deviation principles, in *Soft Methodology and Random Information Systems*, Miguel Lopéz-Díaz *et al.* (eds.), 32-44, Springer, 2004.

[11] M. L. Puri and D. A. Ralescu, Fuzzy random variables, *J. Math. Anal. Appl.*, **114** (1986), 409-422.

[12] G. Wei and Y. Wang, Characterize the Matheron Topology Using Hausdorff Metric of the Alexandroff Compatification, *preprint*.

A General Law of Large Numbers, with Applications

Pedro Terán[1] and Ilya Molchanov[2]

[1] Facultad de Ciencias Económicas y Empresariales. Unidad Docente de Métodos Estadísticos. Grupo Decisión Multicriterio Zaragoza. Universidad de Zaragoza. Gran Vía, 2. E-50005 Zaragoza, Spain
teran@unizar.es

[2] Department of Mathematical Statistics and Actuarial Science. University of Bern. Sidlerstrasse 5. CH-3012 Bern, Switzerland
ilya.molchanov@stat.unibe.ch

Summary. We present a general law of large numbers in a (separable complete) metric space endowed with an abstract convex combination operation. Spaces of fuzzy sets are shown to be particular cases of that framework. We discuss the compatibility of the usual definition of expectation with the abstract one. We close the paper with two applications to the theory of fuzzy random variables (fuzzy random variables and level-2 fuzzy random variables in a *metric, maybe non-Banach* space).

1 Convex Combination (CC) Spaces

Let $(\mathbf{E}, d)$ be a metric space with a *convex combination operation* $[\cdot, \cdot]$

$$(\lambda_1, \ldots, \lambda_n, u_1, \ldots, u_n) \mapsto [\lambda_1, u_1; \ldots; \lambda_n, u_n]$$

(where $n \geq 2$, $\lambda_i > 0$, $\sum \lambda_i = 1$, $u_i \in \mathbf{E}$) yielding an element of $\mathbf{E}$ for each tuple of weights and points of $\mathbf{E}$. Note that $[\lambda_1, u_1; \ldots; \lambda_n, u_n]$ and the shorthand $[\lambda_i, u_i]_{i=1}^n$ have the same intuitive meaning as the more familiar $\lambda_1 u_1 + \ldots + \lambda_n u_n$ and $\sum_{i=1}^n \lambda_i u_i$, but $\mathbf{E}$ is not assumed to have an addition.

We will say that $\mathbf{E}$ is a *convex combination space* (CC space) if the following axioms are satisfied:

(i) (Commutativity) $[\lambda_i, u_i]_{i=1}^n = [\lambda_{\sigma(i)}, u_{\sigma(i)}]_{i=1}^n$ for every permutation σ of $\{1, \ldots, n\}$;

(ii) (Associativity) $[\lambda_i, u_i]_{i=1}^{n+2} = [\lambda_1, u_1; \ldots; \lambda_n, u_n;$
$\lambda_{n+1} + \lambda_{n+2}, [\frac{\lambda_{n+j}}{\lambda_{n+1}+\lambda_{n+2}}; u_{n+j}]_{j=1}^2];$

(iii) (Continuity) If $u, v \in \mathbf{E}$ and $\lambda^{(k)} \to \lambda \in (0, 1)$ as $k \to \infty$, then

$$[\lambda^{(k)}, u; 1 - \lambda^{(k)}, v] \to [\lambda, u; 1 - \lambda, v];$$

(iv) (Negative curvature) For all $u_1, u_2, v_1, v_2 \in \mathbf{E}$ and $\lambda \in (0, 1)$,

$$d([\lambda, u_1; 1 - \lambda, u_2], [\lambda, v_1; 1 - \lambda, v_2]) \leq \lambda d(u_1, v_1) + (1 - \lambda) d(u_2, v_2)$$

P. Terán and I. Molchanov: *A General Law of Large Numbers, with Applications*, Advances in Soft Computing **6**, 153–160 (2006)
www.springerlink.com   © Springer-Verlag Berlin Heidelberg 2006

(v) (Convexification) For each $u \in \mathbf{E}$, there exists $\lim_{n \to \infty} [n^{-1}, u]_{i=1}^n$, which will be denoted by Ku (or $K_{\mathbf{E}}u$ if there is ambiguity).

We prove in [9] that these axioms suffice to construct an expectation operator E which satisfies the law of large numbers.

Theorem 1. *Let $\mathbf{E}$ be a separable complete CC space. Then $\mathbf{E}$ admits an expectation operator E such that*

(a) If $X = \sum_j I_{\Omega_j} u_j$ is a simple function, then $EX = [P(\Omega_j), Ku_j]_j$;
(b) E extends by continuity to all Borel functions $X \in L_{\mathbf{E}}^1$ (i.e. such that $Ed(u, X) < \infty$ for some $u \in \mathbf{E}$);
(c) If $X \in L_{\mathbf{E}}^1$ and $\{X_i\}_{i \in \mathbb{N}}$ are pairwise i.i.d. as X, then $[n^{-1}, X_i]_{i=1}^n \to EX$ almost surely.

With this theorem, it suffices to check axioms (i) through (v) in order to prove a law of large numbers in a concrete space. These axioms are considerably weaker than those defining a Banach space, and include some spaces of sets, fuzzy sets, probability measures and metric spaces with geometric averages (see [9, Section 9] for detailed examples).

Theorem 2. *([9], Theorem 6.2) Let $\mathbf{E}$ be a separable complete CC space. Then, the space $\mathcal{K} = \mathcal{K}(\mathbf{E})$ with the convex combinations*

$$[\lambda_i, A_i]_{i=1}^n = \{[\lambda_i, u_i]_{i=1}^n \mid u_i \in A_i \text{ for all } i\}$$

and the Hausdorff metric d_H is a separable complete CC space where $K_{\mathcal{K}}$ is given by

$$K_{\mathcal{K}} A = \operatorname{co} K_{\mathbf{E}}(A) = \operatorname{co}\{K_{\mathbf{E}} u \mid u \in A\}$$

and the SLLN holds.

Example 1. Let X be a random compact set in a separable Banach space $\mathbf{E}$. Then $K_{\mathbf{E}}$ is the identity so $K_{\mathcal{K}}$ is the convex hull mapping. From the uniqueness of the limit in the law of large numbers, its expectation in the CC sense reduces to the Aumann expectation of $\operatorname{co} X$.

2 Spaces of Fuzzy Sets are CC Spaces

If $\mathbf{E}$ is a CC space, the operations in $\mathcal{K}$ can be further uplifted to $\mathcal{F}$ in the usual manner, and the larger space can be endowed with the metrics d_p and d_∞. In this section we prove that those are CC spaces again.

Theorem 3. *Let $\mathbf{E}$ be a CC space. Then, the space $\mathcal{F} = \mathcal{F}(\mathbf{E})$ with the convex combination operation given by*

$$([\lambda_i, u_i]_{i=1}^n)_\alpha = [\lambda_i, (u_i)_\alpha]_{i=1}^n, \ \alpha \in (0, 1]$$

and the metric

$$d_\infty(u,v) = \sup_{\alpha \in (0,1]} d_H(u_\alpha, v_\alpha)$$

is a CC space where $K_{\mathscr{F}}$ is given by

$$(K_{\mathscr{F}}u)_\alpha = K_{\mathscr{K}}u_\alpha = \mathrm{co}K_E(u_\alpha), \ \alpha \in (0,1].$$

Proof. Properties (i) and (ii) are immediate.

As to (iii), we have to show that

$$\sup_{\alpha \in (0,1]} d_H([\lambda^{(k)},u_\alpha;1-\lambda^{(k)},v_\alpha],[\lambda,u_\alpha;1-\lambda,v_\alpha]) \to 0.$$

It clearly suffices to prove

$$\sup_{A \in \mathscr{K}(u_0), C \in \mathscr{K}(v_0)} d_H([\lambda^{(k)},A;1-\lambda^{(k)},C],[\lambda,A;1-\lambda,C]) \to 0.$$

Reasoning by contradiction, assume the contrary. Then there exists a subsequence $\{k'\}_k$, some $\varepsilon > 0$ and sets $A_{k'} \in \mathscr{K}(u_0), C_{k'} \in \mathscr{K}(v_0)$ such that

$$\varepsilon < d_H([\lambda^{(k')},A_{k'};1-\lambda^{(k')},C_{k'}],[\lambda,A_{k'};1-\lambda,C_{k'}]).$$

By the compactness of u_0 and v_0, the family $\mathscr{K}(u_0) \times \mathscr{K}(v_0)$ is compact in the Vietoris topology and so in the Hausdorff metric. Then we take a further subsequence $\{k''\}_k$ such that $A_{k''} \to A, C_{k''} \to C$ for some $A \in \mathscr{K}(u_0), C \in \mathscr{K}(v_0)$. By the triangle inequality,

$$\varepsilon < d_H([\lambda^{(k'')},A_{k''};1-\lambda^{(k'')},C_{k''}],[\lambda^{(k'')},A;1-\lambda^{(k'')},C])+$$

$$d_H([\lambda^{(k'')},A;1-\lambda^{(k'')},C],[\lambda,A;1-\lambda,C])+$$

$$d_H([\lambda,A;1-\lambda,C],[\lambda,A_{k'};1-\lambda,C_{k'}]) = (I)+(II)+(III).$$

But

$$(I) \le \lambda^{(k'')}d_H(A_{k''},A)+(1-\lambda^{(k'')})d_H(C_{k''},C) \to 0,$$

similarly $(III) \to 0$ and finally $(II) \to 0$, using the CC properties (iii), (iv) of $\mathscr{K}$.

This is a contradiction, so (iii) is proven.

Negative curvature of $\mathscr{F}$ follows easily from that of $\mathscr{K}$:

$$d_\infty([\lambda,u_1;1-\lambda,u_2],[\lambda,v_1;1-\lambda,v_2]) =$$

$$\sup_{\alpha \in (0,1]} d_H([\lambda,(u_1)_\alpha;1-\lambda,(u_2)_\alpha],[\lambda,(v_1)_\alpha;1-\lambda,(v_2)_\alpha]) \le$$

$$\sup_{\alpha \in (0,1]} (\lambda d_H((u_1)_\alpha,(v_1)_\alpha)+(1-\lambda)d_H((u_2)_\alpha,(v_2)_\alpha)) \le$$

$$\le \lambda d_\infty(u_1,v_1)+(1-\lambda)d_\infty(u_2,v_2).$$

Finally, in order to prove the existence of $K_{\mathscr{F}}$ we need to prove

$$d_{\infty}([n^{-1},u]_{i=1}^n,K_{\mathscr{F}}u) \to 0.$$

It suffices to show that

$$\sup_{A \in \mathscr{K}(u_0)} d_H([n^{-1},A]_{i=1}^n,K_{\mathscr{K}}A) \to 0.$$

By a similar compactness argument, we obtain a value $\varepsilon > 0$, a subsequence $\{n''\}_n$ and sets $A_{n''},A \in \mathscr{K}(u_0)$ such that

$$d_H(A_{n''},A) \to 0$$

and

$$\varepsilon < d_H([(n'')^{-1},A_{n''}]_{i=1}^{n''},K_{\mathscr{K}}A_{n''}) \le$$
$$d_H([(n'')^{-1},A_{n''}]_{i=1}^{n''},[(n'')^{-1},A]_{i=1}^{n''})+d_H([(n'')^{-1},A]_{i=1}^{n''},K_{\mathscr{K}}A])$$
$$+d_H(K_{\mathscr{K}}A,K_{\mathscr{K}}A_{n''}) = (IV)+(V)+(VI).$$

Then

$$(IV) \le n'' \cdot (n'')^{-1}d_H(A_{n''},A) \to 0,$$

summand (V) goes to 0 by the CC property (v) of $\mathscr{K}$ and

$$(VI) \le d_H(A,A_{n''}) \to 0$$

using [9, Proposition 3.6].

Remark 1. Notice that property (v) is a generalization of [1, Lemma 6].

As regards the metric d_p, the result is as follows.

Theorem 4. *Let* **E** *be a CC space. Then, the space $\mathscr{F}$ with the same convex combination operation and the metric*

$$d_p(u,v) = \left(\int_0^1 d_H(u_\alpha,v_\alpha)^p d\alpha\right)^{1/p}$$

is a CC space where $K_{\mathscr{F}}$ is given by

$$(K_{\mathscr{F}}u)_\alpha = K_{\mathscr{K}}u_\alpha.$$

Proof. Properties (i), (ii), (iii), (v) follow from those of $(\mathscr{F},[\cdot,\cdot],d_\infty)$. It just remains to prove (iv), but

$$d_p([\lambda,u_1;1-\lambda,u_2],[\lambda,v_1;1-\lambda,v_2]) =$$
$$\|d_H([\lambda,(u_1).;1-\lambda,(u_2).],[\lambda,(v_1).;1-\lambda,(v_2).])\|_p \le$$
$$\le \|\lambda d_H((u_1).,(v_1).)+(1-\lambda)d_H((u_2).,(v_2).)\|_p \le$$
$$\le \lambda\|d_H((u_1).,(v_1).)\|_p+(1-\lambda)\|d_H((u_2).,(v_2).)\|_p$$
$$= \lambda d_p(u_1,v_1)+(1-\lambda)d_p(u_2,v_2).$$

The topological properties of $\mathscr{F}_c$ with those structures are the same as in the Banach case, namely

1. $(\mathscr{F}_c,d_\infty)$ is non-separable in general, but it is complete if **E** is so.
2. $(\mathscr{F}_c,d_p)$ is separable if **E** is so, but it is not complete.

3 On the Notion of Expectation

Our definition of expectation in a CC space requires the space to be separable and complete, so that every random element can be approximated by simple functions and limits of expectations of simple functions are well-behaved.

Since $\mathscr{F}_c$ fails to have both properties with the usual metrics, it may be more convenient to define an expectation à la Puri-Ralescu [5], namely as a (necessarily unique) solution $EX \in \mathscr{F}$ to the equations

$$(EX)_\alpha = EX_\alpha, \ \alpha \in (0,1].$$

We will say that X is Puri-Ralescu integrable if its Puri-Ralescu expectation exists. A mapping satisfying (i) and (ii) below will be called a *fuzzy random variable*.

Theorem 5. *Let* **E** *be a separable complete CC space. Let* X *be an* $\mathscr{F}$-*valued mapping on a measurable space. Then, the following conditions are equivalent:*

(i) X_α *is* d_H-*Borel for every* $\alpha \in (0,1]$,
(ii) X *is* d_p-*Borel.*

Moreover, if the Puri-Ralescu expectation of a fuzzy random variable X *exists, then* $X \in L^1_{(\mathscr{F}_c, d_p)}$ *and its CC expectation is the same.*

Proof. The proof of the first part is the same as in [2, Theorem 6.4]. For the second part, we use the notations E_{PR} and E_{CC} with obvious meaning.

If $E_{PR}X$ exists, in particular $X_0 \in L^1_{\mathscr{K}}$. Taking an arbitrary $u \in \mathbf{E}$, we have

$$Ed_p(X, I_{\{u\}}) = E[\int_0^1 d_H(X_\alpha, \{u\})^p d\alpha)^{1/p}] \le Ed_H(X_0, \{u\}) < \infty$$

so indeed $X \in L^1_{(\mathscr{F}, d_p)}$.

As regards expectations, assume first that $X = \sum_{j=1}^r I_{\Omega_j} u_j$ is a simple function. Then

$$E_{CC}X = [P(\Omega_j), K_{\mathscr{F}} u_j]_{j=1}^r$$

so

$$(E_{CC}X)_\alpha = [P(\Omega_j), K_{\mathscr{K}}(u_j)_\alpha]_{j=1}^r = EX_\alpha$$

and we deduce $E_{CC}X = E_{PR}X$.

Let now X be arbitrary and $\{X_n\}_n$ be a sequence of simple functions converging to X in the sense of $L^1_{(\mathscr{F}, d_p)}$, namely $Ed_p(X_n, X) \to 0$.

By construction, $d_p(E_{CC}X_n, E_{CC}X) \to 0$. But $E_{CC}(X_n) = E_{PR}(X_n)$ and we know that

$$d_p(E_{PR}X_n, E_{PR}X) = (\int_0^1 d_H((E_{PR}X_n)_\alpha, (E_{PR}X)_\alpha)^p d\alpha)^{1/p} =$$

$$= (\int_0^1 d_H(E(X_n)_\alpha, EX_\alpha)^p d\alpha)^{1/p}$$

By the non-expansiveness of the expectation in the CC space $\mathscr{K}$, the latter is bounded above by

$$\left(\int_0^1 [Ed_H((X_n)_\alpha, X_\alpha)]^p d\alpha\right)^{1/p} \le \left(\int_0^1 E[d_H((X_n)_\alpha, X_\alpha)]^p d\alpha\right)^{1/p}$$

$$= Ed_p(X_n, X) \to 0.$$

The uniqueness of the limit yields $E_{PR}X = E_{CC}X$.

Remark 2. It should be noted that $L^1_{(\mathscr{F}, d_p)}$ must be understood as the family of all $\mathscr{F}$-valued elements of the L^1 space constructed using *the completion* of $(\mathscr{F}, d_p)$. A direct construction is not possible because $(\mathscr{F}, d_p)$ is not a complete CC space itself.

A fuzzy random variable can be integrable in the completion of $(\mathscr{F}, d_p)$ without being Puri-Ralescu integrable.

4 SLLN for Fuzzy Random Variables in Non-Banach Spaces

Some versions of the Strong Law of Large Numbers have been proven for fuzzy random variables with the uniform metric d_∞, see e.g. [1, 4, 3, 6]. However all those results ultimately rely on the SLLN for Banach spaces. The results we have proven so far show that one does not really have to depend on Banach space results to prove the SLLN.

Theorem 6. *Let* **E** *be a separable complete CC space. Let* $\{X_n\}_n$ *be a sequence of pairwise independent fuzzy random variables, identically distributed as X. Then,*

$$[n^{-1}, X_i]_{i=1}^n \to EX$$

almost surely, in the sense of d_∞.

Proof. The proof uses the same device as in [4, Theorem 2] to reduce the proof to the $\mathscr{K}$-valued case. The latter follows from Theorem 2.

Remark 3. If **E** is a Banach space, then K_E is trivially the identity mapping and we recover the LLNs in [1, 4].

5 SLLN for Level-2 Fuzzy Random Variables

For a related application, consider level-2 fuzzy sets, namely elements of $\mathscr{F}(\mathscr{F})$. Level-2 fuzzy sets are increasingly used in database modelling and other applications (see [10] and references therein) but no formalization of level-2 fuzzy random variables exists in the literature. In fact, since $\mathscr{F}$ does not embed into a Banach space, the usual tools cannot provide limit theorems in this case.

What we are doing here is to take $\mathscr{F}$ itself as the underlying CC space. Its structure is thus just a level more complicated than that of $\mathscr{F}$: instead of uplifting the operations and metric of $(\mathbf{E}, d)$ to $(\mathscr{F}, d_\infty)$, we use the obtained operations of $\mathscr{F}$ and the metric d_∞ and uplift them in the same way. We denote the uplifted d_∞ metric by $d_{\infty,\infty}$.

Remark 4. The convex hull at the second level (convex hull of subsets of $\mathscr{F}$) is *not at all* obtained by taking the fuzzy convex hull of each element. As an example, take $A = \{I_{\{0\}}, I_{\{2\}}\}$: although A is formed by two convex fuzzy sets, it is not convex itself because, for instance, $I_{\{1\}}$ appears as a convex combination of them.

Theorem 7. *Let X be a $d_{\infty,\infty}$-Borel level-2 fuzzy random variable and $\{X_n\}_n$ be pairwise independent and identically distributed as X. Then,*

$$\sum_{i=1}^{n} n^{-1} X_n \to EX$$

almost surely in the sense of $d_{\infty,\infty}$.

Proof. If X is Borel, then under the Continuum Hypothesis its range is essentially separable, see [8, 7]. Since $(\mathscr{F}(\mathscr{F}), d_{\infty,\infty})$ is complete, we can modify X in a null set so that the closure of its range becomes a separable complete CC space. Then Theorem 1 applies.

Acknowledgement

The first author's research has been partially funded by the Spanish *Ministerio de Ciencia y Tecnología* and *Ministerio de Educación y Ciencia*, under their research grants BFM2002-03263, MTM2005-02254 and TSI2005-02511, and the *Gobierno de Aragón*, under its research grant PM2004-052.

References

[1] A. Colubi, M. López-Díaz, J. S. Domínguez-Menchero, M. A. Gil (1999). A generalized strong law of large numbers. *Probab. Theory Related Fields* **114**, 401–417.

[2] V. Krätschmer (2001). A unified approach to fuzzy random variables *Fuzzy Sets and Systems* **123**, 1–9.

[3] S. Li, Y. Ogura, V. Kreinovich (2002). *Limit theorems and applications of set-valued and fuzzy set-valued random variables.* Kluwer, Dordrecht.

[4] I. S. Molchanov (1999). On strong laws of large numbers for random upper semicontinuous functions. *J. Math. Anal. Appl.* **235**, 349–355.

[5] M. L. Puri, D. A. Ralescu (1986). Fuzzy random variables. *J. Math. Anal. Appl.* **114**, 409–422.

[6] P. Terán (2003). A strong law of large numbers for random upper semicontinuous functions under exchangeability conditions. *Statist. Probab. Lett.* **65**, 251–258.

[7] P. Terán (2006). When levelwise fuzzy random variables are also d_∞-Borel. *Proc. 11th IPMU Int. Conf.*, to appear.

[8] P. Terán (200x). On Borel measurability and large deviations for fuzzy random variables. *Fuzzy Sets and Systems*, to appear.

[9] P. Terán, I. Molchanov (200x). The law of large numbers in a metric space with a convex combination operation. *J. Theoret. Probab.*, to appear.

[10] G. de Tré, R. de Caluwe (2003). Level-2 fuzzy sets and their usefulness in object-oriented database modelling. *Fuzzy Sets and Systems* **140**, 29–49.

Part IV

Applications and Modelling of Imprecise Operators

Fuzzy Production Planning Model for Automobile Seat Assembling *

J. Mula[1†], R. Poler[1], and J.P. Garcia-Sabater[2]

[1] CIGIP(Research Centre on Production Management and Engineering) Polytechnic University of Valencia (SPAIN)
    fmula@cigip.upv.es, rpoler@cigip.upv.es
[2] Business Management Department. Polytechnic University of Valencia (SPAIN)
    jpgarcia@cigip.upv.es

1 Introduction

Production planning consists of the simultaneous determination of the production, inventory and capacity levels of a company on a finite planning horizon with the objective of minimizing the total costs generated by production plans. Fuzzy set theory has been used to model systems that are difficult to define accurately (Bellman and Zadeh 1970; Dubois and Prade 1980; Zimmermann 1996). This theory represents an attractive tool to support the production planning research when the dynamics of the manufacturing environment limits the specification of the model objectives, constraints and parameters. Guiffrida and Nagi (1998) provide an exhaustive literature survey on the fuzzy set theory applications in production management research.

It is necessary to distinguish between flexibility in constraints and goals and uncertainty of the data or epistemic uncertainty. Flexibility is modelled by fuzzy sets and may reflect the fact that constraints or goals are linguistically formulated, their satisfaction is a matter of tolerance and degrees or fuzziness (Bellman and Zadeh 1970). Epistemic uncertainty is concerned with ill-known parameters modelled by fuzzy intervals in the setting of possibility theory (Zadeh 1978; Dubois and Prade 1988). This paper aims to formulate a model for mid-term production planning problem in a multi-product, multi-level and multi-period manufacturing environment with fuzziness in demand and available capacity. An approach widely studied in the literature that allows avoiding the imprecision is the use of Soft Computing

---

[†] This research has been carried out in the framework of a project funded by the Polytechnic University of Valencia, titled 'Development of fuzzy mathematical programming models for production planning in context of uncertainty. Application to an industrial company of the automobile sector'.

[†] Corresponding author. Josefa Mula, Dpto. Organización de Empresas, Economía Financiera y Contabilidad, Escuela Politécnica Superior de Alcoy, Plaza Ferrándiz y Carbonell, 2, 03801 Alcoy (Alicante), SPAIN. Tel.: + 34 96 652 84 23. Fax: + 34 96 652 84 65. E-mail address: fmula@cigip.upv.es

methodologies, and in particular, the use of flexible constraints in the problem formulation. This work uses an approach of fuzzy linear programming based on fuzzy constraints to generate an optimal fuzzy solution. In this context, the survey work by Kacprzyk and Orlovsky (1987) and Delgado et al. (1994) show some possibilities of how fuzziness can be accommodated within linear programming. In this paper, the main goal is to determine the master production schedule, stock levels, delayed demand, and capacity usage levels over a given planning horizon in such a way as to hedge against the fuzziness of demand and capacity constraints. Therefore, we focus on the demands and capacities, it is clear that the uncertainty can be the result of a certain imprecision in satisfying the constraints. The contribution of this paper to the modelling and optimization domain is a practical application of known flexible programming. Other applications of flexible programming in production planning problems can be found in Miller et al. (1997), Pendharkar (1997), Dubois et al. (2003), Itoh et al. (2003), Melian and Verdegay (2005) and Mula et al. (2006).

The rest of this paper is organized as follows. In Section 2, a model for mid-term production planning with fuzzy constraints is presented. In Section 3, the fuzzy model is transformed into an equivalent crisp model. Section 4 uses a real-numerical example to illustrate the fuzzy model proposed. The conclusions are discussed in Section 5.

2 Formulation of the Problem

A linear programming model for the capacity constrained material requirement planning (MRP) problem originally proposed in Mula et al. (2006) and called MRPDet is adopted as the basis of this work. MRPDet is a model for the optimization of the mid-term production planning problem in a multi product, multi level and multi period manufacturing environment. Let us consider the following fuzzy formulation of the MRPDet model. Decision variables and parameters for the model are defined in Table 1.

Minimize

$$z = \sum_{i=1}^{I}\sum_{t=1}^{T}(cp_iP_{it} + ci_iINVT_{it} + crd_iRd_{it}) + \sum_{r=1}^{R}\sum_{t=1}^{T}(ctoc_{rt}Toc_{rt} + ctex_{rt}Tex_{rt}) \quad (1)$$

Subject to

$$INVT_{i,t-1} + P_{i,t-TS_i} + RP_{it} - INVT_{i,t} - Rd_{i,t-1} - \sum_{j=1}^{I}\alpha_{ij}(P_{jt} + RP_{jt})$$
$$+ Rd_{it} \lesssim d_{it} \quad i = 1\ldots I, t = 1\ldots T \quad (2)$$

$$-INVT_{i,t-1} - P_{i,t-TS} - RP_{it} + INVT_{i,t} + Rd_{i,t-1} + \sum_{j=1}^{I}\alpha_{ij}(P_{jt} + RP_{jt})$$
$$- Rd_{it} \lesssim d_{it} \quad i = 1\ldots I, t = 1\ldots T \quad (3)$$

$$\sum_{i=1}^{I} AR_{ir}P_{it} + Toc_{rt} - Tex_{rt} \lesssim CAP_{rt} \qquad r = 1 \ldots R, t = 1 \ldots T \qquad (4)$$

$$-\sum_{i=1}^{I} AR_{ir}P_{it} - Toc_{rt} + Tex_{rt} \lesssim -CAP_{rt} \qquad r = 1 \ldots R, t = 1 \ldots T \qquad (5)$$

$$Rd_{iT} = 0 \qquad\qquad i = 1 \ldots I \qquad (6)$$

$$P_{it}, \; INVT_{it}, \; Rd_{it}, \; Toc_{rt}, \; Tex_{rt} \geq 0 \qquad i = 1 \ldots I, r = 1 \ldots R, t = 1 \ldots T \qquad (7)$$

Eq. (1) shows the total costs to be minimized. The balance equations for the inventory are given by the group of constraints (2). These equations take into account the backlogs of the demand which behave as a negative inventory. It is important to highlight the consideration of the parameter RP_{it} that guarantees the continuity of the MRP along the successive explosions carried out during a given planning horizon. The production in every period is limited by the availability of a group of shared resources. The Eq. (3) considers the limits of capacity of these resources. The symbol $\lesssim$ represents the fuzzy version of $\leq$ and means "essentially less than or similar to". These constraints show that the planner wants to make the left hand side of the constraints smaller or similar to the right hand side "if it is possible". A constraint has also been added (4) to finish with the delays in the last period (T) of the planning horizon. The model also contemplates the non negativity constraints (5) for the decision variables. Finally, the decision variables P_{it}, $INVT_{it}$ and Rd_{it} will be defined as continuous or integer variables depending on the manufacturing environment where the model is applied.

Table 1. Decision variables and model parameters

Sets			
T	Set of periods ($t = 1 \ldots T$)		
I	Set of products ($i = 1 \ldots I$)		
J	Set the parent products in the bill of materials ($j = 1 \ldots J$)		
R	Set of resources ($r = 1 \ldots R$)		
Decision Variables		*Data*	
P_{it}	Quantity to produce of i on t	d_{it}	Demand of i on t
$INVT_{it}$	Inventory of i at the end of t	α_{ij}	Quantity of i to produce a unit of j
Rd_{it}	Delayed demand of i at the end of t	TS_i	Lead time of the product i
Toc_{rt}	Undertime hours of r on t	$INVT_{i0}$	Inventory of i on period 0
Tex_{rt}	Overtime hours of r on t	Rd_{i0}	Delayed demand of i on period 0
		RP_{it}	Programmed receptions of i on t
Objetive Function cost coefficients		*Technological coefficients*	
cp_i	Production cost of a unit of i	AR_{ir}	Time of r for unit of production of i
ci_i	Inventory cost of a unit of i	AR_{ir}	Available capacity of r on t
crd_i	Delayed demand cost of a unit of i		
$ctoc_{rt}$	Undertime hour cost of r on t		
$ctex_{rt}$	Overtime hour cost of r on t		

3 Fuzzy Linear Programming Model

This model attempts to solve the mid-term production planning problem where the market demand and the available capacity are considered fuzzy data.

In order to determine the fuzzy set of the solution, Orlovski (1977) suggests to calculate, for all the α-levels of the solution space, the optimal values corresponding to the objective function and to consider as the fuzzy set of the solution the optimal values of the function objective, with the degree of membership similar to the α-level of the solution space.

Werners (1987) defines the α-level sets of the solution space as $R_\alpha = \{x | x \in X, \mu_R(x) \geq \alpha\}$ and the set of optimal solutions for each set of the α-level as $N(\alpha) = \{x | x \in R_\alpha, \ f(x) = \sup_{x' \in R_\alpha} f(x')\}$. The fuzzy set of the solution is defined by the membership function:

$$\mu_{opt}(x) = \begin{cases} \sup \alpha & if \ x \in UN(\alpha) \\ & \alpha > 0 \\ & x \in N(\alpha) \\ 0 & Otherwise \end{cases} \qquad (8)$$

where $0 \leq \alpha \leq 1$ is a cut value. To solve the problem α is settled down parametrically to obtain the value of the objective function for each one of those $\alpha \lesssim [0, 1]$. Thus, α is equal to 1 when the constraint is perfectly accomplished (no violation) and decreases to zero according to greater violations. For not admissible violations the accomplishment degree will be zero. This membership function can be formulated as follows:

$$\mu_i(x) = \begin{cases} 0 & if \ \sum_{i=1}^{n} B_i x > d_i + p_i \\ 1 - ((\sum_{i=1}^{n} B_i x - d_i)/p_i) & if \ p_i \leq \sum_{i=1}^{n} B_i x \leq d_i + p_i \\ 1 & if \ \sum_{i=1}^{n} B_i x \leq d_i \end{cases} \qquad (9)$$

The membership function of the fuzzy set of "the optimal values of the objective function" is defined as:

$$\mu_f(r) = \begin{cases} \sup \mu_{opt}(x) & if \ r \in R_1 \wedge f^{-1}(r) \neq ext O \\ x \in f^{-1}(r) \\ 0 & Otherwise \end{cases} \qquad (10)$$

where $f(x)$ is the objective function with functional values r.

In the case of the linear programming, the determination of the values r and $\mu_{opt}(x)$ can be obtained by using parametric programming (Chanas, 1983). Given the fuzzy model in Section 3, the equivalent parametric linear programming model is formulated as follows:

Minimize

$$z = \sum_{i=1}^{I} \sum_{t=1}^{T} (cp_i P_{it} + ci_i INVT_{it} + crd_i Rd_{it}) + \sum_{r=1}^{R} \sum_{t=1}^{T} (ctoc_{rt} Toc_{rt} + ctex_{rt} Tex_{rt}) \qquad (11)$$

Subject to

$$INVT_{i,t-1} + P_{i,t-TS} + RP_{it} - INVT_{i,t} - Rd_{i,t-1} - \sum_{j=1}^{I} \alpha_{ij}(P_{jt} + RP_{jt})$$

$$+Rd_{it} \leq d_{it} + p2_{it} - p2_{it}\alpha \qquad i = 1\ldots I, t = 1 \ldots T \qquad (12)$$

$$-INVT_{i,t-1} - P_{i,t-TS} - RP_{it} + INVT_{i,t} + Rd_{i,t-1} + \sum_{j=1}^{I} \alpha_{ij}(P_{jt} + RP_{jt})$$

$$-Rd_{it} \leq -d_{it} + p2_{it} - p2_{it}\alpha \qquad i = 1\ldots I, t = 1 \ldots T \qquad (13)$$

$$\sum_{i=1}^{I} AR_{ir} P_{it} + Toc_{rt} - Tex_{rt} \leq CAP_{rt} + p3rt - p3rt\alpha$$

$$r = 1\ldots R, t = 1 \ldots T \qquad (14)$$

$$-\sum_{i=1}^{I} AR_{ir} P_{it} - Toc_{rt} + Tex_{rt} \leq -CAP_{rt} + p3rt - p3rt\alpha$$

$$r = 1\ldots R, t = 1 \ldots T \qquad (15)$$

$$Rd_{iT} = 0 \qquad i = 1\ldots I \qquad (16)$$

$$P_{it}, \ INVT_{it}, \ Rd_{it}, \ Toc_{rt}, \ Tex_{rt} \geq 0 \qquad i = 1\ldots I, r = 1\ldots R, t = 1\ldots T \qquad (17)$$

In the Eqs. (12) and (13), d_{it} is an estimated value which corresponds to the inferior limit of the tolerance interval for the demand of the product i in the period t. While $p2_{it}$ represents the maximum extension of d_{it} in the tolerance interval of the demand. In the Eqs. (14) and (15), CAP_{rt} is an estimated value which corresponds to the inferior limit of the tolerance interval for the available capacity of the resource r in the period t. On the other hand, $p3_{rt}$ represents the maximum extension of CAP_{rt} in the tolerance interval of the available capacity.

The result of this model is, however, a fuzzy set and the planner have to decide that pair (α, z) considers optimal if he wants to obtain a crisp solution.

If we substitute the membership function (9) by the following (Hamacher et al. 1978):

$$\mu_i(x) = \begin{cases} 1 - \dfrac{t_i}{p_i} & if \ \sum_{i=1}^{n} B_i x = d_i + t_i \ and \ t_i \geq 0 \\[3mm] 1 & if \ \sum_{i=1}^{n} B_i x \leq d_i \end{cases} \qquad (18)$$

where p_i is the acceptable maximum value of the violation t_i in the fuzzy constraint i. The Eqs. (12), (13), (14) and (15) can also be written as (19), (20), (21) and (22), respectively:

$$INVT_{i,t-1} + P_{i,t-TS} + RP_{it} - INVT_{i,t} - Rd_{i,t-1} - \sum_{j=1}^{I} \alpha_{ij}(P_{jt} + RP_{jt})$$

$$+Rd_{it} \leq d_{it} + \theta p2_{it} \qquad i = 1\ldots I, t = 1 \ldots T \qquad (19)$$

$$-INVT_{i,t-1} - P_{i,t-TS} - RP_{it} + INVT_{i,t} + Rd_{i,t-1} + \sum_{j=1}^{I} \alpha_{ij}(P_{jt} + RP_{jt}) \tag{20}$$
$$-Rd_{it} \leq -d_{it} + \theta p2_{it} \qquad\qquad i = 1 \ldots I, t = 1 \ldots T$$

$$\sum_{i=1}^{I} AR_{ir}P_{it} + Toc_{rt} - Tex_{rt} \leq CAP_{rt} + \theta p3_{rt} \tag{21}$$
$$r = 1 \ldots R, t = 1 \ldots T$$

$$-\sum_{i=1}^{I} AR_{ir}P_{it} - Toc_{rt} + Tex_{rt} \leq -CAP_{rt} + \theta p3_{rt} \tag{22}$$
$$r = 1 \ldots R, t = 1 \ldots T$$

4 Numerical Example

The model has been implemented with a high level language for mathematical programming models, the modelling language MPL (2004). Resolution has been carried out with the optimization solver CPLEX (1994). Finally, the input data and outputs of the models are managed through a Microsoft Access 2000 database.

This section uses a real example to illustrate the performance of the fuzzy model. The proposed model is applied in a company dedicated to the assembly of seats for automobiles. The hypotheses to carry out the computational experiment are summarized as follows: the study considers a single part; the decision variables P_{it}, $INVT_{it}$, and Rd_{it} are integer; the external demand only exists for the final product; delayed demand for the final product is considered; an only productive resource restricts the production: the assembly line; the parameters $p2$ and $p3$ required by the fuzzy model have been defined following company criteria; it has been considered a six months planning horizon with a weekly period planning; the performance measures are: production, inventory, delayed demand and overtime costs.

The company receives weekly from the automobile assembler the demand information with a planning horizon for six months. However, these demand forecasts are rarely precise (see Mula et al. 2005). Therefore, this section will validate if the fuzzy model for production planning, proposed in this paper, can be a useful tool for the decision making process of the production planners.

Table 2 summarizes the evaluation results, with a fuzziness level in the demands and available capacities equal to α, according to a group of parameters defined in Mula et al. (2006): (i) the service level; (ii) the levels of inventory; (iii) the planning nervousness respect to the planned period and the planned quantity; and (iv) the total costs.

Table 3 shows the computational efficiency of the crisp model, *MRPDet* (see Mula et al. 2006), and the fuzzy MRP model proposed in this work. Both of the models can obtain the optimal solution of the mixed integer linear programming with a null number of iterations in the first MRP execution (*Release* 1). Obviously, the number of iterations can change in the rest of MRP executions depending on the input data. On the other hand, the fuzzy MRP model has a larger number of constrains but same number of variables and integer variables, which is not implying a

Table 2. Evaluation of the results

α ($\theta = 1 - \alpha$)	Service level %	Number of minimun inventory levels	Planning nervousness (Period)	Planning nervousness (Quantity)	Total costs €
0	99.25%	46	0.5	13.25	4111577.90
0.1	99.27%	9	0.5	14.65	4113786.06
0.2	99.28%	9	0.5	14.85	4112890.28
0.3	99.30%	9	0.5	14.9	4098756.63
0.4	99.32%	9	0.5	15.05	4083810.06
0.5	99.34%	9	0.4	5.05	4065984.25
0.6	99.37%	9	0.4	15.15	4045561.22
0.7	99.39%	9	0.35	15.1	028525.18
0.8	99.40%	9	0.35	14.8	4016677.67
0.9	99.42%	8	0.35	14.8	4004564.22
1	99.44%	8	0.35	14.05	3988276.07

Table 3. Efficiency of the computational experiments for a MRP execution (first week)

Model	Iterations	Variables	Integer	Constraints	Elements non Zero	Array density (%)	CPU time (Secons)
MRPDet	0	4237	5612	2797	8239	0.07	0.86
Fuzzy MRP	0	4237	5612	2857	8477	0.07	3.63*

*This value corresponds to one α

greater requirements of information storage. Finally, the fuzzy model has not caused an explosive growth of the CPU time.

5 Conclusions

In many manufacturing environments, such as the automobile industry, the production planning decisions have to be made under conditions of uncertainty in parameters as important as market demand or capacity data. Due to the unavailability and fuzziness of information it is extremely difficult to exactly generate production plans under a dynamic environment. For improving these production plans a fuzzy production planning model is developed in this paper. The main advantages of the fuzzy MRP model proposed are: (i) a fuzzy complete decision considers other election possibilities besides the maximization variant, which can help managers to foresee the effects of the uncertainty in demands and capacities; and (ii) the size of the modeling problem has not been increased with respect to the crisp model, *MRPDet*, which has not caused a decrement of the computational efficiency. Finally, the testing of the proposed fuzzy model for all the products and resources of the company will be the aim of a forthcoming work.

References

[1] Bellman R, Zadeh L (1970) Decision making in a fuzzy environment Management Science 17: 141-164.

[2] Chanas S (1983) The use of parametric programming in fuzzy linear programming. Fuzzy Sets and Systems 11: 243-251.

[3] CPLEX Optimization Inc., Using the CPLEX callable library (1994).

[4] Delgado M, Kacprzyk J, Verdegay JL, Vila MA (1994) (Eds.) Fuzzy Optimization: "Recent Ad- vances", Physica Verlag,Heidelberg.

[5] Dubois D, Prade H (1980) Fuzzy Sets and Systems: Theory and Applications. New York London, Toronto.

[6] Dubois D, Prade H (1988) Possibility Theory, Plenum Press, New York.

[7] Dubois D, Fargier H, Fortemps P (2003) Fuzzy scheduling: Modelling flexible constraints vs. cop-ing with incomplete knowledge. European Journal of Operational Research 147: 231-252.

[8] Guiffrida AL, Nagi R(1998) Fuzzy set theory applications in production management research: a literature survey. Journal of Intelligent Manufacturing 9: 39-56.

[9] Hamacher H, Leberling H, Zimmermann HJ (1978) Sensitivity analysis in fuzzy linear program- ming. Fuzzy Sets and Systems 1: 269-281.

[10] Itoh T, Ishii H, Nanseki T (2003) A model of crop planning under uncertainty in agricultural management, International Journal of Production Economics 81-82: 555-558.

[11] Kacprzyk J, Orlovski SA (1987) Fuzzy optimization and mathematical program-ming: a brief introduction and survey. In J. Kacprzyk and S.A. Orlovski (Eds.): Optimization Models Us- ing Fuzzy Sets and Possibility Theory, Reidel, Boston, pp. 50 72.

[12] Maximal Software Incorporation (2004) MPL modeling system 4.2, USA.

[13] Melian B, Verdegay JL (2005) Fuzzy optimization problems in wavelength division multiple-xing (WDM) networks. Proceedings of the 4th conference of the EUS-FLAT-LFA, September 7-9, Barcelona, Spain.

[14] Miller WA, Leung LC, Azhar TM, Sargent S (1997) Fuzzy production planning model for fresh tomato packing, International Journal of Production Economics 53: 227-238.

[15] Mula J, Poler R, Garcia JP (2006) MRP with flexible constraints: a fuzzy mathematical pro- gramming approach. Fuzzy Sets and Systems 157: 74-97.

[16] Mula J, Poler R, Garcia JP, Ortiz A (2005) Demand uncertainty effects on first tier suppliers of an automobile industry supply chain. The ICFAI Journal of Supply Chain Management 2 (3): 19-40.

[17] Orlovsky SA (1977) On programming with fuzzy constraint sets. Kybernetes 6: 197-201.

[18] Pendharkar PC (1997) A fuzzy linear programming model for production planning in coal mines. Computers Operations Research 24: 1141-1149.

[19] Werners B (1987) An interactive fuzzy programming system. Fuzzy Sets and Systems 23: 131-147.

[20] Zadeh LA (1978) Fuzzy sets as a basis for a theory of possibility. Fuzzy Sets and Systems 1: 3-28.

[21] Zimmermann HJ (1996) Fuzzy set theory and its applications. Third Edition. Ed.Kluwer Academic Publishers.

Optimal Selection of Proportional Bounding Quantifiers in Linguistic Data Summarization

Ingo Glöckner

FernUniversität in Hagen, 58084 Hagen, Germany
iglockner@web.de

Summary. Proportional bounding quantifiers like "Between p_1 and p_2 percent" are potentially useful for expressing linguistic summaries of data. Given p_1, p_2, existing methods for data summarization based on fuzzy quantifiers can be used to assign a quality score to the summary. However, the problem remains how the optimal choice of p_1, p_2 in the range $0 \leq p_1 \leq p_2 \leq 100\%$ can be established. Moreover, the proposed quality indicators are rather heuristic in nature. The paper presents a method for computing the optimal bounding quantifier which best summarizes the given data. Specifically, the most specific quantifier will be chosen which results in the highest validity score of the summary given a constraint on the the percentage range $p_2 - p_1$. The method not only assigns validity scores to the quantifiers of interest but also determines the best choice of quantifier in $\mathscr{O}(N \log m)$ time, where N is the size of the base set and m the number of different membership grades in the fuzzy arguments.

1 Introduction

A framework for generating linguistic summaries from imprecise data and for evaluating their usefulness has been developed by Yager [8, 10, 7] and refined by Kacprzyk and Strykowski [6]. We assume a finite base set $E \neq \varnothing$ of individuals of interest, and a description of these individuals by fuzzy sets $X \in \widetilde{\mathscr{P}}(E)$. In practice, the data will likely be described by a set A of attributes $a : E \longrightarrow V_a$ which assign an attribute value $a(e)$ to each individual. The fuzzy sets on E, in turn, will only indirectly be given by fuzzy sets declared on the attribute values. Thus, a fuzzy set $X' \in \widetilde{\mathscr{P}}(V_a)$ declared on the attribute range of $a \in A$ gives rise to the associated fuzzy set $X \in \widetilde{\mathscr{P}}(E)$ defined by $\mu_X(e) = \mu_{X'}(a(e))$. The attribute-based description of the data in a database will not be of relevance in this paper, however, so that we drop it for simplicity. From this simplified viewpoint, then, a linguistic summary has the form "Q objects are X's" or "Q X_1's are X_2's" (with an associated 'validity' or 'truthfulness' score $\tau \in [0, 1]$). $Q \in \{almost\,all, many, \dots\}$ is called the 'quantity in agreement' of the summary.

While existing data summarization systems are mostly based on Zadeh's Σ-count or FG-count approach [11] or on Yager's OWA operators [9], we will assume a broader framework which avails us with a uniform analysis of all kinds of linguistic

I. Glöckner: *Optimal Selection of Proportional Bounding Quantifiers in Linguistic Data Summarization*, Advances in Soft Computing **6**, 173–181 (2006)
www.springerlink.com

Table 1. Main types of linguistic quantifiers

Type	Example	Definition
absolute unrestrictive	There are more than 3 Y's	$Q(Y) = q(|Y|)$
absolute	More than 3 Y_1's are Y_2's	$Q(Y_1, Y_2) = q(|Y_1 \cap Y_2|)$
exception	All except 3 Y_1's are Y_2's	$Q(Y_1, Y_2) = q(|Y_1 \setminus Y_2|)$
proportional	Two of three Y_1's are Y_2's	$Q(Y_1, Y_2) = \begin{cases} f(\frac{|Y_1 \cap Y_2|}{|Y_1|}) & |Y_1| > 0 \\ v_0 & \text{else} \end{cases}$
cardinal comparative	More Y_1's than Y_2 are Y_3	$Q(Y_1, Y_2, Y_3) = q(|Y_1 \cap Y_3|, |Y_2 \cap Y_3|)$

quantifiers [4, 3, 5]. This framework is inspired by the linguistic Theory of Generalized Quantifiers (TGQ) [1]. It extends the notion of a (two-valued) generalized quantifier to fuzzy arguments and gradual quantification results in the obvious way:

Definition 1. *An nary* fuzzy quantifier *on a base set $E \neq \emptyset$ is a mapping $\widetilde{Q}$:* $\widetilde{\mathscr{P}}(E)^n \longrightarrow [0,1]$ *(E needs not be finite).*

Expressing an NL quantifier of interest in terms of a fuzzy quantifier is not an easy task, though – mainly because a simple cardinality-based definition is no longer possible when the arguments of the quantifier are fuzzy. We therefore introduce so-called semi-fuzzy quantifiers which serve as a simplified description of the target quantifier.

Definition 2. *An nary* semi-fuzzy quantifier *on a base set $E \neq \emptyset$ is a mapping* $Q : \mathscr{P}(E)^n \longrightarrow [0,1]$.

Semi-fuzzy quantifiers are easier to define than fuzzy quantifiers because one needs not describe their interpretation for fuzzy arguments. The specification of an NL quantifier in terms of a semi-fuzzy quantifier will be linked to the target fuzzy quantifier (which also accepts fuzzy arguments) by a quantifier fuzzification mechanism.

Definition 3. *A quantifier fuzzification mechanism (QFM) $\mathscr{F}$ assigns a fuzzy quantifier $\mathscr{F}(Q) : \widetilde{\mathscr{P}}(E)^n \longrightarrow [0,1]$ to each semi-fuzzy quantifier $Q : \mathscr{P}(E)^n \longrightarrow [0,1]$.*

Table 1 lists the main types of semi-fuzzy quantifiers which are also of interest for data summarization. In this paper, we will restrict attention to *proportional bounded quantifiers*, i.e. quantifiers defined by

$$rate_{[r_1, r_2]}(Y_1, Y_2) = \begin{cases} 1 & : \quad Y_1 \neq \emptyset \wedge \frac{|Y_1 \cap Y_2|}{|Y_1|} \in [r_1, r_2] \\ 0 & : \quad \text{else} \end{cases}$$

These quantifiers are of apparent utility for expressing summaries like "between p_1 and p_2 percent of the X_1's are X_2's" ($r_1 = p_1/100$, $r_2 = p_2/100$) or "at least p percent of the X_1's are X_2's" ($r_1 = p/100$, $r_2 = 1$).

From the perspective of natural language use (i.e. pragmatics), truthfulness of $Q(Y_1, Y_2)$ only indicates a possible use of Q – which might represent a very unusual case of applying Q, however. In other words, the truth score $\tau = \mathscr{F}(Q)(X_1, X_2)$ which judges the validity of the summary "Q X_1's are X_2's", is not restrictive enough to guide quantifier selection to the most appropriate choice of Q. Consider *at least eighty percent*, for example. Clearly the corresponding quantifier should also be true if *all* X_1's are X_2's. However, only the quantifier *all* is appropriate for describing this situation, while *at least eighty percent* has a very low appropriateness grade in this case. Existing approaches to linguistic data summarization have introduced various quality indicators for quantifier selection to solve this problem. While Yager [10] uses only the validity score and a metric for informativeness, Kacprzyk and Strykowski [6] use a multi-dimensional measure based on the degree of truth, the degree of imprecision, the degree of covering, the degree of 'appropriateness' and the length of the summary. The proposed quality indicators are rather heuristic in nature, though, while in this paper, we target at a more principled solution.

Hence let $X_1, X_2 \in \widetilde{\mathscr{P}}(E)$ be given. We further assume a fixed model of fuzzy quantification, i.e. a QFM $\mathscr{F}$ (see below). The summary generation is controlled by a constraint $\delta_{\max} \in [0, 1)$ which specifies the maximal admissible percentage range $r_2 - r_1$ (a summary "between 20% and 90% of the X_1's are X_2's" might be pretty odd, for example). We search for an optimal choice of $Q^* = rate_{[r_1^*, r_2^*]}$, $r_1^* \leq r_2^*$, with

a. $r_2^* - r_1^* \leq \delta_{\max}$.
b. For all $r_1' \leq r_2'$ with $r_2' - r_1' \leq \delta_{\max}$, $\mathscr{F}(rate_{[r_1', r_2']})(X_1, X_2) \leq \mathscr{F}(Q^*)(X_1, X_2)$.
c. If $\mathscr{F}(rate_{[r_1', r_2']})(X_1, X_2) = \mathscr{F}(Q^*)(X_1, X_2)$, then $r_1' \leq r_1^*$ and $r_2^* \leq r_2'$.

The first condition asserts that the percentage span $r_2^* - r_1^*$ of the optimal quantifier is within the limits given by $\delta_{\max}$, i.e. the summary is sufficiently specific. The second condition asserts that Q^* is an optimal choice of such a quantifier, which results in the highest validity score. The third condition ensures that Q^* is the narrowest choice of all quantifiers optimal with respect to the validity score. Let me now explain how Q^* can actually be computed. In order to simplify presentation, we will further assume that $\mathscr{F}(Q^*)(X_1, X_2) > \frac{1}{2}$, i.e. summaries with low truth scores will be ignored.

2 Choosing the Model $\mathscr{F}$

We need a concrete, well-motivated choice of $\mathscr{F}$ to determine the validity score $\tau = \mathscr{F}(Q)(X_1, X_2)$ of the summary. The following construction will assign a suitable choice of fuzzy connectives to the given QFM.

Definition 4. *Let $\mathscr{F}$ be a QFM and $f : \{0, 1\}^n \longrightarrow [0, 1]$ a (semi-fuzzy) truth function. The induced fuzzy truth function $\widetilde{\mathscr{F}}(f) : [0, 1]^n \longrightarrow [0, 1]$ is defined by $\widetilde{\mathscr{F}}(f) = \mathscr{F}(f \circ \eta^{-1}) \circ \widetilde{\eta}$, where $\eta : \{0, 1\}^n \longrightarrow \mathscr{P}(\{1, \ldots, n\})$ and $\widetilde{\eta} : [0, 1]^n \longrightarrow \widetilde{\mathscr{P}}(\{1, \ldots, n\})$ are defined by $\eta(y_1, \ldots, y_n) = \{i : y_i = 1\}$ and $\mu_{\widetilde{\eta}(x_1, \ldots, x_n)}(i) = x_i$, respectively.*

The fuzzy set operations $\widetilde{\cup} : \widetilde{\mathscr{P}}(E)^2 \longrightarrow \widetilde{\mathscr{P}}(E)$ (fuzzy union) and $\widetilde{\neg} : \widetilde{\mathscr{P}}(E) \longrightarrow \widetilde{\mathscr{P}}(E)$ (fuzzy complement) will be defined element-wise in terms of fuzzy disjunction $\widetilde{\vee} = \widetilde{\mathscr{F}}(\vee)$ and fuzzy negation $\widetilde{\neg} = \widetilde{\mathscr{F}}(\neg)$. Based on these induced fuzzy connectives and fuzzy set operations, we can define a class of well-behaved models.

Definition 5. *A QFM $\mathscr{F}$ is called a* determiner fuzzification scheme *(DFS) if it satisfies the following conditions for all semi-fuzzy quantifiers $Q : \mathscr{P}(E)^n \longrightarrow [0,1]$ and fuzzy arguments $X_1, \ldots, X_n \in \widetilde{\mathscr{P}}(E)$:*

(a) $\mathscr{F}(Q) = Q$ if $n = 0$;

(b) $\mathscr{F}(Q)(Y) = Q(Y)$ for crisp $Y \in \mathscr{P}(E)$, $n = 1$;

(c) $\mathscr{F}(\pi_e) = \widetilde{\pi}_e$ for all $E \neq \emptyset$, $e \in E$, where $\pi_e(Y) = 1$ iff $e \in Y$ and $\widetilde{\pi}_e(X) = \mu_X(e)$;

(d) $\mathscr{F}(Q')(X_1, \ldots, X_n) = \widetilde{\neg} \mathscr{F}(Q)(X_1, \ldots, X_{n-1}, \widetilde{\neg} X_n)$ whenever the semi-fuzzy quantifier Q' is defined by $Q'(Y_1, \ldots, Y_n) = \widetilde{\neg} Q(Y_1, \ldots, Y_{n-1}, \neg Y_n)$ for all crisp Y_i;

(e) $\mathscr{F}(Q')(X_1, \ldots, X_{n+1}) = \mathscr{F}(Q)(X_1, \ldots, X_{n-1}, X_n \widetilde{\cup} X_{n+1})$ whenever Q' is defined by $Q'(Y_1, \ldots, Y_{n+1}) = Q(Y_1, \ldots, Y_{n-1}, Y_n \cup Y_{n+1})$ for all crisp Y_i;

(f) $\mathscr{F}(Q)(X_1, \ldots, X_n) \geq \mathscr{F}(Q)(X_1, \ldots, X_{n-1}, X'_n)$ if $X_n \subseteq X'_n$ given that $Q(Y_1, \ldots, Y_n) \geq Q(Y_1, \ldots, Y_{n-1}, Y'_n)$ for all crisp Y_i, $Y_n \subseteq Y'_n$;

(g) $\mathscr{F}(Q \circ \times_{i=1}^n \widehat{\mathscr{F}}(f_i)) = \mathscr{F}(Q) \circ \times_{i=1}^n \widehat{f_i}$ for all $f_i : E' \longrightarrow E$, where $\widehat{f}(Y) = \{f(e) : e \in Y\}$ for all crisp Y and $\mu_{\widehat{\mathscr{F}}(f)(X)}(e) = \mathscr{F}(\pi_e \circ \widehat{f})(X)$.

The choice of postulates (a) through (g) was based on a large catalogue of semantic desiderata from which a minimal (independent) system of core requirements was then distilled. The total list of desiderata validated by these models is discussed in [5]. A DFS will be called a *standard DFS* if it induces the standard set of fuzzy connectives min, max and $1 - x$.

Table 2 lists three general constructions of models which result in the classes of $\mathscr{F}_\Omega$, $\mathscr{F}_\xi$ and $\mathscr{M}_{\mathscr{B}}$ DFSes.[1] The $\mathscr{F}_\Omega$-DFSes form the broadest class of standard DFSes currently known. All practical $\mathscr{F}_\Omega$-DFSes belong to the more regular $\mathscr{F}_\xi$ class, though. The $\mathscr{M}_{\mathscr{B}}$-DFSes can be characterized as the subclass of those $\mathscr{F}_\xi$ models which propagate fuzziness (in the sense of being compatible with a natural fuzziness order). The most prominent example is the following standard DFS $\mathscr{M}_{CX}$, which generalizes the Zadeh's FG-count approach:

$$\mathscr{M}_{CX}(Q)(X_1, \ldots, X_n) = \begin{cases} \frac{1}{2} + \frac{1}{2} \sup\{\gamma : \bot(\gamma) \geq \frac{1}{2} + \frac{1}{2}\gamma\} & : \quad \bot(0) > \frac{1}{2} \\ \frac{1}{2} - \frac{1}{2} \sup\{\gamma : \top(\gamma) \leq \frac{1}{2} - \frac{1}{2}\gamma\} & : \quad \top(0) < \frac{1}{2} \\ \frac{1}{2} & : \quad \text{else} \end{cases}$$

abbreviating $\top(\gamma) = \top_{Q,X_1,\ldots,X_n}(\gamma)$ and $\bot(\gamma) = \bot_{Q,X_1,\ldots,X_n}(\gamma)$. In the following, it is sufficient to consider the model $\mathscr{F} = \mathscr{M}_{CX}$ only because all standard DFSes coincide

---

[1] Here, $X_{\geq \alpha}$ denotes the α-cut and $X_{>\alpha}$ the strict α-cut, respectively. Moreover, $\mathrm{med}_{1/2}(x, y)$ is the fuzzy median, i.e. the second-largest of the three values x, y, $\frac{1}{2}$.

Table 2. Known classes of standard models: an overview

Type	Construction
$\mathscr{F}_\Omega$-DFS	From supervaluation results of three-valued cuts:

$$X_\gamma^{\min} = \begin{cases} X_{\geq \frac{1}{2} + \frac{1}{2}\gamma} & \gamma \in (0,1] \\ X_{> \frac{1}{2}} & \gamma = 0 \end{cases} \qquad X_\gamma^{\max} = \begin{cases} X_{> \frac{1}{2} - \frac{1}{2}\gamma} & \gamma \in (0,1] \\ X_{\geq \frac{1}{2}} & \gamma = 0 \end{cases}$$

$$\mathscr{T}_\gamma(X_i) = \{Y : X_\gamma^{\min} \subseteq Y \subseteq X_\gamma^{\max}\}$$
$$S_{Q,X_1,\dots,X_n}(\gamma) = \{Q(Y_1,\dots,Y_n) : Y_i \in \mathscr{T}_\gamma(X_i), i = 1,\dots,n\}$$
$$\mathscr{F}_\Omega(Q)(X_1,\dots,X_n) = \Omega(S_{Q,X_1,\dots,X_n})$$

$\mathscr{F}_\xi$-DFS	From suprema and infima of supervaluations:

$$\top_{Q,X_1,\dots,X_n}(\gamma) = \sup S_{Q,X_1,\dots,X_n}(\gamma) \qquad \bot_{Q,X_1,\dots,X_n}(\gamma) = \inf S_{Q,X_1,\dots,X_n}(\gamma)$$
$$\mathscr{F}_\xi(Q)(X_1,\dots,X_n) = \xi(\top_{Q,X_1,\dots,X_n},\bot_{Q,X_1,\dots,X_n})$$

$\mathscr{M}_\mathscr{B}$-DFS	From fuzzy median of supervaluation results:

$$Q_\gamma(X_1,\dots,X_n) = \text{med}_{1/2}(\top_{Q,X_1,\dots,X_n}(\gamma),\bot_{Q,X_1,\dots,X_n}(\gamma))$$
$$\mathscr{M}_\mathscr{B}(Q)(X_1,\dots,X_n) = \mathscr{B}(Q_\gamma(X_1,\dots,X_n)_{\gamma \in [0,1]})$$

with $\mathscr{M}_{CX}$ for two-valued quantifiers, and because proportional bounding quantifiers are two-valued quantifiers.

3 Implementation of Proportional Bounding Quantifiers

In order to describe the method for optimal quantifier selection, we must recall the computational analysis of quantitative quantifiers in $\mathscr{F}_\xi$ models developed in [3]. In the following, we assume a finite base set $E \neq \varnothing$ of cardinality $|E| = N$. For given fuzzy arguments $X_1,\dots,X_n \in \widetilde{\mathscr{P}}(E)$, the set of relevant cutting levels is given by $\Gamma(X_1,\dots,X_n) = \{2\mu_{X_i}(e) - 1 : \mu_{X_i}(e) \geq \frac{1}{2}\} \cup \{1 - 2\mu_{X_i}(e) : \mu_{X_i}(e) < \frac{1}{2}\} \cup \{0,1\}$. The computation of quantifiers will be based on an ascending sequence of cutting levels $0 = \gamma_0 < \gamma_1 \cdots < \gamma_{m-1} < \gamma_m = 1$ with $\{\gamma_1,\dots,\gamma_m\} \supseteq \Gamma(X_1,\dots,X_n)$ (usually we will have an equality here). For $\gamma = 0,\dots,m-1$, we abbreviate $\bar{\gamma}_j = \frac{\gamma_j + \gamma_{j+1}}{2}$. We further let $\top_j = \top_{Q,X_1,\dots,X_n}(\bar{\gamma}_j)$ and $\bot_j = \bot_{Q,X_1,\dots,X_n}(\bar{\gamma}_j)$. As a prerequisite for implementing the quantifier selection, let us rewrite $\mathscr{M}_{CX}$ as a function of the finite sample $\Gamma(X_1,\dots,X_n)$ of (three-valued) cut levels.

Proposition 1. *Let* $Q : \mathscr{P}(E)^n \longrightarrow [0,1]$, $X_1,\dots,X_n \in \widetilde{\mathscr{P}}(E)$ *and* $0 = \gamma_0 < \gamma_1 < \cdots < \gamma_{m-1} < \gamma_m = 1$ *be given,* $\Gamma(X_1,\dots,X_n) \subseteq \{\gamma_0,\dots,\gamma_m\}$. *For* $j \in \{0,\dots,m-1\}$ *let* $B_j = 2\bot_j - 1$ *if* $\bot_0 \geq \frac{1}{2}$ *and* $B_j = 1 - 2\top_j$ *otherwise. Further let*

$$\hat{J} = \{j \in \{0,\dots,m-1\} : B_j \leq \gamma_{j+1}\}, \quad \hat{j} = \min \hat{J}.$$

Then

$$\mathscr{M}_{CX}(Q)(X_1,\ldots,X_n) = \begin{cases} \frac{1}{2} + \frac{1}{2}\max(\gamma_{\hat{j}}, B_{\hat{j}}) & : \quad \top_0 > \frac{1}{2} \\ \frac{1}{2} - \frac{1}{2}\max(\gamma_{\hat{j}}, B_{\hat{j}}) & : \quad \top_0 < \frac{1}{2} \\ \frac{1}{2} & : \quad \text{else.} \end{cases}$$

These formulas enable us to evaluate fuzzy quantifications in $\mathscr{M}_{CX}$ (and for two-valued quantifiers like $rate_{[r_1,r_2]}$, in arbitrary standard DFSes). The computation of $\top_j$ and $\perp_j$ must be optimized, though, because a naive implementation which considers each $Y_i \in \mathscr{T}_{\bar{\gamma}_j}(X_i)$ will not achieve sufficient performance. To this end, we observe that for proportional bounding quantifiers, $\top_j$ and $\perp_j$ can be rewritten as

$$\top_j = \max\{q(c_1,c_2) : (c_1,c_2) \in R_j\} \qquad \perp_j = \min\{q(c_1,c_2) : (c_1,c_2) \in R_j\}$$

$$R_j = \{(|Y_1|, |Y_1 \cap Y_2|) : Y_i \in \mathscr{T}_\gamma(Y_i)\} \quad q(c_1,c_2) = \begin{cases} 1 & : \quad c_1 > 0 \wedge \frac{c_2}{c_1} \in [r_1,r_2] \\ 0 & : \quad \text{else.} \end{cases}$$

In numeric terms, the relation R_j can be precisely described as follows:

$$R_j = \{(c_1,c_2) : \ell_1 \leq c_1 \leq u_1, \max(\ell_2, c_1 - u_3) \leq c_2 \leq \min(u_2, c_1 - \ell_3)\}, \quad (1)$$

where $\ell_s = |Z_s|_\gamma^{\min} = |(Z_s)_\gamma^{\min}|$ and $u_s = |Z_s|_\gamma^{\max} = |(Z_s)_\gamma^{\max}|$, $\gamma = \bar{\gamma}_j$, depend on $Z_1 = X_1$, $Z_2 = X_1 \cap X_2$ and $Z_3 = X_1 \cap \neg X_2$, assuming the standard fuzzy intersection and complement.[2] We conclude that

$$\top_j = \max\{q(c_1,c_2) : \ell_1 \leq c_1 \leq u_1, \max(\ell_2, c_1 - u_3) \leq c_2 \leq \min(u_2, c_1 - \ell_3)\}$$
$$= \max\{q'(c_1) : \ell_1 \leq c_1 \leq u_1\}$$
$$\perp_j = \min\{q(c_1,c_2) : \ell_1 \leq c_1 \leq u_1, \max(\ell_2, c_1 - u_3) \leq c_2 \leq \min(u_2, c_1 - \ell_3)\}$$
$$= \begin{cases} 1 & : \quad \ell_1 > 0 \wedge r_1 \leq \frac{\ell_2}{\ell_2 + u_3} \wedge \frac{u_2}{u_2 + \ell_3} \leq r_2 \\ 0 & : \quad \text{else} \end{cases}$$

where

$$q'(c_1) = \begin{cases} q(c_1, \min(u_2, c_1 - \ell_3)) & : \quad \min(u_2, c_1 - \ell_3) < r_1 \\ q(c_1, \max(\ell_2, c_1 - u_3)) & : \quad \max(\ell_2, c_1 - u_3) > r_1 \\ q(c_1, r_1) & : \quad \text{else.} \end{cases}$$

In order to compute a quantification result based on this formula, one must consider every choice of j (i.e. m cutting levels) and (at worst) $N = |E|$ choices of c_1. Thus, the complexity of evaluating a proportional bounding quantifier is $\mathcal{O}(Nm)$.

4 Optimal Quantifier Selection

Given $X_1, X_2 \in \widetilde{\mathscr{P}}(E)$, we define rate bound mappings $r^{\min}, r^{\max} : [0,1] \longrightarrow [0,1]$:

---

[2] The efficient computation of $\ell_s(j)$ and $u_s(j)$ from the histogram of Z_s is explained in [5, 3].

$$r^{\min}(\gamma) = \begin{cases} \frac{\ell_2}{\ell_2 + u_3} & : \quad \ell_1 > 0 \\ 0 & : \quad \ell_1 = 0 \end{cases} \qquad r^{\max}(\gamma) = \begin{cases} \frac{u_2}{u_2 + \ell_3} & : \quad \ell_1 > 0 \\ 1 & : \quad \ell_1 = 0 \end{cases}$$

where $\ell_s = \ell_s(j)$ and $u_s = u_s(j)$ are defined as (1), and $j = \max\{j : \gamma_j \leq \gamma\}$ for $\{\gamma_0, \ldots, \gamma_m\} = \Gamma(X_1, \ldots, X_n), 0 = \gamma_0 < \gamma_1 < \cdots < \gamma_{m-1} < \gamma_m = 1$. Abbreviating $r_j^{\min} = r^{\min}(\overline{\gamma}_j)$, $r_j^{\max} = r^{\max}(\overline{\gamma}_j)$, it is easily shown from the above analysis of $\top_j$ and $\bot_j$ that for a proportional bounding quantifier $rate_{[r_1, r_2]}$ with $r_2 - r_1 \in [0, 1)$,

$$\bot_j = \begin{cases} 1 & : \quad r_1 \leq r_j^{\min} \wedge r_j^{\max} \leq r_2 \\ 0 & : \quad \text{else} \end{cases} \tag{2}$$

In particular, choosing $r_1 = r_j^{\min}$ and $r_2 = r_j^{\max}$ will result in $\bot_j = 1 > \frac{1}{2} + \frac{1}{2}\overline{\gamma}_j$, i.e. $\mathcal{M}_{CX}(rate_{[r_j^{\min}, r_j^{\max}]})(X_1, X_2) \geq \frac{1}{2} + \frac{1}{2}\gamma_{j+1} > \frac{1}{2}$. Hence let $j_* = \max\{j = 0, \ldots, m - 1 : r_j^{\max} - r_j^{\min} \leq \delta_{\max}\}$, $r_1^* = r_{j_*}^{\min}$ and $r_2^* = r_{j_*}^{\max}$. If $r_2^* - r_1^* > \delta_{\max}$, then no summarization based on bounding quantifiers is possible because X_1, X_2 are too fuzzy. In the normal case that $r_2^* - r_1^* \leq \delta_{\max}$, however, the choice of r_1^* and r_2^* will be optimal. To see this, suppose that $\mathcal{M}_{CX}(rate_{[r_1, r_2]})(X_1, X_2) \geq \mathcal{M}_{CX}(Q^*)(X_1, X_2)$, $Q^* = rate_{[r_1^*, r_2^*]}$. Because $\mathcal{M}_{CX}(Q^*)(X_1, X_2) > \frac{1}{2}$, we also have $\mathcal{M}_{CX}(Q')(X_1, X_2) > \frac{1}{2}$, where $Q' = rate_{[r_1, r_2]}$. Now consider $\bot^* = \bot_{Q^*, X_1, X_2}$ and $\bot' = \bot_{Q', X_1, X_2}$. Since Q^* and Q' are two-valued, $\bot^*$ and $\bot'$ are also two-valued. Moreover $\bot^*$ and $\bot'$ are monotonically non-increasing. Thus $\mathcal{M}_{CX}(rate_{[r_1, r_2]})(X_1, X_2) \geq \mathcal{M}_{CX}(Q^*)(X_1, X_2)$ is only possible if $\bot' \geq \bot^*$, since $\mathcal{M}_{CX}$ preserves inequations of $\bot'$ and $\bot^*$. We see from $\bot' \geq \bot^*$ that $\bot'_{j_*} \geq \bot^*_{j_*} = 1$, i.e. $r_1 \leq r_1^*$ and $r_2^* \leq r_2$. Now suppose that $\mathcal{M}_{CX}(Q')(X_1, X_2) > \mathcal{M}_{CX}(Q^*)(X_1, X_2)$. Then $\mathcal{M}_{CX}(Q') = \frac{1}{2} + \frac{1}{2}\gamma'$ for $\gamma' > \gamma_{j_*+1}$. In particular, $\bot'_{j_*+1} = 1$, i.e. $r_1 \leq r_{j_*+1}^{\min}$ and $r_{j_*+1}^{\max} \leq r_2$. By definition of j_*, $r_2 - r_1 \geq r_{j_*+1}^{\max} - r_{j_*+1}^{\min} > \delta_{\max}$, i.e. Q' exceeds the $\delta_{\max}$ limit. Thus in fact $\mathcal{M}_{CX}(Q')(X_1, X_2) = \mathcal{M}_{CX}(Q^*)(X_1, X_2)$ and $r_1 \leq r_1^*, r_2^* \leq r_2$, confirming the optimality of Q^*.

Apparently, the result can be computed in at most m steps ($j = 0, \ldots, m - 1$). Computation of $\bot_j$ given j is possible in constant time, see (2). Therefore, the optimal choice of r_1 and r_2 can be computed in $\mathcal{O}(m)$ time. However, the computation of the cardinality coefficients must also be considered. The method for determining $\ell_s(j)$ and $u_s(j)$ described in [5, 3] requires a pre-computation of the histograms of X_1, X_2, which has complexity $\mathcal{O}(N \log m)$. In practice, this is sufficient to compute optimal bounding quantifiers even for large base sets.

Finally we consider an example of quantifier selection. Let us assume that $E = \{a, b, c, d, e, f, g, h, i\}$, $X_1 = 1/a + 0.9/b + 0.8/c + 0.8/d + 0.7/e + 0.7/f + 0.1/g + 1/h + 0.9/i$, $X_2 = 0.05/a + 0.9/b + 0.7/c + 1/d + 1/e + 0.8/f + 0.1/g + 0.5/h + 0.1/i$. The corresponding minimum and maximum rates and the results of the quantifier selection method for several choices of $\delta_{\max}$ are shown in Figure 1. Here, $\tau = \mathcal{M}_{CX}(Q^*)(X_1, X_2)$ is the resulting validity score of the summary. Notice that a summary like "Between 62.5% and 75% of the X_1's are X_2's" might be misleading because it gives an illusion of precision which is not justified by the data. We therefore suggest a subsequent *linguistic fitting* of the selected quantifier which

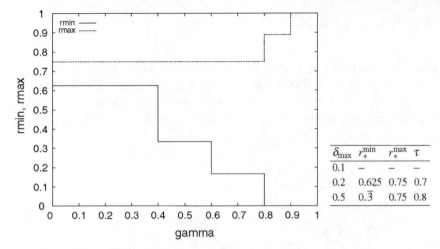

Fig. 1. Example plot of $r^{\min}$ and $r^{\max}$ and results of quantifier selection

should be adapted to a reasonable granularity scale. For example, assuming a 5% granularity scaling, the result for $\delta_{\max} = 0.2$ should better be expressed as "Between 60% and 75% of the X_1's are X_2's".

5 Conclusions

This paper was concerned with the problem of quantifier selection in fuzzy data summarization. For the important class of proportional bounding quantifiers, we have presented an efficient algorithm which computes the optimal quantifier in $\mathcal{O}(N \log m)$ time. Improving upon existing work on fuzzy data summarization, the new method uses a straightforward optimality criterion rather than heuristic quality indicators; the method is guaranteed to determine the optimal quantifier; and the optimal selection can be established very quickly. The resulting quantifier should be fitted to the quality of the data to improve the linguistic appropriateness of the summary. The relative proportion coefficients $r^{\min}$ and $r^{\max}$ on which the computation of the optimal quantifier is based, are interesting in their own right as a graphical representation of relative proportions found in imprecise data.

References

[1] J. Barwise and R. Cooper. Generalized quantifiers and natural language. *Linguistics and Philosophy*, 4:159–219, 1981.

[2] D. Dubois, H. Prade, and R.R. Yager, editors. *Fuzzy Information Engineering*. Wiley, New York, 1997.

[3] I. Glöckner. Evaluation of quantified propositions in generalized models of fuzzy quantification. *International Journal of Approximate Reasoning*, 37(2):93–126, 2004.

[4] I. Glöckner. DFS – an axiomatic approach to fuzzy quantification. TR97-06, Technical Faculty, University Bielefeld, 33501 Bielefeld, Germany, 1997.

[5] Ingo Glöckner. *Fuzzy Quantifiers: A Computational Theory*, volume 193 of *Studies in Fuzziness and Soft Computing*. Springer, Berlin, 2006.

[6] J. Kacprzyk and P. Strykowski. Linguistic summaries of sales data at a computer retailer: A case study. In *Proc. IFSA '99*, pages 29–33, 1999.

[7] D. Rasmussen and R.R. Yager. A fuzzy SQL summary language for data discovery. In Dubois et al. [2], pages 253–264.

[8] R.R. Yager. A new approach to the summarization of data. *Information Sciences*, 28:69–86, 1982.

[9] R.R. Yager. On ordered weighted averaging aggregation operators in multicriteria decisionmaking. *IEEE Trans. on Systems, Man, and Cybernetics*, 18(1):183–190, 1988.

[10] R.R. Yager. On linguistic summaries of data. In W. Frawley and G. Piatetsky-Shapiro, editors, *Knowledge Discovery in Databases*, pages 347–363. AAAI/MIT Press, 1991.

[11] L.A. Zadeh. A computational approach to fuzzy quantifiers in natural languages. *Computers and Mathematics with Applications*, 9:149–184, 1983.

A Linguistic Quantifier Based Aggregation for a Human Consistent Summarization of Time Series

Janusz Kacprzyk[1], Anna Wilbik[1], and Sławomir Zadrożny[1,2]

[1] Systems Research Institute, Polish Academy of Sciences, ul. Newelska 6, 01-447 Warsaw, Poland
[2] Warsaw Information Technology (WIT) ul. Newelska 6, 01-447 Warsaw, Poland
`{kacprzyk,wilbik,zadrozny}@ibspan.waw.pl`

1 Introduction

Dynamics, or variability over time, is crucial in virtually all real world processes. Among many formal approaches to the description of dynamic behavior is the use of time series, notably those composed of a sequence of real numbers that represent how values of a quantity, variable, etc. evolve over time. Time series are then used for many diverse purposes exemplified by decision making, prediction, etc. However, in all these situations first we have to grasp the very meaning of a particular time series in the sense of what is going on with the quantity or variable whose values it represents.

Unfortunately, this is a difficult task. First, nontrivial time series are usually long and maybe beyond human comprehension. Second, real time series rarely exhibit an obvious behavior exemplified by a steady growth. Usually, their behavior is variable and complicated like, e.g., after a long period of a strong growth there is a short period of an almost steady value, then a very long period of a slight decrease, etc. Notice that such a description of time series does carry much information to a human being but is, first, difficult to represent using traditional formal means (e.g. statistical), and, second, is difficult to derive by using conventional tools and techniques. Needless to say that such a description of trends is a kind of summarization.

We propose how to derive a human consistent summarization of time series in terms of trends observed in the data. Although we relate to some traditional approaches to the discovery and assessment of trends, exemplified by various concepts of variability that are somehow rooted in statistics, we employ a different approach.

Basically, we follow the philosophy of the so-called paradigm of computing with words and perceptions due to Zadeh and Kacprzyk [6] which postulates that since for a human being the only fully natural means of communication and articulation is natural language, then the use of natural language based tools, techniques, reports,

J. Kacprzyk et al.: *A Linguistic Quantifier Based Aggregation for a Human Consistent Summarization of Time Series*, Advances in Soft Computing **6**, 183–190 (2006)
`www.springerlink.com`

etc. should be more human consistent than the use of "artificial" numbers, functions, numerical algorithms, etc., and hence might help solve even complicated problems.

In our context we advocate (fuzzy) linguistic data summaries in the sense of Yager [12] (cf. also Kacprzyk and Yager [4], and Kacprzyk, Yager and Zadrożny [5]). Such summaries, e.g. "*most* employees are *middle aged*" for a personnel database, make it possible to grasp the very meaning of even a huge data set by summarizing it by a short natural language like sentence. Fuzzy logic with linguistic quantifiers is here indispensable. We show how to derive (fuzzy) linguistic summaries of time series trends exemplified by "*most* trends are *increasing*", "*most* of *slowly decreasing* trends have a *large variability*", etc. We use Zadeh's [13] simple fuzzy logic based calculus of linguistically quantified propositions, using our modification of Sklansky and Gonzalez's [11] technique to extract elementary trends to be aggregated.

It seems that the method proposed can often be, on the one hand, a viable technique for the description of trends that can effectively and efficiently supplement well established traditional techniques with roots in statistics. On the other hand, since it is heavily based on the use of natural language, it can provide simple, easily comprehensive and human-consistent description of trends in time series. The technique proposed can be easily implemented in human centric systems.

2 Characterization of Time Series

In our approach a time series $\{(x_i, y_i)\}$ is approximated by a piecewise linear function f such that for a given $\varepsilon > 0$, there holds $|f(x_i) - y_i| \leq \varepsilon$, for each i.

Among many algorithms that find such approximations (cf. [1, 3]), the Sklansky and Gonzalez [11] one seems to be a good choice due to its simplicity and efficiency. We modified it as follows. The algorithm constructs the intersection of cones starting from a point p_i of the time series and including a circle of radius ε around the subsequent data points p_{i+j}, $j = 1, \ldots$ until this intersection becomes empty. If for p_{i+k} the intersection is empty, then the points $p_i, p_{i+1}, \ldots, p_{i+k-1}$ are approximated by a straight line segment, and to approximate the remaining points we construct a new cone starting at p_{i+k-1}. Figure 1 presents the idea of the algorithm. The family of possible solutions, i.e., straight line segments to approximate points is indicated with a gray area. This method is fast as it requires only a single pass through the data.

We characterize the trends, meant as the straight line segments of the above described uniform ε-approximation, using the following three features: (1) duration, (2) dynamics of change, and (3) variability.

In what follows we will briefly discuss these factors.

Dynamics of change

Under the term *dynamics of change* we understand the speed of changes. It can be described by the slope of a line representing the trend, (cf. α in Fig. 1). Thus, to quantify dynamics of change we may use the interval of possible angles $\alpha \in \langle -90; 90 \rangle$.

However it might be impractical to use such a scale directly while describing trends. Therefore we may use a fuzzy granulation in order to meet the users' needs

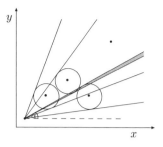

Fig. 1. An illustration of the algorithm [11] for an uniform ε-approximation

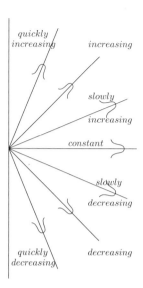

Fig. 2. Example of membership function describing the term "long" concerning the trend duration

Fig. 3. A visual representation of angle granules defining dynamics of change

and task specificity. The user may construct a scale of linguistic terms corresponding to various directions of a trend line as, e.g.: *quickly decreasing, decreasing, slowly decreasing, constant, slowly increasing, increasing, quickly increasing.*

Figure 3 illustrates the lines corresponding to the particular linguistic terms. In fact, each term represents a fuzzy granule of directions. There are many methods of constructing such a fuzzy granulation (cf. [2]), alternatively the user may define a membership functions of particular linguistic terms depending on his or her needs.

We map a single value α (or the whole interval of the angles corresponding to the gray area in Figure 1) characterizing the dynamics of change of a trend identified using modified Sklansky and Gonzalez method, into a fuzzy set best matching given angle. Then we will say that a given trend is, e.g., "decreasing to a degree 0.8", if $\mu_{decreasing}(\alpha) = 0.8$, where $\mu_{decreasing}$ is the membership function of a fuzzy set representing the linguistic term "decreasing" that is a best match for the angle α characterizing the trend under consideration.

Duration

Duration describes the length of a single trend corresponding to a line segment of the linear uniform ε-approximation. Again we will treat it as a linguistic variable. An example of its linguistic labels is "long" defined as a fuzzy set, whose membership function might be assumed as in Figure 2, where OX is the axis of time measured with units that are used in the time series data under consideration.

The actual definitions of linguistic terms describing the duration depend on the perspective assumed by the user. He or she, analyzing the data, may adopt this or another time horizon implied by his or her needs. The analysis may be a part of a policy, strategic or tactical planning, and thus, may require a global or local look, respectively.

Variability

Variability refers to how "spread out" (in the sense of values taken on) a group of data is. There are five frequently used statistical measures of variability:

- the range
- the interquartile range (IQR)
- the variance
- the standard deviation
- the mean absolute deviation (MAD)

We propose to measure the variability of a trend as the distance of the data points covered by this trend from their linear uniform ε-approximation that represents given trend. For this purpose we may bisect the cone and then compute the distance between the point and this ray.

Again the measure of variability is treated as a linguistic variable and expressed by linguistic terms (labels) modeled by fuzzy sets.

3 Linguistic Summaries

A linguistic summary, as presented in [9, 10] is meant as a natural language like sentence that subsumes the very essence of a set of data. This set is assumed to be numeric and is usually large, not comprehensible in its original form by the human being. In Yager's approach (cf. Yager [12], Kacprzyk and Yager [4], and Kacprzyk, Yager and Zadrożny [5]) we assume that (1) $Y = \{y_1, \ldots, y_n\}$ is a set of objects (records) in a database, e.g., the set of workers; and (2) $A = \{A_1, \ldots, A_m\}$ is a set of attributes characterizing objects from Y, e.g., salary, age, etc. in a database of workers, and $A_j(y_i)$ denotes a value of attribute A_j for object y_i.

A linguistic summary of a data set D consists of:

- a summarizer S, i.e. an attribute together with a linguistic value (fuzzy predicate) defined on the domain of attribute A_j (e.g."low" for attribute "salary");
- a quantity in agreement Q, i.e. a linguistic quantifier (e.g. most);
- truth (validity) T of the summary, i.e. a number from the interval $[0,1]$ assessing the truth (validity) of the summary (e.g. 0.7); usually, only summaries with a high value of T are interesting;
- optionally, a qualifier R, i.e. another attribute together with a linguistic value (fuzzy predicate) defined on the domain of attribute A_k determining a (fuzzy subset) of Y (e.g. "young" for attribute "age").

Thus, a linguistic summary may be exemplified by

$$T(\textit{most of employees earn low salary}) = 0.7 \tag{1}$$

A richer linguistic summary may include a qualifier (e.g. young) as, e.g.,

$$T(most \text{ of } young \text{ employees earn } low \text{ salary}) = 0.75 \tag{2}$$

Thus, basically, the core of a linguistic summary is a *linguistically quantified proposition* in the sense of Zadeh [13]. A linguistically quantified proposition, corresponding to (1) may be written as

$$Qy\text{'s are } S \tag{3}$$

and the one corresponding to (2) may be written as

$$QRy\text{'s are } S \tag{4}$$

Then, the component of a linguistic summary, T, i.e., its truth (validity), directly corresponds to the truth value of (3) or (4). This may be calculated by using either original Zadeh's calculus of linguistically quantified propositions (cf. [13]), or other interpretations of linguistic quantifiers.

4 Trend Summarization

In order to characterize the summaries of trends we will refer to Zadeh's concept of a protoform (cf., Zadeh [14]). Basically, a protoform is defined as a more or less abstract prototype (template) of a linguistically quantified proposition. Then, summaries mentioned above might be represented by protoforms of the following form:

- We may consider a short form of summaries:

$$Q \text{ trends are } S \tag{5}$$

like e.g., "*Most* of trends have *a large variability*";
- We may also consider an extended form of the summary represented by the following protoform:

$$QR \text{ trends are } S \tag{6}$$

and exemplified by "*Most* of *slowly decreasing trends* have *a large variability*".

Using Zadeh's [13] fuzzy logic based calculus of linguistically quantified propositions, a (proportional, nondecreasing) linguistic quantifier Q is assumed to be a fuzzy set in the interval $[0, 1]$ as, e.g.

$$\mu_Q(x) = \begin{cases} 1 & \text{for } x \geq 0.8 \\ 2x - 0.6 & \text{for } 0.3 < x < 0.8 \\ 0 & \text{for } x \leq 0.3 \end{cases} \tag{7}$$

The truth values (from $[0,1]$) of (5) and (6) are calculated, respectively as

$$\text{truth}(Qy\text{'s are } S) = \mu_Q\left[\frac{1}{n}\sum_{i=1}^{n}\mu_S(y_i)\right] \tag{8}$$

$$\text{truth}(QR\text{y's are } S) = \mu_Q \left[\frac{\sum_{i=1}^{n}(\mu_R(y_i) \wedge \mu_S(y_i))}{\sum_{i=1}^{n} \mu_R(y_i)} \right] \tag{9}$$

Both the fuzzy predicates S and R are assumed above to be of a rather simplified, atomic form referring to just one attribute. They can be extended to cover more sophisticated summaries involving some confluence of various attribute values as, e.g, "slowly decreasing and short".

5 Example

Let us assume that we have discovered from some given data trends listed in Table 1. We assume the granulation of dynamics of change depicted in Figure 3.

Table 1. Example of trends

id	Dynamics of change (α in degrees)	Duration (Time units)	Variability ([0,1])
1	25	15	0.2
2	-45	1	0.3
3	75	2	0.8
4	-40	1	0.1
5	-55	1	0.7
6	50	2	0.3
7	-52	1	0.5
8	-37	2	0.9
9	15	5	0.0

We can consider the following trend summarization:

Most of trends are *decreasing*

In this summary *most* is the linguistic quantifier Q. The membership function is as in (7).

"*Trends are decreasing*" has a summarizer S with the membership function of "decreasing" given in (10).

$$\mu_S(x) = \begin{cases} 0 & \text{for } \alpha \leq -65 \\ 0.066\alpha + 4.333 & \text{for } -65 < \alpha < -50 \\ 1 & \text{for } -50 \leq \alpha \leq -40 \\ -0.05\alpha - 1 & \text{for } -40 < \alpha < -20 \\ 0 & \text{for } \alpha \geq -20 \end{cases} \tag{10}$$

The truth is computed via (8), n is the number of all trends, here $n = 9$:

$$\text{truth}(\textit{Most of the trends are decreasing}) = \mu_Q\left[\tfrac{1}{n}\sum_{i=1}^{n}\mu_S(y_i)\right] = 0.389$$

If we consider an extended form, we may have the following summary:

$$Most \text{ of } short \text{ trends are } decreasing$$

Again, *most* is the linguistic quantifier Q with the membership function defined as (7). *"The trends are decreasing"* is a summarizer S as in the previous example. *"The trend is short"* is the qualifier R. We define the membership function μ_R as follows:

$$\mu_R(t) = \begin{cases} 1 & \text{for } t \leq 1 \\ -\frac{1}{2}t + \frac{3}{2} & \text{for } 1 < t < 3 \\ 0 & \text{for } t \geq 3 \end{cases}$$

The truth is computed via (9):

$$\text{truth}(Most \text{ of } short \text{ trends are } decreasing) =$$

$$\mu_Q \left[\frac{\sum_{i=1}^{n}(\mu_R(y_i) \wedge \mu_S(y_i))}{\sum_{i=1}^{n} \mu_R(y_i)} \right] = 0.892$$

These summaries are based on trend frequency of a given type identified in data. Thus, many short trends can influence or even dominate long, although rare trends. A further research is needed to solve this problem.

6 Concluding Remarks

We showed how to derive (fuzzy) linguistic summaries of time series exemplified by *"most* trends are *increasing"*, *"most* of *slowly decreasing* trends have a *large variability"*, etc. using Zadeh's [13] fuzzy logic based calculus of linguistically quantified propositions as a "soft" aggregation tool of elementary trends extracted by Sklansky and Gonzalez's [11] technique. The method proposed can find applications in all kinds of human centric systems, notably decision support systems.

For the future, one of promising directions would be the use of other aggregation techniques that make it possible to obtain a more sophisticated representation of time series. Moreover, other, more sophisticated linguistic summaries of trends in time series are conceivable and will be a subject of further research.

References

[1] J.F.Baldwin, T.P.Martin, J.M.Rossiter (1998). Time Series Modelling and Prediction using Fuzzy Trend Information. In *Proceedings of Fifth International Conference on Soft Computing and Information/Intelligent Systems*, 499-502.
[2] I. Batyrshin (2002). On Granular Derivatives and the Solution of a Granular Initial Value Problem. In *International Journal of Applied Mathematics and Computer Science*, 12(3): 403-410.

[3] J. Colomer, J. Melendez, J. L. de la Rosa, and J. Augilar. A qualitative/quantitative representation of signals for supervision of continuous systems. In *Proceedings of the European Control Conference -ECC97*, Brussels, 1997.

[4] J. Kacprzyk and R.R. Yager (2001). Linguistic summaries of data using fuzzy logic. In *International Journal of General Systems*, 30:33-154.

[5] J. Kacprzyk, R.R. Yager and S. Zadrożny (2000). A fuzzy logic based approach to linguistic summaries of databases. In *International Journal of Applied Mathematics and Computer Science*, 10: 813-834.

[6] L.A. Zadeh and J. Kacprzyk, Eds. (1999) *Computing with Words in Information/Intelligent Systems. Part 1. Foundations, Part 2. Applications*, Springer–Verlag, Heidelberg and New York.

[7] J. Kacprzyk and S. Zadrożny (1995). FQUERY for Access: fuzzy querying for a Windows-based DBMS. In P. Bosc and J. Kacprzyk (Eds.) *Fuzziness in Database Management Systems*, Springer-Verlag, Heidelberg, 415-433.

[8] J. Kacprzyk and S. Zadrożny (1999). The paradigm of computing with words in intelligent database querying. In L.A. Zadeh and J. Kacprzyk (Eds.) *Computing with Words in Information/Intelligent Systems. Part 2. Foundations*, Springer–Verlag, Heidelberg and New York, 382-398.

[9] J. Kacprzyk, S. Zadrożny (2005). Linguistic database summaries and their protoforms: toward natural language based knowledge discovery tools. In *Information Sciences* 173: 281-304.

[10] J. Kacprzyk, S. Zadrożny (2005). Fuzzy linguistic data summaries as a human consistent, user adaptable solution to data mining. In B. Gabrys, K. Leiviska, J. Strackeljan (Eds.) *Do Smart Adaptive Systems Exist?* Springer, Berlin Heidelberg New York, 321-339.

[11] J. Sklansky and V. Gonzalez (1980) Fast polygonal approximation of digitized curves. In *Pattern Recognition* 12(5): 327-331.

[12] R.R. Yager (1982). A new approach to the summarization of data. *Information Sciences*, 28: 69-86.

[13] L.A. Zadeh (1983). A computational approach to fuzzy quantifiers in natural languages. In *Computers and Mathematics with Applications*, 9: 149-184.

[14] L.A. Zadeh (2002). A prototype-centered approach to adding deduction capabilities to search engines – the concept of a protoform. BISC Seminar, University of California, Berkeley.

Efficient Evaluation of Similarity Quantified Expressions in the Temporal Domain

F. Díaz-Hermida[1], P. Cariñena[2], and A. Bugarín[2]

[1] Department of Computer Science. University of Oviedo
diazfelix@uniovi.es
[2] Department of Electronics and Computer Science. University of Santiago de Compostela
Intelligent Systems Group, University of Santiago de Compostela
{puri, alberto}@dec.usc.es

1 Introduction

Modelling of fuzzy temporal quantified statements is of great interest for real time systems. In [1] use of fuzzy proportional quantifiers to model temporal statements (sentences involving occurrence of events within a time framework) has been proposed. By using these proportional quantifiers a semantics can be associated to expressions like *"medium or high temperature values were measured together to risky high pressure values in the last few seconds"*. Nevertheless evaluation of a number of temporal statements cannot be modelled with these quantifiers, as *"Association between risky high pressures and high temperatures has been very high in the last few seconds"*. We will see how evaluation of these similarity or correlation expressions between two signals can be modelled by using *similarity quantifiers*.

The issue of computational efficiency is also addressed in the paper. It is usual in real time systems that situations of interest are evaluated involving operations on the values of signals in temporal windows made up of even several thousand points. In this case, the algorithms that efficiently evaluate expressions using the quantification model defined in [3] are sketched.

The remainder of the paper is organized as follows. Firstly, we introduce the fuzzy quantification field and the use of fuzzy quantification in temporal reasoning. Secondly, we show the use of similarity quantifiers in the temporal domain. The paper ends with sketching of the algorithms that allow us to evaluate similarity fuzzy temporal expressions and some conclusions.

2 Fuzzy Quantifiers and Fuzzy Temporal Rules

In this paper we follow the framework presented in [4, 5] to evaluate fuzzy quantified sentences *(FQSs)*, defined as sentences involving fuzzy quantifiers and a number of

F. Díaz-Hermida et al.: *Efficient Evaluation of Similarity Quantified Expressions in the Temporal Domain*, Advances in Soft Computing **6**, 191–198 (2006)
www.springerlink.com

fuzzy properties. Evaluation of a FQS consist of calculating a degree of fulfilment of the sentence for a given universe.

In order to evaluate FQSs *fuzzy quantifiers* are needed. An n-ary fuzzy quantifier $\tilde{Q}$ on a base set $E \neq \varnothing$ is a mapping $\tilde{Q} : \widetilde{\mathscr{P}}(E)^n \longrightarrow [0,1]$ which to each choice of $X_1, \ldots, X_n \in \widetilde{\mathscr{P}}(E)$ assigns a gradual result $\tilde{Q}(X_1, \ldots, X_n) \in [0,1]$ ($\widetilde{\mathscr{P}}(E)$ is the fuzzy powerset of E). Fuzzy quantifiers are the generalization of *classic quantifiers* (mappings $Q : \mathscr{P}(E)^n \longrightarrow \{0,1\}$ of the crisp powerset of E into the bivalued logical set $\{0,1\}$) to the fuzzy case.

It is very difficult to achieve consensus for defining a suitable expression to evaluate a FQSs. To overcome this problem *semi-fuzzy quantifiers*, which are a halfway point between classic quantifiers and fuzzy quantifiers, are described [4, 5]. Formally, an n-ary semi-fuzzy quantifier Q on a base set $E \neq \varnothing$ is a mapping $Q : \mathscr{P}(E)^n \longrightarrow [0,1]$ which assigns a gradual result $Q(Y_1, \ldots, Y_n) \in [0,1]$ to each choice of crisp sets $Y_1, \ldots, Y_n \in \mathscr{P}(E)$.

For instance, the semi-fuzzy quantifier associated to the expression *"nearly all"* may be defined as:

$$\mathbf{nearly\_all}(Y_1, Y_2) = \begin{cases} S_{0.6,0.8}\left(\frac{|Y_1 \cap Y_2|}{|Y_1|}\right) & Y_1 \neq \varnothing \\ 1 & Y_1 = \varnothing \end{cases}$$

where $S_{0.6,0.8}(x) \to [0,1]$ is the S's Zadeh function shown in Figure 1.

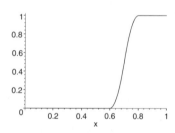

Fig. 1. S Zadeh's function of paramters $0.6, 0.8$.

Semi-fuzzy quantifiers are much more intuitive and easier to define than fuzzy quantifiers, although they do not resolve by their own the problem of evaluating FQSs. In order to evaluate a FQS *quantifier fuzzification mechanisms (QFMs)* [4, 5] are needed that enable us to transform semi-fuzzy quantifiers into fuzzy quantifiers, i.e. mappings with domain in the universe of semi-fuzzy quantifiers and range in the universe of fuzzy quantifiers:

$$\mathscr{F} : (Q : \mathscr{P}(E)^n \to [0,1]) \mapsto \left(\tilde{Q} : \widetilde{\mathscr{P}}(E)^n \to [0,1]\right)$$

As an example of application we will use in this paper the QFM model[3] defined in [3]:

---

[3] This model is not a QFM in the full sense, because the integral can be undefined for some non-finite quantifiers [3].

$$\mathscr{F}^I(Q)(X_1,\ldots,X_n) = \int_0^1 \ldots \int_0^1 Q\left((X_1)_{\geq \alpha_1},\ldots,(X_n)_{\geq \alpha_n}\right) d\alpha_1 \ldots d\alpha_n \quad (1)$$

where $(X_i)_{\geq \alpha_i}$ denotes the alpha cut α_i of the fuzzy set $X_i \in \widetilde{\mathscr{P}}(E)$.

Fuzzy quantification has been proposed to improve the expressiveness of temporal rules in the field of fuzzy temporal reasoning [1]. Temporal rules are used for describing the occurrence of events that are described by means of temporal evolution of signals, temporal relationships among them and/or spatial (as opposed to non-temporal) relationships among the signals values, thus noticeably enhancing the expressivity of usual fuzzy rules in knowledge-based systems. Fuzzy temporal rules have been successfully used in intelligent control for implementation of behaviours in robotics such as mobile obstacles avoidance, wall following or landmarks detection, intelligent monitoring in medical domains and many other fields of application [1].

These propositions describe an event of interest for a particular system. The aim in this context is to calculate a degree of fulfillment for the proposition, based on the historic of signal values, the relationships among them (usually expressed by means of temporal and/or spatial operators) and also the time reference that frames the occurrence of the event of interest (temporal window).

The concept of signal in this context has to be understood in a broad sense. Apart of the usual meaning of signals in monitoring/control processes (*"pressure(t)"*, *"temperature(t)"*, ...) other types of signals are consistent in this framework. For instance, in a given process it may be of interest to consider "*high_pressures(t)*" or "*low_temperatures(t)*" interpreted as the degree of membership to fuzzy sets *"high"/ "low"* of the pressure/temperature value at time point t. The process of calculating these signals from the historic values of "pressure" and "temperature" has been defined as a fuzzy filtering operation (e.g., $high\_pressures(t) = \mu_{high}(pressure(t))$). For the sake of simplicity we will call these *fuzzy signals*, in order to reflect they take values in $[0,1]$.

Proportional quantifiers are used in fuzzy temporal rules to model the *persistence* of events in the temporal window. Examples of temporal propositions or *Fuzzy Temporal Statements* involving *proportional quantifiers* are:

"About the eighty percent or more of the temperature values were high in the last month".

"Medium or high temperature values were associated to risky high pressure values in about the eighty percent or more of the time points in the last two minutes".

Whilst universal quantifier (*"all"*) requests the complete occurrence of the event throughout the temporal window and existential quantifier (*"in"* or *"exist"*) requests the occurrence of the event in a particular point of the temporal window, proportional quantifiers allow a partial occurrence of the event. Modelling of proportional quantifiers can be addressed by applying any QFM endowed with adequate properties to a semi-fuzzy quantifier that adequately describes the quantified expression.

For example, the following proportional semi-fuzzy quantifier can be used for modelling the temporal expression *"about the eighty percent or more of the temperature values were high in the last month"*:

$$Q(T,S) = \begin{cases} S_{0.6,0.8}\left(\frac{|S \cap T|}{|T|}\right) & T \neq \varnothing \\ 1 & T = \varnothing \end{cases}$$

where T represents a crisp temporal reference, S a crisp set of high temperature values, and $S_{0.6,0.8}(x)$ is a Zadeh's S-function that models the quantifier *"about 80% or more"*. In order to construct the fuzzy quantifier from semi-fuzzy quantifier Q to be applied to the fuzzy temporal window *"the last month"* and the fuzzy signal *"high temperatures"* an appropiate QFM must be used (e.g., the one defined in expression 1).

Also, to model the expression *"medium or high temperature values were associated to risky high pressure values in about the eighty percent or more of the instants of the last two minutes"* the following semi-fuzzy quantifier can be used:

$$Q(T,S_1,S_2) = \begin{cases} S_{0.6,0.8}\left(\frac{|S_1 \cap S_2 \cap T|}{|T \cap S_1|}\right) & T \cap S_1 \neq \varnothing \\ 1 & T \cap S_1 = \varnothing \end{cases}$$

where S_1 represents crisp signal *"medium or high temperature"* and S_2 represents crisp signal *"risky high pressure values"*, T represents temporal window *"in the last two minutes"*, and the S function models the quantifier *"about 80% or more"*. By applying a QFM to Q we can construct the fuzzy quantifier that can be used for computing the degree of fulfilment of the Fuzzy Temporal Statement with the fuzzy time window and the fuzzy signals.

We will use the previous example to explain other important aspect of the evaluation of temporal expressions. When we use this kind of expressions in real time systems, the temporal window is usually related to the "current time point "; that is, the temporal window advances with the arrival of new information to the system. In Figure 2 we show an example for fuzzy pressure and fuzzy temperature signals. The time window *"in the last two minutes"* is modelled by a fuzzy number. As time advances, new time points should be considered by the system and also the temporal window advances too. We can think that the pressure and the temperature values are continuosly monitored in a real time system and, in each control iteration, new signal values arrive to the system thus making it necessary to recompute the condition expressed by the fuzzy temporal statement.

3 Using Similarity Quantifiers in Fuzzy Temporal Statements

Althoug the use of fuzzy quantifiers we have shown in previous section is very interesting, there are expressions that do not fit previous semantics that can be very useful for applications. In [2] we have presented a classification of semi-fuzzy quantifiers of undoubted interest for applications. Now a new type of semi-fuzzy quantifiers is being considered: *similarity quantifiers*. These quantifiers let us to measure

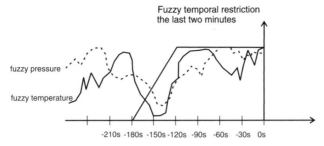

Fig. 2. Fuzzy signals *"riskly high pressure"* and *"medium or high temperature"* and temporal constraint *"the last two minutes"*

the similarity between two sets. When we extend these ideas to the temporal field, the similarity quantifiers can be used to measure the similarity between two signals considered within a temporal window. Some examples of expressions that can be evaluated using this kind of quantifiers are:

"Rainy days and moderately cold days were associated in January"
"High values of the asset market and small interest values were associated in the nineties"
"The correlation between the risky high pressures and the high temperatures has been very high in the last few seconds"

These examples involving similarity quantifiers state an association between the two signals within the temporal window; that is, proportional quantifiers measure the degree to which one of the signals is contained into the other. Similarity quantifiers measure the overlapping between the two signals. The semi-fuzzy quantifiers we are going to use to model these situation help us to make it clearer the requested semantics for similarity operators.

In the following we present two semi-fuzzy quantifiers that can be used to model previous examples:

- Basic temporal similarity semi-fuzzy quantifier:

$$Q(T, S_1, S_2) = \begin{cases} fn\left(\frac{|S_1 \cap S_2 \cap T|}{|(S_1 \cup S_2) \cap T|}\right) & (S_1 \cup S_2) \cap T \neq \varnothing \\ 1 & (S_1 \cup S_2) \cap T = \varnothing \end{cases} \tag{2}$$

where T is a crisp temporal restriction, S_1 and S_2 are crisp signals and $fn : [0,1] \rightarrow [0,1]$ is a fuzzy number. We should note that $\frac{|S_1 \cap S_2|}{|S_1 \cup S_2|}$ is a similarity measure between the crisp signals S_1 and S_2 that is widely used in the fields of signal processing and information retrieval (known as Jaccard index). Temporal window T delimits the range of interest.

In $\frac{|S_1 \cap S_2 \cap T|}{|(S_1 \cup S_2) \cap T|}$ only positive values of the signals are considered. For example, if $|S_1 \cap S_2| = 0$ and $|S_1 \cup S_2| = 1$ in the temporal window T then the two signals are considered completely different when we restrict us to T. For example, if the temporal window T contains one thousand points then S_1 are S_2 are equal in 999 negative values but $\frac{|S_1 \cap S_2 \cap T|}{|(S_1 \cup S_2) \cap T|} = 0$.

- Simple matching temporal semi-fuzzy quantifier:

$$Q(T, S_1, S_2) = \begin{cases} fn\left(\frac{|S_1 \cap S_2 \cap T| + |\overline{S_1} \cap \overline{S_2} \cap T|}{|T|} \right) & T \neq \varnothing \\ 1 & T = \varnothing \end{cases}$$

In this quantifier we simultaneously consider positive and negative signal values to define the similarity between the signals. We should note here that $\frac{|S_1 \cap S_2| + |\overline{S_1} \cap \overline{S_2}|}{|E|}$ is also a similarity measure between crisp signals S_1 and S_2 that is widely used in the fields of signal processing and information retrieval.

By using an appropriate QFM these semi-fuzzy quantifiers can be converted into fuzzy ones. If we use the one defined on expression 2 we obtain:

$$\mathscr{F}^I(Q)(T, S_1, S_2) = \int_0^1 \int_0^1 \int_0^1 Q\left((T)_{\geq \alpha_1}, (S_1)_{\geq \alpha_2}, (S_2)_{\geq \alpha_3} \right) d\alpha_1 d\alpha_2 d\alpha_3 \quad (3)$$

In this way we can evaluate examples like the ones at the beginning of this section, in which a fuzzy temporal window and fuzzy signals are involved.

Expression 3 can be approximated by using numerical integration. By dividing the integral interval $(0, 1]^3$ into the set of intervals:

$$(i \times h, i \times (h+1)] \times (j \times h, j \times (h+1)] \times (k \times h, k \times (h+1)]$$

where h is the size of the integration step, and $0 \leq i, j, k < 1/h$. Let $n = 1/h$ be the number of intervals in the $(0, 1]$ interval. In this way, expression 3 is approximated by

$$\mathscr{F}^I(Q)(T, S_1, S_2) \approx \sum_{i=0}^{n-1} \sum_{j=0}^{n-1} \sum_{k=0}^{n-1} Q\left((T)_{\geq (i+\frac{1}{2}) \times h}, (S_1)_{\geq (j+\frac{1}{2}) \times h}, (S_2)_{\geq (k+\frac{1}{2}) \times h} \right) h^3$$

$$(4)$$

where we have approximated the integral in each interval by its result in the medium point of the interval.

4 Efficient Evaluation of Similarity Temporal Expressions

Now we briefly sketch the ideas that let us to apply efficiently expression 4 when we assume the temporal window is related to the the current time point; that is, the temporal window T "moves forward" when a new time point is processed by the system. We will assume that the fuzzy temporal restriction and the fuzzy signals are stored in sorted vectors, and that the temporal window is defined by using a convex fuzzy number.

Although we focus on the semi-fuzzy temporal quantifier defined on expression 2, similar ideas can be used to develop algorithms for other temporal expressions.

As the semi-fuzzy temporal quantifier defined on expression 2 only depends on the cardinalities $|S_1 \cap S_2 \cap T|$ and $|T \cap (S_1 \cup S_2)|$ a ternary array can be used to store this pair of cardinalities for each integration interval; that is, for each set of alpha-cuts levels $\alpha_i, \alpha_j, \alpha_k$ used to approximate the integral. This cardinality array can be straighforwardly initialized in $O(M \times n^2)$ time, where M is the size of the support of the temporal constraint and n is the number of intervals we are using to approximate the integral (the number of alpha-cuts). Once this array has been calculated evaluation of expression 4 can be trivially done.

This array can be updated in $O(n^3)$ with the arrival of new information. Since the temporal window is related to the current time point (e.g., *"in the last two minutes"*) the temporal window advances just a single time point when a new time point arrives to the system. The finite set of alpha cuts of the temporal window also advances a single time point too. The "oldest point" in each alpha cut of the temporal constraint is removed and a new point is added as a consequence of the time advance.

By checking the signal values in the time point that is removed of the alpha cut and in the time point that is added to the alpha cut the cardinality array can be updated in $O(n^3)$ time and thus temporal expression can also be recalculated in $O(n^3)$.[4]

5 Conclusions

In this paper we have proposed two models for similarity quantifiers that can be used in the field of temporal knowledge representation for modelling similarity expressions between signals. Both models are directly based on two crisp similarity indexes that are based on well-known and widely used definitions in the fields of signal processing and information retrieval.

The approximation we have presented, based on fuzzy quantifiers, expands the set of temporal expressions previously considered in [1], where only proportional quantifiers were used.

By using the QFM approach coherence with the modelling of other temporal quantified expressions is maintained. Moreover, the crisp semantics of semi-fuzzy quantifiers is very clear and easily understandable, and it is easily combined with the studied set of similarity indexes.

Moreover, we have sketched the ideas that let us efficiently evaluate temporal expression when we use these quantifiers. Also generalization of other similarity, dissimilarity or association indexes is straighforward. Although our work has been focused on similarity quantifiers, the ideas we have presented are easily adapted to

---

[4] For example, if we use 5 alpha-cuts to approximate the integral only $5^3 = 125$ operations are needed to recompute the cardinality array. This enables us to eliminate the dependence on the size of the temporal reference in the computation.

other types of quantifiers, and also to some other quantifier fuzzification mechanisms. Therefore, proposal of similar efficient mechanisms for other QFM will be our immediate aim.

References

[1] P. Cariñena. *A model of Fuzzy Temporal Rules for reasoning on dynamic systems*. PhD thesis, Universidade de Santiago de Compostela, 2003.

[2] F. Díaz-Hermida, A. Bugarín, and S. Barro. Definition and classification of semi-fuzzy quantifiers for the evaluation of fuzzy quantified sentences. *International Journal of Approximate Reasoning*, 34(1):49–88, 2003.

[3] F. Díaz-Hermida, A. Bugarín, P. Cariñena, and S. Barro. Voting model based evaluation of fuzzy quantified sentences: a general framework. *Fuzzy Sets and Systems*, 146:97–120, 2004.

[4] I. Glöckner and A. Knoll. A formal theory of fuzzy natural language quantification and its role in granular computing. In W. Pedrycz, editor, *Granular computing: An emerging paradigm*, volume 70 of *Studies in Fuzziness and Soft Computing*, pages 215–256. Physica-Verlag, 2001.

[5] I. Glöckner. *Fuzzy Quantifiers in Natural Language: Semantics and Computational Models*. Der Andere Verlag, 2004.

Part V

Imprecise Probability Theory

Conditional Lower Previsions
for Unbounded Random Quantities

Matthias C. M. Troffaes

Department of Philosophy, Carnegie Mellon University, Department of Mathematical Sciences, Durham University
matthias.troffaes@gmail.com

Summary. In this paper, a theory of conditional coherent lower previsions for arbitrary random quantities, including unbounded ones, is introduced, based on Williams's [13] notion of coherence, and extending at the same time unconditional theories studied for unbounded random quantities known from the literature. We generalize a well-known envelope theorem to the domain of all contingent random quantities. Finally, using this duality result, we prove equivalence between maximal and Bayes actions in decision making for convex option sets.

1 Introduction

Williams's and Walley's theories of conditional lower previsions [13, 12] provide a wide and unifying framework to study various imprecise probability models known in the literature, such as lower and upper probabilities, possibility measures [7, 4], credal sets [8], risk measures [1], and many others [12]. For reasons of mathematical convenience, these theories deal with bounded random quantities only. Nevertheless, unbounded random quantities occur often in practice. To give but a few examples, unbounded costs occur regularly in optimal control theory (see for instance Chevé and Congar [2]). In reliability theory, one typically asks about the time to failure of a component (see for instance Utkin [11]), which is unbounded from above. For applications like these, a generalization to arbitrary random quantities is necessary.

To some extent, such generalizations have already been studied in the literature. Crisma, Gigante and Millossovich [3] studied previsions for arbitrary random quantities, and these previsions may also assume the values $+\infty$ and $-\infty$. Troffaes and De Cooman [9, 10] have constructed an extension for coherent lower previsions defined on bounded random quantities to a larger set of random quantities, using a limit procedure similar to Dunford integration, and also studied how unconditional lower previsions can be handled to encompass arbitrary random quantities, including unbounded ones. However, these theories only deal with unconditional assessments and inferences. Moreover, duality results, which, roughly said, express imprecise models by means of precise ones through an envelope theorem, have been limited to domains where lower previsions do not assume infinite values. The aims of this paper are: (i)

M.C.M. Troffaes: *Conditional Lower Previsions for Unbounded Random Quantities*, Advances in Soft Computing **6**, 201–209 (2006)
www.springerlink.com

to bring conditioning into the picture, and (ii) to arrive at duality results on the full domain of random quantities.

Section 2 recalls Williams's theory and links it to Walley's. Section 3 suggests a new generalization to arbitrary random quantities, also mentioning the duality result and proving an equivalence between maximal and Bayes actions. Due to limitations of space, most proofs have been omitted, and we stick to a summary and brief discussion of the main results.

2 Conditional Lower Previsions

Let's recall the most important definitions and results from Williams's technical report [13], and link them to Walley's behavioral theory [12].

Let ω be a random variable taking values in some set Ω, called the possibility space. Denote by $\wp(\Omega)$ the set of all non-empty subsets of Ω. We shall denote a set and its indicator by the same symbol. A real-valued function of ω is called a random quantity.

Let $\mathscr{L}(\Omega)$ be the set of all bounded Ω–$\mathbb{R}$-mappings, i.e., all bounded random quantities. A real-valued mapping $\underline{P}$ on a subset of $\mathscr{L}(\Omega) \times \wp(\Omega)$ is called a *conditional lower prevision*. For $(f,A) \in \mathrm{dom}\,\underline{P}$, the real value $\underline{P}(f,A)$, or, $\underline{P}(f|A)$ using the more traditional notation, is interpreted as a supremum buying price for f contingent on A: prior to observing ω, we agree to receive $f(\omega)$ and pay any $x < \underline{P}(f|A)$ once ω has been observed and $\omega \in A$, nothing if $\omega \notin A$. Briefly, $A(f-x)$ is an acceptable random quantity for all $x < \underline{P}(f|A)$.

Every conditional lower prevision $\underline{P}$ has a conditional upper prevision $\overline{P}$ associated with it: $\overline{P}(f|A) := -\underline{P}(-f|A)$. So, $\overline{P}$ is defined on $\mathrm{dom}\,\overline{P} = \{(f,A) \colon (-f,A) \in \mathrm{dom}\,\underline{P}\}$. The real value $\overline{P}(f|A)$ is the infimum selling price for f contingent on A.

2.1 Coherence

The following definition of coherence is due to Williams [13] and can be behaviorally motivated using Walley's axioms of desirability [12, S2.2.3];[1] also see Sect. 3.2 further on. By $\sup(f|A)$ and $\inf(f|A)$ we denote the supremum and infimum of $f(\omega)$ over $\omega \in A$.

Definition 1 (Williams [13], p. 5, Eq. (A*)). *Say that $\underline{P}$ is* coherent *if*

$$\sup\left(\sum_{i=1}^{n} \lambda_i B_i\big(f_i - \underline{P}(f_i|B_i)\big) - \lambda_0 B_0\big(f_0 - \underline{P}(f_0|B_0)\big)\,\Big|\,B_0 \cup \cdots \cup B_n\right) \geq 0 \quad (1)$$

for all $n \in \mathbb{N}$, all non-negative $\lambda_0, \ldots, \lambda_n \in \mathbb{R}$, and all $(f_0,B_0), \ldots, (f_n,B_n) \in \mathrm{dom}\,\underline{P}$.

---

[1] … if we add a monotonicity axiom: if f is desirable and $g \geq f$ then g is desirable.

In contradistinction to Walley coherence [12, 7.1.4(b)], Williams coherence does not guarantee $\underline{P}$ to be conglomerable with respect to any partition in $\{B \subseteq \Omega : \exists f \in \mathscr{L}(\Omega) \text{ s.t. } (f,B) \in \operatorname{dom}\underline{P}\}$. Although there are compelling reasons for accepting infinite combinations of gambles contingent on events that form a partition [12, S6.3.3, S6.8.4], for this paper we prefer to use Williams coherence, as the resulting theory is mathematically much easier to handle, especially regarding natural extension. This means however that we should be careful and only accept finite combinations of bets from a conditional lower prevision that is coherent in the sense of Def. 1.

The "only if" part of the following theorem is due to Williams [13, 1.0.4, p. 6]. The "if" part is easily proved. Condition (iv) (which I have formulated slightly differently than Williams) is a generalization of the *generalized Bayes rule* [12, S6.4]. For a field $\mathscr{F}$, let $\mathscr{F}^\circ$ denote $\mathscr{F}$ without $\emptyset$.

Theorem 1. *Suppose that the conditional lower prevision $\underline{P}$ is defined on $\mathscr{K} \times \mathscr{F}^\circ$, where $\mathscr{F}$ is a field, and $\mathscr{K}$ is a linear subspace of $\mathscr{L}(\Omega)$ that contains at least $\mathscr{F}$, and which satisfies $Af \in \mathscr{K}$ whenever $A \in \mathscr{F}$ and $f \in \mathscr{K}$. Then $\underline{P}$ is coherent if and only if*

(i) $\underline{P}(f|A) \geq \inf(f|A)$,
(ii) $\underline{P}(\lambda f|A) = \lambda \underline{P}(f|A)$ for all $\lambda \in \mathbb{R}$, $\lambda \geq 0$,
(iii) $\underline{P}(f+g|A) \geq \underline{P}(f|A) + \underline{P}(g|A)$, and
(iv) $\underline{P}(A(f - \underline{P}(f|A))|B) = 0$,

for all f and g in $\mathscr{K}$, and all A and B in $\mathscr{F}$ such that $\emptyset \neq A \subseteq B$.

2.2 Previsions

When infimum selling prices coincide with supremum buying prices, that is, if $\operatorname{dom}\underline{P} = \operatorname{dom}\overline{P}$ and $\underline{P}(f|A) = \overline{P}(f|A)$ for all $(f,A) \in \operatorname{dom}\underline{P}$, then we say that $\underline{P}$ is *self-conjugate*, denote both $\underline{P}$ and $\overline{P}$ by P (we drop the bar), and call P a *conditional prevision*. They correspond to de Finetti's fair prices [5].

Theorem 2. *Suppose that the conditional prevision P is defined on $\mathscr{K} \times \mathscr{F}^\circ$, where $\mathscr{F}$ is a field, and $\mathscr{K}$ is a linear subspace of $\mathscr{L}(\Omega)$ that contains at least $\mathscr{F}$, and which satisfies $Af \in \mathscr{K}$ whenever $A \in \mathscr{F}$ and $f \in \mathscr{K}$. Then P is coherent if and only if*

(i) $P(f|A) \geq \inf(f|A)$,
(ii) $P(\lambda f|A) = \lambda P(f|A)$ for all $\lambda \in \mathbb{R}$,
(iii) $P(f+g|A) = P(f|A) + P(g|A)$, and
(iv) $P(Af|B) = P(f|A)P(A|B)$,

for all f and g in $\mathscr{K}$, and all A and B in $\mathscr{F}$ such that $\emptyset \neq A \subseteq B$.

Thm. 2 says coherent conditional previsions correspond to conditional expectations: they remain within the appropriate bounds, are linear mappings, and satisfy (a conditional version of) Bayes rule [6].

3 Generalizing to Unbounded Random Quantities

Let $\mathscr{R}(\Omega)$ be the set of all random quantities on Ω, i.e., all Ω–$\mathbb{R}$-mappings. An $\mathbb{R} \cup \{-\infty, +\infty\}$-valued mapping $\underline{P}$ on a subset of $\mathscr{R}(\Omega) \times \wp(\Omega)$ is called an *extended conditional lower prevision*. It is interpreted again as a supremum buying price. Thus, if $\underline{P}(f|A) = -\infty$, then we will never buy f contingent on A, and if $\underline{P}(f|A) = +\infty$, then we will buy f contingent on A for any price.

As before, we associate an *extended conditional upper prevision* $\overline{P}$ with $\underline{P}$, defined on $\operatorname{dom}\overline{P} = \{(f,A): (-f,A) \in \operatorname{dom}\underline{P}\}$, and $\overline{P}(f|A) = -\underline{P}(-f|A)$. We say that $\underline{P}$ is self-conjugate if $\underline{P} = \overline{P}$, in which case we denote it by P and call it an *extended conditional prevision*.

3.1 Avoiding Partial Loss

Following Walley [12] it is instructive not only to introduce a notion of coherence, but also to introduce a notion of avoiding partial loss. For our generalization, we cannot write terms of the form $B_i(f_i - \underline{P}(f_i|B_i))$ as in Def. 1, because if $\underline{P}(f_i|B_i)$ is not finite, $B_i(f_i - \underline{P}(f_i|B_i))$ is not a real-valued random quantity. Fortunately, there's a workaround:

Definition 2. *We say that $\underline{P}$ avoids partial loss if*

$$\sup\left(\sum_{i=1}^n \lambda_i B_i (f_i - x_i) \Big| B_1 \cup \cdots \cup B_n\right) \geq 0 \tag{2}$$

for all $n \in \mathbb{N}$, all non-negative $\lambda_1, \ldots, \lambda_n \in \mathbb{R}$, all $(f_1, B_1), \ldots, (f_n, B_n) \in \operatorname{dom}\underline{P}$ and all $x_i \in \mathbb{R}$, $x_i < \underline{P}(f_i|B_i)$.

Let's explain. Suppose Eq. (2) is violated for some $n \in \mathbb{N}$, non-negative $\lambda_1, \ldots, \lambda_n \in \mathbb{R}$, $(f_1, B_1), \ldots, (f_n, B_n) \in \operatorname{dom}\underline{P}$ and $x_i < \underline{P}(f_i|B_i)$. We are disposed to accept $B_i(f_i - x_i)$, hence, $\sum_{i=1}^n \lambda_i B_i(f_i - x_i)$ as well. But, if Eq. (2) is violated, then this acceptable random quantity is uniformly negative contingent on $B_1 \cup \cdots \cup B_n$, which means that $\sum_{i=1}^n \lambda_i B_i(f_i - x_i)$ incurs a sure loss contingent on $B_1 \cup \cdots \cup B_n$. Because the sure loss is only contingent on a subset, it is called a partial loss, following Walley [12, S7.1.2].

Note that, if $\underline{P}(f_i|B_i) = -\infty$, then there's no $x_i \in \mathbb{R}$, such that $x_i < \underline{P}(f_i|B_i)$, and hence, the corresponding term does not enter the definition. Next, let's generalize Def. 1 to extended conditional lower previsions.

3.2 Coherence

Definition 3. *We say that $\underline{P}$ is coherent if*

$$\sup\left(\sum_{i=1}^n \lambda_i B_i (f_i - x_i) - \lambda_0 B_0 (f_0 - x_0) \Big| B_0 \cup \cdots \cup B_n\right) \geq 0 \tag{3}$$

for all $n \in \mathbb{N}$, all non-negative $\lambda_0, \ldots, \lambda_n \in \mathbb{R}$, all $(f_0, B_0), \ldots, (f_n, B_n) \in \operatorname{dom}\underline{P}$, all $x_i \in \mathbb{R}$, $x_i < \underline{P}(f_i|B_i)$ and all $x_0 \in \mathbb{R}$, $x_0 > \underline{P}(f_0|B_0)$.

Let's explain. Suppose Eq. (3) is violated. The case $\lambda_0 = 0$ has already been argued for (Sec. 3.1). Assume $\lambda_0 > 0$. Again, $\sum_{i=1}^{n} \lambda_i B_i (f_i - x_i)$ is an acceptable random quantity. But, if Eq. (3) is violated, then

$$\sum_{i=1}^{n} \frac{\lambda_i}{\lambda_0} B_i (f_i - x_i) \leq B_0 (f_0 - x_0)$$

which means that also $B_0 (f_0 - x_0)$ is acceptable. But $x_0 > \underline{P}(f_0|B_0)$: we are disposed to increase the supremum buying price for f_0 contingent on B_0.

Theorem 3. *Suppose that the extended conditional lower prevision $\underline{P}$ is defined on $\mathscr{K} \times \mathscr{A}$, where $\mathscr{A}$ is closed under finite unions, and $\mathscr{K}$ is a linear subspace of $\mathscr{R}(\Omega)$ containing at least $\mathscr{A}$, and satisfying $Af \in \mathscr{K}$ whenever $A \in \mathscr{A}$ and $f \in \mathscr{K}$. Then $\underline{P}$ is coherent if and only if*

(i) $\underline{P}(f|A) \geq \inf(f|A)$,
(ii) $\underline{P}(\lambda f|A) = \lambda \underline{P}(f|A)$ for all $\lambda \in \mathbb{R}$, $\lambda \geq 0$,
(iii) $\underline{P}(f+g|A) \geq \underline{P}(f|A) + \underline{P}(g|A)$ whenever the right hand side is well defined, and
(iv) $\underline{P}(A(f-\mu)|B) \begin{cases} \geq 0, & \text{if } \mu < \underline{P}(f|A) \\ \leq 0, & \text{if } \mu > \underline{P}(f|A) \end{cases}$, for all $\mu \in \mathbb{R}$.

for all f and g in $\mathscr{K}$, and all A and B in $\mathscr{A}$ such that $\emptyset \neq A \subseteq B$.

3.3 Previsions

Clearly, an extended conditional prevision avoids partial loss if and only if it is coherent. Also,

Theorem 4. *Suppose that the extended conditional prevision P is defined on $\mathscr{K} \times \mathscr{A}$, where $\mathscr{A}$ is closed under finite unions, and $\mathscr{K}$ is a linear subspace of $\mathscr{R}(\Omega)$ containing at least $\mathscr{A}$, and satisfying $Af \in \mathscr{K}$ whenever $A \in \mathscr{A}$ and $f \in \mathscr{K}$. Then P is coherent if and only if*

(i) $P(f|A) \geq \inf(f|A)$,
(ii) $P(\lambda f|A) = \lambda P(f|A)$ for all $\lambda \in \mathbb{R}$,
(iii) $P(f+g|A) = P(f|A) + P(g|A)$ whenever the right hand side is well defined, and
(iv) $P(Af|B) = P(f|A)P(A|B)$ if $P(f|A) \in \mathbb{R}$ or $P(A|B) > 0$,
$P(Af|B) \geq 0$ if $P(f|A) = +\infty$ and $P(A|B) = 0$, and
$P(Af|B) \leq 0$ if $P(f|A) = -\infty$ and $P(A|B) = 0$.

for all f and g in $\mathscr{K}$, and all A and B in $\mathscr{A}$ such that $\emptyset \neq A \subseteq B$.

It is evident from Thm. 4 that coherent extended conditional previsions also behave as a conditional expectation. The only new constraints are the inequalities in (iv). Still, they can be given a simple intuitive (but non-sense) interpretation in terms of Bayes rule: if $P(f|A)$ is non-negative and very large, and $P(A|B)$ very small, then $P(Af|B) = P(f|A)P(A|B)$ can range anywhere within the set of non-negative real numbers; similar for the other inequality.

3.4 Natural Extension

As in the unconditional case [10], we can again establish that a conditional extended lower prevision has a least-committal coherent extension (or natural extension) to the whole set $\mathscr{R}(\Omega) \times \mathscr{P}^\circ(\Omega)$ if and only if it avoids partial loss. Note that this would *not* hold if we had based our study on Walley's definition of coherence [12, S8.1.2]; see Walley's examples [12, p. 410, ll. 18–22].

Definition 4. *An extended conditional lower prevision $\underline{Q}$ is called a* behavioral extension *of an extended conditional lower prevision $\underline{P}$ if* $\operatorname{dom}\underline{P} \subseteq \operatorname{dom}\underline{Q}$ *and* $\underline{P}(f|B) \leq \underline{Q}(f|B)$ *for any* $(f, B) \in \operatorname{dom}\underline{P}$.

Definition 5. *Let $\underline{P}$ be an extended conditional lower prevision, and let* $\operatorname{dom}\underline{P} \subseteq \mathscr{K} \subseteq \mathscr{R}(\Omega) \times \mathscr{P}^\circ(\Omega)$. *The point-wise smallest coherent behavioral extension of $\underline{P}$ to $\mathscr{K}$, if it exists, is called the* natural extension *of $\underline{P}$ to $\mathscr{K}$, and is denoted by $\mathbf{E}_{\underline{P}}^{\mathscr{K}}$. By $\mathbf{E}_{\underline{P}}$ we denote $\mathbf{E}_{\underline{P}}^{\mathscr{R}(\Omega) \times \mathscr{P}^\circ(\Omega)}$.*

In the next theorem we define another function $\underline{E}$, which coincides with the natural extension exactly when $\underline{P}$ avoids sure loss. We can even make a stronger statement.

Theorem 5. *Let $\underline{P}$ be an extended conditional lower prevision, and let* $\operatorname{dom}\underline{P} \subseteq \mathscr{K} \subseteq \mathscr{R}(\Omega) \times \mathscr{P}^\circ(\Omega)$. *For any $f \in \mathscr{R}(\Omega)$ and $B \in \mathscr{P}^\circ(\Omega)$, define $\underline{E}(f|B)$ as the supremum value of $\alpha \in \mathbb{R}$ for which there are $n \in \mathbb{N}$, non-negative $\lambda_1, \ldots, \lambda_n \in \mathbb{R}$, $(f_0, B_0), \ldots, (f_n, B_n) \in \operatorname{dom}\underline{P}$, and $x_i < \underline{P}(f_i|B_i)$ such that*

$$\sup\left(\sum_{i=1}^{n} \lambda_i B_i (f_i - x_i) - B(f - \alpha) \,\middle|\, B \cup B_1 \cup \cdots \cup B_n \right) < 0 \tag{4}$$

The following conditions are equivalent.

(i) *$\underline{E}$ is a coherent extended conditional lower prevision on $\mathscr{R}(\Omega) \times \mathscr{P}^\circ(\Omega)$.*
(ii) *The natural extension of $\underline{P}$ to $\mathscr{K}$ exists and is equal to $\underline{E}$ restricted to $\mathscr{K}$.*
(iii) *$\underline{P}$ has at least one coherent behavioral extension.*
(iv) *$\underline{P}$ has at least one behavioral extension that avoids partial loss.*
(v) *$\underline{P}$ avoids partial loss.*

3.5 Duality

Let $\mathscr{P}(\Omega)$ denote the set of all coherent extended conditional previsions defined on $\mathscr{R}(\Omega) \times \mathscr{P}^\circ(\Omega)$. The set of coherent extended conditional previsions on $\mathscr{R}(\Omega) \times \mathscr{P}^\circ(\Omega)$ that dominate $\underline{P}$ is denoted by

$$\mathscr{M}(\underline{P}) := \{Q \in \mathscr{P}(\Omega) : \underline{P} \leq Q\} \tag{5}$$

The following result has a rather long proof. Suffice it to say that it can be proved using the same classical technique to prove the Hahn-Banach theorem, in particular, invoking Zorn's lemma on a set of intelligently constructed successive *finite* extensions—only increasing the domain by a finite number of elements. The reason why Zorn renders the existence of finite extensions sufficient for the existence of extensions to any domain is precisely because coherence itself involves only finite sums.

Theorem 6. *The following statements hold.*

(i) $\underline{P}$ avoids partial loss if and only if $\mathcal{M}(\underline{P}) \neq \emptyset$.
(ii) If $\underline{P}$ avoids partial loss, then for all $f \in \mathcal{R}(\Omega)$ and $B \in \wp(\Omega)$,

$$\mathbf{E}_{\underline{P}}(f|B) = \min\{Q(f|B) : Q \in \mathcal{M}(\underline{P})\}. \tag{6}$$

3.6 Equivalence of Bayes and Maximal Actions

For the sake of simplicity, in this section we shall only consider unconditional extended lower previsions: they are defined on subsets of $\mathcal{R}(\Omega) \times \{\Omega\}$, $\underline{P}(\cdot)$ denotes $\underline{P}(\cdot|\Omega)$, $\mathbf{E}_{\underline{P}}(\cdot)$ denotes $\mathbf{E}_{\underline{P}}(\cdot|\Omega)$, and $\overline{\mathbf{E}}_{\underline{P}}(\cdot)$ denotes $-\mathbf{E}_{\underline{P}}(-\cdot)$.

Within a set $\mathcal{K}$ of random quantities, $f \in \mathcal{K}$ is called *maximal* if it is undominated in $\mathcal{K}$ with respect to this partial order: $f > g$ if we wish to pay a strictly positive price to exchange f for g, that is, if $\mathbf{E}_{\underline{P}}(f - g) > 0$. Thus, f is maximal in $\mathcal{K}$ exactly when $\overline{\mathbf{E}}_{\underline{P}}(f - g) \geq 0$ for all $g \in \mathcal{K}$. Similarly, f is called a *Bayes action* if it maximizes Q over $\mathcal{K}$ for at least one Q in $\mathcal{M}(\underline{P})$.

A nice application of the above duality theorem is in decision making: if the set of options is convex, maximal actions are exactly the Bayes actions. This is well-known for coherent lower previsions on bounded random quantities (see for instance Walley [12, S3.9.5]). It holds for extended lower previsions as well.

Theorem 7. *Let $\underline{P}$ be an unconditional extended lower prevision that avoids partial loss. Let $\mathcal{K}$ be a convex subset of $\mathcal{R}(\Omega)$, and let f be a fixed element of $\mathcal{K}$. Then $\overline{\mathbf{E}}_{\underline{P}}(f - g) \geq 0$ for all $g \in \mathcal{K}$ if and only if there is a Q in $\mathcal{M}(\underline{P})$ such that $Q(f - g) \geq 0$ for all $g \in \mathcal{K}$.*

Proof. Define $\underline{R}$ on $\mathcal{R}(\Omega)$ as follows. Let $\mathcal{J}$ be the set

$$\mathcal{J} = \{f - g : g \in \mathcal{K} \text{ and } \mathbf{E}_{\underline{P}}(f - g) < 0\}.$$

Define $\underline{R}(h) = 0$, for all $h \in \mathcal{J}$, and let $\underline{R}(h) = \mathbf{E}_{\underline{P}}(h)$ for all $h \in \mathcal{R}(\Omega) \setminus \mathcal{J}$. Let's first show that $\underline{R}$ avoids partial loss. We must establish that

$$\sup\left[\sum_{i=1}^{n} \lambda_i(h_i - x_i) + \sum_{j=1}^{m} \mu_j(f - g_j - y_j)\right] \geq 0 \tag{7}$$

for all $n, m \in \mathbb{N}$, all non-negative $\lambda_1, \ldots, \lambda_n, \mu_1, \ldots, \mu_m \in \mathbb{R}$, all $h_1, \ldots, h_n \in \mathcal{R}(\Omega) \setminus \mathcal{J}$, all $x_i \in \mathbb{R}$, $x_i < \mathbf{E}_{\underline{P}}(h_i)$, all $g_1, \ldots, g_m \in \mathcal{K}$, and all $y_j \in \mathbb{R}$, $y_j < 0$.

Let $\mu = \sum_{j=1}^{m} \mu_j$. By Thm. 5, $\mathbf{E}_{\underline{P}}$ is coherent, so Equation (7) holds if $\mu = 0$. If $\mu > 0$, let $\alpha_j = \mu_j/\mu$, and rewrite Equation (7) as

$$\sup \left[\sum_{i=1}^{n} \lambda_i (h_i - x_i) - \mu \left(-f + \sum_{j=1}^{m} \alpha_j g_j - \left(-\sum_{j=1}^{m} \alpha_j y_j \right) \right) \right] \geq 0.$$

Again by the coherence of $\underline{\mathbf{E}}_P$ the above inequality must hold, because $\underline{\mathbf{E}}_P(-f + \sum_{j=1}^{m} \alpha_j g_j) = -\overline{\mathbf{E}}_{\underline{P}}(f - \sum_{j=1}^{m} \alpha_j g_j) \leq 0$ by assumption and convexity of $\mathcal{K}$, and $-\sum_{j=1}^{m} \alpha_j y_j > 0$.

Thus, $\underline{R}$ avoids partial loss. By Thm. 6, $\mathcal{M}(\underline{R})$ is non-empty. Let $Q \in \mathcal{M}(\underline{R})$. Then $Q(f - g) \geq \underline{R}(f - g) = 0$ for all $g \in \mathcal{K}$. Also, $Q(h) \geq \underline{R}(h) \geq \underline{\mathbf{E}}_{\underline{P}}(h) \geq \underline{P}(h)$ for all $h \in \mathrm{dom}\,\underline{P}$, and hence $Q \in \mathcal{M}(\underline{P})$.

Acknowledgments

I wish to thank the Belgian American Educational Foundation for funding the research reported in this paper. I'm also indebted to Teddy Seidenfeld for inspiring discussions and comments. Finally, I want to thank two anonymous referees for their useful suggestions.

References

[1] P. Artzner, F. Delbaen, J. M. Eber, and D. Heath. Coherent measures of risk. *Mathematical Finance*, 9(3):203–228, 1999.

[2] M. Chevé and R. Congar. Optimal pollution control under imprecise environmental risk and irreversibility. *Risk Decision and Policy*, 5:151–164, 2000.

[3] L. Crisma, P. Gigante, and P. Millossovich. A notion of coherent prevision for arbitrary random quantities. *Journal of the Italian Statistical Society*, 6(3):233–243, 1997.

[4] G. de Cooman. Possibility theory I–III. *International Journal of General Systems*, 25:291–371, 1997.

[5] B. de Finetti. *Theory of Probability: A Critical Introductory Treatment*. Wiley, New York, 1974–5. Two volumes.

[6] L. E. Dubins. Finitely additive conditional probabilities, conglomerability and disintegrations. *The Annals of Probability*, 3(1):89–99, 1975.

[7] D. Dubois and H. Prade. *Théorie des possibilité*. Masson, Paris, 1985.

[8] I. Levi. *The Enterprise of Knowledge. An Essay on Knowledge, Credal Probability, and Chance*. MIT Press, Cambridge, 1983.

[9] M. C. M. Troffaes and G. de Cooman. Extension of coherent lower previsions to unbounded random variables. In *Proceedings of the 9th International Conference IPMU 2002*, pages 735–42, Annecy, France, July 2002.

[10] M. C. M. Troffaes and G. de Cooman. Lower previsions for unbounded random variables. In Przemyslaw Grzegorzewski, Olgierd Hryniewicz, and Mariá Ángeles Gil, editors, *Soft Methods in Probability, Statistics and Data Analysis*, Advances in Soft Computing, pages 146–155, New York, September 2002. Physica-Verlag.

[11] L. V. Utkin and S. V. Gurov. Imprecise reliability models for the general life-time distribution classes. pages 333–342, Ghent, 1999. Imprecise Probabilities Project.

[12] P. Walley. *Statistical Reasoning with Imprecise Probabilities*. Chapman and Hall, London, 1991.

[13] P. M. Williams. Notes on conditional previsions. Technical report, School of Math. and Phys. Sci., Univ. of Sussex, 1975.

Extreme Lower Probabilities

Erik Quaeghebeur and Gert de Cooman

Ghent University, EESA Department, SYSTeMS Research Group
Technologiepark-Zwijnaarde 914, 9052 Zwijnaarde, Belgium
Erik.Quaeghebeur@UGent.be, Gert.deCooman@UGent.be

Summary. We consider lower probabilities on finite possibility spaces as models for the uncertainty about the state. These generalizations of classical probabilities can have some interesting properties; for example: k-monotonicity, avoiding sure loss, coherence, permutation invariance. The sets formed by all the lower probabilities satisfying zero or more of these properties are convex. We show how the extreme points and rays of these sets – the extreme lower probabilities – can be calculated and we give an illustration of our results.

Key words: Lower probabilities, extreme points, imprecise probabilities.

1 Introduction

We use and work on theories of imprecise probabilities. This means that we use concepts such as lower (and upper) probabilities to represent uncertainty. Calculating them often entails solving optimization problems. These can be hard, sometimes so hard that approximations seem the only option in practice. We are picky about the kind of approximation, however; it must be conservative. This means that the approximating lower (and upper) probabilities must be less precise than – or dominated by – the ones they approximate.

We were – and still are – aware of very few methods for making such conservative approximations. The useful ones are fewer still. One of the ideas for a new approximation approach is what led to the results communicated in this paper – which is *not* about approximations. The idea was, that perhaps we could write an arbitrary lower probability (that is hard to calculate directly) as a series of some special lower probabilities (that should be easier to calculate; breaking off the series would then constitute an approximation). The germ for this idea entered our minds when we read a paper by Maaß [10], where he mentions what in our terminology became *extreme lower probabilities*.

To get started, let us clear up some terminology and introduce the basic concepts, notation, and assumptions.

A *lower probability* $\underline{P}$ is a concept used in almost all the theories that make up the field of imprecise probabilities; it generalizes the classical concept of a probability

E. Quaeghebeur and G. de Cooman: *Extreme Lower Probabilities*, Advances in Soft Computing **6**, 211–221 (2006)

P. Important examples of such theories are the ones described by Dempster [6] and Shafer [13], Fine [7], Walley [15], and Weichselberger [16].

Alhough the definition and interpretation of a lower probability differ between the theories, the idea is similar. Like a probability, it is a real-valued set function defined on the set $\wp(\Omega)$ of all *events* (subsets A, B, C) of a *possibility space* Ω of *states* (elements ω).[1] We are uncertain about the state.

A probability P is (i) positive: for any event A, $\underline{P}(A) \geq 0$, (ii) normed: $P(\Omega) = 1$, and (iii) additive: for any disjunct events B and C, $P(B \cup C) = P(B) + P(C)$. Similarly, a lower probability $\underline{P}$ has to satisfy some properties, but these are weaker than those for probabilities. A lower probability is usually required to be (i) normed, and (iii) super-additive: for any disjunct events B and C, $\underline{P}(B \cup C) \geq \underline{P}(B) + \underline{P}(C)$. In this paper, we do not assume a priori that a lower probability satisfies any property (not even positivity).

A probability P *dominates* a lower probability $\underline{P}$ if $\underline{P}(A) \leq P(A)$ for all events A. A probability P is called *degenerate* when it is 1 on a singleton.

The set of all lower probabilities defined on some possibility space is *convex*. This is also the case for the set of all lower probabilities additionally satisfying some interesting properties. Any closed convex set is fully determined by its *extreme points* and *extreme rays*, and vice-versa [12, Thm 18.5]: all elements of the set can be written as a linear combination of (i) the extreme rays, and (ii) a convex combination of extreme points. The extreme points and extreme rays of a convex set of lower probabilities are its *extreme lower probabilities*.

If you can see that a triangle can be described by its three vertices, you have understood the main idea behind extreme lower probabilities. Of course, we will be talking about things that are a bit more complicated than triangles.

In this paper, we restrict ourselves to finite possibility spaces Ω, of cardinality $|\Omega| = n \in \mathbb{N}$.[2] We will look at sets of lower probabilities satisfying some interesting properties; for example 2-monotonicity: for any events B and C, $\underline{P}(B \cup C) + \underline{P}(B \cap C) \geq \underline{P}(B) + \underline{P}(C)$. It turns out that for finite possibility spaces, not surprisingly, the number of extreme points is finite. We show how the extreme points can be calculated for these cases and illustrate this.

The rest of this paper is structured as follows. In the next section, we will outline the approach we took to calculating extreme lower probabilities: calculating a set of constraints, and using these to compute extreme points. Then, we look at how we can obtain *manageable* sets of constraints for the properties that interest us. Finally, before concluding, we give an illustration of our results, with comments added.

---

[1] An *upper probability* $\overline{P}$ can be defined using its so-called conjugate lower probability $\underline{P}$: for any event A, $\overline{P}(A) = 1 - \underline{P}(\Omega \setminus A)$. Because $\wp(\Omega)$ is closed under complementation, we can – and do so here – work with lower probabilities only.

[2] Notation for number sets: $\mathbb{N}$, $\mathbb{Q}$, and $\mathbb{R}$ respectively denote the nonnegative integers, the rationals, and the reals. To denote common subsets of these, we use predicates as subscripts; e.g., $\mathbb{R}_{>0} = \{r \in \mathbb{R} | r > 0\}$ denotes the strictly positive reals.

2 On Constraints and Vertex Enumeration

A (lower) probability can satisfy a variety of interesting properties; most can be expressed using sets of linear constraints. (We do not consider things like independence and conditioning.) The introduction featured three simple examples: additivity, super-additivity, and 2-monotonicity. In general, these constraints are linear inequalities; equalities are expressed using two inequalities.

Sometimes, not all constraints are necessary: a constraint can be implied by another constraint, or a set of other constraints. A so-called *redundant* constraint can be removed from the set of constraints. If a constraint makes another one redundant, we call the former more *stringent* than the latter.

It is useful to look at this problem in a geometrical framework. Lower probabilities can be viewed as points in a 2^n-dimensional vector space – one dimension per subset in $\wp(\Omega)$. A linear inequality then generally corresponds to a half-space delimited by a hyperplane. With some property, there corresponds a set of half-spaces, and the intersection of these is the convex set of all lower probabilities satisfying that property.

The geometrical approach is illustrated in Fig. 1 below using a toy example. We consider $n = 1$, so the dimension of the vector space is 2 (this is actually the only one we can draw directly). The constraints are given using a set of hyperplanes $\{h_i | i = 1, \ldots, 6\}$, the 'hairs' indicate the half-spaces the constraints correspond to. Constraints h_3 and h_6 are redundant; the former because of h_1 and h_2, the latter because h_5 is more stringent. The set of points satisfying the constraints is filled.

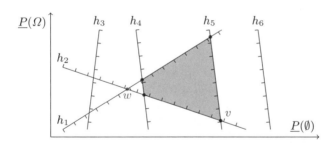

Fig. 1. Illustration of constraints and vertices.

Figure 1 also shows the vertices – v, for example – of the convex set defined by the set of half-spaces (constraints). These are the geometrical equivalent of what we call extreme lower probabilities.

In general, the *vertex enumeration problem* – finding the vertices corresponding to a given set of half-spaces – is hard: no polynomial time algorithm (in dimension, number of constraints, and number of vertices) is known [8]. To get a feeling for the complexity, realize that not only all intersections of nonredundant constraints have to be found, but we must also decide which of these to discard (such as w).

To obtain the set of constraints for different cardinalities and for the various properties we looked at, we have written our own program, constraints [11]. The properties in question will be described in the next section, where we will also give the theorems that form the basis of this program.

We have used publicly available programs (redund, lrs, and cdd) to remove redundant constraints and do vertex enumeration. They are maintained by Avis [1] and Fukuda [9].

3 From Properties to Constraints

Although for some properties – such as avoiding sure loss (see below) and additivity plus positivity plus normalization – it is possible to obtain the corresponding extreme points directly, we were not able to do this for the most interesting ones (k-monotonicity and coherence). For these, we use the vertex enumeration approach described in the previous section: generate the constraints corresponding to the property of interest and then use vertex enumeration to obtain the corresponding extreme lower probabilities. With this approach it is also easy to combine properties; one just has to combine the corresponding sets of constraints. The (big) downside is that it cannot be used in practice for 'large' possibility spaces – large here means $n \geq 5$.

We will now look at how we can obtain the constraints for some interesting properties. At this point we assume nothing about lower probabilities, not even that they are positive, normed, or super-additive.

Most of the results we mention in this section are either not hard to obtain or not entirely new. The most innovative part of this research was the combination of these results with vertex enumeration.

3.1 k-Monotonicity

In the theory of Dempster [6] and Shafer [13] lower probabilities are completely monotone. This is an extreme case of a mathematically interesting type of property: k-monotonicity, where $k \in \mathbb{N}_{>0}$.

A formal definition (adapted from De Cooman et al. [5]): a lower probability $\underline{P}$ is k-monotone if and only if for all $\ell = 1, \ldots, k-1$, any event A and any ℓ-tuple of events $(B_i | i \in \mathbb{N}_{<\ell})$, it holds that $\sum_{I \subseteq \mathbb{N}_{<\ell}} (-1)^{|I|} \underline{P}(A \cap \bigcap_{i \in I} B_i) \geq 0$, where the convention $\bigcap_{i \in \emptyset} B_i = \Omega$ is used. You can see that a k-monotone lower probability is also ℓ-monotone, for $\ell = 1, \ldots, k-1$. We have seen the case $k = 2$ in the introduction (in a different, but equivalent form).

Monotonicity is the same as 1-monotonicity: $\underline{P}(B) \leq \underline{P}(A)$ for any event A and all $B \subseteq A$. *Completely monotone* means k-monotone for all $k \in \mathbb{N}_{>0}$.

The above definition gives rise to a lot of redundant constraints. A lot of constraints are equivalent or are trivially satisfied. Removing them allowed us to formulate the following definition, leading to a more efficient program.

Theorem 1 (constraints for k-monotonicity). *A lower probability $\underline{P}$ is k-monotone if and only if it is monotone, and for all nonempty events A and all $\mathscr{A} \subseteq \wp(A) \setminus \{A, \emptyset\}$ such that*

(i) $0 < |\mathcal{A}| \leq k$,
(ii) $\bigcup_{B \in \mathcal{A}} B = A$, *and such that*
(iii) no event $C \in \mathcal{A}$ *exists for which* $C = \bigcap_{B \in \mathcal{A}} B$,

it holds that $\underline{P}(A) + \sum_{\mathcal{B} \subseteq \mathcal{A}} (-1)^{|\mathcal{B}|} \underline{P}(\bigcap_{B \in \mathcal{B}} B) \geq 0$.

We close this subsection with some remarks. It is a consequence of a result by Chateauneuf and Jaffray [2, Cor. 1] that k-monotonicity for some $k \geq n$ is equivalent to complete monotonicity. Note that $\underline{P}(\emptyset) = 0$ and $\underline{P}(\Omega) = 1$ are not included in k-monotonicity; they are commonly added, however.

3.2 Avoiding Sure Loss

Avoiding sure loss is a property that is useful in the behavioral theory of Walley [15, S2.4.1].[3] A lower probability $\underline{P}$ avoids sure loss if and only if for all $\mathcal{B} \subseteq \wp(\Omega)$, and all $\lambda \in (\mathbb{R}_{\geq 0})^{\mathcal{B}}$, it holds that $\sum_{B \in \mathcal{B}} \lambda_B \underline{P}(B) \leq \sup \sum_{B \in \mathcal{B}} \lambda_B I_B$, where I_B is the indicator function of B, which is 1 for $\omega \in B$ and 0 elsewhere.

If we only require a lower probability to avoid sure loss, the extreme lower probabilities can be determined by reasoning. Because a lower probability avoids sure loss if and only if it is dominated by a probability [15, S3.3.3], the set of extreme lower probabilities consists of the degenerate probabilities as extreme points and all negative main directions as extreme rays.

If we want to use vertex enumeration to obtain the extreme lower probabilities, the problem arises that the definition gives an infinite number of constraints (because λ is real-valued). This is of course unmanageable for any computer. This situation is inevitable when the usual assumption of positivity is made and the extreme lower probabilities cannot be determined by reasoning.

Luckily – by removing redundant constraints –, we can reduce the set of constraints for avoiding sure loss to a finite set. It was shown by Walley [15, SA.3] that in the definition, to get the most stringent constraints, we only need to consider (i) $\mathcal{B}$ such that $\{I_B | B \in \mathcal{B}\}$ is a linearly independent set and $\bigcup_{B \in \mathcal{B}} B = \Omega$, and (ii) λ such that the function on right hand side is the constant function 1.

Walley [15, SA.3] assumes positivity ($\underline{P}$ is in the first orthant). When we want to do the vertex enumeration approach without this assumption (as we do here), we need to add extra constraints for the cases where some (or all) of the components of $\underline{P}$ are negative. This can be done by taking every original constraint, and creating a new one for each of the 2^{2^n} possible orthants $\underline{P}$ can be located in. We do this by setting the λ_B for which $\underline{P}(B) < 0$ to 0.

The above and some other, minor, changes result in the following theorem.

Theorem 2 (constraints for avoiding sure loss). *A lower probability* $\underline{P}$ *avoids sure loss if and only if* $\underline{P}(\emptyset) \leq 0$ *and for all*

(i) $\mathcal{B}$ *such that* $\bigcup_{B \in \mathcal{B}} B = \Omega$ *and* $\{I_B | B \in \mathcal{B}\}$ *is a linearly independent set,*

---

[3] In Walley's theory [15], *lower previsions* – expectation operators – play a central role. Here, we restrict ourselves to the less general lower probabilities.

(ii) $\lambda \in (\mathbb{Q} \cap]0,1])^{\mathscr{B}}$ *such that* $\sum_{B \in \mathscr{B}} \lambda_B I_B = 1$, *and*
(iii) binary masks $\beta \in \{0,1\}^{\mathscr{B}}$,

it holds that $\sum_{B \in \mathscr{B}} \beta_B \lambda_B \underline{P}(B) \leq 1$.

3.3 Coherence

Coherent lower probabilities are at the core of Walley's theory [15, S2.6.4]. Weich-selberger [16] uses the term F-probability for the same concept. A lower proba-bility $\underline{P}$ is coherent if and only if for all events A, all $\mathscr{B} \subseteq \wp(\Omega) \setminus \{A\}$, and all $\lambda \in (\mathbb{R}_{\geq 0})^{\{A\} \cup \mathscr{B}}$, it holds that $\sum_{B \in \mathscr{B}} \lambda_B \underline{P}(B) - \lambda_A \underline{P}(A) \leq \sup(\sum_{B \in \mathscr{B}} \lambda_B I_B - \lambda_A I_A)$. Coherence implies avoiding sure loss and monotonicity, but not 2-monotonicity.

The above definition is similar enough to the one for avoiding sure loss to allow the same techniques for the removal of redundant constraints to be applied, up to some technicalities. Because coherence implies positivity, we need not use binary masks. Working this out results in the following theorem.

Theorem 3 (constraints for coherence). *A lower probability* $\underline{P}$ *is coherent if and only if* $\underline{P}(\emptyset) = 0$, $\underline{P}(\Omega) = 1$, *and*

(a) for all events A it holds that $0 \leq \underline{P}(A) \leq 1$;
(b) for all
  (i) events A and $\mathscr{B} \subseteq \wp(\Omega) \setminus \{A\}$ such that $\{I_B | B \in \mathscr{B}\}$ is a linearly inde-
       pendent set and $\bigcup_{B \in \mathscr{B}} B = A$, and
  (ii) $\lambda \in (\mathbb{Q} \cap]0,1])^{\mathscr{B}}$ such that $\sum_{B \in \mathscr{B}} \lambda_B I_B = I_A$,
  it must hold that $\sum_{B \in \mathscr{B}} \lambda_B \underline{P}(B) \leq \underline{P}(A)$;
(c) for all
  (i) events A and $\mathscr{B} \subseteq \wp(\Omega) \setminus \{A\}$ such that $\{I_A\} \cup \{I_B | B \in \mathscr{B}\}$ is a linearly
       independent set and $\bigcup_{B \in \mathscr{B}} B = \Omega$, and
  (ii) $\lambda \in (\mathbb{Q}_{>0})^{\{A\} \cup \mathscr{B}}$ such that $\sum_{B \in \mathscr{B}} \lambda_B I_B - \lambda_A I_A = 1$,
  it must hold that $\sum_{B \in \mathscr{B}} \lambda_B \underline{P}(B) - \lambda_A \underline{P}(A) \leq 1$.

3.4 Permutation Invariance

As a last property, we look at (weak) *permutation invariance* [4]. It is the odd duck of the lot; whereas all the previous properties allow for lower probabilities that express a quite broad a range of uncertainty models, permutation invariance restricts them to some very specific ones. We mention it to show how easy it can be to add the constraints for another property.

A lower probability is invariant under permutations of the elements of the pos-sibility space if and only if, for any event A and all events B resulting from some permutation, $\underline{P}(A) = \underline{P}(B)$. Let us give an example for $n = 3$: consider the permuta-tion $(1 \to 3, 2 \to 1, 3 \to 2)$, then $A = \{1,2\}$ becomes $B = \{1,3\}$.

We can characterize permutation invariance as follows [16, S4.3.1].

Theorem 4 (constraints for permutation invariance). *A lower probability* $\underline{P}$ *is permutation invariant if and only if for all $k = 1, \ldots, n-1$, any one event A such that $|A| = k$, and for all other events B with $|B| = k$, it holds that $\underline{P}(B) = \underline{P}(A)$.*

4 Results

We are not the first to hunt for extreme lower probabilities. However, as far as we know, we are as of yet the most systematic. ([11] contains a list of results.)

For $n = 4$ Shapley [14] gives a list with 37 of the 41 – he omits the 4 degenerate probabilities – 2-monotone extreme lower probabilities. For $n = 5$, we have found all 117983 for this case. For $n \leq 4$, we have found the extreme (permutation invariant) k-monotone lower probabilities for all k.

In an example, Maaß [10] mentions the 8 extreme coherent lower probabilities for $n = 3$. We give a graphical representation of them in Fig. 2 using their corresponding credal sets. Let us clarify: The set of all probabilities that dominate some lower probability is called its *credal set* (cfr. *core* in game theory [14]). All probabilities can be represented as a point of the $(n-1)$-dimensional unit simplex – a regular triangle for $n = 3$ – and so coherent lower probabilities can be represented by their credal set, which is a convex subset of this unit simplex [15, S3.6.1]. The vertices of the simplex correspond to the degenerate probabilities, so to the states – here a, b, and c.

Fig. 2. The credal sets of the 8 extreme coherent lower probabilities for $n = 3$.

In Fig. 2, we only give the border of the credal sets and show only one element of each permutation class. At the top right, we indicate the following: the total number of elements of the permutation class (e.g., $4\times$), if it is permutation invariant (↻), and k-monotonicity for $k \in \mathbb{N}_{>1}$ (∞ for complete monotonicity, 2 for 2-monotonicity). Along the simplex's left edge, we give a vector that is proportional to the extreme coherent lower probability (component order: $\emptyset$ $\{a\}\{b\}\{c\}$ $\{a,b\}\{a,c\}\{b,c\}$ Ω). Remember that $\underline{P}(\Omega) = 1$.

As convex combinations preserve monotonicity, we can immediately see from Fig. 2 that for $n = 3$, all coherent lower probabilities are 2-monotone. This was already known, but it illustrates how these computational results can help in finding theoretical results. (For $n = 2$, all are completely monotone.)

Once we implemented our program, finding all 402 extreme coherent lower probabilities for $n = 4$ was easy.[4] Figure 3 shows the corresponding credal sets, as well as those for the 16 extreme 3-monotone lower probabilities, in the same way we did for $n = 3$ in Fig. 2. The unit simplex is now a regular tetrahedron, so we had to use a projection from three dimensions to two.

---

[4] For $n = 5$, we know 1 743 093 of the extreme coherent lower probabilities. These have been found by a computer in our lab, after some months of vertex enumeration. A hardware failure cut this gargantuan task short.

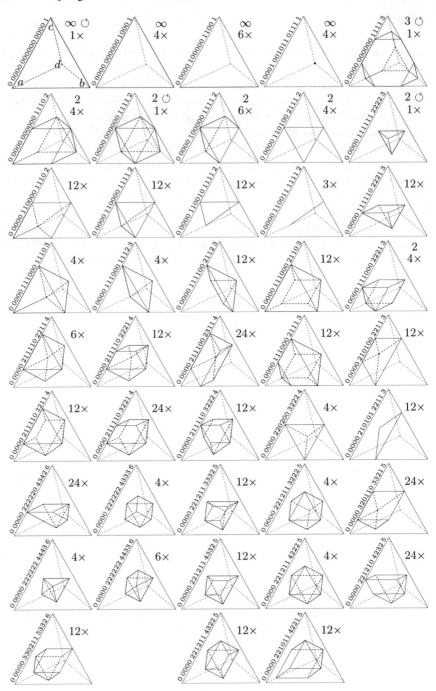

Fig. 3. The credal sets for $n = 4$ of the extreme coherent lower probabilities (all except last of top row) and extreme 3-monotone lower probabilities (top row).

Take $\Omega = \{a,b,c,d\}$. In this case, the component order is $\emptyset$ $\{a\}\{b\}\{c\}\{d\}$ $\{a,b\}\{a,c\}\{a,d\}\{b,c\}\{b,d\}\{c,d\}$ $\{a,b,c\}\{a,b,d\}\{a,c,d\}\{b,c,d\}$ Ω.

With Fig. 3 as a guide, we can give some observations and results.

Using results from Choquet [3, Ch. 7], it can be proven that the extreme completely monotone lower probabilities are the *vacuous* lower probabilities (cfr. *unanimity games* in game theory) with respect to events: $\underline{P}$ is vacuous with respect to A if $\underline{P}(B) = 1$ for $A \subseteq B$ and 0 otherwise.[5] These correspond to the first three (Fig. 2) and first four (Fig. 3) permutation classes shown. The last of these classes corresponds to the degenerate probabilities, which are all the extreme (classical) probabilities.

We have observed that the extreme completely monotone probabilities are always included in the extreme coherent proabilities. This is not the case for all of the extreme k-monotone and permutation invariant lower probabilities. An example for $n = 4$ is the only non-completely monotone lower probability of the extreme 3-monotone lower probabilities (shown in Fig. 3).

Notice that, except for the degenerate probabilities, all credal sets touch all tetrahedron faces. This is so because it can be shown that the degenerate probabilities are the only extreme coherent lower probabilities that can be nonzero in singletons.

5 Conclusions

Although we have not intensively pursued our initial goal – finding conservative approximations for lower and upper probabilities –, it did lead us to this interesting research. Obtaining sets of extreme lower probabilities for many cases and formulating a systematic approach to calculating them are the main results of the research presented in this paper.

Apart from these results, this topic also allows one to become familiar with different ways of looking at lower probabilities and their properties. A lower probability can be seen as a set function satisfying some properties, as a convex combination of some special set functions, as a point of a convex subset of a vector space, and – for coherent ones – as a credal set. They can be k-monotone, avoid sure loss, be coherent, be permutation invariant, etc.

And last but not least, this topic can lead to beautiful figures.

Acknowledgements

We wish to thank M. Troffaes, S. Maaß, A. Wallner, and E. Miranda for many a useful discussion and two referees for useful suggestions. These led to a much clearer view on the subject.

---

[5] In general (Ω can be infinite), the extreme completely monotone lower probabilities are those that take the value 1 on a proper filter of sets and are 0 elsewhere.

This paper presents research results of the Belgian Program on Interuniversity Attraction Poles, initiated by the Belgian Federal Science Policy Office. The scientific responsibility rests with its authors. Erik Quaeghebeur's research is financed by a Ph.D. grant from the Institute for the Promotion of Innovation through Science and Technology in Flanders (IWT-Vlaanderen).

References

[1] Avis, D. (2005). `lrs` homepage. `http://cgm.cs.mcgill.ca/~avis/C/lrs.html`.

[2] Chateauneuf, A. and Jaffray, J.-Y. (1989). Some characterizations of lower probabilities and other monotone capacities through the use of Möbius inversion. *Math. Social Sci.* **17**, 263–283.

[3] Choquet, G. (1954). Theory of Capacities. *Ann. Inst. Fourier* **5**, 131–295.

[4] De Cooman, G. and Miranda, E. (2006). Symmetry of models versus models of symmetry. In: Harper, W. L. and Wheeler, G. R. (Editors) *Probability and Inference: Essays in Honor of Henry E. Keyburg, Jr.* King's College Publications.

[5] De Cooman, G., Troffaes, M. and Miranda, E. (2005). *n*-Monotone lower previsions and lower integrals. In: Cozman, F. G., Nau, R. and Seidenfeld, T. (Editors) *ISIPTA '05: Proceedings of the Fourth International Symposium on Imprecise Probabilities and Their Applications*, pp. 145–154.

[6] Dempster, A. P. (1967). Upper and lower probabilities induced by a multivalued mapping. *Ann. Math. Stat.* **38**, 325–339.

[7] Fine, T. L. (1973). *Theories of probability: An examination of foundations.* Academic Press.

[8] Fukuda, K. (2004). Frequently Asked Questions in Polyhedral Computation. `http://www.ifor.math.ethz.ch/~fukuda/polyfaq/`.

[9] Fukuda, K. (2005). `cdd` and `cddplus` homepage. `http://www.ifor.math.ethz.ch/~fukuda/cdd_home/cdd.html`.

[10] Maaß, S. (2003). Continuous Linear Representations of Coherent Lower Previsions. In: Bernard, J.-M., Seidenfeld, T. and Zaffalon, M. (Editors) *ISIPTA '03: Proceedings of the Third International Symposium on Imprecise Probabilities and Their Applications*, pp. 372–382.

[11] Quaeghebeur, E. (2006). `constraints` homepage. `http://users.UGent.be/~equaeghe/constraints.php`.

[12] Rockafellar, R. T. (1970). *Convex Analysis.* Princeton Univ. Press.

[13] Shafer, G. (1976). *A mathematical theory of evidence.* Princeton Univ. Press.

[14] Shapley, L. S. (1971). Cores of Convex Games. *Int. J. Game Theory* **1**, 11–26.

[15] Walley, P. (1991). *Statistical Reasoning with Imprecise Probabilities.* Chapman and Hall.

[16] Weichselberger, K. (2001). *Elementare Grundbegriffe einer allgemeineren Wahrscheinlichkeitsrechnung,* vol. I: Intervallwarscheinlichkeit als umfassendes Konzept. Physica-Verlag.

Equivalence Between Bayesian and Credal Nets on an Updating Problem [*]

Alessandro Antonucci and Marco Zaffalon

Istituto Dalle Molle di Studi sull'Intelligenza Artificiale (IDSIA), Galleria 2, CH-6928
Manno (Lugano), Switzerland
{alessandro,zaffalon}@idsia.ch

We establish an intimate connection between Bayesian and credal nets. Bayesian nets
are precise graphical models, credal nets extend Bayesian nets to imprecise probability. We focus on traditional belief updating with credal nets, and on the kind of belief
updating that arises with Bayesian nets when the reason for the missingness of some
of the unobserved variables in the net is unknown. We show that the two updating
problems are formally the same.

1 Introduction

Imagine the following situation. You want to use a graphical model to formalize
your uncertainty about a domain. You prefer precise probabilistic models and so you
choose the *Bayesian network* (BN) formalism [5] (see Sect. 2.1). You take care to
precisely specify the graph and all the conditional mass functions required. At this
point you are done with the modelling phase, and start updating beliefs about a target
variable conditional on the observation of some variables in the net. The remaining
variables are not observed, i.e., they are *missing*. You know that some of the missing
variables are simply *missing at random* (MAR), and so they can easily be dealt
with by traditional approaches. Yet, there is a subset of missing variables for which
you do not know the process originating the missingness.

This innocuous-looking detail is going to change the very nature of your model:
while you think you are working with BNs, what you are actually using are credal
networks. *Credal networks* (CNs, see Sect. 2.2) are graphical models that generalize
Bayesian nets to sets of distributions [3], i.e., to *imprecise probability* [6].

The implicit passage from Bayesian to credal nets is based on two steps. First, the
above conditions, together with relatively weak assumptions, give rise to a specific
way to update beliefs called *conservative inference rule* (CIR, see Sect. 3) [7]. CIR
is an imprecise-probability rule: it leads, in general, to imprecise posterior probabilities for the target variable, even if the original model is precise. The second step is

---

[*] This research was partially supported by the Swiss NSF grant 200020-109295/1.

A. Antonucci and M. Zaffalon: *Equivalence Between Bayesian and Credal Nets on an Updating Problem*, Advances in Soft Computing **6**, 223–230 (2006)
www.springerlink.com

done in this paper: we show the formal equivalence between CIR-based updating in BNs, and traditional credal-network updating (see Sect. 4 and App. A) based on the popular notion of *strong independence* [3].

CIR and CNs have been proposed with quite different motivations in the literature: CIR as an updating rule for the case of partial ignorance about the missingness (or incompleteness) process; CNs as a way to relax the strict modelling requirements imposed by precise graphical models. The main interest in our result is just the established connection between two such seemingly different worlds. But the result appears also to be a basis to use algorithms for CNs to solve CIR-based updating problems.

2 Bayesian and Credal Networks

In this section we review the basics of Bayesian networks and their extension to convex sets of probabilities, i.e., credal networks. Both the models are based on a collection of random variables $\mathbf{X}$, which take values in finite sets, and a directed acyclic graph $\mathscr{G}$, whose nodes are associated to the variables of $\mathbf{X}$.

For both models, we assume the *Markov condition* to make $\mathscr{G}$ represent probabilistic independence relations between the variables in $\mathbf{X}$: every variable is independent of its non-descendant non-parents conditional on its parents. What makes BNs and CNs different is a different notion of independence and a different characterization of the conditional mass functions for each variable given the possible values of the parents, which will be detailed next.

Regarding notation, for each $X_i \in \mathbf{X}$, Ω_{X_i} is the possibility space of X_i, x_i a generic element of Ω_{X_i}, $P(X_i)$ a mass function for X_i and $P(x_i)$ the probability that $X_i = x_i$. A similar notation with uppercase subscripts (e.g., X_E) denotes arrays (and sets) of variables in $\mathbf{X}$. The parents of X_i, according to $\mathscr{G}$, are denoted by Π_i and for each $\pi_i \in \Omega_{\Pi_i}$, $P(X_i|\pi_i)$ is the conditional mass function for X_i given the joint value π_i of the parents of X_i.

2.1 Bayesian Networks

In the case of Bayesian networks, the modelling phase involves specifying a conditional mass function $P(X_i|\pi_i)$ for each $X_i \in \mathbf{X}$ and $\pi_i \in \Omega_{\Pi_i}$; and the standard notion of probabilistic independence is assumed in the Markov condition. A BN can therefore be regarded as a joint probability mass function over $\mathbf{X} \equiv (X_1, \ldots, X_n)$, that factorizes as follows: $P(\mathbf{x}) = \prod_{i=1}^{n} P(x_i|\pi_i)$, for each $\mathbf{x} \in \Omega_{\mathbf{X}}$, because of the Markov condition. The updated belief about a queried variable X_q, given some evidence $X_E = x_E$, is:

$$P(x_q|x_E) = \frac{\sum_{x_M} \prod_{i=1}^{n} P(x_i|\pi_i)}{\sum_{x_M, x_q} \prod_{i=1}^{n} P(x_i|\pi_i)}, \tag{1}$$

where $X_M \equiv \mathbf{X} \setminus (\{X_q\} \cup X_E)$, the domains of the arguments of the sums are left implicit and the values of x_i and π_i are consistent with (x_q, x_M, x_E). Despite its hardness in general, Eq. (1) can be efficiently solved for polytree-shaped BNs with standard propagation schemes based on local computations and message propagation [5]. Similar techniques apply also for general topologies with increased computational time.

2.2 Credal Networks

CNs relax BNs by allowing for imprecise probability statements. There are many kinds of CNs. We stick to those consistent with the following:[1]

Definition 1. *Consider a finite set of BNs with the same graph $\mathcal{G}$, over the same variables $\mathbf{X}$, i.e., a pair $\langle \mathcal{G}, \mathbf{P}(\mathbf{X}) \rangle$, where $\mathbf{P}(\mathbf{X})$ is the array of the joint mass functions associated to the set of BNs. Define a* credal network *as the convex hull of such mass functions: i.e., $K(\mathbf{X}) \equiv \mathrm{CH}\{\tilde{P}(\mathbf{X})\}_{\tilde{P} \in \mathbf{P}}$.*

We define a *credal set* as the convex hull of a collection of mass functions over a vector of variables. In this paper we assume this collection to be finite; therefore a credal set can be geometrically regarded as a *polytope*. Such a credal set contains an infinite number of mass functions, but only a finite number of *extreme mass functions*: those corresponding to the *vertices* of the polytope. It is possible to show that inference based on a credal set is equivalent to that based only on its vertices [6]. Clearly $K(\mathbf{X})$ is a credal set over $\mathbf{X}$ [we similarly denote by $K(X)$ a credal set over X]. The vertices of $K(\mathbf{X})$ are generally a subset of the original set of BNs and the CN is said equivalent to this *finite set* of BNs.

Note that $K(\mathbf{X})$ in Def. 1 is not specified via local pieces of probabilistic information, and so we say that the corresponding CN is *globally specified*. When the construction of $K(\mathbf{X})$ emphasizes locality, we talk of *locally specified* CNs. We can specify CNs locally in two ways. In the first, each probability mass function $P(X_i|\pi_i)$ is defined to belong to a finite set of mass functions [whose convex hull $K(X_i|\pi_i)$ is a credal set by definition, which is said to be local]. We talk of *separately specified* credal nets in this case. In the second, the generic probability table $P(X_i|\Pi_i)$, i.e., a function of both X_i and Π_i, is defined to belong to a finite set of tables, denoted by $K(X_i|\Pi_i)$. In this case we talk of *extensive* specification. In both cases, the multiplicity of local mass functions or tables gives rise to a multiplicity of joint mass functions over $\mathbf{X}$ by simply taking all the combinations of the local pieces of knowledge. Such joint mass functions are just those making up the finite set of BNs in Def. 1.

Belief updating with CNs is defined as the computation of the posterior credal set for a queried variable X_q, conditionally on evidence about some other variables X_E:

$$K(X_q|x_E) \equiv \mathrm{CH}\left\{\tilde{P}(X_q|x_E)\right\}_{\tilde{P} \in \mathbf{P}}. \tag{2}$$

---

[1] By Def. 1 we are implicitly assuming the notion of *strong independence* in the Markov condition for CNs, see [3]. $K(\mathbf{X})$ is usually called the *strong extension*.

3 Conservative Inference Rule

The most popular approach to missing data in the literature and in the statistical practice is based on the so-called *missing-at-random* assumption. MAR allows missing data to be neglected, thus turning the incomplete data problem into one of complete data. Unfortunately, MAR embodies the idea that the process responsible for the missingness (i.e., the *missingness process*) is not selective, which is not realistic in many cases. De Cooman and Zaffalon have developed an inference rule based on much weaker assumptions than MAR, which deals with near-ignorance about the missingness process [4]. This result has been expanded by Zaffalon [7] to the case of mixed knowledge about the missingness process: for some variables the process is assumed to be nearly unknown, while it is assumed to be MAR for the others. The resulting updating rule is called *conservative inference rule* (CIR).

To show how CIR-based updating works, we partition the variables in $\mathbf{X}$ in four classes: (i) the queried variable X_q, (ii) the observed variables X_E, (iii) the unobserved MAR variables X_M, and (iv) the variables X_I made missing by a process that we basically ignore. CIR leads to the following credal set as our updated beliefs about the queried variable:

$$K(X_q||^{X_I}x_E) \equiv \mathrm{CH}\left\{P(X_q|x_E,x_I)\right\}_{x_I \in \Omega_{X_I}}, \qquad (3)$$

where the superscript on the double conditioning bar is used to denote beliefs updated with CIR and to specify the set of missing variables X_I assumed to be non-MAR, and clearly $P(X_q|x_E,x_I) = \sum_{x_M} P(X_q,x_M|x_E,x_I)$.

4 Equivalence Between CIR-Based Updating in Bayesian Nets and Credal Nets Updating

In this section we prove the formal equivalence between updating with CIR on BNs and standard updating on CNs, defining two distinct mappings from a generic instance of the first problem in a corresponding instance of the second (see Sect. 4.1) and *vice versa* (see Sect. 4.2). Fig. 1 reports the correspondence scheme with the names of the mappings that will be introduced next. According to CIR assumptions [7], we focus on the case of BNs assigning positive probability to each event.

4.1 From Bayesian to Credal Networks

First let us define the B2C transformation, mapping a BN $\langle \mathscr{G}, P(\mathbf{X}) \rangle$, where a subset X_I of $\mathbf{X}$ is specified, in a CN. For each variable $X \in X_I$, B2C prescribes to: (i) add to X an *auxiliary child node*[2] X', associated to a binary variable with possible values x' and $\neg x'$; and (ii) extensively specify the probability table $P(X'|X)$, to belong to the following set of $|\Omega_X|$ tables:

---

[2] This transformation takes inspirations from Pearl's prescriptions about boundary conditions for propagation [5, Sect. 4.3].

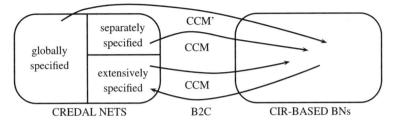

Fig. 1. Relations between updating on CNs and CIR-updating in BNs

$$\left\{\begin{bmatrix} 1\,0\,0\ldots0 \\ 0\,1\,1\ldots1 \end{bmatrix}, \ldots, \begin{bmatrix} 0\ldots010\ldots0 \\ 1\ldots101\ldots1 \end{bmatrix}, \ldots, \begin{bmatrix} 0\,0\,0\ldots01 \\ 1\,1\,1\ldots10 \end{bmatrix}\right\}. \tag{4}$$

Each table in Eq. (4) specifies a conditional probability for the state x' of X' (corresponding to the first row of the table), which is zero conditionally on any state of X except a single one, different for any table. The B2C transformation, clearly linear in the input size, is the basis for the following:

Theorem 1. *Consider a CIR instance on a Bayesian network $\langle \mathcal{G}, P(\mathbf{X}) \rangle$. Let X_I be the array of the unobserved non-MAR variables. Let $K(X_q||^{X_I}x_E)$ be the credal set returned by CIR for a queried variable X_q given the evidence $X_E = x_E$. If $K(X_q|x_E, x'_I)$ is the posterior credal set for X_q in the credal net $\langle \mathcal{G}', P(\mathbf{X}') \rangle$ obtained from $\langle \mathcal{G}, P(\mathbf{X}) \rangle$ by a B2C transformation with the nodes X_I specified, conditional on the evidences $X_E = x_E$ and $X'_I = x'_I$, then:* [3]*

$$K(X_q||^{X_I}x_E) = K(X_q|x_E, x'_I). \tag{5}$$

4.2 From Credal to Bayesian Networks

For globally specified CNs we define a transformation that returns a BN given a CN $\langle \mathcal{G}, P(\mathbf{X}) \rangle$ as follows. The BN is obtained: (i) adding a *transparent* node X'' that is parent of all the nodes in $\mathbf{X}$ (see Fig. 2 left) and such that there is a one-to-one correspondence between the elements of $\Omega_{X''}$ and those of $\mathbf{P}$; and (ii) setting for each $X_i \in \mathbf{X}$ and $x'' \in \Omega_{X''}$: $P(X_i|\Pi_i, x'') \equiv \tilde{P}(X_i|\Pi_i)$, where Π_i are the parents of X_i in the CN and $\tilde{P}$ is the element of $\mathbf{P}$ corresponding to x''.

In the case of locally specified CNs, we consider a slightly different transformation, where: (i) we add a transparent node X''_i for each $X_i \in \mathbf{X}$, that is parent only of X_i (see Fig. 2 right) and such that there is a one-to-one correspondence between the elements of $\Omega_{X''_i}$ and the probability tables $P(X_i|\Pi_i)$ in the extensive[4] specification of $K(X_i|\Pi_i)$; and (ii) we set for each $X_i \in \mathbf{X}$: $P(X_i|\Pi_i, x''_i) \equiv \tilde{P}(X_i|\Pi_i)$, where Π_i are

---

[3] Th. 1 can be extended also to CIR instances modeling incomplete observations where the value of the observed variable is know to belong to a generic subset of the possibility space, rather than missing observations for which the universal space is considered. We skip this case for lack of space.

[4] Separately specified credal sets can be extensively specified, considering all the probability tables obtained from the combinations of the vertices of the original credal sets. Although

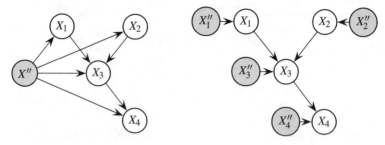

Fig. 2. The Bayesian networks returned by CCM' (left) and CCM (right). Transparent nodes are gray, while the nodes of the original CN are white

the parents of X_i in the CN and $\tilde{P}(X_i|\Pi_i)$ is the probability table of $K(X_i|\Pi_i)$ relative to x_i''. Note that no prescriptions are given about the unconditional mass functions for the transparent nodes in both the transformations, because irrelevant for the results we will obtain. The second is the so-called CCM transformation [1] for CNs, while the first is simply an extension of CCM to the case of globally specified CNs and will be denoted as CCM'. These transformations are the basis for the following:

Theorem 2. *Let $K(X_q|x_E)$ be the posterior credal set of a queried variable X_q, given some evidence $X_E = x_E$, for a CN $\langle \mathscr{G}, \mathbf{P}(\mathbf{X})\rangle$. Let also $\langle \mathscr{G}', P'(\mathbf{X}')\rangle$ be the BN obtained from $\langle \mathscr{G}, \mathbf{P}(\mathbf{X})\rangle$ through CCM' (or CCM if the CN is not globally specified). Denote as $K(X_q||^{X''}x_E)$ the CIR-based posterior credal set for X_q in the BN obtained assuming what follows: the nodes in X_E instantiated to the values x_E, the transparent nodes, denoted as X'' also if CCM is used, to be not-MAR and the remaining nodes MAR. Then:*

$$K(X_q|x_E) = K(X_q||^{X''}x_E). \tag{6}$$

5 Conclusions and Outlook

We have proved the formal equivalence between two updating problems on different graphical models: CIR-based updating on BNs and traditional updating with CNs. The result follows easily via simple transformations of the graphical models. An important consequence of the established link between BNs and CNs is that under realistic conditions of partial ignorance about the missingness process, working with BNs is actually equivalent to working with CNs. This appears to make CNs even more worthy of investigation than before.

The result makes it also possible in principle to solve CIR-based updating on BNs, for which there are no algorithms at presents, by means of algorithms for CNs.

---

correct, this transformation gives rise to an exponential explosion of the number of tables. An alternative transformation, described in [2], might avoid this problem, as suggested by a reviewer.

Unfortunately, the main corpus of algorithms for CNs considers the case of separately specified CNs, while CIR problems on BNs correspond to extensively specified CNs (see Fig. 1). Future work should therefore involve developing generalizations of the existing algorithms for CNs to the extensive case.

A Proofs of the Theorems

Proof of Theorem 1 *According to Eq. (3) and Eq. (2) respectively, we have:*

$$K(X_q||^{X_I}x_E) = \mathrm{CH}\{P(X_q|x_E,\tilde{x}_I)\}_{\tilde{x}_I \in \Omega_{X_I}} \tag{7}$$

$$K(X_q|x_E,x_I') = \mathrm{CH}\{\tilde{P}(X_q|x_E,x_I')\}_{\tilde{P} \in \mathbf{P}}. \tag{8}$$

An obvious isomorphism holds between $\mathbf{P}$ *and* Ω_{X_I}*: that follows from the correspondence, for each* $X_i \in X_I$*, between the conditional probability tables for* $P(X_i'|X_i)$ *as in Eq. (4) and the elements of* Ω_{X_i}*. Accordingly, we denote by* $\tilde{x}_I$ *the element of* Ω_{X_I} *corresponding to* $\tilde{P} \in \mathbf{P}$*. The thesis will be proved by showing, for each* $\tilde{P} \in \mathbf{P}$*,* $\tilde{P}(X_q|x_E) = P(X_q|x_E,\tilde{x}_I)$*. For each* $x_q \in \Omega_{X_q}$*:*

$$P(x_q|x_E,\tilde{x}_I) = \sum_{x_M} P(x_q,x_M|x_E,\tilde{x}_I) \propto \sum_{x_M} P(x_q,x_M,x_E,\tilde{x}_I) \tag{9}$$

$$\tilde{P}(x_q|x_E,x_I') = \sum_{x_M,x_I} \tilde{P}(x_q,x_M,x_I|x_E,x_I') \propto \sum_{x_M,x_I} \tilde{P}(x_q,x_M,x_I,x_E,x_I'). \tag{10}$$

According to the Markov condition:

$$\tilde{P}(x_q,x_M,x_I,x_E,x_I') = \prod_{i:X_i \in X_I} \left[\tilde{P}(x_i'|x_i) \cdot \tilde{P}(x_i|\pi_i)\right] \cdot \prod_{j:X_j \in \mathbf{X}' \setminus (X_I \cup X_I')} \tilde{P}(x_j|\pi_j), \tag{11}$$

with the values of x_i'*,* x_i*,* π_i*,* x_j *and* π_j *consistent with* (x_q,x_M,x_E,x_I,x_I')*.*

According to Eq. (4), $P(x_i'|x_i)$ *is zero for each* $x_i \in \Omega_{X_i}$ *except for the value* $\tilde{x}_i$*, for which is one. The sum over* $x_i \in \Omega_{X_i}$ *of the probabilities in Eq. (11) is therefore reduced to a single non-zero term. Thus, taking all the sums over* X_i *with* $X_i \in X_I$*:*

$$\sum_{x_I} \tilde{P}(x_q,x_M,x_I,x_E,x_I') = \prod_{i:X_i \in X_I} P(\tilde{x}_i|\pi_i) \cdot \prod_{j:X_j \in \mathbf{X} \setminus X_I} P(x_j|\pi_j) = P(x_q,x_M,x_E,\tilde{x}_I), \tag{12}$$

with the values of π_i*,* x_j *and* π_j *consistent with* $(x_q,x_M,x_E,\tilde{x}_I)$*. But Eq. (12) allows us to rewrite Eq. (9) as Eq. (10) and conclude the thesis.* $\square$

Proof of Theorem 2 *Consider a globally specified CN, for which CCM' should be used and* X'' *denotes a single transparent node. According to Eq. (3):*

$$K(X_q||^{X''}x_E) = \mathrm{CH}\{P(X_q|x_E,x'')\}_{x'' \in \Omega_{X''}}. \tag{13}$$

Setting $X_M \equiv \mathbf{X} \setminus (X_E \cup \{X_q\})$, for each $x_q \in \Omega_{X_q}$:

$$P(x_q|x_E,x'') = \sum_{x_M} P(x_q,x_M|x_E,x'') \propto \sum_{x_M} P(x_q,x_M,x_E,x''). \qquad (14)$$

According to the Markov condition and CCM' definition, we have:

$$P(x_q,x_M,x_E,x'') = P(x'') \cdot \prod_{i=1}^{n} P(x_i|\pi_i,x'') \propto \prod_{i=1}^{n} \tilde{P}(x_i|\pi_i) = \tilde{P}(x_q,x_M,x_E), \qquad (15)$$

where $\tilde{P}$ is the element of $\mathbf{P}$ associated to $x'' \in \Omega_{X''}$. The sum over x_M of the probabilities in Eq. (15) is proportional to $\tilde{P}(x_q|x_E)$. Thus, $\tilde{P}(X_q|x_E) = P(X_q|x_E,x'')$ for each $(\tilde{P},x'') \in \mathbf{P} \times \Omega_{X''}$, that proves the thesis. Analogous considerations can be done for locally defined CNs transformed by CCM. $\square$

References

[1] A. Cano, J. Cano, and S. Moral. Convex sets of probabilities propagation by simulated annealing on a tree of cliques. In *Proceedings of the Fifth International Conference (IPMU '94)*, pages 978–983, Paris, 1994.

[2] A. Cano and S. Moral. Using probability trees to compute marginals with imprecise probabilities. *Int. J. Approx. Reasoning*, 29(1):1–46, 2002.

[3] F.G. Cozman. Graphical models for imprecise probabilities. *Int. J. Approx. Reasoning*, 39(2-3):167–184, 2005.

[4] G. de Cooman and M. Zaffalon. Updating beliefs with incomplete observations. *Artificial Intelligence*, 159:75–125, 2004.

[5] J. Pearl. *Probabilistic Reasoning in Intelligent Systems: Networks of Plausible Inference*. Morgan Kaufmann, San Mateo, 1988.

[6] P. Walley. *Statistical Reasoning with Imprecise Probabilities*. Chapman and Hall, New York, 1991.

[7] M. Zaffalon. Conservative rules for predictive inference with incomplete data. In F.G. Cozman, R. Nau, and T. Seidenfeld, editors, *Proceedings of the Fourth International Symposium on Imprecise Probabilities and Their Applications (ISIPTA '05)*, pages 406–415, Pittsburgh, 2005.

Varying Parameter in Classification Based on Imprecise Probabilities*

Joaquín Abellán, Serafín Moral, Manuel Gómez and Andrés Masegosa

Department of Computer Science and Artificial Intelligence, University of Granada, 18071 Granada - Spain
{jabellan,smc,mgomez,andrew}@decsai.ugr.es

Summary. We shall present a first explorative study of the variation of the parameter s of the imprecise Dirichlet model when it is used to build classification trees. In the method to build classification trees we use uncertainty measures on closed and convex sets of probability distributions, otherwise known as credal sets. We will use the imprecise Dirichlet model to obtain a credal set from a sample, where the set of probabilities obtained depends on s. According to the characteristics of the dataset used, we will see that the results can be improved varying the values of s.

1 Introduction

The problem of classification (an important problem in the field of *machine learning*) may generally be defined in the following way: we have a set of observations, called the training set, and we wish to obtain a set of laws in order to assign a value of the variable to be classified (also called class variable) to each new observation. The set used to verify the quality of this set of laws is called the test set. Classification has important applications in medicine, character recognition, astronomy, banking, etc. A classifier may be represented using a Bayesian network, a neural network, a classification tree, etc. Normally, these methods use the probability theory in order to estimate the parameters with a stopping criterion in order to limit the complexity of the classifier.

Our starting point shall be a classification procedure based on the use of classification trees, which use closed and convex sets of probability distributions, otherwise known as credal sets, and uncertainty measures on these. A classification tree is a structure that is easy to understand and an efficient classifier. It has its origin in Quinlan's ID3 algorithm [18]. As a basic reference, we should mention the book by Breiman et al. [11]. We shall apply decision trees for classification and we shall use the imprecise Dirichlet model (IDM) [22] in order to estimate the probabilities of membership to the respective classes defined by the variable to be classified.

* This work has been supported by the Spanish Ministry of Science and Technology under the Algra project (TIN2004-06204-C03-02).

J. Abellán et al.: *Varying Parameter in Classification Based on Imprecise Probabilities*, Advances in Soft Computing **6**, 231–239 (2006)
www.springerlink.com

In Abellán and Moral [2, 4], we studied how to quantify the uncertainty of a set of probabilities by extending the measures which are used in the theory of evidence, Dempster [12] and Shafer [20]. We shall consider two main origins of uncertainty: conflict and non-specificity. In Abellán and Moral [3, 4] and Abellán, Klir and Moral [8], we present functions which verify the fundamental properties associated with these types of functions (Dubois and Prade [13], Klir and Wierman [17]).

In Abellán and Moral [7] an extended method is defined by modifying the tree construction method used in Abellán and Moral [5]. In this work, the relationships of a single variable in the dataset were sought with the variable to be classified. The variable which most reduces the uncertainty of the classification was introduced. If no variable reduces the uncertainty, it stops. In Abellán and Moral [7] instead of searching for the relationships of a single variable, we also consider how each pair of variables in the dataset affects the variable to be classified, introducing the variable which either individually or jointly with another most reduces the uncertainty. In this way, we search for more complex relationships which only come to light when a study is made of how two variables jointly affect the variable to be classified, but are not revealed by either of the two variables separately. Another variation introduced in Abellán and Moral [7] is the use of maximum entropy of a credal set as a measure of total uncertainty. A conflict measure favors branching and with a non-specificity measure the complexity of the model is limited. With this function of total uncertainty we have obtained the best results. In recent results (Abellán, Klir and Moral [8]) we have managed to separate maximum entropy into components which coherently quantify conflict and non-specificity.

In this paper, we carry out a series of experiments to analyze the success of the method exposed in Abellán and Moral [7] varying the parameter s of the IDM. We will apply different values of s for the method on several known data sets and we will see that the variation of success depending of s is different for each data set.

In Section 2 of this article, we shall present necessary notations and definitions for understanding the classification method used. In Section 3, we shall describe the method in detail. In Section 4, we shall check our procedure varying parameters with a series of datasets which are widely used in classification.

2 Notations and Prior Knowledge

The study of the uncertainty measures in the Dempster-Shafer theory is the starting point for the study of these measures in the context of more general theories. In any of these theories, it is justifiable that a measure capable of measuring the uncertainty represented by a credal set must quantify the parts of conflict and non-specificity ([3, 4, 15, 17]).

Recently, in Abellán and Moral [7] and Klir and Smith [16], the authors justified the use of maximum entropy on credal sets as a good measure of total uncertainty. The problem lies in separating these functions into others which really do measure the conflict and non-specificity parts represented by the use of a credal set for representing the information. More recently, in Abellán, Klir and Moral [8], they managed to split maximum entropy into functions capable of coherently measuring the conflict and non-specificity of a credal set $\mathscr{P}$, and they also propose algorithms in order to facilitate their calculation in order-2 capacities, Abellán and Moral [6, 9].

In any classification problem, we must consider that we have a dataset $\mathscr{D}$ with the values of a set $\mathscr{L}$ of discrete or discretized variables $\{X_i | i = 1, \ldots, \eta\}$, also called attribute variables. Each variable has as its states, a set of cases belonging to a finite set $\Omega_{X_i} = \{x_i^1, x_i^2, \ldots, x_i^{|\Omega_{X_i}|}\}$. Our aim will be to create a classification tree from the data $\mathscr{D}$, of the variable to be classified C, with states in the set $\Omega_C = \{c_1, c_2, \ldots, c_k\}$.

Definition 1. *Let* $\{X_i | i = 1, \ldots, \eta\}$ *be a set of discrete variables with states in the finite sets* Ω_{X_i}, *respectively. We shall call any* m-*tuple a configuration of* $\{X_1, \ldots, X_\eta\}$: $(X_{h_1} = x_{h_1}^{t_{h_1}}, X_{h_2} = x_{h_2}^{t_{h_2}}, \ldots, X_{h_m} = x_{h_m}^{t_{h_m}})$, *where* $x_{h_j}^{t_{h_j}} \in \Omega_{X_{h_j}}$, $j \in \{1, \ldots, m\}$, $h_j \in \{1, \ldots, \eta\}$ *and* $h_j \neq h_v$ *with* $j \neq v$. *In other words, a configuration is a set of values of the variables of* $\{X_1, \ldots, X_\eta\}$.

Definition 2. *Given a dataset and a configuration* σ *of the set* $\{X_1, \ldots, X_\eta\}$, *we shall consider a credal set* $\mathscr{P}_C^\sigma$ *for a variable* C *in relation to* σ *defined by the set of probability distributions,* p, *such that* $p_j = p(\{C = c_j\}) \in \left[\frac{n_{c_j}^\sigma}{N+s}, \frac{n_{c_j}^\sigma + s}{N+s} \right]$, *for each* $j \in \{1, \ldots, k\}$, *obtained on the basis of the imprecise Dirichlet model, Walley [22]. Here* $n_{c_j}^\sigma$ *is the number of occurrences of the configuration* $\{C = c_j\} \cap \sigma$ *in the dataset,* N *is the number of observations compatible with the configuration* σ *and* $s > 0$ *is a hyperparameter. We shall denote this interval as* $\left[\underline{P}(c_j | \sigma), \overline{P}(c_j | \sigma) \right]$.

Noting $^s\mathscr{P}$ as the credal set associated with a set of fixed frequencies $\{n_{c_j}, j = 1, \ldots, \eta\}$ for a s value, it can be checked that $s_1 \leq s_2 \Longleftrightarrow {}^{s_1}\mathscr{P} \subseteq {}^{s_2}\mathscr{P}$. The value s represents the strength of the prior ignorance about the probabilities of each state c_j and determines how fast the lower and upper probabilities converge as more data are taken (a greater value of s produces more cautious inferences). For the IDM, Walley [22] suggests a value for s between $s = 1$ and $s = 2$. In Bernard [10] we can find arguments to use values $s > 1$ when $k > 2$.

In the literature, when Dirichlet prior is used in a precise Bayesian approach, different values for s have been proposed in base of some principles: $s \longrightarrow 0$, $s = 1$, $s = \frac{k}{2}$, $s = k$ (see [10]). When one of the last two values are used in the IDM, it does not satisfy the *representation invariance principle* [22], but here we want to study the performance at the light of the results in an empirical study, even if this principle is not verified.

3 The Classification Method

The method starts with a tree with a single node. The procedure shall be described as a recursive algorithm [6], which is started with the root node with no label associated to it. Each node will have a list $\mathscr{L}^*$ of possible labels of variables which can be associated to it. The procedure will initially be started with the complete list of variables.

We will consider that we have two functions implemented: $\text{Inf1}(\sigma, X_i)$ and $\text{Inf2}(\sigma, X_i, X_j)$, computing respectively the values:

$$\text{Inf1}(\sigma, X_i) = \left(\sum_{x_i \in U_i} r_{x_i}^{\sigma} TU(\mathscr{P}^{\sigma \cap (X_i = x_i)}) \right),$$

$$\text{Inf2}(\sigma, X_i, X_j) = \left(\sum_{x_i \in U_i, x_j \in U_j} r_{x_i, x_j}^{\sigma} TU(\mathscr{P}^{\sigma \cap (X_i = x_i, X_j = x_j)}) \right),$$

where $r_{x_i}^{\sigma}$ is the relative frequency with which X_i takes value x_i in $\mathscr{D}[\sigma]$ (the part of the dataset compatible with σ), r_{x_i, x_j}^{σ} is the relative frequency with which X_i and X_j take values x_i and x_j, respectively, in $\mathscr{D}[\sigma]$, $\sigma \cap (X_i = x_i)$ is the result of adding the value $X_i = x_i$ to configuration σ (analogously for $\sigma \cap (X_i = x_i, X_j = x_j)$), and TU is any total uncertainty measure on credal sets.

If No is a node and σ a configuration associated with it, Inf1 tries to measure the weighted average total uncertainty of the credal sets associated with the children of this node if variable X_i is added to it (and there is a child for each one of the possible values of this node). The average is weighted by the relative frequency of each one of the children in the data set. Inf2 is similar, but considers adding two variables in one step: assigning X_i to the first node and then assigning X_j to all the children of the first node. It measures the average of the total uncertainty of the credal sets associated to the grandchildren (the result of this function does not depend on the order).

The basic idea is very simple and it is applied recursively to each one of the nodes we obtain. For each one of these nodes, we consider whether the total uncertainty of the credal set at this node can be decreased by adding one or two nodes. If this is the case, then we add a node with a maximum decrease of uncertainty. If the uncertainty cannot be decreased, then this node is not expanded and it is transformed into a leaf of the resulting tree.

Procedure `BuilTree` (*No*, $\mathscr{L}^*$)

1. If $\mathscr{L}^* = \emptyset$, then `Exit`
2. $\sigma \longleftarrow \sigma \cap \{No\}$
3. Compute $TU(\mathscr{P}^\sigma)$
4. Compute $\alpha = \min_{X_i \in \mathscr{L}^*} \texttt{Inf1}(\sigma, X_i)$, $\beta = \min_{X_i, X_j \in \mathscr{L}^*} \texttt{Inf2}(\sigma, X_i, X_j)$
5. If $\text{Min}\{\alpha, \beta\} \geq TU(\mathscr{P}^\sigma)$ then `Exit`
6. Else
   7. If $\alpha \leq \beta$ then $X_k \longleftarrow arg(\alpha)$
   8. Else
      9. Let $\{X_i, X_j\} = arg(\beta)$
      10. $X_k \longleftarrow arg(Min\{\texttt{Inf1}(\sigma, X_i), \texttt{Inf1}(\sigma, X_j)\})$
11. $\mathscr{L}^* \longleftarrow \mathscr{L}^* - X_k$
12. $No \longleftarrow X_k$
13. For each possible value x_k of X_k
    14. Add a node No_k
    15. Make No_k a child of No
    16. Call `BuilTree` (*No_k*, $\mathscr{L}^*$)

In the above algorithm, X_k is the branching variable of node *No*. The intuitive idea is that when we assign this variable to *No*, we divide the database associated with this node among its different children. In each one of the children, we can have more precise average knowledge about C but based on a smaller sample.

Parameter s can be used to control the complexity of the resulting model. Smaller values of s will produce larger trees. With very large s the resulting tree will be small.

Decision in the Leaves

In order to classify a new case with values of all the variables except the one in C, we start from the tree's root node and we continue the path indicated to us by the values of the new case. If we are at a node with variable X_i and this variable takes state x_i^r in this case, then we choose the offspring corresponding to this value. This process is repeated until we reach a leaf node and using the associated credal set, we can find the corresponding value of the variable to be classified C, using the maximum frequency criterion. Another criterion that could be used is the interval dominance criterion (strong dominance) [5]. This criterion generally implies a partial order and in certain situations it is not possible to specify any value for the variable which is being classified. The state $C = c_h$ shall be chosen having verified that $\forall i \neq h$ $\overline{P}(c_i|\sigma) < \underline{P}(c_h|\sigma)$

When there is no dominant value, the result could be the set of non-dominated states of C, (states c_i for those for which there is no other dominating state according to the previous inequality). In this regard, we obtain a *credal* classifier [23], with which a set of non-dominated states is obtained and a single state or nothing as in our case. The application of this criterion shall avoid the loss of information that we would obtain by leaving unclassified those cases where there are states of C with frequencies which are much higher than the others but which our criterion does not allow to be classified. To compare the results with the ones of other known method to build classification trees, we use the maximum frequency criterion in this paper. As this criterion is used, apart from determining the structure of the tree, parameter s will not have an additional effect in the final classification. If a credal classifier were used, then larger s with give rise to more unclassified cases.

4 Experimentation

We have applied the methods on some known and different datasets, obtained from the *UCI repository of machine learning databases*, available online at ftp://ftp.ics.uci.edu/pub/machine-learning-databases.

The datasets were: *Cmc* (Contraceptive method choice), *Flare2* (astronomy), *German* (financial); *Car* (vehicles); *Tae* (teaching); *Monks1* (artificial); *Tic-tac-toe* (games) and *Balance-scale* (psychological).

Some of the datasets had missing observations and in some cases, they had non-discrete variables. The cases with missing values were eliminated and the continuous variables were discretized using MLC++ software, available at http://www.sgi.com/tech/mlc. The entropy measure was used for discretization. The number of intervals is not fixed, and is obtained following Fayyad and Irani's procedure [14]. Also, previously we have randomized the data when the dataset was ordered by attributes. Table I presents a brief description of these datasets. We can see the number of cases in the training set (Tr), the number in the test set (Ts), the number of attribute variables in the dataset (η), the number of different states of the variable to be classified (k) and the range of states of the attribute variables (R.attr), i.e. the minimum and maximum number of states of the attribute variables.

The algorithms were implemented with Java language (version 1.5) using maximum entropy as the total uncertainty measure on IDM probability intervals. A simple algorithm for the computation of upper entropy can be found in [1]. This measure is chosen on the basis of arguments which have been widely discussed in Klir and Smith [16], Abellán and Moral [7] and Abellán, Moral and Klir [8].

We would also like to compare the results with other standard method, using the same datasets and the same pre-processing. In Table 2, therefore, we present in each column E_s the results obtained on the datasets test sets varying the values of parameter s. We have considered the following set of values for s $\{0.5, 1, 1.5, 2, 2.5, 3, 3.5, 4, 5, 6, 8, \frac{k}{2}\}$. To use as reference, in this table, we have inserted column J48, that corresponds to an improved version of the C4.5 method of Quinlan [19], based on the ID3 [18], which uses a classification tree with classic probabilities[2].

Observing the results of Table 2 we can appreciate that for the datasets (1), (2) and (3), we obtain better results with values of $s \geq 4$. For datasets (4), (6) and (7) the results are better for $s < 1$. This allow us to wonder if it is reasonable for our supervised classification problem to consider values for s in $[1, 2]$ as some authors suggest when the IDM is used ([22, 10]). The best average obtained is for $s = 1.5$ and $s = \frac{k}{2}$. Now, as $s = 1.5$ obtains the greatest performance and it is a fixed value, this

---

[2] J48 method can be obtained via *weka* software, available in http://www.cs.waikato.ac.nz/ ml/weka/

is our actual proposal for s. However, this result is not close to the average obtained taking the best results possible for every dataset (77.1). A possible explanation is that larger values of s are good when the number of attributes is high. Perhaps this is due to the necessity of making a more strict control of the size in this case, due to the possibility of overfitting due to the fact that we have more candidate variables for each node of the tree. This allow us to consider in the future a more general possible relation between the characteristics of each data set (number of attribute variables, number of states of the attribute variables, size of the dataset,..) with the parameter s and to study whether a value of $s = 1.5$ with some additional control about the model size depending of the number of attributes can provide optimal results.

Table 1. Description of the datasets

Dataset	Tr	Ts	η	k	R.attr	Dataset	Tr	Ts	η	k	R.attr
(1)Cmc	986	487	8	3	2-4	(5)Tae	101	50	5	3	2-26
(2)Flare2	714	352	10	9	2-7	(6)Monks1	122	434	6	2	2-4
(3)German	670	330	24	2	2-5	(7)Tic-tac-toe	641	317	9	2	3-3
(4)Car	1157	571	6	4	3-4	(8)Balance-scale	418	207	4	3	5-5

Table 2. Percentages of correct classifications in the test set

Dataset	$J48$	$E_{0.5}$	E_1	$E_{1.5}$	E_2	$E_{2.5}$	E_3	$E_{3.5}$	E_4	E_5	E_6	E_8	$E_{k/2}$
(1)	52.2	47.0	48.9	49.7	50.9	51.3	52.6	52.4	52.4	**56.7**	56.1	56.1	49.7
(2)	85.2	82.1	85.5	85.8	85.8	85.5	85.5	**86.4**	**86.4**	**86.4**	84.7	84.7	**86.4**
(3)	72.9	67.0	71.5	71.2	72.1	72.1	73.0	73.0	72.7	**73.9**	73.6	73.3	71.5
(4)	90.9	**91.2**	88.4	87.4	86.9	85.8	86.2	84.8	79.5	77.6	75.7	75.7	86.9
(5)	42.0	50.0	52.0	**54.0**	50.0	50.0	50.0	42.0	42.0	42.0	42.0	42.0	**54.0**
(6)	80.0	**100.0**	94.4	94.5	91.7	91.7	91.7	91.7	80.4	79.4	72.1	74.7	94.4
(7)	81.4	**88.0**	85.8	86.1	86.1	81.4	79.8	79.8	80.1	79.8	81.1	78.2	85.8
(8)	65.2	53.6	64.3	64.3	**66.2**	59.9	58.9	58.9	58.9	57.0	57.0	57.0	64.3
Average	71.2	72.4	73.9	**74.1**	73.7	72.2	72.2	71.1	69.1	69.1	67.8	67.7	**74.1**

5 Conclusions

We have presented a first explorative study of the results of our classification method varying parameters. We have proved that for each dataset it is possible to improve the result changing the value of the total prior strength. But more studies and experiments are necessary to ascertain the ideal relationship between the value of s and some characteristics of each data set, such that: number of states of the class variable, number of attribute variables, size of dataset, etc.

References

[1] Abellán J. (2006) Uncertainty measures on probability intervals from the imprecise Dirichlet model. To appear in *Int. J. of Gen. Syst.* 35(3).

[2] Abellán J. and Moral, S. (1999) Completing a Total Uncertainty Measure in Dempster-Shafer Theory, *Int. J. of Gen. Syst.*, 28:299-314.

[3] Abellán J. and Moral, S. (2000) A Non-specificity measure for convex sets of probability distributions, *Int. J. of Uncer., Fuzz. and K-B Syst.*, 8:357-367.

[4] Abellán J. and Moral, S. (2003) Maximum entropy for credal sets, *Int. J. of Uncer., Fuzz. and K-B Syst.*, 11:587-597.

[5] Abellán J. and Moral, S. (2003) Using the total uncertainty criterion for building classification trees, *Int. J. of Int. Syst.* 18(12):1215-1225.

[6] Abellán J. and Moral, S. (2005) Maximum difference of entropies as a non-specificty measure for credal sets, *Int. J. of Gen. Syst.*, 34(3):201-214.

[7] Abellán J. and Moral, S. (2005) Upper entropy of credal sets. Applications to credal classification, *Int. J. of Appr. Reas.*, 39:235-255.

[8] Abellán J., Klir, G.J. and Moral, S. (2006) Disaggregated total uncertainty measure for credal sets, *Int. J. of Gen. Syst.*, 35(1):29-44.

[9] Abellán J. and Moral, S. (2006) An algorithm that computes the upper entropy for order-2 capacities, *Int. J. of Uncer., Fuzz. and K-B Syst.*, 14(2):141-154.

[10] Bernard, J.M. (2005) An introduction to the imprecise Dirichlet model for multinomial data, *Int. J. of Appr. Reas.*, 39:123-150.

[11] Breiman, L., Friedman, J.H., Olshen, R.A. and Stone. C.J. (1984) Classification and Regression Trees. Wadsworth Statistics, Probability Series, Belmont.

[12] A.P. Dempster, A.P. (1967) Upper and lower probabilities induced by a multi-valued mapping, *Ann. Math. Stat.*, 38:325-339.

[13] Dubois, D. and Prade, H. (1987) Properties and Measures of Information in Evidence and Possibility Theories. *Fuzz. Sets and Syst.*, 24:183-196.

[14] Fayyad, U.M. and Irani, K.B. (1993) Multi-valued interval discretization of continuous-valued attributes for classification learning. *Proc. of the 13th Inter. Joint Conf. on Art. Int.*, Morgan Kaufmann, San Mateo, 1022-1027.

[15] Klir, G.J. (2006) Uncertainty and Information: Foundations of Generalized Information Theory. John Wiley, Hoboken, NJ.

[16] Klir G.J. and Smith, R.M. (2001) On measuring uncertainty and uncertainty-based information: Recent developments, *Ann. of Math. and Art. Intell.*, 32(1-4):5-33.

[17] Klir G.J. and Wierman, M.J. (1998) *Uncertainty-Based Inf.*, Phisica-Verlag.

[18] Quinlan, J.R. (1986) Induction of decision trees, *Machine Learning*, 1:81-106.

[19] Quinlan, J.R. (1993) *Programs for Machine Learning*. Morgan Kaufmann series in Machine Learning.

[20] Shafer G. (1976) *A Mathematical Theory of Evidence*, Princeton University Press, Princeton.

[21] Walley, P. (1991) *Statistical Reasoning with Imprecise Probabilities*, Chapman and Hall, London.

[22] Walley, P. (1996) Inferences from multinomial data: learning about a bag of marbles, *J. Roy. Statist. Soc. B*, 58:3-57.

[23] Zaffalon, M. (1999) A credal approach to naive classification, *Proc. of the First Int. Symp. on Impr. Prob. and their Appl. '99*, Gent, 405-414.

Comparing Proportions Data with Few Successes

F.P.A. Coolen[1] and P. Coolen-Schrijner[2]

[1] Dept. of Mathematical Sciences, Durham University, Durham DH1 3LE, UK
   Frank.Coolen@durham.ac.uk
[2] Pauline.Schrijner@durham.ac.uk

Summary. We review a recently introduced nonparametric predictive approach for comparison of groups of proportions data, using interval probability. We particularly focus on cases where groups have zero of few successes. These inferences are for events that $m \geq 1$ future observations from a particular group will include more successes than m future observations from each other group.

1 Introduction

We use Coolen's [5] predictive upper and lower probabilities [16, 17] to compare future numbers of successes in Bernoulli trials for $k \geq 2$ independent groups, using data consisting of observed numbers of trials and the numbers of successes in these trials. In [6] we presented nonparametric predictive methods for pairwise and multiple comparisons for such proportions data, the latter restricted to the event that the number of successes out of m future trials for one group is greater than (or equal to) the corresponding number for each of the other groups. We briefly review this approach, and focus on data with few successes as may occur in comparison of highly reliable technical units or of highly effective medicines. In such cases, inference based on established methods is often deemed unsatisfactory. Classical frequentist methods typically test for differences in parameters representing underlying success probabilities, and rarely indicate differences if all groups involved have zero or few successes. Bayesian inference in case of zero or few successes tends to be sensitive to chosen prior distributions [8]. As our method uses upper and lower probabilities, it may not provide inferences which appear to be as strong as those based on established methods, but we claim that the explicit representation of indeterminacy is an advantage of our method, and it is of great relevance to study the way in which our upper and lower probabilities depend on the data and the choice of m, the number of future observations per group considered.

F.P.A. Coolen and P. Coolen-Schrijner: *Comparing Proportions Data with Few Successes*, Advances in Soft Computing **6**, 241–248 (2006)
www.springerlink.com

2 Nonparametric Predictive Comparison of Proportions

Coolen [5] presented and justified upper and lower probabilities for nonparametric prediction of Bernoulli random quantities, Coolen and Coolen-Schrijner [6] used these to derive upper and lower probabilities for nonparametric predictive comparison of proportions. We briefly state the key results from these two papers, after which we focus on the use of these results in cases where the data contain zero or few successes. We refer to [5, 6] for more detailed presentation, justification and discussion of these results, but we wish to remark that this inferential method [5] uses Hill's assumption $A_{(n)}$ [13], and defines direct predictive upper and lower probabilities [1, 11, 12] for future observations, based on available data. The upper and lower probabilities in [5, 6] fit in the framework of nonparametric predictive inference (NPI) [1], and hence have strong internal consistency properties [1, 6].

Suppose that we have a sequence of $n+m$ exchangeable Bernoulli trials [10], each with success and failure as possible outcomes, and data consisting of s successes observed in n trials. Let Y_a^b denote the random quantity representing the number of successes in trials a to b, then a sufficient representation of the data for our inferences is $Y_1^n = s$, due to the exchangeability of all trials. We are interested in the number of successes in trials $n+1$ to $n+m$. Let $R_t = \{r_1, \ldots, r_t\}$, with $1 \leq t \leq m+1$ and $0 \leq r_1 < r_2 < \ldots < r_t \leq m$, and let $\binom{s+r_0}{s} = 0$. Then the NPI upper probability for the event $Y_{n+1}^{n+m} \in R_t$, given data $Y_1^n = s$, for $s \in \{0, \ldots, n\}$, is [5]

$$\overline{P}(Y_{n+1}^{n+m} \in R_t | Y_1^n = s) =$$

$$\binom{n+m}{n}^{-1} \sum_{j=1}^{t} \left[\binom{s+r_j}{s} - \binom{s+r_{j-1}}{s} \right] \binom{n-s+m-r_j}{n-s}$$

The corresponding lower probability is derived via

$$\underline{P}(Y_{n+1}^{n+m} \in R_t | Y_1^n = s) = 1 - \overline{P}(Y_{n+1}^{n+m} \in \{0, 1, \ldots, m\} \backslash R_t | Y_1^n = s)$$

This is justified in [5], and agrees with the fact that these upper and lower probabilities are F-probability in the theory of interval probability [1, 6, 17]. These upper and lower probabilities are based on an assumed underlying representation of Bernoulli data as realisations of real-valued continuously distributed random quantities, analogous to the representation used by Bayes [2], which allows predictive inference to be based on Hill's assumption $A_{(n)}$ [5, 13]. As in [6], it implies that, for our NPI comparisons of proportions, we only need upper probabilities for events $Y_{n+1}^{n+m} \geq y$ and $Y_{n+1}^{n+m} < y$. For $y \in \{0, 1, \ldots, m\}$ and $0 < s < n$,

$$\overline{P}(Y_{n+1}^{n+m} \geq y | Y_1^n = s) =$$

$$\binom{n+m}{n}^{-1} \left[\binom{s+y}{s} \binom{n-s+m-y}{n-s} + \sum_{l=y+1}^{m} \binom{s+l-1}{s-1} \binom{n-s+m-l}{n-s} \right]$$

and for $y \in \{1, \ldots, m+1\}$ and $0 < s < n$,

$$\overline{P}(Y_{n+1}^{n+m} < y | Y_1^n = s) =$$

$$\binom{n+m}{n}^{-1} \left[\binom{n-s+m}{n-s} + \sum_{l=1}^{y-1} \binom{s+l-1}{s-1} \binom{n-s+m-l}{n-s} \right]$$

If the data are all successes ($s = n$) or failures ($s = 0$), then for $y \in \{0, 1, \ldots, m\}$,

$$\overline{P}(Y_{n+1}^{n+m} \geq y | Y_1^n = n) = 1 \quad \text{and} \quad \overline{P}(Y_{n+1}^{n+m} \geq y | Y_1^n = 0) = \frac{\binom{n+m-y}{n}}{\binom{n+m}{n}}$$

and for $y \in \{1, \ldots, m+1\}$,

$$\overline{P}(Y_{n+1}^{n+m} < y | Y_1^n = n) = \frac{\binom{n+y-1}{n}}{\binom{n+m}{n}} \quad \text{and} \quad \overline{P}(Y_{n+1}^{n+m} < y | Y_1^n = 0) = 1$$

In [6] we presented the theory for comparing $k \geq 2$ groups of proportions data within the NPI framework, with interest in the event that the number of successes in m future trials in group i is greater than (or equal to) the maximum of the number of successes in m future trials for each of the $k - 1$ other groups. For $k = 2$ this procedure is called 'pairwise comparison'. For $k \geq 3$ such a simultaneous 'multiple comparison' of one group with all other groups cannot directly be inferred from corresponding pairwise comparisons, and is often advocated for problems where one is explicitly interested in distinguishing a 'best' group. Throughout, we assume the groups to be fully independent [6]. In a straightforward manner, we add an index i to notation when referring to group $i \in \{1, \ldots, k\}$, and we use the notation $j \neq i$ for $j \in \{1, \ldots, k\} \backslash i$ and $(\underline{n}, \underline{s})$ to denote all data for the k groups. In [6] we derived the upper probability

$$\overline{P}(Y_{i,n_i+1}^{n_i+m} > \max_{j \neq i} Y_{j,n_j+1}^{n_j+m} | (\underline{n}, \underline{s})) = \sum_{y=0}^{m} \left[\Delta(y) \prod_{j \neq i} \overline{P}(Y_{j,n_j+1}^{n_j+m} < y | Y_{j,1}^{n_j} = s_j) \right],$$

with $\Delta(y) = \overline{P}(Y_{i,n_i+1}^{n_i+m} \geq y | Y_{i,1}^{n_i} = s_i) - \overline{P}(Y_{i,n_i+1}^{n_i+m} \geq y+1 | Y_{i,1}^{n_i} = s_i)$

The corresponding lower probability and the upper and lower probabilities for this event with '>' replaced by '$\geq$' are derived similarly. In Section 3, we will use $\overline{P}(i > \max_{j \neq i} j)$ as short-hand notation for the upper probability above, whereas for pairwise comparisons between groups i and j we just use $\overline{P}(i > j)$ (and similarly for other upper and lower probabilities).

3 Proportions Data with Few Successes

The upper and lower probabilities for comparisons of proportions data [6], given in Section 2, normally need to be computed numerically, but this is straightforward as it only involves finite sums and products. However, it does complicate analytical study of these upper and lower probabilities, where particular interest is in their dependence on the data and on the choice of m. We mostly illustrate such dependences via examples in [6], where it is shown that the choice of m is highly relevant due to both the inherent randomness involved with sampling Bernoulli random quantities and the fact that imprecision, i.e. the difference between corresponding upper and lower probabilities, tends to increase with m. Also, imprecision in these predictive inferences tends to decrease if the number of available data grows, although one must be careful as imprecision tends to be smaller when upper and lower probabilities are close to zero or one than when these are nearer 0.5. For data with zero or few successes, we illustrate some pairwise and multiple comparisons in the example below. If pairwise comparisons ($k = 2$) involve a group with zero successes, some of these upper and lower probabilities reduce to simple general forms,

$$\overline{P}(Y_{1,n_1+1}^{n_1+m} > Y_{2,n_2+1}^{n_2+m} | Y_{1,1}^{n_1} = 0, Y_{2,1}^{n_2} = 0) = \frac{m}{n_1 + m}$$

$$\underline{P}(Y_{1,n_1+1}^{n_1+m} > Y_{2,n_2+1}^{n_2+m} | Y_{1,1}^{n_1} = 0, Y_{2,1}^{n_2} = s_2) = 0 \text{ for all } s_2 \in \{0, \dots, n_2\}$$

$$\overline{P}(Y_{1,n_1+1}^{n_1+m} \geq Y_{2,n_2+1}^{n_2+m} | Y_{1,1}^{n_1} = s_1, Y_{2,1}^{n_2} = 0) = 1 \text{ for all } s_1 \in \{0, \dots, n_1\}$$

$$\underline{P}(Y_{1,n_1+1}^{n_1+m} \geq Y_{2,n_2+1}^{n_2+m} | Y_{1,1}^{n_1} = 0, Y_{2,1}^{n_2} = 0) = \frac{n_2}{n_2 + m}$$

These results are based on the idea, in line with intuition and the underlying representation of the Bernoulli random quantities [5] as mentioned in Section 2, that if no success has been observed for a particular group, it cannot be excluded that no success can ever be observed for that group. We see that imprecision increases with m.

Example. We illustrate NPI comparison of proportions data with few successes, using data on tumors in mice (Table 1) that were used by Tamura and Young [14] to illustrate an estimator for the Beta-Binomial distribution. These data are from 26 studies ($i = 1, \dots, 26$), n_i is the number of mice in Study i of which s_i tested positive for having a tumor.

Table 1. Tumors in mice data

i (n_i, s_i)	i (n_i, s_i)	i (n_i, s_i)	i (n_i, s_i)
1 (12,0)	8 (17,0)	15 (22,2)	21 (47,4)
2 (12,0)	9 (20,1)	16 (20,2)	22 (54,6)
3 (10,0)	10 (19,1)	17 (20,3)	23 (49,8)
4 (10,1)	11 (19,1)	18 (20,3)	24 (20,2)
5 (20,0)	12 (17,1)	19 (18,3)	25 (49,6)
6 (20,0)	13 (15,1)	20 (20,4)	26 (49,10)
7 (19,0)	14 (25,2)		

For pairwise comparisons, these data are suitable for illustrating the effect of different n_i if the corresponding s_i are zero. Table 2 illustrates this effect, as the

Table 2. (Study 1 vs Study 2) compared to (Study 1 vs Study 5)

m :	1	3	5	10	50		1	3	5	10	50
$\overline{P}(1 > 2)$	0.077	0.200	0.294	0.455	0.806	$\overline{P}(1 \geq 2)$	1	1	1	1	1
$\underline{P}(1 > 2)$	0	0	0	0	0	$\underline{P}(1 \geq 2)$	0.923	0.800	0.706	0.545	0.194
$\overline{P}(1 > 5)$	0.077	0.200	0.294	0.455	0.806	$\overline{P}(1 \geq 5)$	1	1	1	1	1
$\underline{P}(1 > 5)$	0	0	0	0	0	$\underline{P}(1 \geq 5)$	0.952	0.870	0.800	0.667	0.286

data for Study 1 and Study 2 are both $(12, 0)$ and for Study 5 they are $(20, 0)$. We have to keep in mind here that such data imply that it cannot be logically excluded that in such groups there would never appear a mouse which tests positive, which is reflected in some of the lower probabilities being equal to 0 for all m, and some of the upper probabilities being equal to 1 for all m. It is clear that the choice of m greatly influences these predictive upper and/or lower probabilities, which is mostly due to the decreasing chance of equal numbers of successes in the two groups considered when m increases.

The lower probabilities for the event '$1 \geq 2'$ are smaller than the corresponding lower probabilities for '$1 \geq 5'$, for all m, which is caused by $n_2 < n_5$, so imprecision is logically related to the amount of information available. The upper probabilities for the events '$1 > 2'$ and '$1 > 5'$ are identical, for all m. These upper probabilities correspond to the situation where for Studies 2 and 5 there would never be any mice that test positive, hence for both events these are just the NPI-based upper probabilities of at least 1 positive test result out of m further tested mice for Study 1. For the values in Table 2, $\overline{P}(1 > 2)$ is equal to $1 - \underline{P}(1 \geq 2)$, which is caused by the fact that here $\underline{P}(2 \geq 1) = \underline{P}(1 \geq 2)$, as we have precisely the same information from Studies 1 and 2. As Studies 1 and 5 did not give the same observations, there such a relation does not hold, but of course as always we have $\overline{P}(5 > 1) = 1 - \underline{P}(1 \geq 5)$, so we can derive all upper and lower probabilities for such pairwise comparisons between these studies from the entries in Table 2.

Table 3 gives our NPI-based upper and lower probabilities for comparison of Study 3, with data $(10, 0)$, to Study 26, with data $(49, 10)$, and for comparison of Study 5, with data $(20, 0)$, to Study 26. The effect of $n_5 > n_3$ is clearly seen in the upper probabilities for '$3 > 26'$ and '$5 > 26'$, all of which are of course pretty small values. Note that, due to the randomness involved and the decreasing chance of having the same numbers of mice testing positive in two studies when m increases,

Table 3. (Study 3 vs Study 26) compared to (Study 5 vs Study 26)

m :	1	3	5	10	50		1	3	5	10	50
$\overline{P}(3 > 26)$	0.073	0.134	0.156	0.167	0.148	$\overline{P}(3 \geq 26)$	0.818	0.609	0.494	0.354	0.183
$\underline{P}(3 > 26)$	0	0	0	0	0	$\underline{P}(3 \geq 26)$	0.780	0.482	0.304	0.104	0.0002
$\overline{P}(5 > 26)$	0.038	0.072	0.083	0.081	0.044	$\overline{P}(5 \geq 26)$	0.810	0.569	0.428	0.255	0.064
$\underline{P}(5 > 26)$	0	0	0	0	0	$\underline{P}(5 \geq 26)$	0.780	0.482	0.304	0.104	0.0002

the values for these upper probabilities first increase but then decrease if m becomes larger. The corresponding lower probabilities for the events '3 $\geq$ 26' and '5 $\geq$ 26' are identical, which is caused by the fact that these are both equal to the NPI-based lower probability of 0 positives out of m further tested mice for Study 26, as these lower probabilities correspond to the possible situation that, for both Studies 3 and 5, no mice might ever test positive.

Table 4 gives the multiple comparisons upper and lower probabilities for events '$i > \max_{j \neq i} j'$', for $m = 50$. The lower probabilities for studies that revealed no tumors, to lead to the largest future number of mice with tumors, is zero, which agrees with our earlier observations, and relates to the fact that there is no evidence against the possibility that mice in such studies would never develop tumors. The corresponding lower probabilities for events '$i \geq \max_{j \neq i} j'$', which we have not reported here, are non-zero but of course very small, as for these studies this relates only to the event that all studies had zero tumors in m future tests. The values n_i strongly influence the imprecision, which clearly shows when comparing the upper probabilities for Studies 3 (with $n_3 = 10$) and 5 (with $n_5 = 20$). Studies 20 and 26 have the highest upper and lower probabilities, with the smaller number of mice observed in Study 20 ($n_{20} = 20$, $n_{26} = 49$) reflected by larger imprecision. It is also interesting to compare Studies 24 and 25 ($(n_{24}, s_{24}) = (20, 2)$, $(n_{25}, s_{25}) = (49, 6)$), where even though a smaller proportion of mice with tumors was observed in Study 24, its upper and lower probabilities of leading to the maximum future proportion of mice with tumors are greater than for Study 25, which is due to the fact that for Study 25 there is substantially more evidence making it less likely that this study would lead to the maximum future proportion. This occurs because both Studies 24 and 25 are pretty unlikely to lead to the maximum future proportion, Study 24 also has substantially more imprecision than Study 25 due to its smaller number of observations. For smaller values of m (not shown), the larger upper probabilities as in Table 4 are

Table 4. Multiple comparisons with $m = 50$

i	$\underline{P}(i > \max_{j \neq i} j)$	$\overline{P}(i > \max_{j \neq i} j)$	i	$\underline{P}(i > \max_{j \neq i} j)$	$\overline{P}(i > \max_{j \neq i} j)$
1	0	0.0252	14	0.0027	0.0328
2	0	0.0252	15	0.0052	0.0523
3	0	0.0418	16	0.0082	0.0721
4	0.0215	0.1744	17	0.0307	0.1626
5	0	0.0043	18	0.0307	0.1626
6	0	0.0043	19	0.0482	0.2191
7	0	0.0053	20	0.0844	0.3035
8	0	0.0080	21	0.0010	0.0127
9	0.0012	0.0236	22	0.0027	0.0249
10	0.0015	0.0284	23	0.0193	0.1036
11	0.0015	0.0284	24	0.0082	0.0721
12	0.0026	0.0413	25	0.0048	0.0389
13	0.0045	0.0611	26	0.0566	0.2213

smaller, mostly because again it would be more likely for numbers of successes out of m observations, for two or more groups, to be equal.

4 Discussion

As shown in Section 3, the choice of m can make a big difference to the apparent conclusions from our pairwise and multiple comparisons. In practice, we would recommend not to restrict attention to a single value of m, but to present results for a variety of m-values. Of course, if one has a specific interest in a particular value of m, for example when decisions must be made clearly related to such a specific value, then our method can be used in a straightforward manner. A nice feature of our method is that its explicit predictive nature allows it to be used easily for decision making, where costs or utilities can directly be linked to future numbers of successes for the different groups. If one has imperfect information about such costs or utilities, which might well be the case as these are related to future realisations, one could include imprecise values of such costs or utilities in the decision processes, bringing the approach within the realms of 'information-gap decision theory' [4].

In this paper, events of the form $Y_{1,n_1+1}^{n_1+m} > Y_{2,n_2+1}^{n_2+m}$ were considered. Comparisons could also be based on lower and upper previsions [16] for $Y_{1,n_1+1}^{n_1+m} - Y_{2,n_2+1}^{n_2+m}$, which is relatively straightforward in our NPI framework. Because the lower and upper probabilities for these random quantities depend on m, so will such lower and upper previsions. This could well lead to quite different conclusions. It will be interesting to compare both these methods in a further study, whereas in practice it might be sensible to report on the outcomes of both simultaneously.

In [7] we present further related methods for multiple comparisons for proportions data, considering the NPI-based upper and lower probabilities that a selected subset of the k groups contains (all) the 'best' group(s). Such subset selection methods have been widely studied in the statistics literature [3, 15], but rarely from an explicitly predictive perspective as in [7]. Coolen and van der Laan [9] presented NPI-based selection methods for real-valued random quantities. Several possible uses of such inferential methods have been suggested [3], e.g. screening experiments where, to end up with a small number of preferred treatments, one starts with all those available, and after several observations wishes to continue with only a subset of all treatments, which should be very likely to contain (all) the best treatment(s). The specific features in case of data sets containing few successes carry over in a logical manner to such NPI-based subset selection methods for proportions data.

References

[1] Augustin T, Coolen FPA (2004) Nonparametric predictive inference and interval probability. Journal of Statistical Planning and Inference 124: 251-272

[2] Bayes T (1763) An essay towards solving a problem in the doctrine of chances. Philosophical Transactions, Royal Society of London 53: 370-418, 54: 296-325.

[3] Bechhofer RE, Santner TJ, Goldsman DM (1995) Design and analysis of experiments for statistical selection, screening, and multiple comparisons. Wiley, New York

[4] Ben-Haim Y (2001) Information-gap decision theory. Academic Press, San Diego

[5] Coolen FPA (1998) Low structure imprecise predictive inference for Bayes' problem. Statistics and Probability Letters 36: 349-357

[6] Coolen FPA, Coolen-Schrijner P (2006) Nonparametric predictive comparison of proportions. Journal of Statistical Planning and Inference, to appear

[7] Coolen FPA, Coolen-Schrijner P (2006) Nonparametric predictive subset selection for proportions. Statistics and Probability Letters, to appear

[8] Coolen FPA, Coolen-Schrijner P (2006) On zero-failure testing for Bayesian high reliability demonstration. Journal of Risk and Reliability, to appear

[9] Coolen FPA, van der Laan P (2001) Imprecise predictive selection based on low structure assumptions. Journal of Statistical Planning and Inference 98: 259-277

[10] De Finetti B (1974) Theory of probability. Wiley, Chichester

[11] Dempster AP (1963) On direct probabilities. Journal of the Royal Statistical Society B 25: 100-110

[12] Geisser S (1993) Predictive inference: an introduction. Chapman and Hall, London

[13] Hill BM (1968) Posterior distribution of percentiles: Bayes' theorem for sampling from a population. Journal of the American Statistical Association 63: 677-691

[14] Tamura RN, Young SS (1987) A stabilized moment estimator for the Beta-Binomial distribution. Biometrics 43: 813-824

[15] Verheijen JHM, Coolen FPA, van der Laan P (1997) Combining two classical approaches for statistical selection. Communications in Statistics - Theory and Methods 26: 1291-1312

[16] Walley P (1991) Statistical Reasoning with Imprecise Probabilities. Chapman and Hall, London

[17] Weichselberger K (2001) Elementare Grundbegriffe einer allgemeineren Wahrscheinlichkeitsrechnung I. Intervallwahrscheinlichkeit als umfassendes Konzept. Physika, Heidelberg

A Unified View of Some Representations of Imprecise Probabilities

S. Destercke and D. Dubois

Institut de Recherche en Informatique de Toulouse (IRIT)
Université Paul Sabatier, 118 route de Narbonne, 31062 Toulouse, France
desterck@irit.fr and dubois@irit.fr

Summary. Several methods for the practical representation of imprecise probabilities exist such as Ferson's p-boxes, possibility distributions, Neumaier's clouds, and random sets. In this paper some relationships existing between the four kinds of representations are discussed. A cloud as well as a p-box can be modelled as a pair of possibility distributions. We show that a generalized form of p-box is a special kind of belief function and also a special kind of cloud.

1 Introduction

Many uncertainty calculi can be viewed as encoding families of probabilities. Representing such families in a practical way can be a real challenge, and several proposals have been made to do so, under various assumptions. Among these proposals are p-boxes[6], possibility distributions [3], clouds [8] and random sets [1].

Possibility theory, p-boxes, and clouds use nested confidence sets with upper and lower probability bounds. This way of representing imprecise subjective probabilistic knowledge is very natural, and corresponds to numerous situations where an expert is asked for confidence intervals. In this paper, we investigate or recall various links existing between these representations, illustrating the fact that they are all closely related.

Section 2 reviews the different kinds of representations considered in this paper, and generalizes the notion of p-boxes. In section 3, we show that a generalized p-box (which encompasses usual p-boxes) can be encoded by a belief function, and we then give a practical method to build it. Finally, section 4 recalls briefly some results on clouds and possibility theory, before examining the relationship between clouds and generalized p-boxes more closely.

S. Destercke and D. Dubois: *A Unified View of Some Representations of Imprecise Probabilities*, Advances in Soft Computing **6**, 249–257 (2006)
www.springerlink.com

2 Imprecise Probability Representations

2.1 Upper and Lower Probabilities

A family $\mathscr{P}$ of probabilities on X induces lower and upper probabilities on sets A [12]. Namely $\underline{P}(A) = \inf_{P \in \mathscr{P}} P(A)$ and $\overline{P}(A) = \sup_{P \in \mathscr{P}} P(A)$. Let $\mathscr{P}_{\underline{P}, \overline{P}}(A) = \{P | \forall A \subseteq X \text{measurable}, \underline{P}(A) \leq P(A) \leq \overline{P}(A)\}$. It should be noted that $\mathscr{P}_{\underline{P}, \overline{P}}$ is convex and generally larger than the original family $\mathscr{P}$, since lower and upper probabilities are projections of $\mathscr{P}$ on sets A. Representing either $\mathscr{P}$ or $\mathscr{P}_{\underline{P}, \overline{P}}$ on a computer can be tedious, even for one-dimension problems. Simpler representations can be very useful, even if it implies a loss in generality.

2.2 Random Sets

Formally, a random set is a set-valued mapping from a (here finite) probability space to a set X. It induces lower and upper probabilities on X [1]. Here, we use mass functions [10] to represent random sets. A mass function m is defined by a mapping from the power set $\mathscr{P}(X)$ to the unit interval, s.t. $\sum_{A \subseteq X} m(A) = 1$. A set E with positive mass is called a focal set. Each focal set is viewed as the disjunction of its elements and represents a piece of incomplete information. Plausibility and belief functions can then be defined from this mass function :

$$Bel(A) = \sum_{E, E \subseteq A} m(E) \text{ and } Pl(A) = 1 - Bel(A^c) = \sum_{E, E \cap A \neq \emptyset} m(E).$$

The set $\mathscr{P}_{Bel} = \{P | \forall A \subseteq X \text{ measurable}, Bel(A) \leq P(A) \leq Pl(A)\}$ is the special probability family induced by the belief function.

2.3 Quantitative Possibility Theory

A possibility distribution π is a mapping from X to the unit interval (hence a fuzzy set) such that $\pi(x) = 1$ for some $x \in X$. Several set-functions can be defined from a possibility distribution π [3]:

- Possibility measures: $\Pi(A) = \sup_{x \in A} \pi(x)$
- Necessity measures: $N(A) = 1 - \Pi(A^c)$
- Guaranteed possibility measures: $\Delta(A) = \inf_{x \in A} \pi(x)$

Possibility degrees express the extent to which an event is plausible, i.e., consistent with a possible state of the world. Necessity degrees express the certainty of events and Δ-measures the extent to which all states of the world where A occurs are plausible. They apply to so-called guaranteed possibility distributions [3] generally denoted by δ.

A possibility degree can be viewed as an upper bound of a probability degree [4]. Let $\mathscr{P}_\pi = \{P, \forall A \subseteq X \text{ measurable}, P(A) \leq \Pi(A)\}$ be the set of probability measures encoded by π. A necessity (resp. possibility) measure is a special case of belief (resp. plausibility) function when focal sets are nested.

2.4 Generalized Cumulative Cistributions

Let Pr be a probability function on the real line with density p. The *cumulative distribution* of Pr is denoted F^p and is defined by $F^p(x) = \Pr((-\infty, x])$.

Interestingly the notion of cumulative distribution is based on the existence of the natural ordering of numbers. Consider a probability distribution (probability vector) $\alpha = (\alpha_1 \ldots \alpha_n)$ defined over a finite domain X of cardinality n; α_i denotes the probability $\Pr(x_i)$ of the i-th element x_i, and $\sum_{j=1}^{n} \alpha_j = 1$. Then no obvious notion of cumulative distribution exists. In order to make sense of this notion over X one must equip it with a complete preordering $\leq_R$, which is a reflexive, complete and transitive relation. An R-downset is of the form $\{x_i : x_i \leq_R x\}$, and denoted $(x]_R$.

Definition 1 *The generalized R-cumulative distribution of a probability distribution on a finite, completely preordered set $(X, \leq_R)$ is the function $F_R^\alpha : X \to [0,1]$ defined by $F_R^\alpha(x) = \Pr((x]_R)$.*

Consider another probability distribution $\beta = (\beta_1 \ldots \beta_n)$ on X. The corresponding R-dominance relation of α over β can be defined by the pointwise inequality $F_R^\alpha < F_R^\beta$. In other words, a generalized cumulative distribution can always be considered as a simple one, up to a reordering of elements.

In fact any generalized cumulative distribution F_R^α with respect to a weak order $>_R$ on X, of a probability measure Pr, with distribution α on X, can be viewed as a possibility distribution π_R whose associated measure dominates Pr, i.e. $\max_{x \in A} F_R^\alpha(x) \geq \Pr(A), \forall A \subseteq X$. This is because a (generalized) cumulative distribution is constructed by computing the probabilities of events $\Pr(A)$ in a nested sequence of downsets $(x_i]_R$ [2].

2.5 Generalized p-box

A p-box [6] is defined by a pair of cumulative distributions $\underline{F} \leq \overline{F}$ on the real line bounding the cumulative distribution of an imprecisely known probability function with density p. Using the results of section 2.4, we define a generalized p-box as follow

Definition 2 *A R-p-box on a finite, completely preordered set $(X, \leq_R)$ is a pair of R-cumulative distributions $F_R^\alpha(x)$ and $F_R^\beta(x)$, s.t. $F_R^\alpha(x) \leq F_R(x) \leq F_R^\beta(x)$ with β a probability distribution R-dominated by α*

The probability family induced by a R-p-box is $\mathscr{P}_{p-box} = \{P | \forall x, F_R^\alpha(x) \leq F_R(x) \leq F_R^\beta(x)\}$ If we choose R and consider the sets $A_i = (x_i]_R, \forall x_i \in X$ with $x_i \leq_R x_j$ iff $i < j$, we define a family of nested confidence sets $\emptyset \subseteq A_1 \subseteq A_2 \subseteq \ldots \subseteq A_n \subset X$. The family $\mathscr{P}_{p-box}$ can be encoded by the constraints

$$\alpha_i \leq P(A_i) \leq \beta_i \qquad i = 1, \ldots, n \tag{1}$$

with $\alpha_1 \leq \alpha_2 \leq \ldots \leq \alpha_n \leq 1$ and $\beta_1 \leq \beta_2 \leq \ldots \leq \beta_n \leq 1$. If X is the real line and $A_i = (-\infty, x_i]$, it is easy to see that we find back the usual definition of p-boxes.

2.6 Clouds

This section recalls basic definitions and results due to Neumaier [8], cast in the terminology of fuzzy sets and possibility theory. A *cloud* is an Interval-Valued Fuzzy Set F such that $(0,1) \subseteq \bigcup_{x \in X} F(x) \subseteq [0,1]$, where $F(x)$ is an interval $[\delta(x), \pi(x)]$. In the following it is defined on a finite set X or it is an interval-valued fuzzy interval (IVFI) on the real line (then called a cloudy number). In the latter case each fuzzy set has cuts that are intervals. When the upper membership function coincides with the lower one, $(\delta = \pi)$ the cloud is called *thin*. When the lower membership function is identically 0, the cloud is said to be *fuzzy*.

A random variable x with values in X is said to belong to a cloud F if and only if $\forall \alpha \in [0,1]$:

$$P(\delta(x) \geq \alpha) \leq 1 - \alpha \leq P(\pi(x) > \alpha) \tag{2}$$

under all suitable measurability assumptions. Obviously, a fuzzy cloud is a possibility distribution.

If X is a finite set of cardinality n, a *cloud* can be defined by the following constraints :

$$P(B_i) \leq 1 - \alpha_{i+1} \leq P(A_i) \text{ and } B_i \subseteq A_i \quad i = 1, \ldots, n \tag{3}$$

Where $1 = \alpha_1 > \alpha_2 > \ldots > \alpha_n = 0$ and $A_1 \subseteq A_2 \subseteq \ldots \subseteq A_n; B_1 \subseteq B_2 \subseteq \ldots \subseteq B_n$. The confidence sets A_i and B_i are respectively the α-cuts of fuzzy sets π and δ ($A_i = \{x_i, \pi(x_i) > \alpha_{i+1}\}$ and $B_i = \{x_i, \delta(x_i) \geq \alpha_{i+1}\}$).

3 Generalized p-boxes are Belief Functions

In this section, we show that $\mathcal{P}_{p-box}$, the probability family described in section 2.5 can be encoded by a belief function. In order to achieve this, we reformulate the constraints given by equations (1).
Consider the following partition of $X : E_1 = A_1, E_2 = A_2 \setminus A_1, \ldots, E_n = A_n \setminus A_{n-1}, E_{n+1} = X \setminus A_n$
The constraints on the confidence sets A_i can be rewritten

$$\alpha_i \leq \sum_{k=1}^{i} P(E_i) \leq \beta_i \quad i = 1, \ldots, n \tag{4}$$

The proof that a belief function encoding $\mathcal{P}_{p-box}$ exists follows in four points

a. The family $\mathcal{P}_{p-box}$ is always non-empty
b. Constraints induce $\underline{P}(\bigcup_{k=i}^{j} E_k) = \max(0, \alpha_j - \beta_{i-1})$
c. Construction of a belief function s.t. $Bel(\bigcup_{k=i}^{j} E_k) = \underline{P}(\bigcup_{k=i}^{j} E_k)$
d. For any subset A of X, $Bel(A) = \underline{P}(A)$, then $\mathcal{P}_{p-box} = \mathcal{P}_{Bel}$ follows.

3.1 $\mathscr{P}$ is Non-empty

Consider the case where $\alpha_i = \beta_i, \quad i = 1, \ldots, n$ in equation (4). A probability distribution s.t. $P(E_1) = \alpha_1; P(E_2) = \alpha_2 - \alpha_1; \ldots; P(E_n) = \alpha_n - \alpha_{n-1}; P(E_{n+1}) = 1 - \alpha_n$ always exists and is in $\mathscr{P}_{p-box}$. Hence, $\mathscr{P}_{p-box} \neq \emptyset$. Every other cases being a relaxation of this one, $\mathscr{P}_{p-box}$ always contains at least one probability.

3.2 Lower Probabilities on Sets $(\bigcup_{k=i}^{j} E_k)$

Using the partition given in section 3, we have $P(\bigcup_{k=i}^{j} E_k) = \sum_{k=i}^{j} P(E_k)$. Equations (4) induce the following lower and upper bounds on $P(\bigcup_{k=i}^{j} E_k)$

Proposition 1 $\underline{P}(\bigcup_{k=i}^{j} E_k) = \max(0, \alpha_j - \beta_{i-1}); \overline{P}(\bigcup_{k=i}^{j} E_k) = \beta_j - \alpha_{i-1}$

Proof To obtain $\underline{P}(\bigcup_{k=i}^{j} E_k)$, we must minimize $\sum_{k=i}^{j} P(E_k)$. From equation (4), we have

$$\alpha_j \leq \sum_{k=1}^{i-1} P(E_k) + \sum_{k=i}^{j} P(E_k) \leq \beta_j \text{ and } \alpha_{i-1} \leq \sum_{k=1}^{i-1} P(E_k) \leq \beta_{i-1}$$

Hence $\sum_{k=i}^{j} P(E_k)) \geq \max(0, \alpha_j - \beta_{i-1})$ and this lower bound $\max(0, \alpha_j - \beta_{i-1})$ is always reachable : if $\alpha_j > \beta_{i-1}$, take P s.t. $P(A_{i-1}) = \beta_{i-1}, P(\bigcup_{k=i}^{j} E_k) = \alpha_j - \beta_{i-1}, P(\bigcup_{k=j+1}^{n+1} E_k) = 1 - \alpha_j$. If $\alpha_j \leq \beta_{i-1}$, take P s.t. $P(A_{i-1}) = \beta_{i-1}, P(\bigcup_{k=i}^{j} E_k) = 0, P(\bigcup_{k=j+1}^{n+1} E_k) = 1 - \beta_{i-1}$. Proof for $\overline{P}(\bigcup_{k=i}^{j} E_k) = \beta_j - \alpha_{i-1}$ follows the same line.

3.3 Building the Belief Function

We now build a belief function s.t. $Bel(\bigcup_{k=i}^{j} E_k) = \underline{P}(\bigcup_{k=i}^{j} E_k)$, and in section 3.4, we show that this belief function is equivalent to the lower envelope of $\mathscr{P}_{p-box}$. We rank the α_i and β_i increasingly and rename them as

$$\alpha_0 = \beta_0 = \gamma_0 = 0 \leq \gamma_1 \leq \ldots \leq \gamma_{2n} \leq 1 = \gamma_{2n+1} = \beta_{n+1} = \alpha_{n+1}$$

and the successive focal elements F_l with $m(F_l) = \gamma_l - \gamma_{l-1}$. The construction of the belief function can be summarized as follow :

$$\text{If } \gamma_{l-1} = \alpha_i, \text{ then } F_l = F_{l-1} \cup E_{i+1} \tag{5}$$

$$\text{If } \gamma_{l-1} = \beta_i, \text{ then } F_l = F_{l-1} \setminus E_i \tag{6}$$

equation (5) means that elements in E_{i+1} are added to the previous focal set after reaching α_i, and equation (6) means that elements in E_i are deleted from the previous focal set after reaching β_i.

3.4 $\mathscr{P}_{Bel}$ is Equivalent to $\mathscr{P}_{p-box}$

To show that $\mathscr{P}_{Bel} = \mathscr{P}_{p-box}$, we show that $Bel(A) = \underline{P}(A) \ \forall A \subseteq X$

Lower probability on sets A_i

Looking at equations (5,6) and taking $\gamma_l = \alpha_i$, we see that focal elements $F_1, \ldots, F_l$ only contain sets E_k s.t. $k \leq i$, hence $\forall j = 1, \ldots, l, F_j \subset A_i$. After γ_l, the focal elements $F_{l+1}, \ldots, F_{2n}$ contain at least one element E_k s.t. $k > i$. Summing the weights $m(F_1), \ldots, m(F_l)$, we have $Bel(A_i) = \gamma_l = \alpha_i$.

Sets of the type $P(\bigcup_{k=i}^{j} E_k)$

From section 3.2, we have $\underline{P}(\bigcup_{k=i}^{j} E_k) = \max(0, \alpha_j - \beta_{i-1})$. Considering equations (5,6) and taking $\gamma_l = \alpha_j$, we have that focal elements $F_{l+1}, \ldots, F_{2n}$ contain at least one element E_k s.t. $k > j$, hence the focal elements $F_j \not\subset \bigcup_{k=i}^{j} E_k$ for $j = l+1, \ldots, 2n$. Taking then $\gamma_m = \beta_{i-1}$, we have that the focal elements $F_1, \ldots, F_m$ contain at least one element E_k s.t. $k < i$, hence the focal elements $F_j \not\subset \bigcup_{k=i}^{j} E_k$ for $j = 1, \ldots, m$.

If $m < l$ (i.e. $\gamma_l = \alpha_j \geq \beta_{i-1} = \gamma_m$), then, for $j = m+1, \ldots, l$, the focal elements $F_j \subset \bigcup_{k=i}^{j} E_k$, and we have $Bel(\bigcup_{k=i}^{j} E_k) = \gamma_l - \gamma_m = \alpha_j - \beta_{i-1}$. Otherwise, there is no focal element F_l, $l = 1, \ldots, 2n$ s.t. $F_l \subset \bigcup_{k=i}^{j} E_k$ and we have $Bel(\bigcup_{k=i}^{j} E_k) = \underline{P}(\bigcup_{k=i}^{j} E_k) = 0$.

Sets made of non-successive E_k

Consider a set of the type $A = (\bigcup_{k=i}^{i+l} E_k \cup \bigcup_{k=i+l+m}^{j} E_k)$ with $m > 1$ (i.e. there's a "hole" in the sequence, since at least $E_{i+l+1} \notin A$).

Proposition 2 $\underline{P}(\bigcup_{k=i}^{i+l} E_k \cup \bigcup_{k=i+l+m}^{j} E_k) = Bel(\bigcup_{k=i}^{i+l} E_k)) + Bel(\bigcup_{k=i+l+m}^{j} E_k)$

Sketch of proof The following inequalities gives us a lower bound on $\underline{P}$

$$\inf\left(P(\bigcup_{k=i}^{i+l} E_k \cup \bigcup_{k=i+l+m}^{j} E_k)\right) \geq \inf P(\bigcup_{k=i}^{i+l} E_k) + \inf P(\bigcup_{k=i+l+m}^{j} E_k)$$

we then use a reasoning similar to the one of section 3.2 to show that this lower bound is always reachable. The result can then be easily extended to a number n of "holes" in the sequence of E_k. This completes the proof and shows that $Bel(A) = \underline{P}(A)\ \forall A \in X$, so $\mathscr{P}_{Bel} = \mathscr{P}_{p-box}$.

4 Clouds and Generalized p-boxes

Let us recall the following result regarding possibility measures (see [2]):

Proposition 3 $P \in \mathscr{P}_\pi$ if and only if $1 - \alpha \leq P(\pi(x) > \alpha), \forall \alpha \in (0, 1]$

Consider a cloud (δ, π), and define $\bar{\pi} = 1 - \delta$. Note that $P(\delta(x) \geq \alpha) \leq 1 - \alpha$ is equivalent to $P(\bar{\pi} \geq \beta) \geq 1 - \beta$, letting $\beta = 1 - \alpha$. So it is clear from equation (2) that probability measure P is in the cloud (δ, π) if and only if it is in $\mathscr{P}_\pi \cap \mathscr{P}_{\bar{\pi}}$. So a cloud is a family of probabilities dominated by two possibility distributions (see [5]). It follows that

Proposition 4 *A generalized p-box is a cloud*

Consider the definition of a generalized p-box and the fact that a generalized cumulative distribution can be viewed as a possibility distribution π_R dominating the probability distribution Pr (see section 2.4). Then, the set of constraints $(P(A_i) \geq \alpha_i)_{i=1,n}$ from equation (1) generates a possibility distribution π_1 and the set of constraints $(P(A_i^c) \geq 1 - \beta_i)_{i=1,n}$ generates a possibility distribution π_2. Clearly $\mathscr{P}_{p-box} = \mathscr{P}_{\pi_1} \cap \mathscr{P}_{\pi_2}$, and corresponds to the cloud $(1 - \pi_2, \pi_1)$. The converse is not true.

Proposition 5 *A cloud is a generalized p-box iff $\{A_i, B_i, i = 1, \ldots, n\}$ form a nested sequence of sets (i.e. there's a complete order with respect to inclusion)*

Assume the sets A_i and B_j form a globally nested sequence whose current element is C_k. Then the set of constraints defining a cloud can be rewritten in the form $\gamma_k \leq P(C_k) \leq \beta_k$, where $\gamma_k = 1 - \alpha_{i+1}$ and $\beta_k = \min\{1 - \alpha_{j+1} : A_i \subseteq B_j\}$ if $C_k = A_i$; $\beta_k = 1 - \alpha_{i+1}$ and $\gamma_k = \max\{1 - \alpha_{j+1} : A_j \subseteq B_i\}$ if $C_k = B_i$.

Since $1 = \alpha_1 > \alpha_2 > \ldots > \alpha_n = 0$, these constraints are equivalent to those of a generalized p-box. But if $\exists B_j, A_i$ with $j > i$ s.t. $B_j \not\subseteq A_i$ and $A_i \not\subseteq B_j$, then the cloud is not equivalent to a p-box.

In term of pairs of possibility distributions, a cloud is a p-box iff π_1 and π_2 are comonotonic.

When the cloud is thin $(\delta = \pi)$, cloud constraints reduce to $P(\pi(x) \geq \alpha) = P(\pi(x) > \alpha) = 1 - \alpha$. On finite sets these constraints are contradictory. The closest approximation corresponds to the generalized p-box such that $\alpha_i = P(A_i), \forall i$. It allocates fixed probability weights to elements E_i of the induced partition. In the continuous case, a thin cloud is non trivial. A cumulative distribution function defines a thin cloud containing the only random variable having this cumulative distribution. A continuous unimodal possibility distribution π on the real line induces a thin cloud $(\delta = \pi)$ which can be viewed as a generalized p-box and is thus a (continuous) belief function with uniform mass density, whose focal sets are doubletons of the form $\{x(\alpha), y(\alpha)\}$ where $\{x : \pi(x) \geq \alpha\} = [x(\alpha), y(\alpha)]$. It is defined by the Lebesgue measure on the unit interval and the multimapping $\alpha \longrightarrow \{x(\alpha), y(\alpha)\}$. It is indeed clear that $Bel(\pi(x) \geq \alpha) = 1 - \alpha$. There is an infinity of probability measures dominating this belief function.

5 Conclusions and Open Problems

There are several concise representations of imprecise probabilities. This paper highlights some links existing between clouds, possibility distributions, p-boxes and belief functions. We generalize p-boxes and show that they can be encoded by a belief function (extending results from [7, 9]). Another interesting result is that generalized p-boxes are a particular case of clouds, which are themselves equivalent to a pair of possibility distributions.

This paper shows that at least some clouds can be represented by a belief function. Two related open questions are : can a cloud be encoded by a belief function as well? can a set of probabilities dominated by two possibility measures be encoded by a belief function ? and if not, can we find inner or outer approximations following a principle of minimal commitment? Another issue is to extend these results to the continuous framework of Smets [11].

References

[1] A. Dempster. Upper and lower probabilities induced by a multivalued mapping. *Annals of Mathematical Statistics*, 38:325–339, 1967.

[2] D. Dubois, L. Foulloy, G. Mauris, and H. Prade. Probability-possibility transformations, triangular fuzzy sets, and probabilistic inequalities. *Reliable computing*, 10:273–297, 2004.

[3] D. Dubois, P. Hajek, and H. Prade. Knowledge-driven versus data-driven logics. *Journal of Logic, Language and Information*, 9:65–89, 2000.

[4] D. Dubois and H. Prade. When upper probabilities are possibility measures. *Fuzzy Sets and Systems*, 49:65–74, 1992.

[5] D. Dubois and H. Prade. Interval-valued fuzzy sets, possibility theory and imprecise probability. In *Proceedings of International Conference in Fuzzy Logic and Technology (EUSFLAT'05)*, Barcelona, September 2005.

[6] S. Ferson, L. Ginzburg, V. Kreinovich, D. Myers, and K. Sentz. Construction probability boxes and Dempster-Shafer structures. Technical report, Sandia National Laboratories, 2003.

[7] E. Kriegler and H. Held. Utilizing belief functions for the estimation of future climate change. *International Journal of Approximate Reasoning*, 39:185–209, 2005.

[8] A. Neumaier. Clouds, fuzzy sets and probability intervals. *Reliable Computing*, 10:249–272, 2004.

[9] H. Regan, S. Ferson, and D. Berleant. Equivalence of methods for uncertainty propagation of real-valued random variables. *International Journal of Approximate Reasoning*, 36:1–30, 2004.

[10] G. Shafer. *A mathematical Theory of Evidence*. Princeton University Press, 1976.

[11] P. Smets. Belief functions on real numbers. *International Journal of Approximate Reasoning*, 40:181–223, 2005.

[12] P. Walley. *Statistical Reasoning with Imprecise Probabilities*. Chapman and Hall, 1991.

Part VI

Possibility, Evidence and Interval Methods

Estimating an Uncertain Probability Density

Yakov Ben-Haim

Yitzhak Moda'i Chair in Technology and Economics
Technion-Israel Institute of Technology, Haifa 32000 Israel
yakov@technion.ac.il

1 Two Foci of Uncertainty

In the first years of the 19th century Gauss and Legendre independently invented least-squares estimation in order to estimate planetary orbits. Based on complete confidence in Newtonian dynamics, they overcame the challenge of noisy and inconsistent astronomic observations [4]. Least-squares estimation is the paradigm of optimal estimation and system identification.

Gauss and Legendre where justified in focussing entirely on data error because Newtonian celestial mechanics has such tremendous fidelity to the truth, which is fortunate for the history of statistics. In contrast, modern estimation problems in vast domains of social and technological sciences are characterized by egregiously incomplete, mis-specified or simply erroneous models. For instance, in macro-economic modelling

> there is genuine uncertainty about how good a model is, even within the sample. Moreover, since the economy if evolving, we can take it for granted that the data generation process will change in the forecast period, causing any model of it to become mis-specified over that period, and this is eventually the main problem in economic forecasting. [1, p.246]

Two foci of uncertainty are present in these estimation problems: noisy data as well as fundamental errors in model structure. The least-squares paradigm of optimality – maximize fidelity of model to data by minimizing an error function – is not directly applicable to this situation.

A basic theorem of info-gap theory asserts the irrevocable trade-off between enhancing fidelity of a model to data, and ameliorating the structural errors in the model itself [2]. Robustness to model error decreases as the analyst demands greater fidelity to the data; maximal fidelity entails minimal robustness to model mis-specification.

What this means is that parameter estimation cannot, realistically, be as good as the data themselves suggest, when models are wrong in unknown ways. Thus the key insight of Gauss and Legendre – let the data themselves dictate the fidelity

Y. Ben-Haim: *Estimating an Uncertain Probability Density*, Advances in Soft Computing **6**, 261–265 (2006)
www.springerlink.com   © Springer-Verlag Berlin Heidelberg 2006

to the model – is inappropriate when the model structures which underlie the estimation are uncertain. The implication is that fidelity to data should be satisfied rather than optimized. Satisficed (sub-optimal) fidelity rarely entails a unique estimate, so there remains an additional degree of freedom in the estimation process which can be devoted to maximizing the robustness to model uncertainty. In this paper we demonstrate these ideas for a particular class of problems, based on info-gap decision theory.

2 Info-Gap Robust Estimation

In many situations one wishes to estimate the parameters of a probability density function (pdf) based on observations. A common approach is to select those parameter values which maximize the likelihood function for the class of pdfs in question. We will develop a simple example, based on info-gap decision theory [2], to show how to deal with the situation in which the form of the pdf is uncertain. This is a special case of system identification when the structure of the system model is uncertain. A simple example is found in [3, section 11.4].

Consider a random variable x for which a random sample has been obtained, $X = (x_1, \ldots, x_N)$. Let $\widetilde{p}(x|\lambda)$ be a pdf for x, whose parameters are denoted by λ. The likelihood function is the product of the pdf values at the observations because the observations are statistically independent of one another:

$$L(X, \widetilde{p}) = \prod_{i=1}^{N} \widetilde{p}(x_i|\lambda) \tag{1}$$

The maximum likelihood estimate of the parameters is the value of λ which maximizes $L(X, \widetilde{p})$:

$$\lambda^\star = \arg\max_{\lambda} L(X, \widetilde{p}) \tag{2}$$

But now suppose that the form of the pdf is not certain. Let $\widetilde{p}(x|\lambda)$ be the most reasonable choice of the form of the pdf, for instance $\widetilde{p}$ might be the normal or exponential distribution, but the actual form of the pdf is unknown. We will still estimate the parameters λ of the nominal pdf $\widetilde{p}(x|\lambda)$, but we wish to choose those parameters to *satisfice* the likelihood and to be *robust* to the info-gaps in the shape of the actual pdf which generated the data, or which might generate data in the future.

Let $\mathscr{P}$ be the set of all normalized and non-negative pdfs on the domain of x. Thus the actual pdf must belong to $\mathscr{P}$. Let $\mathscr{U}(\alpha, \widetilde{p})$ denote an info-gap model for uncertainty in the actual form of the pdf. For instance the envelope-bound info-gap model is the following unbounded family of nested sets of pdfs:

$$\mathscr{U}(\alpha, \widetilde{p}) = \{p(x): \ p(x) \in \mathscr{P}, \ |p(x) - \widetilde{p}(x|\lambda)| \leq \alpha \psi(x)\}, \quad \alpha \geq 0 \tag{3}$$

where $\psi(x)$ is the known envelope function and the horizon of uncertainty, α, is unknown. At any horizon of uncertainty, $\mathscr{U}(\alpha, \widetilde{p})$ is a set of pdfs. These sets are nested by α and become more inclusive as α increases. The family of these nested sets is unbounded so there is no worst case or most extreme pdf.

Now the question is, given the random sample X, and the info-gap model for uncertainty in the form of the pdf, how should we choose the parameters of the nominal pdf $\widetilde{p}(x|\lambda)$?

We would like to choose parameter values for which the likelihood is high. However, since the form (not only the parameters) of the pdf is uncertain, we wish to choose λ so that the likelihood is robust to the info-gaps in the shape of the pdf. The robustness of parameter values λ is the greatest horizon of uncertainty α up to which all pdfs in $\mathscr{U}(\alpha, \widetilde{p})$ have at least a critical likelihood L_c:

$$\widehat{\alpha}(\lambda, L_c) = \max \left\{ \alpha : \left(\min_{p \in \mathscr{U}(\alpha, \widetilde{p})} L(X, p) \right) \geq L_c \right\} \tag{4}$$

A large value of $\widehat{\alpha}(\lambda, L_c)$ implies that fidelity at least as good as L_c will be obtained with parameters λ even if the form of the estimated pdf, $\widetilde{p}(x|\lambda)$, errs greatly. On the other hand, a small value of $\widehat{\alpha}(\lambda, L_c)$ means that the fidelity could be less than L_c if $\widetilde{p}(x|\lambda)$ errs even a little.

The basic trade-off relation referred to earlier states that robustness decreases ($\widehat{\alpha}$ gets smaller) as fidelity improves (L_c gets smaller):

$$L_c < L'_c \text{ implies } \widehat{\alpha}(\lambda, L_c) \leq \widehat{\alpha}(\lambda, L'_c) \tag{5}$$

Furthermore, as mentioned earlier, the fidelity anticipated from the best model and the data has zero robustness to model error:

$$L_c = L[X, \widetilde{p}(x, |\lambda)] \text{ implies } \widehat{\alpha}(\lambda, L_c) = 0 \tag{6}$$

This is true for *any* choice of parameters λ, so it is true for the direct optimal estimate $\lambda^\star$ in eq.(2).

To develop an expression for the robustness, define $\mu(\alpha)$ as the inner minimum in eq.(4). For the info-gap model in eq.(3) we see that $\mu(\alpha)$ is obtained for the following choices of the pdf at the data points X:

$$p(x_i) = \begin{cases} \widetilde{p}(x_i) - \alpha \psi(x_i) & \text{if } \alpha \leq \widetilde{p}(x_i)/\psi(x_i) \\ 0 & \text{else} \end{cases} \tag{7}$$

Define:

$$\alpha_{\max} = \min_i \frac{\widetilde{p}(x_i)}{\psi(x_i)} \tag{8}$$

Since $\mu(\alpha)$ is the product of the densities in eq.(7) we find:

$$\mu(\alpha) = \begin{cases} \prod_{i=1}^{N} [\widetilde{p}(x_i) - \alpha \psi(x_i)] & \text{if } \alpha \leq \alpha_{\max} \\ 0 & \text{else} \end{cases} \tag{9}$$

According to the definition of the robustness in eq.(4), the robustness of likelihood-aspiration L_c is the greatest value of α at which $\mu(\alpha) \geq L_c$. Since $\mu(\alpha)$ strictly decreases as α increases, we see that the robustness is the solution of $\mu(\alpha) = L_c$. In other words, $\mu(\alpha)$ is the inverse of $\widehat{\alpha}(\lambda, L_c)$:

$$\mu(\alpha) = L_c \quad \text{implies} \quad \widehat{\alpha}(\lambda, L_c) = \alpha \tag{10}$$

Consequently a plot of $\mu(\alpha)$ vs. α is the same as a plot of L_c vs. $\widehat{\alpha}(\lambda, L_c)$. Thus, eq.(9) provides a convenient means of calculating robustness curves.

3 Example

Robustness curves are shown in Fig. 1 based on eqs. (9) and (10). The nominal pdf is exponential, $\widetilde{p}(x|\lambda) = \lambda \exp(-\lambda x)$, and the envelope function is constant, $\psi(x) = 1$. An exponentially distributed random sample containing $N = 20$ data points is generated with $\lambda = 3$. The maximum-likelihood estimate (MLE) of λ, based on eq. (2), is $\lambda^\star = 1/\bar{x}$ where $\bar{x} = (1/N)\sum_{i=1}^{N} x_i$ is the sample mean. Robustness curves are shown for three values of λ, namely, $0.9\lambda^\star$, $\lambda^\star$, and $1.1\lambda^\star$.

Given a sample, X, the likelihood function for exponential coefficient λ is $L[X, \widetilde{p}(x|\lambda)]$. Each robustness curve in Fig. 1, $\widehat{\alpha}(\lambda, L_c)$ vs. L_c, reaches the horizontal axis when L_c equals the likelihood, as expected from eq. (6). In other words, the robustness of the estimated likelihood is zero for any value of λ.

$\lambda^\star$ is the MLE of the exponential coefficient. Consequently, for any λ, $L[X, \widetilde{p}(x|\lambda^\star)] \geq L[X, \widetilde{p}(x|\lambda)]$. Thus $\widehat{\alpha}(\lambda^\star, L_c)$ reaches the horizontal axis to the right of $\widehat{\alpha}(\lambda, L_c)$.

For the specific random sample whose robustness curves are shown in Fig. 1, the robustness of the MLE, $\lambda^\star$, is greater than the robustness of the lower value, $0.9\lambda^\star$, at all likelihood aspirations L_c. In other words, $\lambda^\star$ is a better robust-satisficing choice

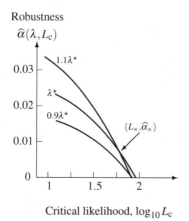

Fig. 1. Robustness curves. $\lambda^\star = 3.4065$.

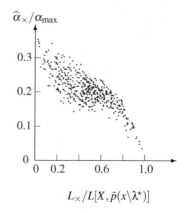

Fig. 2. Loci of intersection of robustness curves $\widehat{\alpha}(\lambda^\star, L_c)$ and $\widehat{\alpha}(1.1\lambda^\star, L_c)$.

than $0.9\lambda^\star$ at any L_c. However, the robustness curves for $\lambda^\star$ and $1.1\lambda^\star$ cross at $(L_\times, \widehat{\alpha}_\times)$, indicating that the MLE is preferable at large L_c and low robustness, while $1.1\lambda^\star$ is preferred elsewhere. The crossing of robustness curves entails the reversal of preference between $\lambda^\star$ and $1.1\lambda^\star$.

This pattern is repeated for all of 500 random samples: $\widehat{\alpha}(\lambda^\star, L_c)$ dominates $\widehat{\alpha}(0.9\lambda^\star, L_c)$, while $\widehat{\alpha}(\lambda^\star, L_c)$ and $\widehat{\alpha}(1.1\lambda^\star, L_c)$ cross. The coordinates of the intersection of $\widehat{\alpha}(\lambda^\star, L_c)$ and $\widehat{\alpha}(1.1\lambda^\star, L_c)$ are plotted in Fig. 2 for 500 robustness curves, each generated from a different 20-element random sample. In each case, $\lambda^\star$ is the MLE of that sample. The vertical axis is the robustness at the intersection, $\widehat{\alpha}_\times$, divided by the maximum robustness for that sample, $\alpha_{\max}$, defined in eq. (8). The horizontal axis is the likelihood-aspiration at the intersection, $L_\times$, divided by the maximum likelihood for the sample, $L[X, \widetilde{p}(x|\lambda^\star)]$.

The center of the cloud of points in Fig. 2 is about $(0.5, 0.2)$. What we learn from this is that the robustness curves for $\lambda^\star$ and $1.1\lambda^\star$ typically cross at a likelihood aspiration of about half the best-estimated value, and at a robustness of about 20% of the maximum robustness. We also see that curve-crossing can occur at much higher values of L_c, and that this tends to be at very low robustness. This happens when $L[X, \widetilde{p}(x|1.1\lambda^\star)]$ is only slightly less than $L[X, \widetilde{p}(x|\lambda^\star)]$. Curve-crossing can also occur at much lower L_c and higher robustness, typically because $L[X, \widetilde{p}(x|1.1\lambda^\star)]$ is substantially less than $L[X, \widetilde{p}(x|\lambda^\star)]$.

Note that the data in this example are generated from an exponential distribution, so there is nothing in the data to suggest that the exponential distribution is wrong. The motivation for the info-gap model of eq. (3) is that, while the *past* has been exponential, the *future* may not be. The robust-satisficing estimate of λ accounts not only for the historical evidence (the sample X) but also for the future uncertainty about the relevant family of distributions.

In short, the curve-crossing shown in Fig. 1 is typical, and info-gap robust-satisficing provides a technique for estimating the parameters of a pdf when the form of the pdf is uncertain.

References

[1] Bardsen, Gunnar, Oyvind Eitrheim, Eilev S. Jansen, and Ragnar Nymoen, 2005, *The Econometrics of Macroeconomic Modelling*, Oxford University Press.

[2] Ben-Haim, Yakov, 2001, *Information-Gap Decision Theory: Decisions Under Severe Uncertainty*, Academic Press, San Diego.

[3] Ben-Haim, Yakov, 2005, Info-gap Decision Theory For Engineering Design. Or: Why 'Good' is Preferable to 'Best', appearing as chapter 11 in *Engineering Design Reliability Handbook*, Edited by Efstratios Nikolaidis, Dan M.Ghiocel and Surendra Singhal, CRC Press, Boca Raton.

[4] Stigler, Stephen M., 1986, *The History of Statistics: The Measurement of Uncertainty before 1900*. The Belknap Press of Harvard University Press.

Theory of Evidence with Imperfect Information

J. Recasens

Sec. Matematiques i Informatica
ETS Arquitectura Valles
Universitat Politecnica de Catalunya
Pere Serra 1-15
08190 Sant Cugat del Valles
Spain
j.recasens@upc.edu

1 Introduction

In the Theory of Evidence it is assumed that there is an exact amount of information that makes the evidences of the focal sets (i.e. their masses) sum to 1. But it is not difficult to think of situations where this is not the case and the mass assignments sum less or more than 1.

The first situation happens when there is a lack of information. In the special case when the focal sets are nested (consonant) we obtain non-normalized possibility distributions and there are several standard ways to normalize them which in fact are equivalent to reassigning the masses of evidence to the subsets of our universe [2]. The general case deserves a deeper attention.

When the mass of the focal sets sum more than 1, we have to handle an excess of information. This can happen, for example, if we get it from different sources, since we can obtain redundant or contradictory information in this way.

In these cases it would be interesting to have some reasonable procedures to add or remove information or equivalently to relocate the excess or defect of masses of the focal sets.

For this purpose, in this paper a plausibility distribution on a universe X (with perfect mass assignment) will be split into a family of (non-normalized) possibility distributions. In the case of imperfect mass assignment, we can deal with pieces of information separately and we can add or remove some or some parts of them to obtain a mass summing to 1.

In the next Section the fact that Possibility Theory fuzzifies the case when there is only one focal set (with mass 1) in our universe and that a plausibility distribution can be seen as a sum of (non-normalized) possibility distributions will be shown. The results of this Section will be used to normalize mass assignments that do not add 1 in Section 4. Previously, in Section 3 normalization of possibility distributions will be revisited from this point of view. Section 5 contains a possible application of

J. Recasens: *Theory of Evidence with Imperfect Information*, Advances in Soft Computing **6**, 267–274 (2006)
www.springerlink.com © Springer-Verlag Berlin Heidelberg 2006

the previous results to Approximate Reasoning and especially to interpolation and defuzzfication of fuzzy outputs in fuzzy control.

2 Plausibility and Possibility Distributions

A plausibility distribution Pl can be split in possibility ones in a way that it can be thought as the union of them in a logic system based on the Lukasiewicz connectives. Let us recall that the t-norm T_L and t-conorm S_L of Lukasiewicz T_L are defined for all $x, y \in [0,1]$ by $T_L(x,y) = Max(x+y-1,0)$ and $S_L(x,y) = Min(x+y,1)$.

As usual in fuzzy logic, the semantics for the connectives "and" and "or" will be given by Lukasiewicz t-norm and t-conorm, the implication by the residuation $(\hat{T}_L(x|y) = Min(1-x+y,1))$ of Lukasiewicz t-norm, the negation by the strong negation $n(x) = 1-x$ and the quantifers $\exists$ and $\forall$ by the suprem and infimum respectively.

Possibility distributions generalize (fuzzify) plausibility distributions when there is only a focal set B in the universe of discourse X.

Indeed, in this case $Pl(A) = 0$ or $1 \; \forall A \subset X$. Considering the set $\mathscr{P}l$ of subsets of X that are plausible we have

$$A \in \mathscr{P}l \text{ if and only if } \exists x \in X \mid x \in B \wedge x \in A.$$

If instead of B we consider a normalized fuzzy subset μ_B of X, then the previous formula becomes $Pl(A) = sup_{x \in X} T_L(\mu_B(x), A(x))$ where $A(x)$ is the characteristic function of A. Since A is a crisp set, last formula is equivalent to

$$Pl(A) = sup_{x \in A} \mu_B(x)$$

which is the possibility of A wrt μ_B.

In a similar way it can be shown that the necessity of A fuzzifies the belief of A when there is only one focal set.

Let us see what happens when there are more than one focal set. We will study the case with two focal sets that can be easily generalized to any finite number. Let B and C be the two focal sets in X. Then $m(B) + m(C) = 1$. B and C are pieces of evidence and therefore a set will be plausible when some of its elements satisfy at least on of them. This can be written

$$A \in \mathscr{P}l \text{ if and only if } (\exists x \in X \mid x \in B \wedge x \in A) \vee (\exists y \in X \mid y \in C \wedge y \in A).$$

If we fuzzify the previous formula associating to B and C the fuzzy subsets μ_B and μ_C defined as follows respectively,

$$\mu_B(x) = \begin{cases} m(B) & \text{if } x \in B \\ 0 & \text{otherwise} \end{cases} \qquad \mu_C(x) = \begin{cases} m(C) & \text{if } x \in C \\ 0 & \text{otherwise.} \end{cases}$$

we get $Pl(A) = sup_{x \in X} T_L(\mu_B, A(x)) + sup_{y \in X} T_L(\mu_C, A(y))$, which is indeed the definition of $Pl(A)$.

In a similar way, the belief of A can be thought from this point of view.

Given a focal set B of X, $sup_{x \in A} \mu_B(x)$ can be interpreted as the possibility of A wrt B (or conditioned to B). This is, knowing the piece of evidence B, $sup_{x \in A} \mu_B(x)$ gives the possibility of A. The total plausibility of A is then the sum of all the possibilities wrt all the focal sets B of X and the sum can be viewed as the 'or' connective modeled by the Lukasiewicz t-conorm.

More general, if we consider a (non-normalized) possibility distribution Pos_μ generated by a fuzzy subset μ of X, then $Pos_\mu(A)$ is the possibility of A wrt μ. If we have n fuzzy subsets $\mu_1, \mu_2, , \mu_n$ of heights $h_{\mu_1}, h_{\mu_2}, , h_{\mu_n}$ with $\sum_{i=1}^{n} h_{\mu_i} = 1$, we can define a plausibility distribution by simple adding (i.e. applying the disjunction):

$$Pl(A) = Pos_{\mu_1}(A) + Pos_{\mu_2}(A) + ... + Pos_{\mu_n}(A).$$

We can consider the question in the opposite way: A given plausibility distribution can be split in a sum of possibility ones (in many different ways). Indeed, let F be the set of focal elements of X and let us consider a (maximal) chain c_1 in F. Since the elements of c_1 are nested, they generate a non-normalized possibility distribution Pos_{c_1}. Now remove the elements of c_1 from F and select another (maximal) chain c_2 from $F - c_1$. Repeating this process we will get a family of chains $c_1, c_2, , c_k$ partitioning F ($c_i \cap c_j = \emptyset$ if $i \neq j$, and $c_1, \cup c_2 \cup \cup c_k = F$). Then the following proposition can be trivially proved.

Proposition 1. $Pl(A) = \sum_{i=1}^{k} Pos_{c_k}(A) \ \forall A \subset X$.

This permits us to define the *dimension* of a random set.

Definition 1. *The minimum number of chains needed in the previous process will be called the dimension of (X, F).*

Then (X, F) is equivalent to give k fuzzy subsets of X (or, equivalently, k non-normalized possibility distributions).

This concept can help us to understand the structure of (X, F).

Proposition 2. *The dimension of (X, F) is 1 if and only if F is nested or equivalently if it generates a possibility distribution.*

Proposition 3. *If F generates a probability distribution, then the dimension of (X, F) is equal to the cardinality of F.*

3 Normalizing Possibility Distributions

It is well-known that any normal fuzzy subset of a universe X generates a possibility distribution and in this paper the terms possibility distribution and (normalized) fuzzy subset will be indistinctly used. Nevertheless, in many cases there is a need of dealing with non-normalized fuzzy subsets that generate non-normalized possibility distributions. In these cases, it is assumed that there is a lack of information or evidence, which implies that the masses assigned to the focal sets sum a value y_1 smaller than 1. The lack of mass $1 - y_1$ is usually assigned to the empty set. If we want to obtain a normalized possibility distribution from a non-normalized one μ, we must relocate the mass $1 - y_1$, which is equivalent to normalize the fuzzy subset μ. There are many ways to do that (see e.g. [2]).

If μ is a non-normalized possibility distribution on a set X with non-zero values $y_1, y_2, ..., y_n$ with $y_i > y_{i+1}$, the focal sets are $M_i = \{x \in X | \mu(x) \geq y_i\}$ for $i = 1, ..., n$ and the mass assigned to M_i is $m_i = y_i - y_{i+1}$ and the empty set M_0 has mass $m_0 = 1 - y_1$.

In [2] a *valid* normalization of μ is defined as a distribution corresponding to a reassignment of masses to the focal elements such that the new masses of M_i are $m_i + u_i$ for all $i = 1, 2, ..., n$ with $\sum_{i=1}^{n} u_i = m_0$ and $0 \leq u_i \leq 1$ for all $i = 1, 2, ...n$.

Following the ideas of the last section, this means that we add to our possibility distribution the one corresponding to the values u_i.

Probably the three most used normalization procedures for possibility distributions are the following ones.

Definition 2. *The minimal normalization of a fuzzy subset μ of X with greatest value y_1 is the fuzzy subset μ of X defined for all $x \in X$ by*

$$\hat{\mu}(x) = \begin{cases} \mu(x) & \text{if } \mu(x) \neq y_1 \\ 1 & \text{otherwise.} \end{cases}$$

In other words, $\hat{\mu}$ is obtained from μ by replacing the greatest value y_1 by 1.

Definition 3. *The maximal normalization $\hat{\mu}$ of μ is obtained by adding $1 - y_1$ (the lack of evidence) to all non-zero values y_i of μ (i.e.: $\hat{y}_i = y_i + 1 - y_1$ for all $i = 1, 2, ..., n$).*

Definition 4. *The product-related normalization $\hat{\mu}$ of μ is obtained by dividing all its values by the greatest one y_1.*

Taking the last Section into account, these methods correspond to adding the following possibility distributions to our universe.

Proposition 4. *Let μ be a fuzzy subset of X with greatest value y_1. The minimal normalization of μ is obtained by adding the fuzzy subset μ' of X defined by*
$$\mu'(x) = \begin{cases} 1 - y_1 & \text{if } \mu(x) = y_1 \\ 0 & \text{otherwise.} \end{cases}$$

Proposition 5. *Let μ be a fuzzy subset of X with greatest value y_1. The maximal normalization of μ is obtained by adding the fuzzy subset μ' of X defined by*
$$\mu'(x) = \begin{cases} 1 - y_1 & \text{if } x \text{ belongs to the support of } \mu \\ 0 & \text{otherwise.} \end{cases}$$

Proposition 6. *Let μ be a fuzzy subset of X with greatest value y_1. The product related normalization of μ is obtained by adding the fuzzy subset μ' of X defined for all $x \in X$ by $\mu'(x) = \frac{1 - y_1}{y_1} \mu(x)$.*

4 Normalizing Plausibility Distributions with Imperfect Information

Let us generalize the results of the last Section when there is a mass assignment to the focal sets that does not sum to 1. Let F be the set of focal sets on a universe X. If B is a focal set let $m(B)$ be its mass and let $m_0 = 1 - \sum_{B \subset F} m(B)$ be the lack of information. Let Pl be the associated plausibility measure. According to Proposition 3.1., there exist k possibility distributions $Pos_1, ..., Pos_k$ such that for all $A \subset X$ $Pl(A) = Pos_1(A) + ... + Pos_k(A)$. Let h_i be the height of the fuzzy subset μ_i of X associated to Pos_i for all $i = 1, ..., k$ and $h = \sum_{i=1}^{k} h_i$.

4.1 Lack of Information

Let us first study when the masses sum less than 1.

Definition 5. *With the previous notations a valid normalization $\hat{Pl}$ of Pl when $m_0 > 0$ is a plausibility measure corresponding to a reassignment of masses to the focal sets such that the new masses of $B \subset F$ are $m(B) + m'(B)$ with $\sum_{B \subset F} m'(B) = m_0$ and for all $B \subset F$ $0 \leq m'(B) \leq 1$.*

The minimal, maximal and product-related normalizations of possibility distributions can be then generalized to this case obtaining valid normalizations of plausibility measures.

Definition 6. *With the previous notations, the minimal normalization of Pl is obtained by shifting the greatest values of the fuzzy subsets μ_i corresponding to the possibility distributions Pos_i to $h_i + 1 - h$*

Proposition 7. *With the previous notations, the minimal normalization of Pl is obtained by adding the possibility measures corresponding to the fuzzy subsets μ_i', $i = 1, ..., k$ defined by* $\mu_i'(x) = \begin{cases} \frac{1-h}{k} & \text{if } \mu_i(x) = h_i \\ 0 & \text{otherwise.} \end{cases}$

Definition 7. *With the previous notations, the maximal normalization of Pl is obtained by adding $1 - h$ to all the values of the supports of the fuzzy subsets μ_i.*

Proposition 8. *With the previous notations, the maximal normalization of Pl is obtained by adding the possibility measure corresponding to the fuzzy subset μ' defined by* $\mu'(x) = \begin{cases} 1 - h & \text{if } x \text{ is in the support of some } \mu_i \ i = 1, ..., k \\ 0 & \text{otherwise.} \end{cases}$

Definition 8. *With the previous notations, the product related of Pl is obtained by dividing the fuzzy subsets μ_i by h.*

Proposition 9. *With the previous notations, the product related normalization of Pl is obtained by adding the possibility measures corresponding to the fuzzy subsets μ_i' defined for all $x \in X$ by $\mu_i'(x) = \frac{1-h}{h} \mu_i(x)$.*

It interesting to note that the minimal, maximal and product-related normalizations are valid ones.

Also it is worth noticing that after a valid normalization the dimension of X remains unchanged since the set of focal sets is the same.

4.2 Excess of Information

If the amount of mass is greater than 1, we have to remove some of it to normalize our plausibility measure.

The procedure is very similar to the one used in the previous subsection, but we have to be aware that some possibility distributions in which Pl is split can contain less mass than is needed to remove. Let us suppose for example that Pl is split into three possibility distributions of heights 0.8, 0.7 and 0.1. This makes $h = 1.6$, which means that in the minimal case, for instance, we would need to lower the heights of these fuzzy sets by removing $\frac{1.6-1}{3} = 0.2$ which exceeds the height of the third fuzzy subset. Therefore, in the case of the maximal and minimal normalization this situation has to be taken into account. (Note that this can not happen in the product-related normalization).

The definition of valid normalization is dual to Definition 5.

Definition 9. *With the previous notations a valid normalization $\hat{Pl}$ of Pl when $m_0 < 0$ is a plausibility measure corresponding to a reassignment of masses to the focal sets such that the new masses of $B \subset F$ are $m(B) - m'(B)$ with $\sum_{B \subset F} m'(B) = m_0$ and for all $B \subset F$ $0 \le m'(B) \le 1$.*

The minimal and maximal normalizations with excess of information can be done in this way

Definition 10. *Let Pl be a plausibility measure split into k possibility distributions $Pos_1, ..., Pos_k$ corresponding to fuzzy subset $\mu_1, ..., \mu_k$ of X with heights $h_1, ..., h_k$ with $h = \sum_{i=1}^{k} h_i > 1$. The minimal normalization of Pl is obtained in the following way: Remove all possibility distributions with $h_i < (h-1)/k$. Calculate the sum of heights h' of the remaining fuzzy subsets $\mu_{i_1}, ... \mu_{i_l}$ and replace them by the fuzzy subsets $\mu'_{i_j}(x) = Min(\frac{h'-1}{l}, \mu_{i_j})$ $j = 1, ..., l$.*

Definition 11. *Let Pl be a plausibility measure split into k possibility distributions $Pos_1, ..., Pos_k$ corresponding to fuzzy subset $\mu_1, ..., \mu_k$ of X with heights $h_1, ..., h_k$ with $h = \sum_{i=1}^{k} h_i > 1$. The maximal normalization of Pl is obtained in the following way: Remove all possibility distributions with $h_i < (h-1)/k$. Calculate the sum of heights h' of the remaining fuzzy subsets $\mu_{i_1}, ... \mu_{i_l}$ and replace them by the fuzzy subsets $\mu'_{i_j}(x) = Max(\mu_{i_j} - \frac{h'-1}{l}, 0)$ $j = 1, ..., l$.*

5 An Application to Interpolative Reasoning

In this Section we will develop an approach to Interpolative Reasoning using the results of the previous ones.

Let us suppose that the input and output spaces of a rule-based model are intervals $[a,b]$ and $[c,d]$ of the real line partitioned into n and m triangular fuzzy subsets $\mu_1,...,\mu_n$ and $\nu_1,...,\nu_m$ respectively such that there are $n+2$ points $a = x_0 = x_1 < x_2 < ... < x_n = x_{n+1} = b$ with $\mu_i = [x_{i-1},x_i,x_{i+1}]$ and $m+2$ points $c = y_0 = y_1 < y_2 < ... < y_n = y_{n+1} = d$ with $\nu_i = [y_{i-1},y_i,y_{i+1}]$. Let us consider a set of rules $R_1,...,R_n$ of the form

$$\text{Rule } R_i : \text{ If } X \text{ is } \mu_i \text{ Then } y \text{ is } \nu_{r(i)} \text{ where } r:\{1,2,...,n\} \to \{1,2,...m\}.$$

In order to assure continuity in the reasoning, we will assume that $|r(i) - r(i-1)| \leq 1 \; \forall i = 1,...,n$ (i.e., the fuzzy subsets of the outputs of two correlative rules are not disjoint).

Let x be a point in $[x_i,x_{i+1}]$. Applying the i-th Rule to x, we obtain the fuzzy subset $Min(\nu_{r(i)},\mu_i(x))$. The two fuzzy subsets $Min(\nu_{r(i)},\mu_i(x))$ and $Min(\nu_{r(i+1)},\mu_{i+1}(x))$ generate a perfect plausibility measure Pl since $\mu_i(x) + \mu_{i+1}(x) = 1$.

Proposition 10. *If $\nu_{r(i)} \neq \nu_{r(i+1)}$, there exists exactly one point y in $[c,d]$ with $Pl(y) = 1$.*

The rules R_i are rough local descriptions of a map f. In this case it seems reasonable to take the value of y with $Pl(y) = 1$ as the image of x.

Proposition 11. *If $\nu_{r(i)} \neq \nu_{r(i+1)}$, then the map f is linear in the interval $[i,i+1]$.*

If $\nu_{r(i)} = \nu_{r(i+1)}$, then there is not only a point y in $[c,d]$, but an interval of points y with $Pl(y) = 1$.

Example 1. Let us consider the following rule-base system: $[a,b] = [0,3]$ $[c,d] = [3,5]$, $[0,3]$ partitioned into the fuzzy subsets $S = [0,0,1]$, $M = [0,1,2]$, $L = [1,2,3]$, $XL = [2,3,3]$ and $[3,5]$ into $S = [3,3,4]$, $M = [3,4,5]$, $L = [4,5,5]$ with the following rules:

$$\text{Rule } R_1 : \text{ If } x \text{ is } S \text{ Then } y \text{ is } S$$

$$\text{Rule } R_2 : \text{ If } x \text{ is } M \text{ Then } y \text{ is } M$$

$$\text{Rule } R_3 : \text{ If } x \text{ is } L \text{ Then } y \text{ is } M$$

$$\text{Rule } R_4 : \text{ If } x \text{ is } XL \text{ Then } y \text{ is } L$$

The following figure shows for every value $x \in [0,3]$ the corresponding values y in $[3,5]$ with $Pl(y) = 1$.

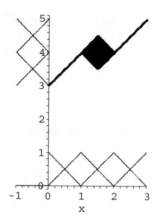

6 Concluding Remarks

Plausibility distributions have been split into a families of non-normalized possibility distributions. This permits to add or remove some information in case the mass assignment do not sum to 1. The normalization of possibility distributions has been revisited and generalized to plausibility ones.

It would be interesting to find some way to determine "good" splits and easy algorithms to generate them.

Also Possibility Theory has been reinterpreted as the fuzzification of Evidence Theory when there is only one focal set with mass 1.

References

[1] Dempster, A. P. (1967) Upper and lower probabilities induced by a multivalued mapping. *Annals of Math. Statistics* **38** 325-339.
[2] Lawry, J. (2001) Possibilistic Normalization and Reasoning under Partial Inconsistency *IJUFKBS*, **9** 413-436.
[3] Zadeh, L. A. (1978) Fuzzy Sets as a Basis for a Theory of Possibility. *FS&S* **1** 3-28.

Conditional IF-probability

Katarína Lendelová

Faculty of Natural Sciences, Matej Bel University, Department of Mathematics, Tajovského 40, SK-974 01 Banská Bystrica
lendelov@fpv.umb.sk

The aim of this paper is to define the product operation on the family of IF-events and the notion of joint IF-observable. We formulate the version of conditional IF-probability on IF-events, too.

1 Introduction

In recent years the theory of IF-sets introduced by Atanassov ([1]) has been studied by many authors. An *IF-set A* on a space Ω is a couple (μ_A, ν_A), where $\mu_A : \Omega \to [0,1]$, $\nu_A : \Omega \to [0,1]$ are functions such that $\mu_A(\omega) + \nu_A(\omega) \leq 1$ for each $\omega \in \Omega$ (see [1]). The function μ_A is called the membership function, the function ν_A is called the non membership function. In [3] Grzegorzewski and Mrówka defined the probability on the family of IF-events $\mathcal{N} = \{(\mu_A, \nu_A) \; ; \; \mu_A, \nu_A \text{ are } \mathcal{S} - measurable \text{ and } \mu_A + \nu_A \leq 1\}$ as a mapping $\mathcal{P}$ from the family $\mathcal{N}$ to the set of all compact intervals in **R** by the formula $\mathcal{P}((\mu_A, \nu_A)) = \left[\int_\Omega \mu_A \, dP, 1 - \int_\Omega \nu_A \, dP \right]$, where $(\Omega, \mathcal{S}, P)$ is probability space. This IF-probability was axiomatically characterized by B. Riečan (see[11]).

More general situation was studied in [12], where the author introduced the notion of IF-probability on the family

$$\mathscr{F} = \{(f,g) \; ; \; f,g \in \mathscr{T}, \mathscr{T} \text{ is a Lukasiewicz tribe and } f+g \leq 1\}$$

as a mapping $\mathcal{P}$ from the family $\mathscr{F}$ to the family $\mathscr{J}$ of all closed intervals $\langle a, b \rangle$ such that $0 \leq a \leq b \leq 1$. Variant of Central limit theorem and Weak law of large numbers were proved as an illustration of method applied on these IF-events. It can see in the papers [8], [9].

More general situation was used in [7]. The authors defined the IF-probability on the family $\mathscr{M} = \{(a,b) \in M, a+b \leq u\}$, where M is σ-complete MV-algebra, which can be identified with the unit interval of a unique ℓ-group G with strong unit u, in symbols,

$$M = \Gamma(G, u) = (\langle 0, u \rangle, 0, u, \neg, \oplus, \odot)$$

K. Lendelová: *Conditional IF-probability*, Advances in Soft Computing **6**, 275–283 (2006)
www.springerlink.com © Springer-Verlag Berlin Heidelberg 2006

where $\langle 0, u \rangle = \{a \in G \, ; \, 0 \leq a \leq u\}$, $\neg a = u - a$, $a \oplus b = (a+b) \wedge u$, $a \odot b = (a+b-u) \vee 0$ (see [14]). We say that G is the ℓ-group (with strong unit u) corresponding to M.

By an ℓ-group we shall mean a lattice-ordered Abelian group. For any ℓ-group G, an element $u \in G$ is said to be a strong unit of G, if for all $a \in G$ there is an integer $n \geq 1$ such that $nu \geq a$.

The independence of IF-observables, the convergence of IF-observables and the Strong law of large numbers were studied on this family of IF-events, see [5], [6].

In this paper we define the product operation on the family $\mathscr{F}$ of IF-events

$$\mathscr{F} = \{(f,g) \, ; \, f, g \in \mathscr{T}, \mathscr{T} \text{ is Lukasiewicz tribe and } f + g \leq 1\}$$

and formulate the version of conditional IF-probability on this family. In *Section 2* we introduce the operations on $\mathscr{F}$ and $\mathscr{J}$, where $\mathscr{J}$ is the family of all closed intervals $\langle a, b \rangle$ such that $0 \leq a \leq b \leq 1$.

2 Basic Notions

Now we introduce operations on $\mathscr{F}$. Let $A = (a_1, a_2)$, $B = (b_1, b_2)$. Then we define

$$A \oplus B = (a_1 \oplus b_1, a_2 \odot b_2) = \big((a_1 + b_1) \wedge 1, (a_2 + b_2 - 1) \vee 0 \big),$$

$$A \odot B = (a_1 \odot b_1, a_2 \oplus b_2) = \big((a_1 + b_1 - 1) \vee 0, (a_2 + b_2) \wedge 1 \big).$$

If $A_n = (a_{n1}, a_{n2})$, then we write

$$A_n \nearrow A \Longleftrightarrow a_{n1} \nearrow a_1, \, a_{n2} \searrow a_2.$$

IF-probability $\mathscr{P}$ on $\mathscr{F}$ is a mapping from $\mathscr{F}$ to the family $\mathscr{J}$ of all closed intervals $\langle a, b \rangle$ such that $0 \leq a \leq b \leq 1$. Here we define

$$\langle a, b \rangle + \langle c, d \rangle = \langle a + c, b + d \rangle,$$

$$\langle a_n, b_n \rangle \nearrow \langle a, b \rangle \Longleftrightarrow a_n \nearrow a, \, b_n \nearrow b.$$

By an **IF-probability on** $\mathscr{F}$ we understand any function $\mathscr{P} : \mathscr{F} \to \mathscr{J}$ satisfying the following properties:

(i) $\mathscr{P}((1,0)) = \langle 1, 1 \rangle = 1$; $\mathscr{P}((0,1)) = \langle 0, 0 \rangle = 0$;
(ii) if $A \odot B = (0,1)$ and $A, B \in \mathscr{F}$, then $\mathscr{P}(A \oplus B) = \mathscr{P}(A) + \mathscr{P}(B)$;
(iii) if $A_n \nearrow A$, then $\mathscr{P}(A_n) \nearrow \mathscr{P}(A)$.

IF-probability $\mathscr{P}$ is called **separating**, if

$$\mathscr{P}\big((a_1, a_2)\big) = \langle \mathscr{P}^{\flat}(a_1), 1 - \mathscr{P}^{\sharp}(a_2) \rangle,$$

where the functions $\mathscr{P}^{\flat}, \mathscr{P}^{\sharp} : \mathscr{T} \to [0,1]$ are probabilities. The next important notion is notion of IF-observable.

By **IF-observable on** $\mathscr{F}$ we understand any mapping $x : \mathscr{B}(\mathbf{R}) \to \mathscr{F}$ satisfying the following conditions:

(i) $x(\mathbf{R}) = (1,0)$;
(ii) if $A \cap B = O$, then $x(A) \odot x(B) = (0,1)$ and $x(A \cup B) = x(A) \oplus x(B)$;
(iii) if $A_n \nearrow A$, then $x(A_n) \nearrow x(A)$.

If we denote $x(A) = \left(x^{\flat}(A), 1 - x^{\sharp}(A)\right)$ for each $A \in \mathscr{B}(\mathbf{R})$, then $x^{\flat}, x^{\sharp} : \mathscr{B}(\mathbf{R}) \to \mathscr{T}$ are observables, see [9].

3 The Family of IF-events with Product

We introduce the notion of product operation on the family of IF-events $\mathscr{F}$ and show an example of this operation.

Definition 1. *We say that a binary operation · on $\mathscr{F}$ is product if it satisfying the following conditions:*

(i) $(1,0) \cdot (a_1, a_2) = (a_1, a_2)$ for each $(a_1, a_2) \in \mathscr{F}$;
(ii) the operation · is commutative and associative;
(iii) if $(a_1, a_2) \odot (b_1, b_2) = (0,1)$ and $(a_1, a_2), (b_1, b_2) \in \mathscr{F}$, then

$$(c_1, c_2) \cdot \left((a_1, a_2) \oplus (b_1, b_2)\right) = \left((c_1, c_2) \cdot (a_1, a_2)\right) \oplus \left((c_1, c_2) \cdot (b_1, b_2)\right)$$

and

$$\left((c_1, c_2) \cdot (a_1, a_2)\right) \odot \left((c_1, c_2) \cdot (b_1, b_2)\right) = (0,1)$$

for each $(c_1, c_2) \in \mathscr{F}$;
(iv) if $(a_{1n}, a_{2n}) \searrow (0,1)$, $(b_{1n}, b_{2n}) \searrow (0,1)$ and $(a_{1n}, a_{2n}), (b_{1n}, b_{2n}) \in \mathscr{F}$, then $(a_{1n}, a_{2n}) \cdot (b_{1n}, b_{2n}) \searrow (0,1)$.

Now we show an example of product operation on $\mathscr{F}$.

Theorem 1. *The operation · defined by*

$$(x_1, y_1) \cdot (x_2, y_2) = (x_1 \cdot x_2, y_1 + y_2 - y_1 \cdot y_2)$$

for each $(x_1, y_1), (x_2, y_2) \in \mathscr{F}$ is product operation on $\mathscr{F}$.

Proof. (i) Let (a_1, a_2) is an element of family $\mathscr{F}$. Then

$$(1,0) \cdot (a_1, a_2) = (1.a_1, 0 + a_2 - 0 \cdot a_2) = (a_1, a_2).$$

(ii) The operation · is commutative and associative.
(iii) Let $(a_1, a_2) \odot (b_1, b_2) = (0,1)$. Then

$$
\begin{aligned}
(c_1, c_2) \cdot \left((a_1, a_2) \oplus (b_1, b_2)\right) &= (c_1, c_2) \cdot (a_1 \oplus b_1, a_2 \odot b_2) = \\
&= (c_1, c_2) \cdot \left((a_1 + b_1) \wedge 1, (a_2 + b_2 - 1) \vee 0\right) = \\
&= (c_1, c_2) \cdot (a_1 + b_1, a_2 + b_2 - 1) = \\
&= (a_1 c_1 + b_1 c_1, a_2 + b_2 + 2c_2 - 1 - a_2 c_2 - b_2 c_2)
\end{aligned}
$$

and

$$((c_1,c_2)\cdot(a_1,a_2)) \ \oplus \ ((c_1,c_2)\cdot(b_1,b_2)) =$$
$$= (c_1a_1,c_2+a_2-c_2a_2)\oplus(c_1b_1,c_2+b_2-c_2b_2) =$$
$$= (c_1a_1\oplus c_1b_1,(c_2+a_2-c_2a_2)\odot(c_2+b_2-c_2b_2)) =$$
$$= ((c_1a_1+c_1b_1)\wedge 1,(a_2+b_2+2c_2-1-a_2c_2-b_2c_2)\vee 0) =$$
$$= (a_1c_1+b_1c_1,a_2+b_2+2c_2-1-a_2c_2-b_2c_2).$$

Hence $(c_1,c_2)\cdot((a_1,a_2)\oplus(b_1,b_2)) = ((c_1,c_2)\cdot(a_1,a_2))\oplus((c_1,c_2)\cdot(b_1,b_2))$. Moreover

$$((c_1,c_2)\cdot(a_1,a_2)) \ \odot \ ((c_1,c_2)\cdot(b_1,b_2)) =$$
$$= (c_1a_1,c_2+a_2-c_2a_2)\odot(c_1b_1,c_2+b_2-c_2b_2) =$$
$$= (c_1a_1\odot c_1b_1,(c_2+a_2-c_2a_2)\oplus(c_2+b_2-c_2b_2)) = (0,1).$$

(iv) Let $(a_{1n},a_{2n}) \searrow (0,1)$, $(b_{1n},b_{2n}) \searrow (0,1)$. Since $a_{1n}\searrow 0$, $a_{2n}\nearrow 1$, $b_{1n}\searrow 0$ and $b_{2n}\nearrow 1$, then

$$(a_{1n},a_{2n})\cdot(b_{1n},b_{2n}) = (a_{1n}b_{1n},a_{2n}+b_{2n}-a_{2n}b_{2n}) \searrow (0,1).$$

The next important notion is a notion of joint observable.

Definition 2. *Let $x,y : \mathscr{B}(\mathbf{R}) \to \mathscr{F}$ be two IF-observables. The joint IF-observable of the IF-observables x,y is a mapping $h : \mathscr{B}(\mathbf{R}^2) \to \mathscr{F}$ satisfying the following conditions:*

(i) $h(\mathbf{R}^2) = (1,0)$;
(ii) if $A,B \in \mathscr{B}(\mathbf{R}^2)$ and $A\cap B = \emptyset$, then $h(A\cup B) = h(A)\oplus h(B)$ and $h(A)\odot h(B) = (0,1)$;
(iii) if $A,A_1,\ldots \in \mathscr{B}(\mathbf{R}^2)$ and $A_n \nearrow A$, then $h(A_n) \nearrow h(A)$;
(iv) $h(C\times D) = x(C)\cdot y(D)$ for each $C,D \in \mathscr{B}(\mathbf{R})$.

Theorem 2. *To each two IF-observables $x,y : \mathscr{B}(\mathbf{R}) \to \mathscr{F}$ there exists their joint IF-observable.*

Proof. In [13] *Theorem 3.3.*

4 Conditional IF-probability

In classical probability space $(\Omega, \mathscr{S}, P)$ we have two possibilities to define conditional probability of $A \in \mathscr{S}$

a. $P(A|\mathscr{S}_0)$, where $\mathscr{S}_0 \subset \mathscr{S}$ is a σ-algebra, is a version of conditional probability of A with respect to $\mathscr{S}_0$, if $P(A|\mathscr{S}_0) : \Omega \to \mathbf{R}$ is $\mathscr{S}_0$-measurable and

$$\int_B P(A|\mathscr{S}_0) \, dP = P(A\cap B)$$

for every $B \in \mathscr{S}_0$.

b. $p(A|\xi)$, where $\xi : \Omega \rightarrow \mathbf{R}$ is a random variable, is a version of conditional probability of A with respect to ξ, if $p(A|\xi) : \mathbf{R} \rightarrow \mathbf{R}$ is a Borel function such that

$$\int_B p(A|\xi) \, dP_\xi = P(A \cap \xi^{-1}(B))$$

for every $B \in \mathscr{B}(\mathbf{R})$.

We take the second case for definition conditional IF-probability on family $\mathscr{F}$ of IF-events with product $\cdot$ defined by

$$(x_1, y_1) \cdot (x_2, y_2) = (x_1 \cdot x_2, y_1 + y_2 - y_1 \cdot y_2)$$

for each $(x_1, y_1), (x_2, y_2) \in \mathscr{F}$. We show some properties of conditional IF-probability in this section, too.

Definition 3. *Let (a_1, a_2) be an element of family $\mathscr{F}$ of IF-events with product $\cdot$ given by formula*

$$(x_1, y_1) \cdot (x_2, y_2) = (x_1 \cdot x_2, y_1 + y_2 - y_1 \cdot y_2)$$

for $(x_1, y_1), (x_2, y_2) \in \mathscr{F}$. Let $y : \mathscr{B}(\mathbf{R}) \rightarrow \mathscr{F}$ be an IF-observable, $y(A) = \left(y^\flat(A), 1 - y^\sharp(A) \right)$, $A \in \mathscr{B}(\mathbf{R})$. A function $\mathbf{p}\big((a_1, a_2)|y\big) : \mathscr{B}(\mathbf{R}) \rightarrow \mathscr{J}$ is a version of the conditional IF-probability of (a_1, a_2) with respect to y, if there exists the Borel functions $\mathbf{p}^\flat(a_1|y^\flat), \mathbf{p}^\sharp(1 - a_2|y^\sharp) : \mathbf{R} \rightarrow \mathbf{R}$ such that

$$\left\langle \int_B \mathbf{p}^\flat(a_1|y^\flat) \, d\mathscr{P}^\flat, \int_B \mathbf{p}^\sharp(1 - a_2|y^\sharp) \, d\mathscr{P}^\sharp \right\rangle = \mathscr{P}\big((a_1, a_2) \cdot y(B)\big)$$

for every $B \in \mathscr{B}(\mathbf{R})$. Then

$$\mathbf{p}\big((a_1, a_2)|y\big)(B) = \left\langle \mathbf{p}^\flat(a_1|y^\flat)(B), \mathbf{p}^\sharp(1 - a_2|y^\sharp)(B) \right\rangle.$$

Theorem 3. *The function $\mathbf{p}^\flat(a_1|y^\flat) : \mathbf{R} \rightarrow \mathbf{R}$ from Definition 3 is a version of conditional probability of a_1 with respect $y^\flat$ and the function $\mathbf{p}^\sharp(1 - a_2|y^\sharp)$ is a version of conditional probability of $1 - a_2$ with respect $y^\sharp$.*

Proof. By Definition 3 we obtain

$$\left\langle \int_B \mathbf{p}^\flat(a_1|y^\flat) \, d\mathscr{P}^\flat, \int_B \mathbf{p}^\sharp(1 - a_2|y^\sharp) \, d\mathscr{P}^\sharp \right\rangle = \mathscr{P}\big((a_1, a_2) \cdot y(B)\big) =$$

$$= \mathscr{P}\big((a_1, a_2) \cdot (y^\flat(B), 1 - y^\sharp(B))\big) = \mathscr{P}\big((a_1 y^\flat(B), 1 - (1 - a_2) y^\sharp(B))\big) =$$

$$= \left\langle \mathscr{P}^\flat(a_1 y^\flat(B)), \mathscr{P}^\sharp\big((1 - a_2) y^\sharp(B)\big) \right\rangle.$$

Hence

$$\int_B \mathbf{p}^\flat(a_1|y^\flat) \, d\mathscr{P}^\flat = \mathscr{P}^\flat(a_1 y^\flat(B)),$$

$$\int_B \mathbf{p}^\sharp(1-a_2|y^\sharp) \, d\mathscr{P}^\sharp = \mathscr{P}^\sharp((1-a_2)y^\sharp(B)).$$

Lemma 1. *The function* $\mathbf{p}((a_1,a_2)|y)$ *from Definition 3 exists and if g,h are two versions of conditional IF-probability of* (a_1,a_2) *with respect to y, then* $g = h$ $\mathscr{P} \circ y$-*almost everywhere.*

Proof. Existence of functions $\mathbf{p}^\flat(a_1|y^\flat) : \mathbf{R} \to \mathbf{R}$ and $\mathbf{p}^\sharp(1-a_2|y^\sharp)$ implies existence of function $\mathbf{p}((a_1,a_2)|y)$.

Let $f, g : \mathscr{B}(\mathbf{R}) \to \mathscr{J}$, $f(A) = \langle f^\flat(A), f^\sharp(A) \rangle$, $g(A) = \langle g^\flat(A), g^\sharp(A) \rangle$ be two versions of conditional IF-probability, where $A \in \mathscr{B}(\mathbf{R})$. From *Theorem 3* and from definition and properties of conditional probability we have

$$f^\flat = g^\sharp \quad \mathscr{P}^\flat - \text{almost everywhere}$$
$$f^\sharp = g^\sharp \quad \mathscr{P}^\sharp - \text{almost everywhere}$$

Hence $f = g$ $\mathscr{P} \circ y$-almost everywhere.

Theorem 4. *Let* $\mathscr{F}$ *be family of IF-events and* $\cdot$ *be the product given by*

$$(x_1,y_1) \cdot (x_2,y_2) = (x_1 \cdot x_2, y_1 + y_2 - y_1 \cdot y_2)$$

for each $(x_1,y_1),(x_2,y_2) \in \mathscr{F}$. *Let* $(a_1,a_2) \in \mathscr{F}$, $y : \mathscr{B}(\mathbf{R}) \to \mathscr{F}$ *be an IF-observable. Then* $\mathbf{p}((a_1,a_2)|y)$ *has the following properties:*

(i) $\mathbf{p}((0,1)|y) = \langle 0,0 \rangle = 0$, $\mathbf{p}((1,0)|y) = \langle 1,1 \rangle = 1$ $\mathscr{P} \circ y$-*almost everywhere;*

(ii) $0 \le \mathbf{p}((a_1,a_2)|y) \le 1$ $\mathscr{P} \circ y$-*almost everywhere;*

(iii) if $\bigodot_{i=1}^{\infty}(a_{1i},a_{2i}) = (0,1)$, *then* $\mathbf{p}\left(\bigoplus_{i=1}^{\infty}(a_{1i},a_{2i})\Big|y\right) = \left\langle \sum_{i=1}^{\infty}\mathbf{p}^\flat(a_{1i}|y^\flat), \sum_{i=1}^{\infty}\mathbf{p}^\sharp(1 - a_{2i}|y^\sharp)\right\rangle$ $\mathscr{P} \circ y$-*almost everywhere.*

Proof.
By *Definition 3* we have $\mathscr{P}((a_1,a_2) \cdot y(B)) = \left\langle \int_B \mathbf{p}^\flat(a_1|y^\flat) \, d\mathscr{P}^\flat, \int_B \mathbf{p}^\sharp(1-a_2|y^\sharp) \, d\mathscr{P}^\sharp \right\rangle$.

(i) If $(a_1,a_2) = (0,1)$, then

$$\mathscr{P}((0,1) \cdot y(B)) = \mathscr{P}((0,1)) = \langle 0,0 \rangle = \left\langle \int_B 0 \, d\mathscr{P}^\flat, \int_B 0 \, d\mathscr{P}^\sharp \right\rangle.$$

Hence $\mathbf{p}((0,1)|y) = \langle 0,0 \rangle = 0$.

If $(a_1,a_2) = (1,0)$, then

$$\mathscr{P}\big((1,0)\cdot y(B)\big) = \mathscr{P}(y(B)) = (\mathscr{P}\circ y)(B) = \left\langle \int_B 1\,\mathrm{d}\mathscr{P}^\flat, \int_B 1\,\mathrm{d}\mathscr{P}^\sharp \right\rangle.$$

(ii) If $B \in \mathscr{B}(\mathbf{R})$, then

$$0 = \langle 0,0 \rangle = \mathscr{P}\big((a_1,a_2)\cdot y(\emptyset)\big) \le \mathscr{P}\big((a_1,a_2)\cdot y(B)\big) =$$

$$= \left\langle \int_B \mathbf{p}^\flat(a_1|y^\flat)\,\mathrm{d}\mathscr{P}^\flat, \int_B \mathbf{p}^\sharp(1-a_2|y^\sharp)\,\mathrm{d}\mathscr{P}^\sharp \right\rangle \le \mathscr{P}\big((a_1,a_2)\cdot y(\mathbf{R})\big) \le \langle 1,1 \rangle = 1$$

and

$$(\mathscr{P}\circ y)\big(\{t \in \mathbf{R}; \mathbf{p}((a_1,a_2)|y) < \langle 0,0 \rangle = 0\}\big) = (\mathscr{P}\circ y)(B_0) = \langle 0,0 \rangle = 0,$$

$$(\mathscr{P}\circ y)\big(\{t \in \mathbf{R}; \mathbf{p}((a_1,a_2)|y) > \langle 1,1 \rangle = 1\}\big) = (\mathscr{P}\circ y)(B_1) = \langle 0,0 \rangle = 0.$$

In the reverse cases

$$(\mathscr{P}\circ y)(B_0) > 0, \ (\mathscr{P}\circ y)(B_1) > 0$$

we get contradictions

$$\int_{B_0} \mathbf{p}^\flat(a_1|y^\flat)\,\mathrm{d}\mathscr{P}^\flat < 0 \ , \ \int_{B_1} \mathbf{p}^\flat(a_1|y^\flat)\,\mathrm{d}\mathscr{P}^\flat > 1,$$

$$\int_{B_0} \mathbf{p}^\sharp(1-a_2|y^\sharp)\,\mathrm{d}\mathscr{P}^\sharp < 0 \ , \ \int_{B_1} \mathbf{p}^\sharp(1-a_2|y^\sharp)\,\mathrm{d}\mathscr{P}^\sharp > 1$$

respectively.

(iii) Let $\bigodot\limits_{i=1}^{\infty} (a_{1i},a_{2i}) = (0,1)$. Then

$$\mathscr{P}\left(\left(\bigoplus_{i=1}^{\infty}(a_{1i},a_{2i})\right)\cdot y(B)\right) = \mathscr{P}\left(\bigoplus_{i=1}^{\infty}\big((a_{1i},a_{2i})\cdot y(B)\big)\right) =$$

$$= \sum_{i=1}^{\infty} \mathscr{P}\big((a_{1i},a_{2i})\cdot y(B)\big) =$$

$$= \sum_{i=1}^{\infty} \left\langle \int_B \mathbf{p}^\flat(a_{1i}|y^\flat)\,\mathrm{d}\mathscr{P}^\flat, \int_B \mathbf{p}^\sharp(1-a_{2i}|y^\sharp)\,\mathrm{d}\mathscr{P}^\sharp \right\rangle =$$

$$= \left\langle \int_B \sum_{i=1}^{\infty}\mathbf{p}^\flat(a_{1i}|y^\flat)\,\mathrm{d}\mathscr{P}^\flat, \int_B \sum_{i=1}^{\infty}\mathbf{p}^\sharp(1-a_{2i}|y^\sharp)\,\mathrm{d}\mathscr{P}^\sharp \right\rangle.$$

5 Conclusion

The paper is concerned in the probability theory on IF-events with product. We define the notion of product operation · and joint IF-observable. We formulate the version of conditional IF-probability on family of IF-events.

Acknowledgements

This paper was supported by Grant VEGA 1/2002/05.

References

[1] K. Atanassov (1999). Intuitionistic Fuzzy sets : Theory and Applications. In *Physica Verlag*, New York.

[2] A. Dvurečenskij - S. Pulmannová (2000). New Trends in Quantum Structures. In *Kluwer*, Dordrecht.

[3] P. Grzegorzewski - E. Mrówka (2002). Probability of intuistionistic fuzzy events. In *Soft Metods in Probability, Statistics and Data Analysis (P. Grzegorzewski et al. eds.)*, Physica Verlag, New York, 105-115.

[4] K. Lendelová (2005). IF-probability on MV-algebras. In *Notes on Intuitionistic Fuzzy Sets, Vol.* **11**, Number 3, 66-72.

[5] K. Lendelová. Convergence of IF-observables. (submitted to Proceeding of IWIFSGN'2005, Warszawa)

[6] K. Lendelová. Strong law of large numbers for IF-events. (accepted in Proceeding of IPMU'2006)

[7] K. Lendelová - J. Petrovičová (2005). Representation of IF-probability on MV-algebras. In *Soft Computing - A Fusion of Foundation, Methodologies and Applications*, Springer-Verlag (Online).

[8] K. Lendelová - B. Riečan (2004). Weak law of large numbers for IF-events. In *Current Issues in Data and Knowledge Engineering (Bernard De Baets et al. eds.)*, EXIT, Warszawa, 309-314.

[9] J. Petrovičová - B. Riečan (2005). On the central limit theorem on IFS-events. In *Mathware Soft Computing* **12**, 5-14.

[10] P. Pták - S. Pulmannová (1991). Orthomodular Structures as Quantum Logics. In *Kluwer Acad. Publ.*, Dordrecht.

[11] B. Riečan (2003). A descriptive definition of the probability on intuitionistic fuzzy sets. In *EUSFLAT '2003 (M. Wagenecht, R. Hampet eds.)*, Zittau-Goerlitz Univ. Appl. Sci., 263-266.

[12] B. Riečan (2004). Representation of Probabilities on IFS Events. In *Soft Methodology and Random Information Systems (López-Diáz et al. eds.)*, Springer, Berlin Heidelberg New York, 243-248.

[13] B. Riečan. On the probability and random variables on IF events. In *Applied Artifical Inteligence. World Scientific 2006* (accepted)

[14] B. Riečan - D. Mundici (2002). Probability in MV-algebras. In *Handbook of Measure Theory (E. Pap ed.)*, Elsevier, Amsterdam, pages 869-909.

[15] B. Riečan - T. Neubrunn (1997). Integral, Measure, and Ordering. In *Kluwer Academic Publishers, Dordrecht and Ister Science*, Bratislava.

[16] M. Vrábelová (2000). On the conditional probability in product MV-algebras. In *Soft Comput.* **4**, 58-61.

On Two Ways for the Probability Theory on IF-sets

Beloslav Riečan

[1] M. Bel University Banská Bystrica
    riecan@fpv.umb.sk
[2] Math. Inst. Slovak Academy of Sciences
    riecan@mat.savba.sk

One of the important problems of IF-sets theory ([1])is the creation of the probability theory on the family $\mathscr{F}$ of all IF-events. We present here two ways how to solve this problem. The first one is an embedding of $\mathscr{F}$ to a convenient MV-algebra. The second is a substituting the notions of an MV-algebra by a more general notions of so-called L-lattice.

1 IF-events

Consider a measurable space $(\Omega, \mathscr{S})$, $\mathscr{S}$ be a σ-algebra, $\mathscr{T}$ be the family of all $\mathscr{S}$-measurable functions $f : \Omega \to [0,1]$,

$$\mathscr{F} = \{(\mu_A, \nu_A); \mu_A, \nu_A : \Omega \to [0,1], \mu_A, \nu_A \text{ are } \mathscr{S}\text{-measurable}, \mu_A + \nu_A \le 1\}.$$

There are at least two ways how to define probability on $\mathscr{F}$. They were discovered independently (see the following examples)

Example 1(Grzegorzewski, Mrowka [5]). The probability $\mathscr{P}(A)$ of an IF event A is defined as the interval

$$\mathscr{P}(A) = [\int_{\Omega} \mu_A dp, 1 - \int_{\Omega} \nu_A dp].$$

Example 2(Gersternkorn, Manko [4]). The probability $\mathscr{P}(A)$ of an IF - event A is defined as the number

$$\mathscr{P}(A) = \frac{1}{2}(\int_{\Omega} \mu_A dp + 1 - \int_{\Omega} \nu_A dp).$$

Based on Example 1 the following axiomatic definition of probability was introduced ([8]) as a function from $\mathscr{F}$ to the family $\mathscr{J}$ of all compact subintervals of the unit interval

Definition . An IF - probability on $\mathscr{F}$ is a mapping $\mathscr{P} : \mathscr{F} \to \mathscr{J}$ satisfying the following conditions:

B. Riečan: *On Two Ways for the Probability Theory on IF-sets*, Advances in Soft Computing **6**, 285–290 (2006)
www.springerlink.com © Springer-Verlag Berlin Heidelberg 2006

(i) $\mathscr{P}((0,1)) = [0,0], \mathscr{P}((1,0)) = [1,1];$

(ii) $\mathscr{P}((\mu_A, \nu_A)) + \mathscr{P}((\mu_B, \nu_B)) = \mathscr{P}((\mu_A, \nu_A) \oplus (\mu_B, \nu_B)) + \mathscr{P}((\mu_A, \nu_A) \odot (\mu_B, \nu_B))$ for any $(\mu_A, \nu_A), (\mu_B, \nu_B) \in \mathscr{F}$;

(iii) $(\mu_{A_n}, \nu_{A_n}) \nearrow (\mu_A, \nu_A) \Longrightarrow \mathscr{P}((\mu_{A_n}, \nu_{A_n})) \nearrow \mathscr{P}((\mu_A, \nu_A)).$

(Recall that $[a,b] + [c,d] = [a+c, b+d]$, and $[a_n, b_n] \nearrow [a,b]$ means that $a_n \nearrow a, b_n \nearrow b$. On the other hand $(\mu_{A_n}, \nu_{A_n}) \nearrow (\mu_A, \nu_A)$ means that $\mu_{A_n} \nearrow \mu_A$, and $\nu_{A_n} \searrow \nu_A$.)

In the present time there is known the general form of all IF-probabilities ([9]). Of course, the representation of the probability $\mathscr{P} : \mathscr{F} \to \mathscr{J}$ is considered in the form

$$\mathscr{P}(A) = [f(\int \mu_A dP, \int \nu_A dP), g(\int \mu_A dP, \int \nu_A dP)].$$

Theorem 1. ([10]) To any probability $\mathscr{P} : \mathscr{F} \to \mathscr{J}$ there exist $\alpha, \beta \in [0,1], \alpha \le \beta$ such that

$$\mathscr{P}((\mu_A, \nu_A)) = [(1-\alpha) \int \mu_A dP + \alpha(1 - \int \nu_A dP), (1-\beta) \int \mu_A dP + \beta(1 - \int \nu_A dP)].$$

It is easy to see that the examples 1 and 2 are special cases of the preceding characterization. In the first case $\alpha = 0$ and $\beta = 1$, in the second example $\alpha = \beta = \frac{1}{2}$.

Of course, in the probability theory on MV-algebras some mappings are considered with the set R of all real numbers as the range. Therefore we must to decompose an $\mathscr{F}$-valued function into R-valued functions.

Definition. A function $p : \mathscr{F} \to [0,1]$ will be called a state if the following conditions are satisfied:

(i) $p((0,1)) = 0, p((1,0)) = 1;$

(ii) $p((\mu_A, \nu_A) \oplus (\mu_B, \nu_B)) + p((\mu_A, \nu_A) \odot (\mu_B, \nu_B)) = p((\mu_A, \nu_A)) + p((\mu_B, \nu_B))$ for any $(\mu_A, \nu_A), (\mu_B, \nu_B) \in \mathscr{F}$;

(iii) $(\mu_{A_n}, \nu_{A_n}) \nearrow (\mu_A, \nu_A) \Longrightarrow p((\mu_{A_n}, \nu_{A_n})) \nearrow p((\mu_A, \nu_A)).$

Theorem 2. Let $\mathscr{P} : \mathscr{F} \to \mathscr{J}$ be a mapping. Denote $\mathscr{P}(A) = [\mathscr{P}^\flat(A), \mathscr{P}^\sharp(A)]$ for any $A \in \mathscr{F}$. Then $\mathscr{P}$ is a probability if and only if $\mathscr{P}^\flat, \mathscr{P}^\sharp$ are states.

Now we are able to work with R-valued functions and then to return to $\mathscr{J}$-valued functions.

2 Embedding

An MV-algebra ([2], [3], [11], [12]) is a system $(M, \oplus, \odot, \neg, 0, 1)$ (where $\oplus, \odot$, are binary operations, $\neg$ is a unary operation, $0, 1$ are fixed elements) such that the following identities are satisfied: $\oplus$ is commutative and associative, $a \oplus 0 = a, a \oplus 1 = 1, \neg(\neg a) = a, \neg 0 = 1, a \oplus (\neg a) = 1, \neg(\neg a \oplus b) \oplus b = \neg(a \oplus \neg b) \oplus a, a \odot b = \neg(\neg a \oplus \neg b)$. Every MV-algebra is a distributive lattice, where $a \vee b = a \oplus (\neg(a \oplus \neg b))$, 0 is the least element, and 1 is the greatest element of M.

Example 3. An instructive example is the unit interval $[0,1]$ endowed with the operations $a \oplus b = (a+b) \wedge 1, a \odot b = (a+b-1) \vee 0, \neg a = 1-a$.

A very important tool for studying MV-algebras is the Mundici theorem ([2, 3]) stating that to any MV-algebra M there exists a lattice ordered group G with a strong unit u (i.e. to any $a \in G$ there exists $n \in N$ such that $na \geq u$) such that

$$M = [0, u],$$
$$a \oplus b = (a+b) \wedge u,$$
$$a \odot b = (a+b-u) \vee 0,$$
$$\neg a = u - a.$$

In Example 3 the corresponding l-group is $(R, +, \leq)$ where $+$ is the usual sum of real numbers, and $\leq$ is the usual ordering, $u = 1$.

Theorem 3. Define $\mathcal{M} = \{(\mu_A, \nu_A); \mu_A, \nu_A \text{ are } \mathcal{S}\text{-measurable}, \mu_A, \nu_A : \Omega \to [0,1]\}$ together with operations

$$(\mu_A, \nu_A) \oplus (\mu_B, \nu_B) = (\mu_A \oplus \mu_B, \nu_A \odot \nu_B),$$
$$(\mu_A, \nu_A) \odot (\mu_B, \nu_B) = (\mu_A \odot \mu_B, \nu_A \oplus \nu_B),$$
$$\neg(\mu_A, \nu_A) = (1 - \mu_A, 1 - \nu_A).$$

Then the system $(\mathcal{M}, \oplus, \odot, \neg, 0, 1)$ is an MV - algebra.

Proof. Consider the set $\mathcal{G} = \{(f,g); f, g : \Omega \to R, f, g \text{ are measurable }\}$. The ordering $\leq$ is induced by the IF-ordering, hence $(f, g) \leq (h, k) \iff f \leq h, g \geq k$. Evidently $(\mathcal{G}, \leq)$ is a lattice, $(f, g) \vee (h, k) = (f \vee h, g \wedge k), (f, g) \wedge (h, k) = (f \wedge h, g \vee k)$. Now we shall define the group operation $+$ by the following formula:

$$(f, g) + (h, k) = (f + h, g + k - 1).$$

It is not difficult to see that $+$ is commutative and associative, and $(0, 1)$ is the neutral element. The inverse element to (f, g) is the couple $(-f, 2 - g)$, since

$$(f, g) + (-f, 2 - g) = (f - f, g + 2 - g - 1) = (0, 1),$$

therefore

$$(f, g) - (h, k) = (f, g) + (-h, 2 - k) = (f - h, g - k + 1).$$

If we put $u = (1, 0)$, then $\mathcal{M} = \{(f, g) \in \mathcal{G}; (0, 1) \leq (f, g) \leq (1, 0)\} = \{(f, g) \in \mathcal{G}; 0 \leq f \leq 1, 0 \leq g \leq 1\}$ with the MV-algebra operations, i.e.

$$(f, g) \oplus (h, k) = ((f, g) + (h, k)) \wedge (1, 0) = (f + h, g + k - 1) \wedge (1, 0) = ((f + h) \wedge 1, (g + g - 1) \vee 0) = (f \oplus h, g \odot k),$$

and similarly

$$(f, g) \odot (h, k) = (f \odot h, g \oplus k).$$

Theorem 4. To any state $p : \mathcal{F} \to [0, 1]$ there exists exactly one state $\bar{p} : \mathcal{M} \to [0, 1]$ such that $\bar{p} | \mathcal{F} = p$.

Proof. Given p and $(f, g) \in \mathcal{M}$ put $\bar{p}((f, g)) = p((f, 0)) - p((0, 1 - g))$. It is not difficult to prove that $\bar{p}$ is a state on $\mathcal{M}$. The rest of the proof is then clear.

By a similar technique the problem of observables can be solved.

Definition. An IF-observable is a mapping $x : \mathcal{B}(R) \to \mathcal{J}$ ($\mathcal{B}(R)$ being the family of all Borel subsets of R) satisfying the following properties:

(i) $x(R) = (1_\Omega, 0_\Omega)$;

(ii) $A, B \in \mathcal{B}(R), A \cap B = \emptyset \implies x(A) \odot x(B) = (0, 1), x(A \cup B) = x(A) \oplus x(B)$;

(iii) $A_n \nearrow A \Longrightarrow x(A_n) \nearrow x(A)$.

Definition. The joint IF observable of IF observables $x, y : \mathscr{B}(R) \to \mathscr{F}$ is a mapping $h : \mathscr{B}(R^2) \to \mathscr{F}$ satisfying the following conditions
   (i) $h(R^2) = (1_\Omega, 0_\Omega)$;
   (ii) $A, B \in \mathscr{B}(R^2), A \cap B = \emptyset \Longrightarrow x(A) \odot h(B) = (0,1), h(A \cup B) = h(A) \oplus h(B)$;
   (iii) $A_n \nearrow A \Longrightarrow h(A_n) \nearrow h(A)$.
   (iv) $h(C \times D) = x(C).y(B)$
   for any $C, D \in \mathscr{B}(R)$. (Here $(f,g).(h,k) = (f.h, g.k)$.)
   Proof.

Theorem 5. To any two IF observables $x, y : \mathscr{B}(R) \to \mathscr{F}$ there exists their joint IF observable.
   Proof. Put $x(A) = (x^\flat(A), 1 - x^\sharp(A)), y(B) = (y^\flat(B), 1 - y^\sharp(B))$, and for fixed $\omega \in \Omega$

$$\lambda_\omega^\flat(A) = x^\flat(A)(\omega), \lambda_\omega^\sharp(A) = x^\sharp(A)(\omega),$$

$$\kappa_\omega^\flat(B) = y^\flat(B)(\omega), \kappa_\omega^\sharp(B) = y^\sharp(B)(\omega).$$

Then $\lambda_\omega^\flat, \lambda_\omega^\sharp, \kappa_\omega^\flat, \kappa_\omega^\sharp : \mathscr{B}(R) \to [0,1]$ are probability measures. For $C \in \mathscr{B}(R^2)$ define

$$h(C) = (h^\flat(C), 1 - h^\sharp(C)),$$

where

$$h^\flat(C)(\omega) = \lambda_\omega^\flat \times \kappa_\omega^\flat(C),$$

$$h^\sharp(C)(\omega) = \lambda_\omega^\sharp \times \kappa_\omega^\sharp(C)$$

For to prove that $h(C) \in \mathscr{F}$ it is necessary to show that $h^\flat(C) + 1 - h^\sharp(C) \le 1$, hence $h^\flat(C) \le h^\sharp(C)$. Of course, we know $x^\flat(A) \le x^\sharp(A), y^\flat(B) \le y^\sharp(B)$, hence $\lambda_\omega^\flat \le \lambda_\omega^\sharp, \kappa_\omega^\flat \le \kappa_\omega^\sharp$ for any $\omega \in \Omega$. We have

$$h^\flat(C)(\omega) = \lambda_\omega^\flat \times \kappa_\omega^\flat(C) = \int_R \kappa_\omega^\flat(C^u) d\lambda_\omega^\flat(u) \le$$

$$\le \int_R \kappa_\omega^\sharp(C^u) d\lambda_\omega^\flat(u) \le \int_R \kappa_\omega^\sharp(C^u) d\lambda_\omega^\sharp(u) = h^\sharp(C)(\omega).$$

The existence of the joint observable $h_n, (n = 1, 2, ...)$ can be used for the defining of functions of observables $x_1, ..., x_n$, e.g. $\frac{1}{n} \sum_{i=1}^n (n = 1, 2, ...)$. Namely

$$\frac{1}{n} \sum_{i=1}^n = h_n \circ g_n^{-1}$$

where $g_n(u_1, ..., u_n) = \frac{1}{n} \sum_{i=1}^n u_i$ The motivation is taken from random vectors, where $(\frac{1}{n} \sum_{i=1}^n \xi_i)^{-1}(A) = (g_n \circ T_n)^{-1}(A) = T_n^{-1}(g_n^{-1}(A)), T_n(\omega) = (\xi_1(\omega), ..., \xi_n(\omega))$.

Since $\mathscr{F} \subset \mathscr{M}$, the whole probability theory on MV-algebras is now applicable to our system $\mathscr{F}$.

3 L-lattices

All basic facts in this section has been presented in the paper [6] and the thesis [7]. The used methods are surprising generalizations of the methods used in [26].

Definition. An L-lattice (Lukasiewicz lattice) is a structure $L = (L, \leq, \oplus, \odot, 0_L, 1_L)$, where $(L, \leq)$ is a lattice, 0_L is the least and 1_L the greatest element of the lattice L, and $\oplus, \odot$ are binary operations on L.

Example 4. Any MV-algebra is an L-lattice.

Example 5. The set $\mathscr{F}$ of all IF-events defined on a measurable space $(\Omega, \mathscr{S})$ is an L-lattice.

Definition. A probability on an L-lattice L is a mapping $p : L \to 0, 1]$ satisfying the following three conditions:
   (i) $p(1_L) = 1, p(0_L) = 0$;
   (ii) if $a \odot b = 0_L$, then $p(a \oplus b) = p(a) + p(b)$;
   (iii) if $a_n \nearrow a$, then $p(a_n) \nearrow p(a)$.
An observable is a mapping $x : \mathscr{B}(R) \to L$ satisfying the following conmditions:
   (i) $x(R) = 1_L$;
   (ii) if $A \cap B = \emptyset$, then $x(A) \odot x(B) = 0_L$ and $x(A \cup B) = x(A) \oplus x(B)$;
   (iii) if $A_n \nearrow A$, then $x(A_n) \nearrow x(A)$.
   It is quite surprising that there are given no conditions about the binary operations $\oplus, \odot$. Of course, here it is used the fact that the main importance in the probability theory has the probability distribution of a random variable (see the following theorem).

Theorem 6. Let $x : \mathscr{B}(R) \to L$ be an observable, $p : L \to [0, 1]$ a probability. Then the composite mapping $p \circ x : \mathscr{B}(R) \to [0, 1]$ is a probability measure on $\mathscr{B}(R)$.

   The key to the possibility to successfully create the probability theory on L-lattices is in the notion of independency.

Definition. Observables $x_1, ..., x_n$ are independent, if there exists and n-dimensional observable $h_n : \mathscr{B}(R) \to L$ such that

$$(p \circ h_n)(A_1 \times ... \times A_n) = (p \circ x_1)(A_1) \cdot ... \cdot (p \circ x_n)(A_n)$$

for all $A_1, ..., A_n \in \mathscr{B}(R)$.

   Now the existence of the joint distribution can be used similarly as in MV-algebras. Of course, in the MV-algebra $\mathscr{M}$ the existence of the joint observable can be proved, here we only assume that it exits.

4 Conclusion

We presented here two effective methods how to reach new results of the probability theory on IF - events. The first one (Section 2) was successfully applied in [9], the second one (Section 3) in [6] and [7].

Acknowledgement

The paper was supported by Grant VEGA 1/2002/05.

References

[1] Atanassov, K.: Intuitionistic Fuzzy sets: Theory and Applications. Physica Verlag, New York 1999.

[2] Cignoli, L. O., D'Ottaviano, M. L., Mundici, D.: Algebraic Foundations of Many-valued Reasoning. Kluwer Academic Publishers, Dordrecht 2000.

[3] Dvurečenskij, A., Pulmannová, S., : New Trends in Quantum Structures. Kluwer, Dordrecht 2000.

[4] Gerstenkorn, T., Manko, J.: Probabilities of intuitionistic fuzzy events. In: Issues in Intelligent Systems: Paradigms (O.Hryniewicz et al. eds.). EXIT, Warszawa, 63 - 58.

[5] Grzegorzewski, P. - Mrowka, E.: Probability of intuitionistic fuzzy events. In: Soft Methods in Probability, Statistics and Data Analysis (P. Grzegorzewski et al. eds.). Physica Verlag, New York 2002, 105-115.

[6] Lendelová, K.: Measure theory on multivalued logics and its applications, PhD thesis. M. Bel University Banská Bystrica 2005.

[7] Lendelová, K.: A mathematical model of probability on IF-events. Soft Computing (submitted).

[8] Riečan, B.: Representation of probabilities on IFS events. Advances in Soft Computing, Soft Methodology and Random Information Systems (M. Lopez - Diaz et al. eds.) Springer, Berlin 2004, 243-246.

[9] Riečan, B.: On the entropy of IF dynamical systems. In: Fuzzy Sets, Intuitionistic Fuzzy Sets, Generalized Nets, and Related Topics (K. Atanassov et al. eds.) EXIT, Warszawa 2005, 328-336.

[10] Riečan, B.: On a problem of Radko Mesiar: general form of IF - probabilities. Accepted to Fuzzy sets and Systems.

[11] Riečan, B., Mundici, D.: Probability on MV-algebras. In: Handbook on Measure Theory (E.Pap ed.), Elsevier, Amsterdam 2002.

[12] Riečan, B., Neubrunn, T.: Integral, Measure, and Ordering. Kluwer, Dordrecht 1997.

A Stratification of Possibilistic Partial Explanations

Sara Boutouhami[1] and Aicha Mokhtari[2]

[1] Institut d'Informatique, USTHB, BP 32, EL Alia, Alger, Algeria
   s_boutouhami@yahoo.fr
[2] Institut d'Informatique, USTHB, BP 32, EL Alia, Alger, Algeria
   mokhtari_aissani@yahoo.fr

Summary. Several problems are connected, in the literature, to causality: prediction, explanation, action, planning and natural language processing... In a recent paper, Halpern and Pearl introduced an elegant definition of causal (partial) explanation in the structural-model approach, which is based on their notions of weak and actual cause [5]. Our purpose in this paper is to partially modify this definition, rather than to use a probability (quantitative modelisation) we suggest to affect a degree of possibility (a more qualitative modelisation) which is nearer to the human way of reasoning, by using the possibilistic logic. A stratification of all possible partial explanations will be given to the agent for a given request, the explanations in the first strate are more possible than those belonging to the other strates. We compute the complexity of this strafication.

1 Introduction

Causation is a deeply intuitive and familiar relation, gripped powerfully by common sense, or so it seems. But as is typical in philosophy, deep intuitive familiarity has not led to any philosophical account of causation that is at once clean and precise [3]. A source of difficulties seems to be that the notion of causality is bound to other ideas like that of explanation. In a recent paper, Halpern and Pearl propose a new definition of explanation and partial explanation, using structural equations to model counterfactuals, the definition is based on the notion of actual cause. Essentially, an explanation is a fact that is not known for certain, but if found true, would constitute an actual cause of the fact to be explained, regardless of the agent's initial uncertainty [4, 5].

Our purpose in this paper is to partially modify this definition, i. e., rather than to use a probability (quantitative modelisation) we suggest to affect a degree of possibility (qualitative modelisation) which is nearer to the human reasoning [7]. A stratification of all possible partial explanations will be given to the agent for a given request (the explanations will be ordered in a set of strates), the explanations in the first strate are more possible than those belonging to the other strates. We compute the complexity of this stratification.

S. Boutouhami and A. Mokhtari: *A Stratification of Possibilistic Partial Explanations*, Advances in Soft Computing **6**, 291–298 (2006)
www.springerlink.com   © Springer-Verlag Berlin Heidelberg 2006

The paper is organized as follows. We present in the section 2, the structural approach, the definition of actual cause and the definition of the explanation. In section 3 we suggest to affect a degree of possibility to the definition advocated by Halpern and Pearl and then we carry out a more qualitative reasoning. We propose a stratification of all possible partial explanations; this stratification reflects a hierarchy of priority between partial explanations.In Section 4, we analyze the complexity of the algorithm of stratification. Finally, in section 5, we conclude and we give some perspectives of this work.

2 Structural Approach

Halpern and Pearl propose a definition of cause (*actual cause*) within the framework of structural causal models. Specifically, they express stories as a structural causal model (or more accurately, a causal world), and then provide a definition for when one event causes another, given this model of the story [4, 5]. Structural models are a system of equations over a set of random variables. We can divide the variables into two sets: endogenous (each of which has exactly one structural equation that determines their value) and exogenous (whose values are determined by factors outside the model, and thus have no corresponding equation). Capital letters X, Y, etc. will denote variables and sets of variables, and the lower-case letters x, y, etc. denote values of the sets of variables X, Y. Formally, *a signature S* is a tuple (U, V, R), where U is a set of exogenous variables, V is a set of endogenous variables, and R associates with every variable $Y \in U \cup V$ a nonempty set $R(Y)$ of possible values for Y (that is, the set of values over which Y ranges).

A *causal model* (or *structural model*) over signature S is a tuple $M = (S, F)$, where F associates with each variables $X \in V$ a function denoted F_X such that $F_X :$ $(\times_{u \in U} R(U)) \times (\times_{Y \in V - \{X\}} R(Y)) \rightarrow R(X)$.$F_X$ determines the values of X given the values of all the other variables in $U \in V$. Causal models can be depicted as a *causal diagram*: a directed graph whose nodes correspond to the variables in V with an edge from X to Y if F_Y depends on the value of X. Given a causal model $M = (S, F)$, a (possibly empty) vector X of variable in V, and vectors x and u of values for the variables in X and U, respectively, we can define a new causal model denoted $M_{X \leftarrow x}$ over the signature $S_X = (U, V - X, R_{|V-X})$. $M_{X \leftarrow x}$ is called *a submodel* of M by [6], $R_{|V-X}$ is the restriction of R to the variables in $V - X$. Intuitively, this is the causal model that results when the variables in X are set to x by some external action that effects only the variables in X. Formally $M_{X \leftarrow x} = (S_X, F^{X \leftarrow x})$, where $F_Y^{X \leftarrow x}$ is obtained from F_Y by setting the values of the variables in X to x.

Given a signature $S = (U, V, R)$, a formula of the form $X = x$, for $X \in V$ and $x \in R(X)$, is called *primitive event*. A basic *causal formula* (over S) is one of the form $[Y_1 \leftarrow y_1, ..., Y_k \leftarrow y_k] \varphi$ where : φ is a Boolean combination of primitive events, $Y_1, ..., Y_k$ are distinct variables in V, $y_i \in R(Y_i)$. Such formula is abbreviated as $[Y \leftarrow y] \varphi$. A basic causal formula is a boolean combination of basic formulas. A causal formula ψ is true or false in a causal model, given a context. We write $(M, u) \models \psi$ if ψ is true in the causal model M given the context u.

Definition 1. *Let $M = (U, V, F)$, be a causal model. Let $X \subseteq V$, $X = x$ is an actual cause of φ if the following three conditions hold:*

- *(AC1): $(M, u) \models X = x \wedge \varphi$ (that is, both $X = x$ and φ are true in the actual world).*
- *(AC2): There exists a partition (Z, W) of V with $X \subseteq V, W \subseteq V \backslash X$ and some setting (x', w') of the variables in (X, W) such that if $(M, u) \models Z = z^*$, then both of the following conditions hold :*
  - a. *$(M, u) \models [X \leftarrow x', W \leftarrow w'] \neg \varphi$. In worlds, changing (X, W) from (x, w) to (x', w') changes φ from true to false.*
  - b. *$(M, u) \models [X \leftarrow x, W \leftarrow w', Z' \leftarrow z^*] \varphi$. for all subsets Z' of Z.*
- *(AC3): X is minimal.*

2.1 Explanation

Essentially, an explanation is a fact that is not known to be certain but, if found to be true, would constitute an actual cause of the fact to be explained, regardless of the agent's initial uncertainty. An explanation is relative to the agent's epistemic state, in that case, one way of describing an agent's state is by simply describing the set of contexts the agent considers possible [4, 5].

Definition 2. *(Explanation) Given a structural model M, $X = x$ is an explanation of φ relative to a set K of contexts if the following conditions hold:*

- *EX1: $(M, u) \models \varphi$ for each $u \in K$. (that is, φ must hold in all contexts the agent considers possible. The agent considers what he is trying to explain as an established fact).*
- *EX2: $X = x$ is a weak cause (without the minimal condition AC3) of φ in (M, u) for each $u \in K$ such that $(M, u) \models X = x$.*
- *EX3: X is minimal; no subset of X satisfies EX2.*
- *EX4: $(M, u) \models \neg(X = x)$ for some $u \in K$ and $(M, u') \models (X = x)$ for some $u' \in K$.*

Halpern and Pearl propose a sophisticated definition for actual causality based on structural causal models, however although this definition works on many previously problematic examples, it still does not fit with intuition on all examples, moreover the explanation proposed in this approach is not qualitative. To handle this problem, we propose an improvement of this definition in the next section.

3 Possibilistic Explanation

Possibilistic logic offers a convenient tool for handling uncertain or prioritized formulas and coping with inconsistency [1]. Propositional logic formulas are thus associated with weight belonging to a linearly ordered scale. In this logic, at the semantic level, the basic notion is a possibility distribution denoted by π, which is a mapping from a set of informations Ω to the interval $[0, 1]$. $\pi(\omega)$ represents the possibility degree of the interpretation ω with the available beliefs. From a possibility distribution π, two measures defined on a set of propositional or first order formulas can be determined: one is the possibility degree of formula φ, denoted $\Pi(\varphi) = max\{\pi(\omega) : \omega \models \Omega\}$, the other is the necessity degree of formula φ is defined as $N(\varphi) = 1 - \Pi(\neg\varphi)$, for more details see [7, 8].

In order to give a more qualitative character to the previous explanation, we suggest to affect a degree of possibility rather than a degree of probability. A new definition of explanation using the possibilistic logic is proposed. It offers an ordering set of possible explanations. The agent's epistemic state will be represented by describing the set of the interpretations that the agent considers possible.

Definition 3. *(Possibilistic explanation) Let ω be an interpretation that the agent considers possible ($\omega \in \Omega$). Given a structural model $M, X = x$ is an explanation of φ relative to a set Ω of possible interpretations if the following conditions hold:*

- *$EX1'$: $(M, \omega) \models \varphi$ for each $\omega \in \Omega$. (that is, φ must be satisfied in all interpretation the agent considers possible).*
- *$EX2'$: $X = x$ is a weak cause of φ in (M, ω) for each $\omega \in \Omega$ such that $(M, \omega) \models X = x$.*
- *$EX3'$: X is minimal; no subset of X satisfies $EX2'$.*
- *$EX4'$: $(M, \omega) \models \neg(X = x)$ for some $\omega \in \Omega$ and $(M, \omega') \models X = x$ for some $\omega' \in \Omega$.*

Not all explanations are considered equally good. Some explanations are more plausible than others. We propose to define the goodness of an explanation by introducing a degree of possibility (by including priority levels between explanations). The measure of possibility of an explanation is given by:

$$\Pi(X = x) = max\{\pi(\omega) : \omega \models X = x, \omega \in \Omega\}$$

There is a situations where we can't find a complete explanation of an event (relative to Ω). But we can find a complete explanation relative to a sub-set Ω' of Ω. That explanation is a partial explanation relative to Ω In the next section we give our definition a partial explanation and it's goodness.

Definition 4. *(partial explanation) Let π be a possibility distribution, i.e., a mapping from a set of interpretations Ω that the agent considers possible into the interval $[0, 1]$. Let $\Omega_{X=x,\varphi}$ be the largest subset such that $X = x$ is an explanation of φ (it consists of all interpretations in Ω except those where $X = x$ is true but is not a weak cause of φ).*

$$\Omega_{X=x,\varphi} = \Omega - \{\omega : \omega \in \Omega | \omega \models X = x, \ \omega \models \varphi \text{ and } X = x \text{ is not a weak cause of } \varphi\}$$

- *$X = x$ is a partial explanation of φ with the goodness $\Pi(\Omega_{X=x,\varphi} | X = x) = max\{\pi(\omega) : \omega \models X = x, \omega \in \Omega_{X=x,\varphi}\}$.*
- *$X = x$ is a $\alpha-partial$ explanation of φ relative to π and Ω, if $\Omega_{X=x,\varphi}$ exists and $\Pi(\Omega_{X=x,\varphi} | X = x) \geq \alpha$.*

- $X = x$ is an partial explanation of φ relative to π and Ω, iff $X = x$ is a α−partial explanation of φ and $\alpha \geq 0$.

Partial explanations will be ordered, in a set of strates $S_{\alpha 1} \cup ... \cup S_{\alpha n}$ for a given request.

- The $S_{\alpha 1}$ will contain the complete explanations if there exists,
- $X = x$ is in the strate $S_{\alpha i}$, if $\Pi(\Omega_{X=x,\varphi}|X = x) = \alpha_i$,
- Let $X = x$ be a partial explanation in the strate $S_{\alpha i}$ and $Y = y$ a partial explanation in the strate $S_{\alpha j}$. $X = x$ is a partial explanation more plausible than the partial explanation $Y = y$, if $\Pi(\Omega_{X=x,\varphi}|X = x) = \alpha_i > \Pi(\Omega_{Y=y,\varphi}|X = x) = \alpha_j$.

Example 1. Suppose I see that Victoria is tanned and I seek an explanation. Suppose that the causal model includes variables for "Victoria took a vacation", "It is sunny in the Canary Islands", "Victoria went to a tanning". The set of Omega includes interpretations for all settings of these variables compatible with Victoria being tanned. Note that, in particular, there is an interpretation where Victoria both went to the Canaries (and didn't get tanned there, since it wasn't sunny) and to a tanning salon. Victoria taking a vacation is not an explanation (relative to Omega), since there is an interpretation where Victoria went to the Canary Islands but it was not sunny, and the actual cause of her tan is the tanning salon, not the vacation. However, intuitively it is "almost" satisfied, since it is satisfied by every interpretation in Omega, in which Victoria goes to the Canaries. "Victoria went to the Canary Islands" is a partial explanation of "Victoria being tanned". There is a situation where we can't find a complete explanation (it is inexplicable).

The usual definition of a conditional distribution of possibility is:

$$\pi(\omega|\varphi) = \begin{cases} 1 \text{ if } \Pi(\varphi) = \pi(\omega) \\ \pi^-(\varphi) \text{ if } = \pi^-(\omega) < \Pi(\varphi) \text{ and } \neg(\omega \models \varphi) \\ 0 \text{ else} \end{cases}$$

Conditioner with φ consists on a revision of degree of possibility associated to different interpretations, after having the certain information φ. (φ is a certain information, so interpretations that falsifie φ are impossibles).
We propose the measure of *explanatory power* of $X = x$ to be $\Pi^-(\Omega_{X=x,\varphi}|X = x) = max\{\pi^-(\omega) : \omega \models X = x, \omega \in \Omega_{X=x,\varphi}\}$.

3.1 Algorithm of Generation of Strates

The main idea of our algorithm is to provide a set of choices of ordered partial explanations for a given request of the agent.

Let φ be a request for which the agent seeks an explanation. Let V be the set of endogenous variables and let $X \subseteq V - \{Y_i\}$, $\forall Y_i \in \varphi$ be a set of possible variables that may formulate the explanation. For all subset X' of X, decide if there exists an attribution of values which makes it a partial explanation. If it is the case, then compute $\Pi(\Omega_{X'=x',\varphi}|X'=x')$. Once that is done, add this partial explanation to the appropriate strate if it exists. If not, create a new strate which will contain this partial explanation. Finally, insert the new strate in its appropriate order according to the existing strates. This algorithm gives us all the partial explanations. This structure facilitates the search of a new explanation when we have a new consideration of the agent as an adaptation with the evolution of the agent believes.

Algorithm of Generation of strates

Input $\{S = \varphi, V, \varphi, \Omega, R(X)\}$
begin

   a. $X = V - \{Y_j\}$, $\forall Y_j$, Y_j is a variable in φ
   b. **for all** $X' \subseteq X$ **do**
      begin
        a) Decide if there exist $x' \in R(X')$, such that $X' = x'$ is an α-partial explanation of φ
          relative to Ω.
        b) **if** $X' = x'$ is an α-partial explanation **then**
          begin
            i. Compute $\Pi(\Omega_{X'=x',\varphi}|X'=x')$; Let $\alpha_i = \Pi(\Omega_{X'=x',\varphi}|X'=x')$
            ii. **If** the strate S_{α_i} exists **then**   Add $\{X' = x'\}$ to the strate S_{α_i}
               else
               begin
                 A. Create a new strate S_{α_i}
                 B. Add $\{X' = x'\}$ to the strate S_{α_i}
                 C. Insert the strate S_{α_i} in the good order
                 D. $S = S \cup S_{\alpha_i}$
               end
        end
      end

end
Output $S = \cup S_{\alpha_i}$

---

4 Complexity of Stratification of Possibilistic Explanations

The complexity of our algorithm is driven from the results given by Eiter and Lukasiewicz [2]. An analysis of the computational complexity of Halpern and Pearl's (causal) explanation in the structural approach is given in a recent paper by Eiter and Lukasiewicz [2].

    An explanation of an observed event φ is a minimal conjunction of primitive events that causes φ even when there is uncertainty about the actual situation at hand. The main idea of the stratification is to compute all the possible partial explanations. This problem can be reduced to that of computing the set of all partial explanations which is equivalent to computing the set of all valid formulas among a Quantified Boolean Formulas $QBF = \exists A\, \forall C\, \exists D\, y$, where $\exists A\, \forall C\, \exists D\, y$ is a reduction of guessing some $X' \subseteq X$ and $x' \in R(x)$ and deciding whether $X' = x'$ is α-partial explanation. All complexity results from the two propositions:

Proposition 1. *For all $X, Y \in V$ and $x \in R(X)$, the values F_Y and $F_Y^{X \leftarrow x}$, given an interpretation $\omega \in \Omega$, are computable in polynomial time.*

Proposition 2. *Let $X \subseteq V$ and $x \in R(X)$. Given $\omega \in \Omega$ and an event φ, deciding whether $(M, \omega) \models \varphi$ and $(M, \omega) \models [X \leftarrow x]\varphi$ (given x) hold can be done in polynomial time.*

Given $M = (U, V, F)$, $X \subseteq V$, an event φ, a set of interpretations Ω such that $(M, \omega) \models \varphi$ for all interpretations $\omega \in \Omega$, for all $X' \subseteq X$ guessing an attribution of values x' of X' ($x' \in R(X')$)such that $X' = x'$ is a partial explanation of φ. After that we compute the explanatory power of the partial explanation $X' = x'$, once that done we insert it in the appropriate strate. Computing the set of strates is $FP_\parallel^{\Sigma_2^P}$-Complete. Recall that $X' = x'$ is a partial explanation of φ iff (a) $X' = x'$ is an explanation of φ relative to $\Omega_{X'=x'}^{\varphi}$ and (b) $\Pi(\Omega_{X'=x'}|X' = x') \geq 0$; To recognize partial explanation, we need to know the set of interpretations $\Omega_{X'=x'}^{\varphi}$. $\Omega_{X'=x'}^{\varphi}$ is the set of all $\omega \in \Omega$ such that either (i) $(M, \omega) \models \neg(X' = x')$ or (ii) $(M, \omega) \models (X' = x')$ and $X' = x'$ is a weak cause of φ under ω. Deciding (i) is polynomial, and deciding (ii) is in NP, $\Omega_{X'=x'}^{\varphi}$ can be computed efficiently with parallel calls to a $NP-oracle$, computing $\Omega_{X=x}^{\varphi}$ is in $P_\parallel^{NP}$. Once $\Omega_{X'=x'}^{\varphi}$ is given, deciding (a) is possible with two $NP-oracle$ calls and deciding (b) is polynomial. Hence, the problem is in $P_\parallel^{NP}$. Deciding whether $X' = x'$ is an α-partial explanation of φ is in $P_\parallel^{NP}$. Hence, guessing some $X' \subseteq X$ and $x' \in R(X')$ and deciding whether $X' = x'$ is an α-partial explanation of φ is in Σ_2^P.

The complexity of our algorithm is inherited from the complexity of guessing a partial explanation (is a Σ_2^P-complete) and of the complexity of the explanatory power ($P_\parallel^{NP}$), this complexity is lies to the problem of computing $\Omega_{X'=x',\varphi}$. The calculus of strate, is a problem of guessing all $X' \subseteq X$ and verifying the existence of partial explanation which is $FP_\parallel^{\Sigma_2^P}$-complete, and computing there explanatory power, so that the stratification problem is $FP_\parallel^{\Sigma_2^P}$-complete.

5 Conclusion and Perspectives

In this paper we have presented a partial modification of the notion of explanation related to the counterfactual idea. We have suggested the use of the possibilistic logic which provides a priority level between the explanations. We prefer the use of possibility instead of probability because possibility reflects better the human reasoning, which is rather qualitative than quantitative.

We have proposed a stratification of all partial explanation for a given request. This stratification facilitates the task of searching a new explanation when we have a new consideration of the agent (an evolution of the agent beliefs). We gave an analysis of the computational complexity of this stratification. As perspectives, we plan to extend this work to deal with the problem of responsibility and blame.

Acknowledgments

This work was partially supported by CMEP project (06MDU687) entitled: *Developpement des reseaux causaux possibilistes : application a la securite informatique*. The authors are indebted to Daniel Kayser for his helpful remarks.

References

[1] S. Benferhat, S. Lagrue, O. Papini. A possibilistic handling of partially ordered information. In *Proceedings UAI'03*, pages 29–36, 2003
[2] T. Eiter, T. Lukasiewicz. Complexity results for explanations in the structural-model approach. *Artificial Intelligence*, 154(1-2):145–198, 2004
[3] N. Hall, L. A. Paul. Causation and counterefactuals. *Edited by John Collins, Ned Hall and L. A. Paul*, Cloth/june, 2004
[4] J.Y. Halpern, J. Pearl. Causes and Explanations: A Structural-model Approach. *British Journal for Philosophy of Science*, To appear
[5] J.Y. Halpern, J. Pearl. Causes and explanations: A structural-model approach, Part II: Explanations. In *Proceedings IJCAI'01*, pages 27–34, Seattle, WA, 2001
[6] J. Pearl. Causality Models, Reasoning, and Inference. Cambridge University Press, New York, 2000
[7] H. Prade, D. Dubois. Possibility theory: An approach to computerized, processing of uncertainty. Plenum Press, New York, 1988
[8] L. A. Zadeh. Fuzzy sets as a basic for a theory of possibility. In *Fuzzy Sets and Systems*, 1:3–28, 1978

Finite Discrete Time Markov Chains
with Interval Probabilities

Damjan Škulj

Faculty of Social Sciences, University of Ljubljana
Kardeljeva ploščad 5, 1000 Ljubljana, Slovenia
Damjan.Skulj@Fdv.Uni-Lj.Si

Summary. A Markov chain model in generalised settings of interval probabilities is presented. Instead of the usual assumption of constant transitional probability matrix, we assume that at each step a transitional matrix is chosen from a set of matrices that corresponds to a structure of an interval probability matrix. We set up the model and show how to obtain intervals corresponding to sets of distributions at consecutive steps. We also state the problem of invariant distributions and examine possible approaches to their estimation in terms of convex sets of distributions, and in a special case in terms of interval probabilities.

1 Introduction

Interval probabilities present a generalised probabilistic model where classical single valued probabilities of events are replaced by intervals. In our paper we refer to Weichselberger's theory [4]; although, several other models also allow interval interpretation of probabilities.

An approach to involve interval probabilities to the theory of Markov chains was proposed by Kozine and Utkin [1]. They assume a model where transitional probability matrix is constant but unknown. Instead of that, only intervals belonging to each transitional probability are known.

In this paper we attempt to relax this model. We do this in two directions. First, we omit the assumption of the transitional probability matrix being constant, and second, instead of only allowing intervals to belong to single atoms, we allow them to belong to all subsets.

Allowing non-constant transitional probability matrix makes Markov chain model capable of modeling real situations where in general it is not reasonable to expect exactly the same transitional probabilities at each step. They can, however, be expected to belong to some set of transitional probabilities. In interval probability theory such sets are usually obtained as structures of interval probabilities. Our assumption is thus that transitional probability at each step is an arbitrary member of a set of transitional probability matrices generated by an interval probability matrix.

D. Škulj: *Finite Discrete Time Markov Chains with Interval Probabilities*, Advances in Soft Computing **6**, 299–306 (2006)
www.springerlink.com   © Springer-Verlag Berlin Heidelberg 2006

A similar relaxation is also made to the initial distribution. Instead of a single distribution, we allow a set of distributions forming a structure of an interval probability.

Our goal is to estimate the interval probabilities after a number of steps and to find an invariant set of distributions. Unfortunately, it turns out that interval probabilities are not always sufficient to represent distributions obtained after some steps, which can form very general sets of distributions, that may not be easy to represent. The method based on interval probabilities can thus only approximate the true sets of distributions. To overcome this drawback, we provide a method to at least in principle approximate the corresponding sets of distributions with convex sets of probability distributions with arbitrary precision. In those settings, interval probabilities only present an easy to handle special case.

2 Basic Definitions and Model Setup

First we introduce basic elements of interval probability theory due to Weichselberger [4], but some of them are used here in a simplified form. Let Ω be a nonempty set and $\mathscr{A}$ a σ-algebra of its subsets. The term *classical probability* or *additive probability* will denote any set function $p\colon \mathscr{A} \to \mathbb{R}$ satisfying Kolmogorov's axioms. Let L and U be set functions on $\mathscr{A}$, such that $L \le U$ and $L(\Omega) = U(\Omega) = 1$. The interval valued function $P(\,.\,) = [L(\,.\,), U(\,.\,)]$ is called an *interval probability*.

To each interval probability P we associate the set $\mathscr{M}$ of all additive probability measures on the measurable space $(\Omega, \mathscr{A})$ that lie between L and U. This set is called the *structure* of the interval probability P. The basic class of interval probabilities are those whose structure is non-empty. Such an interval probability is denoted as an *R-field*. The most important subclass of interval probabilities, *F-fields*, additionally assumes that both lower bound L and upper bound U are strict according to the structure:

$$L(A) = \inf_{p \in \mathscr{M}} p(A) \quad \text{and} \quad U(A) = \sup_{p \in \mathscr{M}} p(A) \quad \text{for every } A \in \mathscr{A}. \quad (1)$$

The above property is in a close relation to *coherence* in Walley's sense (see [3]), in fact, in the case of finite probability spaces both terms coincide. Because of the requirement (1) only one of the bounds L and U is needed. Usually we only take the lower one. Thus, an F-field is sufficiently determined by the triple $(\Omega, \mathscr{A}, L)$, and therefore, we will from now on denote F-fields in this way.

Now we introduce the framework of our Markov chain model. Let Ω be a finite set with elements $\{\omega_1, \ldots, \omega_m\}$ and $\mathscr{A} := 2^\Omega$ be the algebra of its subsets. Further let

$$X_0, X_1, \ldots, X_n, \ldots \quad (2)$$

be a sequence of random variables such that

$$P(X_0 = \omega_i) = q^{(0)}(\omega_i) =: q_i^0,$$

where $q^{(0)}$ is a classical probability measure on $(\Omega, \mathscr{A})$ such that

$$L^{(0)} \leq q^{(0)}, \tag{3}$$

where $Q^{(0)} = (\Omega, \mathscr{A}, L^{(0)})$ is an F-probability field. Thus $q^{(0)}$ belongs to the structure $\mathscr{M}^{(0)}$ of $Q^{(0)}$.

Further, suppose that

$$P\left(X_{n+1} = \omega_j \mid X_n = \omega_i, X_{n-1} = \omega_{k_{n-1}}, \ldots, X_0 = \omega_{k_0}\right) = p_i^{n+1}(\omega_j) =: p_{ij}^{n+1}, \tag{4}$$

where p_{ij}^{n+1} is independent of $X_0, \ldots, X_{n-1}$ and

$$L_i \leq p_i^{n+1}, \tag{5}$$

where $P_i = (\Omega, \mathscr{A}, L_i)$, for $1 \leq i \leq m$, is an F-probability field. Thus p_{ij}^{n+1} are transitional probabilities at time $n+1$; however, they do not need to be constant, but instead, on each step they only satisfy (5), where L_i are constant. Thus, the transitional probabilities are not constant in the usual sense but only in the sense of interval probabilities.

Now we shall generalise the concept of stochastic matrices to interval probabilities. Let $P = [P_1, \ldots, P_m]^T$, where P_i are F-fields for $i = 1, \ldots, m$. We will call such P an *interval stochastic matrix*. The *lower bound* of an interval stochastic matrix is simply $P_L := [L_1, \ldots, L_m]$, where L_i is the lower bound P_i and the *structure* of an interval stochastic matrix is the set $\mathscr{M}(P)$ of stochastic matrices $p = (p_{ij})$ such that $p_i \geq L_i$, where p_i, for $i = 1, \ldots, m$, is the classical probability distribution on $(\Omega, \mathscr{A})$, generated by $p_i(\omega_j) = p_{ij}$ for $j = 1, \ldots, m$.

Thus, the transitional probabilities are given in terms of interval stochastic matrices. Under the above conditions, the probability distribution of each X_n will be given in terms of an F-field $Q^{(n)} = (\Omega, \mathscr{A}, L^{(n)})$. Thus

$$P(X_n = \omega_i) = q^{(n)}(\omega_i) =: q_i^n,$$

where $q^{(n)}$ is a classical probability measure on $(\Omega, \mathscr{A})$ such that

$$L^{(n)} \leq q^{(n)}.$$

We will call a sequence (2) with the above properties an *interval Markov chain*. An advantage of presenting sets of probability measures with interval probabilities is that only one value has to be given for each set to determine an interval probability. Usually, this is the lower probability $L(A)$ of an event A. In general this requires $m(2^m - 2)$ values for the transitional matrix and $2^m - 2$ values for the initial distribution. We demonstrate this by the following example.

Example 1. Take $\Omega = \{\omega_1, \omega_2, \omega_3\}$. The algebra $\mathscr{A} = 2^\Omega$ contains six non-trivial subsets, which we denote by $A_1 = \{\omega_1\}, A_2 = \{\omega_2\}, A_3 = \{\omega_3\}, A_4 = \{\omega_1, \omega_2\}, A_5 = \{\omega_1, \omega_3\}, A_6 = \{\omega_2, \omega_3\}$. Thus, besides $L(\emptyset) = 0$ and $L(\Omega) = 1$ we have to give the values $L(A_i)$ for $i = 1, \ldots, 6$. Let the lower probability L of an interval probability Q be represented through the n-tuple

$$L = (L(A_1), L(A_2), L(A_3), L(A_4), L(A_5), L(A_6)), \tag{6}$$

and take $L = (0.1, 0.3, 0.4, 0.5, 0.6, 0.7)$. Further we represent the interval transitional matrix P by a matrix with three rows and six columns, each row representing an element ω_i of Ω and the values in the row representing the interval probability P_i through its lower probability L_i. Take for example the following matrix:

$$P_L = \begin{pmatrix} 0.5 & 0.1 & 0.1 & 0.7 & 0.7 & 0.4 \\ 0.1 & 0.4 & 0.3 & 0.6 & 0.5 & 0.8 \\ 0.2 & 0.2 & 0.4 & 0.5 & 0.7 & 0.7 \end{pmatrix}. \tag{7}$$

In the next section we will show how how to obtain the lower probability at the second step, given the lower bounds for Q and P.

3 Calculating Distributions at n-th Step

The main advantage of Markov chains is that knowing the probability distribution at time n we can easily calculate the distribution at the next time. This is done simply by multiplying the given distribution with the transitional matrix.

In the generalised case we consider a set of probability distributions and a set of transitional matrices, given as structures of the corresponding interval probabilities. The actual distribution as well as the actual transitional probability matrix can be any pair of members of the two sets. Let $q^{(0)}$ be an initial distribution, thus satisfying (3), and p^1 a transitional probability, satisfying (5). According to the classical theory, the probability at the next step is $q^{(1)} = q^{(0)}p^1$. Thus, the corresponding set of probability distributions on the next step must contain all the probability distributions of this form. Consequently, in the most general form, the set of probability distributions corresponding to X_k would be

$$\mathcal{M}_k := \{q^{(0)}p^1 \ldots p^k \mid q^{(0)} \in \mathcal{M}(Q^{(0)}), p^i \in \mathcal{M}(P) \text{ for } i = 1, \ldots, k\}. \tag{8}$$

But these sets of probability distributions are in general not structures of interval probability measures. Thus, they can not be observed in terms of interval probabilities. However, a possible approach using interval probabilities is to calculate the lower and the upper envelope of the set of probabilities obtained at each step and do the further calculations with this interval probability and its structure. The resulting set of possible distributions at n-th step is then in general larger than $\mathcal{M}_k$, and could only be regarded as an approximate to the true set of distributions.

The advantage of the approach in terms of interval probabilities is that the calculations are in general computationally less difficult and that some calculations, such as the calculation of invariant distributions, can be done directly through systems of linear equations. As we shall see, the level of precision of estimates is very flexible and can be adjusted depending on our needs and the imprecision of the data.

Now we give such a procedure for a direct calculation of the lower bound $L^{(n+1)}$ under the assumption that the set of probabilities at n-th step is given in terms of an interval probability $Q^{(n)}$. Let π_A be a permutation on the set $\{1,\dots,m\}$ such that $L_{\pi_A(i)}(A) \geq L_{\pi_A(i+1)}(A)$ for $1 \leq i < m-1$ and denote $A_i := \bigcup_{k=1}^{i}\{\omega_{\pi_A(k)}\}$ where $A_0 = \emptyset$. Define the probability measure

$$q_{\pi_A(i)}^{\pi_A} = q^{\pi_A}(\omega_{\pi_A(i)}) := L^{(n)}(A_i) - L^{(n)}(A_{i-1}). \qquad (9)$$

The set function $L^{(n+1)}$ is then the infimum of the set of all distributions from the structure of $Q^{(n)}$ multiplied by all members of $\mathcal{M}(P)$. It turns out that it can be directly calculated as

$$L^{(n+1)}(A) = \sum_{i=1}^{m} q_i^{\pi_A} L_i(A). \qquad (10)$$

Example 2. Let us calculate the second step probability distribution on the data of Example 1. Let the lower bound $L^{(0)}$ of $Q^{(0)}$ be as in the previous example, $L^{(0)} = (0.1, 0.3, 0.4, 0.5, 0.6, 0.7)$ and let the transitional probability be given by its lower bound P_L from the same example. Further, let $L^{(1)}$ be the lower bound of the interval probability distribution at step 1, $Q^{(1)}$. By (10) we get

$$L^{(1)} = (0.19, 0.23, 0.28, 0.56, 0.62, 0.64).$$

4 Invariant Distributions

4.1 The Invariant Set of Distributions

One of the main concepts in the theory of Markov chains is the existence of an invariant distribution. In the classical theory, an invariant distribution of a Markov chain with transitional probability matrix P is any distribution q such that $qP = q$. In the case of *ergodic* Markov chain an invariant distribution is also the limit distribution.

In our case, a single transitional probability matrix as well as initial distributions are replaced by sets of distributions given by structures of interval probabilities. Consequently, an invariant distribution has to be replaced by a set of distributions, which is invariant for the interval transitional probability matrix P. It turns out, that there always exists a set $\mathcal{M}$ such that

$$\mathcal{M} = \{qp \mid q \in \mathcal{M}, p \in \mathcal{M}(P)\} \qquad (11)$$

and that for every initial set of probability distributions $\mathcal{M}_0$ late enough members of sequence (8) converge to $\mathcal{M}$.

For simplicity we may always assume the initial distribution to be the set of all probability measures on $(\Omega, \mathscr{A})$, which is equal to the structure of the interval probability $Q_0 = [0,1]$. Thus, from now on, let $\mathcal{M}_0 := \{q \mid q$ is a probability measure on $(\Omega, \mathscr{A})\}$. Clearly, the sequence (8) with initial set of distributions $\mathcal{M}_0$ includes all sequences with any other initial set of distributions.

Consider the following sequence of sets of probability measures:

$$\mathcal{M}_{i+1} := \{qp \mid q \in \mathcal{M}_i, p \in \mathcal{M}(P)\}, \tag{12}$$

z starting with $\mathcal{M}_0$. The above sequence corresponds to sequence (8) with initial set of distributions equal to $\mathcal{M}_0$.

It is easy to see that the sequence (12) is monotone: $\mathcal{M}_{i+1} \subseteq \mathcal{M}_i$, and thus we can define the limiting set of distributions by

$$\mathcal{M}_\infty := \bigcap_{i=1}^{\infty} \mathcal{M}_i. \tag{13}$$

The above set is clearly non-empty, since it contains all eigenvectors of all stochastic matrices from $\mathcal{M}(P)$ corresponding to eigenvalue 1. It is well known that such eigenvectors always exist. Besides, this set clearly satisfies the requirement (11). Thus, we will call the set (13) *the invariant set of distributions* of an interval Markov chain with the interval transitional probability matrix P.

The above definition of the invariant set of an interval Markov chain gives its construction only in terms of limits, but it says nothing about its nature, such as, whether it is representable in terms of the structure of some interval probability or in some other way. However, it is important that such a set always exists.

4.2 Approximating the Invariant Set of Distributions with Convex Sets of Distributions

Since the invariant set of distributions of an interval Markov chain in general does not have a representation in terms of a structure of an interval probability or maybe even in terms of a convex set, we have to find some methods to at least approximate it with such sets.

For every closed convex set $\mathcal{M}$ of probability distributions on $(\Omega, \mathcal{A})$ there exists a set of linear functionals $\mathcal{F}$ and a set of scalars $\{l_f \mid f \in \mathcal{F}\}$ such that

$$\mathcal{M} = \{p \mid p \text{ is a probability measure on } (\Omega, \mathcal{A}), f(p) \geq l_f \; \forall f \in \mathcal{F}\}. \tag{14}$$

Example 3. If the set of functionals is equal to the natural embedding of the algebra $\mathcal{A}$ then the resulting set of distributions forms the structure of an F-probability field: $f_A(p) := p(A)$, $l_{f_A} := L(A)$ and $P = (\Omega, \mathcal{A}, L)$.

Moreover, the set of functionals may correspond to even a smaller set, like a proper subset of $\mathcal{A}$, such as the set of atoms in $\mathcal{A}$ yielding a structure of an interval probability with additional properties.

Thus, every structure of an interval probability may be represented by a set of distributions of the form (14).

Now fix a set of functionals $\mathcal{F}$ and an interval stochastic matrix P and define the following sequence of sets of probability distributions, where $\mathcal{M}_0$ is the set of all probability measures on $(\Omega, \mathcal{A})$:

$$\mathcal{M}_{0,\mathscr{F}} := \mathcal{M}_0;$$
$$\mathcal{M}'_{i+1,\mathscr{F}} := \{qp \mid q \in \mathcal{M}_{i,\mathscr{F}}, p \in \mathcal{M}(P)\}$$
$$\mathcal{M}_{i+1,\mathscr{F}} := \{q \mid f(q) \geq \inf_{q' \in \mathcal{M}'_{i+1,\mathscr{F}}} f(q') \ \forall f \in \mathscr{F}\}.$$

The way the set $\mathcal{M}_{i+1,\mathscr{F}}$ is obtained from $\mathcal{M}'_{i+1,\mathscr{F}}$ is similar to the concept of *natural extension* for a set of lower previsions (see e.g. [3]).

The idea of the above sequence is to replace the sets $\mathcal{M}_{i,\mathscr{F}}$, which are difficult to handle, with sets of distributions representable by linear functionals in $\mathscr{F}$. In the special case from Example 3 such a set is the structure of some interval probability.

The following properties are useful:

(i) If $\mathscr{F}' \subset \mathscr{F}$ then $\mathcal{M}_{i,\mathscr{F}'} \supseteq \mathcal{M}_{i,\mathscr{F}} \supseteq \mathcal{M}_i$ holds for every $i \in \mathbb{N} \cup \{0\}$, where $\mathcal{M}_i$ is a member of the sequence (12).

(ii) The inclusion $\mathcal{M}_{i+1,\mathscr{F}} \subseteq \mathcal{M}_{i,\mathscr{F}}$ holds for every $i \in \mathbb{N}$. Thus, the sequence $(\mathcal{M}_{i,\mathscr{F}})$ is monotone and this implies existence of a limiting set of distributions for every set of functionals $\mathscr{F}$:

$$\mathcal{M}_{\infty,\mathscr{F}} := \bigcap_{i \in \mathbb{N}} \mathcal{M}_{i,\mathscr{F}}.$$

The sets of distributions $\mathcal{M}_{\infty,\mathscr{F}}$ all comprise the set $\mathcal{M}_{\infty}$ and can be in some important cases found directly through a system of linear equations.

(iii) The set $\mathcal{M}_{\infty,\mathscr{F}}$ is a maximal set among all sets $\mathcal{M}$ with the property:

$$\inf_{q \in \mathcal{M}} f(q) = \inf_{\substack{q \in \mathcal{M} \\ p \in \mathcal{M}(P)}} f(q \cdot p) \ \forall f \in \mathscr{F}. \tag{15}$$

While the sets $\mathcal{M}_{\infty,\mathscr{F}}$ only approximate the invariant set of distributions $\mathcal{M}_{\infty}$ from below, it can clearly be approximated from above by the set $\mathcal{M}_e$ containing all eigenvectors of the stochastic matrices from the structure $\mathcal{M}(P)$.

4.3 Approximating Invariant Distributions with Interval Probabilities

The important special case of convex sets of probabilities is the case of structures of interval probabilities. For this case the conditions (15) translate to a system of linear equations with $2^m - 2$ unknowns. We obtain this case by considering the linear functionals on the set of probability measures on $(\Omega, \mathscr{A})$ of the form f_A, where $f_A(q) = q(A)$ for every probability measure q:

$$\mathscr{F}_{\mathscr{A}} = \{f_A \mid A \in \mathscr{A}\}.$$

The set of inequalities (15) can now be rewritten in terms of lower probabilities L and L_i to obtain:

$$L(A) = \sum_{i=1}^{m} q_i^{\pi_A} L_i(A) \ \forall A \in \mathscr{A}. \tag{16}$$

Recall that $q_i^{\pi_A}$ are expressible in terms of L as given by (9). The invariant set of distributions is then the structure of the F-field $Q_\infty = [L, U]$, where L is the minimal solution of the above system of linear equations, as follows from (iii).

Example 4. We approximate the invariant set of distributions of the Markov chain with interval transitional probability matrix given by the lower bound (7). We obtain the following solution to the system of equations (16):

$$L^{(\infty)} = (0.232, 0.2, 0.244, 0.581, 0.625, 0.6),$$

where $L^{(\infty)}$ is of the form (6).

As we have pointed out earlier, the above lower bound is only an approximation of the true lower bound for the invariant set of distributions. For comparison we give the lower bound of the set of eigenvalues of 100,000 random matrices dominating P_L:

$$(0.236, 0.223, 0.275, 0.587, 0.628, 0.608),$$

which is an approximation from above. Thus, the lower bound of the true invariant set of distributions lies between the above approximations.

References

[1] I. Kozine and L. V. Utkin. Interval-valued finite markov chains. *Reliable Computing*, 8(2):97–113, 2002.

[2] J. Norris. *Markov Chains*. Cambridge University Press, Cambridge, 1997.

[3] P. Walley. *Statistical reasoning with imprecise probabilities*. Chapman and Hall, London, New York, 1991.

[4] K. Weichselberger. *Elementare Grundbegriffe einer allgemeineren Wahrscheinlichkeitsrechnung I – Intervallwahrscheinlichkeit als umfassendes Konzept*. Physica-Verlag, Heidelberg, 2001.

Evidence and Compositionality

Wagner Borges[1] and Julio Michael Stern[2]

[1] Mackenzie Presbiterian University
  wborges@mackenzie.com.br
[2] University of São Paulo
  jstern@ime.usp.br

Summary. In this paper, the mathematical apparatus of significance testing is used to study the relationship between the truthness of a complex statistical hypothesis, H, and those of its constituents, H^j, $j = 1 \ldots k$, within the independent setup and under the principle of compositionality.

Key words: Bayesian models; Complex hypotheses; Compositionality; Mellin convolution; Possibilistic and probabilistic reasoning; Significance tests; Truth values, functions and operations.

1 Introduction

According to Wittgenstein [17], (2.0201, 5.0, 5.32):

- The truthness of every complex statement can be infered from the thuthness of its elementary constituents.

- The truthness of a statement results from the statements' truth-function (Wahrheitsfunktionen).

- Truth-functions of complex statements results from successive applications of a finite number of truth-operations (Wahrheitsoperationen) to elementary constituents.

This is known as the principle of Compositionality, which plays a central role in analytical philosophy, see [3], and is related to the semantic theory of truth, developed by Alfred Tarski.

The principle of compositionality also exists in far more concrete mathematical contexts, such as in reliability engineering, see [1] and [2], (1.4):

"One of the main purposes of a mathematical theory of reliability is to develop means by which one can evaluate the reliability of a structure when the reliability of its components are known. The present study will be concerned with this kind of mathematical development. It will be necessary for this purpose to rephrase our intuitive concepts of structure, component, reliability, etc. in more formal language, to restate carefully our assumptions, and to introduce an appropriate mathematical apparatus."

W. Borges and J.M. Stern: *Evidence and Compositionality*, Advances in Soft Computing **6**, 307–315 (2006)
www.springerlink.com   © Springer-Verlag Berlin Heidelberg 2006

When brought into the realm of parametric statistical hypothesis testing, the principle states that a meaningful complex statement, H, concerning $\theta = (\theta^1, \ldots, \theta^k) \in \Theta = (\Theta^1 \times \ldots \times \Theta^k)$ is a logical composition of statements, $H^1, \ldots, H^k$, concerning the elementary components, $\theta^1 \in \Theta^1, \ldots, \theta^k \in \Theta^k$, respectively. Within this setting, means to evaluate the truthness of H, as well as that of each of its elementary components, $H^1, \ldots, H^k$, is provided by the mathematical apparatus of the Full Bayesian Significance Test (FBST) procedure, a coherent Bayesian significance test for sharp hypotheses introduced in [9]. For detailed definitions, interpretations, implementation and applications, see the authors' previous articles, [8] and [16]. Further general references on the subject include [7-12] and [15].

It is of interest, however, to know what can be said about the truthness of H from the knowledge of the truethness of each of its elementary components, $H^1, \ldots H^k$. This is precisely what the authors explore in the present paper, within the independent setup.

2 The FBST Structure

By a FBST *Structure*, we mean a quintuple $M = \{\Theta, H, p_0, p_n, r\}$, where

- Θ is the parameter space of an underlying statistical model $(S, \Sigma(S), P_\theta)$; The Full Bayesian Significance Test (FBST) has been introduced by Pereira and Stern [9], as a coherent Bayesian significance test for sharp hypotheses. For detailed definitions, interpretations, implementation and applications, see the authors' previous articles.

In this paper we analyze the relationship between the credibility, or truth value, of a complex hypothesis, H, and those of its elementary constituents, H^j, $j = 1 \ldots k$.

- $H : \theta \in \Theta_H = \{\theta \in \Theta \mid g(\theta) \leq \mathbf{0} \wedge h(\theta) = \mathbf{0}\}$ is the *Hypothesis*, stating that the parameter lies in the (null) set Θ_H, defined by inequality and equality constraints in terms of vector functions g and h in the parameter space. In the present context, however, the statement H will be considered sharp or precise, in the sense that $\dim(\Theta_H) < \dim(\Theta)$, i.e., with at least one equality constraint in force. We shall often write, for simplicity, H for the set Θ_H.

- p_0, p_n and r are the *Prior*, the *Posterior* and the *Reference* probability densities on Θ, all with respect to the same σ-finite measure μ on a measurable space $(\Theta, \Sigma(\Theta))$.

Within a FBST structure, the following definitions are essential:

- The posterior *Surprise* function, $s(\theta)$, relative to the structure's reference density, $r(\theta)$, and its *constrained* and *unconstraind* suprema are defined as:

$$s(\theta) = \frac{p_n(\theta)}{r(\theta)} , \quad s^* = s(\theta^*) = \sup_{\theta \in H} s(\theta) , \quad \widehat{s} = s(\widehat{\theta}) = \sup_{\theta \in \Theta} s(\theta) .$$

- The *Highest Relative Surprise Set* (HRSS) at level v, $\overline{T}(v)$, and its complement, $T(v)$, are defined as:

$$T(v) = \{\theta \in \Theta \mid s(\theta) \leq v\} , \quad \overline{T}(v) = \Theta - T(v) .$$

- The *Truth Function* of M, $W : R_+ \mapsto [0,1]$, and the *Truth Value* of H in M, $\mathrm{ev}(H)$, are defined as:

$$W(v) = \int_{T(v)} p_n(\theta)\mu(d\theta) , \quad \mathrm{ev}(H) = W(s^*) .$$

Since $\mathrm{ev}(H) = W(s^*)$, the pair (W, s^*) will be referred to as the *Truth Summary* of the structure M. The truth function, W, plays the role of a posterior cumulative surprise distribution and is sometimes refered as that. The function $\overline{W}(v) = 1 - W(v)$, is refered to as the *Untruth Function* of M, and $\overline{\mathrm{ev}}(H) = \overline{W}(s^*) = 1 - \mathrm{ev}(H)$ is refered to as the *Untruth Value*, or *evidence-value* against H.

The tuth value of H plays the role of a quantitative measurement of the truthness of H. Since the *Tangential Set*, $\overline{T} = T(s^*)$, contains the points of the parameter space with higher surprise, relative to the reference density, than any point in H, small values of $\overline{\mathrm{ev}}(H)$ indicate that the hypothesis traverses high density regions, favoring the hypothesis. When $r(\theta)$ is the (possibly improper) uniform density, i.e., $r(\theta) \propto 1$, $\overline{T}$ is the Posterior's Highest Density Probability Set (HDPS) tangential to H. In the statistical jargon $\mathrm{ev}(H)$ is commonly refered to as the evidence value, or *e-value*, supporting H.

The role of the reference density in the FBST is to make $\mathrm{ev}(H)$ implicitly invariant under suitable transformations of the coordinate system of the parameter space. The natural choices for reference density are an uninformative prior, interpreted as a representation of no information in the parameter space, or the limit prior for no observations, or the neutral ground state for the Bayesian operation. Standard (possibly improper) uninformative priors include the uniform and maximum entropy densities, see [5] for a detailed discussion.

As we mentioned in the introduction, our results concern complex hypotheses in independent setups. The precise meaning of this framework is that the FBST structures $M = \{\Theta, H, p_0, p_n, r\}$ and $M^j = \{\Theta^j, H^j, p_0^j, p_n^j, r^j\}$, $j = 1, \ldots k$, bear the following relationships between their elements:

- the parameter space, Θ, of the underlying statistical model, $(S, \Sigma(S), P_\theta)$, is the product space $\Theta^1 \times \Theta^2 \times \ldots \times \Theta^k$;

- H, is a logical composition (conjunctions/disjunctions) of $H^1, H^2 \ldots H^k$;

- p_n and r, are probability densities with respect to the product measure $\mu = \mu^1 \times \mu^2 \times \ldots \times \mu^k$ on $(\Theta, \Sigma(\Theta))$, where μ^j denote the σ-finite measure on $(\Theta^j, \Sigma(\Theta^j))$ with respect to which p_0^j, p_n^j and r^j are densities ; and

- the probability densities p_n and r are such that

$$p_n(\theta) = \prod_{j=1}^{k} p_n^j(\theta^j) \text{ and } r(\theta) = \prod_{j=1}^{k} r^j(\theta^j) , \quad \theta = (\theta^1, \ldots, \theta^k) \in \Theta .$$

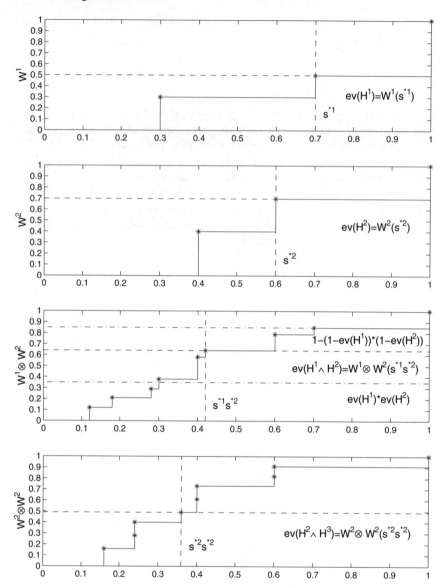

Fig. 1. Truth functions $W(v)$, $v \in [0, \widehat{s}]$, normalized s.t. $\widehat{s} = 1$ Subplots 1,2: W^j, s^{*j}, and $ev(H^j)$, for $j = 1, 2$; Subplot 3: $W^1 \otimes W^2$, $s^{*1}s^{*2}$, $ev(H^1 \wedge H^2)$ and bounds; Subplot 4: Structure M^3 is an independent replica of M^2, $ev(H^1) < ev(H^2)$, but $ev(H^1 \wedge H^3) > ev(H^2 \wedge H^3)$.

3 Inequalities for the Truth-Values of Conjunctions

In this section we consider the case of a conjunctive composite hypothesis, that is, the case in which H is equivalent to $H^1 \wedge H^2 \wedge \ldots \wedge H^j$, and show that in the independent setup an answer to the question of whether the truth value of H can be expressed in terms of the truth values of its elementary constituents can only be given in the form of upper and lower bounds. The following lemmas will be needed to prove the main result of this section:

Lemma 1: For any conjunctin $H = H^1 \wedge H^2 \wedge \ldots H^k$, we have

$$s^* = \sup_{\theta \in H} s(\theta) = \prod_{j=1}^{k} \sup_{\theta^j \in H^j} s^j(\theta^j) = \prod_{j=1}^{k} s^{*j} .$$

Proof: Since for $\theta \in H$, $s^j(\theta^j) \le s^{*j}$, for $1 \le j \le k$, $s(\theta) = \prod_{j=1}^{k} s^j(\theta^j) \le \prod_{j=1}^{k} s^{*j}$ so that $s^* \le \prod_{j=1}^{k} s^{*j}$. However, for $\varepsilon > 0$ and $s = \prod_{j=1}^{k} (s^{*j} - \varepsilon)$, there must exist $\theta \in \bigwedge_{j=1}^{k} H^j$ such that $s(\theta) = \prod_{j=1}^{k} s^j(\theta^j) > \prod_{j=1}^{k} (s^{*j} - \varepsilon)$. So, $\sup_{\theta \in H} s(\theta) > \prod_{j=1}^{k} (s^{*j} - \varepsilon)$, and the result follows by making $\varepsilon \to 0$.

Lemma 2: If W^j, $1 \le j \le k$, and W are the truth functions of $M^j, 1 \le j \le k$, and M, respectively, the following inequality holds:

$$\prod_{j=1}^{k} W^j(v^j) \le W\left(\prod_{j=1}^{k} v^j\right) ,$$

Proof: Let $G : R_+^k \mapsto [0,1]$ be defined as

$$G(v^1, \ldots, v^k) = \int_{\{s^1(\theta^1) \le v^1, \ldots, s^k(\theta^k) \le v^k\}} p_n(\theta) \mu(d\theta) .$$

Since $s = \prod_{j=1}^{k} s^j$, $\mu = \prod_{j=1}^{k} \mu^j$, and $\{s^1(\theta^1) \le v^1, \ldots, s^k(\theta^k) \le v^k\} \subseteq \{\prod_{j=1}^{k} s^j(\theta^j) \le \prod_{j=1}^{k} v^j\} = \{s(\theta) \le \prod_{j=1}^{k} v^j\}$, it follows that $\prod_{j=1}^{k} W^j(v^j) = G(v^1, \ldots, v^k) \le W(\prod_{j=1}^{k} v^j)$.

Lemma 3: For any conjunction $H = H^1 \wedge H^2 \wedge \ldots \wedge H^k$, we have

$$\prod_{j=1}^{k} \mathrm{ev}(H^j) \le \mathrm{ev}(H^1 \wedge H^2 \wedge \ldots \wedge H^k) \quad \text{and}$$

$$\prod_{j=1}^{k} \overline{\mathrm{ev}}(H^j) \le \overline{\mathrm{ev}}(H^1 \wedge H^2 \wedge \ldots \wedge H^k) .$$

Proof: In the inequality of Lemma 2, replacing each v^j by s^{*j}, $1 \le j \le k$, and then using Lemma 1, the first result follows. The same argument proves the other assertion.

Lemma 3 give us lower bounds for the truth and untruth values of any conjunction $H = H^1 \wedge H^2 \wedge \ldots H^k$, respectively in terms of the truth and untruth values of the elementary constituent hypotheses. Upper bound for the same values are easily obtained. More precisely,

Theorem 1: For any conjunction $H = H^1 \wedge H^2 \wedge \ldots \wedge H^k$, we have
$\prod_{j=1}^{k} \text{ev}(H^j) \leq \text{ev}(H^1 \wedge H^2 \wedge \ldots \wedge H^k) \leq 1 - \prod_{j=1}^{k}(1 - \text{ev}(H^j))$ and
$\prod_{j=1}^{k} \overline{\text{ev}}(H^j) \leq \overline{\text{ev}}(H^1 \wedge H^2 \wedge \ldots \wedge H^k) \leq 1 - \prod_{j=1}^{k}(1 - \overline{\text{ev}}(H^j))$.

In the null-or-full support case, that is, when, for $1 \leq j \leq k$, $s^{*j} = 0$ or $s^{*j} = \hat{s}^j$, and the truth values of the simple constituent hypotheses are either 0 or 1, the bounds in proposition 2 are sharp. In fact, it is not hard to see that the composition rule of classical logic holds, that is,

$$\text{ev}(H^1 \wedge \ldots \wedge H^k) = \begin{cases} 1 \;, & \text{if } s^{*1} = \hat{s}^1 \ldots s^{*k} = \hat{s}^k \;; \\ 0 \;, & \text{if, for some } j = 1 \ldots k, \; s^{*j} = 0 \;. \end{cases}$$

In the example below, illustrated by Figure 1, we show that the inequality in theorem 1 can, in fact be strict. Appendix A presents a Matlab function giving thr Mellin convolution of discretized (stepwise) distributions, used to generate all examples.

Example 1: In the third, first and second subplots of Figure 1, we have the graphs of truth functions corresponding, respectively, to the complex hypothesis $H^1 \wedge H^2$ and to its elementary constituents, H^1 and H^2. Note that while $\text{ev}(H^1) = 0.5$ and $\text{ev}(H^2) = 0.7$, $\text{ev}(H^1 \wedge H^2) = 0.64$, which is strictly grater than $\text{ev}(H^1)\text{ev}(H^2) = 0.35$.

4 The Truth Operation for Conjunctions

In this section we will also consider the case of a conjunctive composite hypothesis, $H = H^1 \wedge H^2 \wedge \ldots \wedge H^j$., within an independent setup. The investigation, however, concerns the question of whether the truth function of the FBST structure corresponding to H can be obtained from the truth functions of the FBST structures corresponding to its elementary constituents.

Given two probability distribution functions $G^1 : R_+ \mapsto [0, 1]$ and $G^2 : R_+ \mapsto [0, 1]$, their *Mellin convolution*, $G^1 \otimes G^2$, is the distribution function defined by

$$G^1 \otimes G^2(v) = \int_0^\infty \int_0^{v/y} G^1(dx)G^2(dy) = \int_0^\infty G^1(v/y)G^2(dy) \;.$$

The Mellin convolution output, $G^1 \otimes G^2$ is the probability distribution function of the product of two independent random variables, X and Y, with distribution functions, G^1 and G^2, respectively, see [11] and [16]. Consequently, commutativeness and associativeness of Mellin convolution, $\otimes$, follows immediately.

Lemma 4: For any conjunction $H = H^1 \wedge H^2 \wedge \ldots H^k$,

$$W = \bigotimes_{1 \leq j \leq k} W^j = W^1 \otimes W^2 \otimes \ldots \otimes W^k(v) \;.$$

Proof: This lemma follows straight from the definition of W.

Due to the above result, we shall refer to the Mellin convolution, in the present context, as the *Truth Operation*.

Together with the truth operation, the elementary truth summaries, (W^j, s^{*j}), $1 \leq j \leq k$, efficiently synthetize the independent setup information, in the sense that the truth value of any conjunction $H = H^1 \wedge H^2 \wedge \ldots H^k$ can be obtained. More precisely:

Theorem 2: If (W^j, s^{*j}), $1 \leq j \leq k$, are the truth summaries of the elementary contituents of a conjunction,

$$H = \bigwedge_{1 \leq j \leq k} H^j \text{ , then } ev(H) = W(s^*) = \bigotimes_{1 \leq j \leq k} W^j \left(\prod_{j=1}^k s^{*j} \right) .$$

Proof: Immediate, from Lemmas 2 and 4.

5 Disjunctive Normal Form

Let us now consider the case where H is *Homogeneous* and expressed in *Disjunctive Normal Form*, that is:

$$H = \bigvee_{i=1}^q \bigwedge_{j=1}^k H^{(i,j)} , \quad M^{(i,j)} = \{ \Theta^j, H^{(i,j)}, p_0^j, p_n^j, r^j \} .$$

Let us also define $s^{*(i,j)}$ and $\hat{s}^{(i,j)}$ as the respective constrained and unconstrained suprema of $s(\theta^{(i,j)})$ on the elementary hypotheses $H^{(i,j)}$.

Theorem 3:

$$ev(H) = ev \left(\bigvee_{i=1}^q \bigwedge_{j=1}^k H^{(i,j)} \right) = W \left(\sup_{i=1}^q \prod_{j=1}^k s^{*(i,j)} \right) =$$

$$\max_{i=1}^q W \left(\prod_{j=1}^k s^{*(i,j)} \right) = \max_{i=1}^q ev \left(\bigwedge_{j=1}^k H^{(i,j)} \right) .$$

Proof: Since the supremum of a function over the (finite) union of q sets, is the maximum of the suprema of the same function over each set, and W is non-decreasing, the result follows.

Theorem 3 discloses the *Possibilistic* nature of the FBST truth value, i.e., the e-value of a disjunction is the maximum e-value of the disjuncts, [12-14].

6 Final Remarks

It is worth mentioning that the present article does not abridge the most general composition cases of nested or heterogeneous (independent) structures, within which composite hypotheses are simultaneously assessed in heterogeneous sub-structures of (possibly) different dimensions. The following example indicates that this is not a trivial matter:

Example 2: Let $m = \arg\max_{j=1,2} \text{ev}(H^j)$ and H be equivalent to $(H^1 \vee H^2) \wedge H^3$. Is it true that $\text{ev}(H) = \max\{\text{ev}(H^1 \wedge H^3), \text{ev}(H^2 \wedge H^3)\} = \text{ev}(H^m \wedge H^3)$? Interestingly the answer is in the negative. In the third and forth subplots of Figure 1 we have the graphs of the Truth Functions corresponding, respectively, to the complex hypothesis $H^1 \wedge H^3$ and $H^2 \wedge H^3$, where the structure M^3 is an independent replica of M^2. Observe that $\text{ev}(H^1) = 0.5 < \text{ev}(H^2) = 0.7$, while $\text{ev}(H^1 \wedge H^3) = 0.64 > \text{ev}(H^2 \wedge H^3) = 0.49$.

Forthcomming papers extend the results obtained herein to the conditional independence setup. Such estensions allow us to develop efficient significance testing procedures for multinomial Dirichlet and Bayesian networks models, see [4] and [10]. Forthcomming papers also detail the implementation of Markov Chain Monte Carlo computational procedures for estimating the truth function, $W(v), 0 \leq v \leq \hat{s}$. Such procedures require only minor adaptations, with small computational overhead, in the MCMC procedures for estimating $\text{ev}(H) = W(s^*)$, see [6].

References

[1] Barlow, R.E, Prochan, F. (1981). *Statistical Theory of Reliability and Life Testing.* Silver Spring: To Begin With.

[2] Birnbaum, Z.W, Esary, J.D, Saunders, S.C. (1961). Multicomponent Systems and Structures, and their Reliability. *Technometrics, 3*, 55-77.

[3] Conde, M.L. (1998). *Wittgenstein: Linguagem e Mundo.* São Paulo: Annablume.

[4] Cozman, F.G. (2000). Generalizing Variable Elimination in Bayesian Networks. 7th IBERAMIA / 15th SBIA workshop proceedings, 27-32. São Paulo: Tec Art.

[5] Kapur, J.N. (1989). *Maximum Entropy Models in Science & Engineering.* Wiley.

[6] Lauretto, M, Pereira, C.A.B, Stern, J.M, Zacks, S. (2003). Comparing Parameters of Two Bivariate Normal Distributions Using the Invariant Full Bayesian Significance Test. *Brazilian Journal of Probability and Statistics, 17*, 147-168.

[7] Lauretto, M, Stern, J.M. (2005). FBST for Mixture Model Selection. MaxEnt 2005, *American Institute of Physics Conference Proceedings, 803*, 121-128.

[8] Madruga, M.R, Pereira, C.A.B, Stern, J.M. (2003). Bayesian Evidence Test for Precise Hypotheses. *Journal of Statistical Planning and Inference,* 117,185-198.

[9] Pereira, C.A.B, Stern, J.M. (1999). Evidence and Credibility: Full Bayesian Significance Test for Precise Hypotheses. *Entropy Journal, 1*, 69-80.

[10] Pereira, C.A.B, Stern, J.M. (2005). *Inferência Indutiva com Dados Discretos: Uma Visão Genuinamente Bayesiana*. XV-COMCA. University of Antofagasta.

[11] Springer, M.D. (1979). *The Algebra of Random Variables*. NY: Wiley.

[12] Stern, J.M. (2003). Significance Tests, Belief Calculi, and Burden of Proof in Legal and Scientific Discourse. Laptec'03, *Frontiers in Artificial Intell.and its Applications*, 101, 139–147.

[13] Stern, J.M. (2004). Paraconsistent Sensitivity Analysis for Bayesian Significance Tests. SBIA'04, *Lecture Notes Artificial Intelligence*, 3171, 134–143.

[14] Stern, J.M. (2005). Cognitive Constructivism, Eigen–Solutions, and Sharp Statistical Hypotheses. FIS2005 - Foundations of Information Science. 61, 1–23.

[15] Stern, J.M, Zacks, S. (2002). Testing Independence of Poisson Variates under the Holgate Bivariate Distribution. *Statistical and Probability Letters*, 60, 313–320.

[16] Williamson, R.C. (1989). *Probabilistic Arithmetic*. University of Queensland.

[17] Wittgenstein, L. (1921). *Tractatus Logico Philosophicus*. Ed.1999, NY: Dover.

Appendix A: Mellin Convolution Matlab Function

```
function [z,kk]= mellc(x,y,ii,jj);    *    if (i==ii & j==jj)
%z(1,j)= coord in [0,max_t s(t)]      *        skk= z(1,k); end
%z(1,kk)= s* ,   max surprise over H  * end   end   %for_i for_j
%z(2,j)= prob mass at z(1,j),      M  * z(3,:)= z(2,:);
%z(3,j)= cumulative distribution, W   * [s,ind]= sort(z(1,1:nm)');
n= size(x,2); m=size(y,2); nm= n*m;   * z= z(1:3,ind);   kk= 1;
z= zeros(3,nm); k=0; skk=0;           * for k=2:nm
for i=1:n    for j=1:m                 *    z(3,k)= z(3,k)+z(3,k-1);
  k= k+1;                             *    if ( z(1,k)<=skk )
  z(1,k)= x(1,i)*y(1,j);              *       kk= k; end
  z(2,k)= x(2,i)*y(2,j);              * end %for_k
```

High Level Fuzzy Labels for Vague Concepts

Zengchang Qin[*] and Jonathan Lawry

Artificial Intelligence Group, Department of Engineering Mathematics, University of Bristol, BS8 1TR, UK.
{z.qin; j.lawry}@bris.ac.uk

1 Introduction

Vague or imprecise concepts are fundamental to natural language. Human beings are constantly using imprecise language to communicate each other. We usually say 'John is tall and strong' but not 'John is exactly 1.85 meters in height and he can lift 100kg weights'. Humans have a remarkable capability to perform a wide variety of physical and mental tasks without any measurements. This capability partitionsof objects into granules, with a granule being a clump of objects drawn together by indistinguishability, similarity, proximity or function [8]. We will focus on developing an understanding of how we can use vague concepts to convey information and meaning as part of a general strategy for practical reasoning and decision making.

We may notice that *labels* are used in natural language to describe what we see, hear and feel. Such labels may have different degrees of vagueness. For example, when we say Mary is *young* and she is *female*, the label *young* is more vague than the label *female* because people may have more widely different opinions on being *young* than being *female*. For a particular concept, there could be more than one label that is appropriate for describing this concept, and some labels could be more appropriate than others. A random set framework, *Label Semantics*, was proposed to interpret these facts [3]. In such a framework, linguistic expressions or labels such as *small*, *medium* and *large* are used for modelling. These labels are usually defined by overlapping fuzzy sets which are used to cover the universes of continuous variables. Different from Computing with Words [9], fuzzy labels are usually predefined and used for building intelligent systems such as decision tree [4, 5], naive Bayes learning [7] and rule induction systems [6] without involving the computing of semantic meanings of these labels.

In this paper, we extended the label semantics framework with high level fuzzy labels. In previous research of label semantics, fuzzy labels are used to describe a numerical data element and the corresponding appropriateness degree for using a

---

[*] Current Address: Berkeley Initiative in Soft Computing, Electrical Engineering and Computer Sciences Department, University of California, Berkeley, CA 94720, USA.

Z. Qin and J. Lawry: *High Level Fuzzy Labels for Vague Concepts*, Advances in Soft Computing **6**, 317–324 (2006)

particular fuzzy label is just the membership of this data element belonging to the fuzzy label. Due to the vagueness and impreciseness of the real-world, numerical values are not always available. Here, we extend the label semantics framework to use higher level labels to describe some vague concepts which are also defined by intervals or fuzzy sets. The rest of the paper is structured as follows. Section 2 introduces the label semantics framework, based on which, the idea of high level fuzzy labels is disussed and supported with an example in section 3.

2 Label Semantics For Uncertainty Modeling

Label semantics is a methodology of using linguistic expressions or fuzzy labels to describe numerical values. For a variable x into a domain of discourse Ω we identify a finite set of fuzzy labels $\mathscr{L} = \{L_1, \cdots, L_n\}$ with which to label the values of x. Then for a specific value $x \in \Omega$ an individual I identifies a subset of $\mathscr{L}$, denoted D_x^I to stand for the description of x given by I, as the set of labels with which it is appropriate to label x. If we allow I to vary across a population V with prior distribution P_V, then D_x^I will also vary and generate a random set denoted D_x into the power set of $\mathscr{L}$ denoted by $\mathscr{S}$. We can view the random set D_x as a description of the variable x in terms of the labels in $\mathscr{L}$. The frequency of occurrence of a particular label, say S, for D_x across the population then gives a distribution on D_x referred to as a mass assignment on labels[2]. More formally,

Definition 1 (*Label Description*) *For $x \in \Omega$ the label description of x is a random set from V into the power set of $\mathscr{L}$, denoted D_x, with associated distribution m_x, which is referred to as mass assignment:*

$$\forall S \subseteq \mathscr{L}, \quad m_x(S) = P_V(\{I \in V | D_x^I = S\}) \tag{1}$$

where $m_x(S)$ is called the mass associated with a set of labels S and

$$\sum_{S \subseteq \mathscr{L}} m_x(S) = 1 \tag{2}$$

Intuitively mass assignment is a distribution on appropriate label sets and $m_x(S)$ quantifies the evidence that S is the set of appropriate labels for x.

For example, an expression such as '*the score on a dice is small*', as asserted by individual I, is interpreted to mean $D_{SCORE}^I = \{small\}$, where $SCORE$ denotes the value of the score given by a single throw of a particular dice. When I varies across a population V, different sets of labels could be given to describe the variable $SCORE$, so that we obtain the random set of D_{SCORE} into the power set of $\mathscr{L}$.

---

[2] Since $\mathscr{S}$ is the power set of L, the logical representation $S \in \mathscr{S}$ can be written as $S \subseteq \mathscr{L}$. The latter representation will be used through out this thesis. For example, given $\mathscr{L} = \{L_1, L_2\}$, we can obtain $\mathscr{S} = \{\emptyset, \{L_1\}, \{L_2\}, \{L_1, L_2\}\}$. For every element in $\mathscr{S}: S \in \mathscr{S}$, the relation $S \subseteq \mathscr{L}$ will hold.

In this framework, *appropriateness degrees* are used to evaluate how appropriate a label is for describing a particular value of variable x. Simply, given a particular value α of variable x, the appropriateness degree for labeling this value with the label L, which is defined by fuzzy set F, is the membership value of α in F. The reason we use the new term 'appropriateness degrees' is partly because it more accurately reflects the underlying semantics and partly to highlight the quite distinct calculus based on this framework [3]. This definition provides a relationship between mass assignments and appropriateness degrees.

Definition 2 (*Appropriateness Degrees*)

$$\forall x \in \Omega, \ \forall L \in \mathscr{L} \quad \mu_L(x) = \sum_{S \subseteq \mathscr{L}:L \in S} m_x(S)$$

For example, given a set of labels defined on the temperature outside: $\mathscr{L}_{Temp} = \{low, medium, high\}$. Suppose 3 of 10 people agree that '*medium* is the only appropriate label for the temperature of $15°$ and 7 agree 'both *low* and *medium* are appropriate labels'. According to def. 1, the mass assignment for $15°$ is $m_{15}(medium) = 0.3$, and $m_{15}(low, medium) = 0.7$ or formally:

$$m_{15} = \{medium\} : 0.3, \ \{low, medium\} : 0.7$$

More details about the theory of mass assignment can be found in [1]. In this example, we have that the appropriateness of *medium* as a description of $15°$ is $\mu_{medium}(15) = 0.7 + 0.3 = 1$, and that of *low* is $\mu_{low}(15) = 0.7$.

It is certainly true that a mass assignment on D_x determines a unique appropriateness degree for μ_L for any $L \in \mathscr{L}$, but generally the converse does not hold. For example, given $\mathscr{L} = \{L_1, L_2, L_3\}$ and $\mu_{L_1} = 0.3$ and $\mu_{L_2} = 1$. We could obtain an infinite family of mass assignments:

$$\{L_1, L_2\} : \alpha, \ \{L_2\} : \beta, \ \{L_2, L_3\} : 0.7 - \beta, \ \{L_1, L_2, L_3\} : 0.3 - \alpha$$

for any α and β satisfying: $0 \leq \alpha \leq 0.3, \quad 0 \leq \beta \leq 0.7$. Hence, the first assumption we make is that the mass assignment m_x are consonant and this allows us to determine m_x uniquely from the appropriateness degrees on labels as follows:

Definition 3 (*Consonant Mass Assignments on Labels*) *Let* $\{\beta_1, \cdots, \beta_k\} = \{\mu_L(x) | L \in \mathscr{L}, \mu_L(x) > 0\}$ *ordered such that* $\beta_t > \beta_{t+1}$ *for* $t = 1, 2, \cdots, k-1$ *then:*

$$m_x = M_t : \beta_t - \beta_{t-1}, \ for \ t = 1, 2, \cdots, k-1,$$

$$M_k : \beta_k, \quad M_0 : 1 - \beta_1$$

where $M_0 = \emptyset$ *and* $M_t = \{L \in \mathscr{L} | \mu_L(x) \geq \beta_t\}$ *for* $t = 1, 2 \ldots, k$.

For the previous example, given $\mu_{L_1}(x) = 0.3$ and $\mu_{L_2}(x) = 1$, we can calculate the consonant mass assignments as follows: The appropriateness degrees are ordered as $\{\beta_1, \beta_2\} = \{1, 0.3\}$ and $M_1 = \{L_2\}, M_2 = \{L_1, L_2\}$. We then can obtain

$$m_x = \{L_2\} : \beta_1 - \beta_2, \{L_1, L_2\} : \beta_2 = \{L_2\} : 0.7, \{L_1, L_2\} : 0.3$$

Because the appropriateness degrees are sorted in def. 3 the resulting mass assignments are "nested". Clearly then, there is a unique consonant mapping to mass assignments for a given set of appropriateness degree values. The justification of the consonance assumption can be found in [1, 3]. Notice that in some cases we may have non-zero mass associated with the empty set This means that some voters believe that x cannot be described by any labels in $\mathscr{L}$. For example, if we are given $\mu_{L_1}(x) = 0.3$ and $\mu_{L_2}(x) = 0.8$, then the corresponding mass assignment is:

$$\{L_2\} : 0.5, \{L_1, L_2\} : 0.3, \emptyset : 0.2$$

where the associated mass for the empty set is obtained by $1 - \beta_1 = 0.2$.

3 High Level Label Description

In this section, we will consider how to use a high level fuzzy label to describe another fuzzy label. Here the term *high level* does not mean a hiacrhial structure. We will actually consider two set of fuzzy labels which are independently defined on the same universe. If the cardinality of a set of labels $\mathscr{L}$ is denoted by $|\mathscr{L}|$. We then can say $\mathscr{L}_1$ higher level labels of $\mathscr{L}_2$ if $\mathscr{L}_1 < \mathscr{L}_2$. We will acutally consider the methodology of using one set of fuzzy labels to represent the other set of fuzzy labels.

For example, a fuzzy concept *about_m* is defined by an interval on [a, b] (see the left-hand side figure of Fig. 1), so that the appropriateness degree of using fuzzy label *small* to label *about_m* is:

$$\mu_{small}(about\_m) = \frac{1}{b-a} \int_a^b \mu_{small}(u) du \tag{3}$$

If the vagueness of the concept *about_m* depends on the interval denoted by δ where the length of the interval $|\delta| = b - a$. We then can obtain:

$$\mu_{small}(about\_m) = \frac{1}{|\delta|} \int_{u \in \delta} \mu_{small}(u) du \tag{4}$$

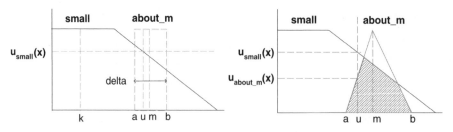

Fig. 1. The appropriateness degree of using *small* to label vague concept *about_m* is defined by the ratio of the area covered by both labels to the area covered by *about_m* only.

If *about_m* is defined by other fuzzy labels rather than an interval, for example, a triangular fuzzy set (e.g., the right-hand side figure of Fig. 1). How can we define the appropriateness degrees?

We begin by considering a data element $x \in [a, b]$, the function $\mu_{about\_m}(x)$ represents the degree of x belonging to the fuzzy label F. Function $\mu_{small}(x)$ defines the appropriateness degrees of using label *small* to describe x [3]. We essentially hope to obtain the appropriateness degrees of using *small* to label *about_m*. We then consider the each elements belonging to *about_m*. If $\mu_{about\_m}(x) = 1$, which means x is absolutely belonging to *about_m*, then the appropriateness degree is just $\mu_{small}(x)$. However, if $\mu_{about\_m} < \mu_{small}(x)$, we can only say it is belonging to *about_m* in certain degrees. Logically, fuzzy operation AND is used, and in practical calculation, the min($\cdot$) function is employed. The appropriateness is then defined by:

$$\mu_{small}(about\_m) = \frac{\int_{u \in \delta} \min(\mu_{small}(u), \mu_{about\_m}(u)) du}{\int_{u' \in \delta} \mu_{about\_m}(u') du'} \tag{5}$$

where function $\min(x, y)$ returns the minimum value between x and y. Equation 4 is a special case of equation 5 where the following equations always hold:

$$\mu_{small}(u) = \min(\mu_{small}(u), \mu_{about\_m}(u))$$

$$|\delta| = \int_{u \in \delta} \mu_{about\_m}(u) du$$

Definition 4 *Given a vague concept (or a fuzzy label) F and a set of labels $\mathscr{L} = \{L_1, \ldots, L_m\}$ defined on a continuous universe Ω. The appropriateness degrees of using label L ($L \in \mathscr{L}$) to describe F is:*

$$\mu_L(F) = \frac{\int_{u \in \delta} \min(\mu_L(u), \mu_F(u)) du}{\int_{u' \in \delta} \mu_F(u') du'} \tag{6}$$

where δ is the universe covered by fuzzy label F.

Given appropriateness degrees, the mass assignment can be obtained from the appropriateness degrees by the consonance assumption. Equation 5 is a general form for all kinds of fuzzy sets which are not limited to an interval or a triangular fuzzy sets.

Example 1. Figure 1 gives a set of isosceles triangular fuzzy labels $F_1, \ldots, F_8$ and two high level fuzzy label *small* and *large* defined on the same universe. The membership functions (the non-zero part) for F_5, F_6 and *small* are defined as follows:

$$PS \rightarrow y = \frac{5}{2}x - 3, \quad PT \rightarrow y = -\frac{5}{2}x + 5$$

$$QR \rightarrow y = \frac{5}{2}(x - 0.4) - 3, \quad QU \rightarrow y = -\frac{5}{2}(x - 0.4) + 5$$

$$OU \rightarrow y = -\frac{5}{6}x + 2$$

---

[3] Here we interpret $\mu(\cdot)$ in different manners: membership function and appropriateness degrees, though they are mathematically the same.

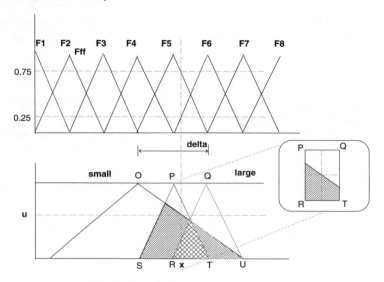

Fig. 2. The relations between fuzzy labels.

As we can see from Fig. 2: $\mu_{F_5}(x) = 0.75$ and $\mu_{F_6}(x) = 0.25$ given $x = 1.7$. According to definition 4 we can obtain:

$$\mu_{small}(F_5) = 0.8, \ \mu_{large}(F_5) = 1$$

$$\mu_{small}(F_6) = 0.5, \ \mu_{large}(F_6) = 1$$

So that the corresponding consonant mass assignments (see definition 3) are:

$$m_{F_5} = \{small, large\} : 0.8, \{large\} : 0.2$$

$$m_{F_6} = \{small, large\} : 0.5, \{large\} : 0.5$$

High level labels *small* and *large* can be used to describe $x = 1.7$ by the following steps.

$$m_x = \{F_5, F_6\} : 0.25, \{F_5\} : 0.5, \emptyset : 0.25$$

F_5 and F_6 can be represented by the mass assignments of high level fuzzy labels: *small* and *large*. Considering the term $\{F_5, F_6\}$, it means that both two labels F_5 and F_6 are appropriate for labeling x with a certain degree. It defines a area covered both by F_5 and F_6 (see Fig. 2) which is an interval between R and T. Therefore, according to def. 4 we can obtain the mass assignment for $\{F_5, F_6\}$:

$$m_{\{F_5, F_6\}} = \{small, large\} : 0.5, \{large\} : 0.5$$

Finally, we obtain:

$$m_x = (\{small, large\} : 0.5, \{large\} : 0.5) : 0.25,$$
$$(\{small, large\} : 0.8, \{large\} : 0.2) : 0.5, \emptyset : 0.25$$
$$= \{small, large\} : 0.525, \{large\} : 0.225, \emptyset : 0.25$$

From the above example, if we use *small* and *large* to describe x directly. By the function of *small* we can obtain $u = \frac{7}{12}$ so that the mass assignments are:

$$m_x = \{small, large\} : \frac{7}{12}, \{large\} : \frac{5}{12}$$

which is different from the result presented in example 1. It is because precision is lost by using two level of fuzzy labels. In our example, x is firstly repressed by F_5 and F_6 which is precise. However, the description of x by *small* and *large* through F_5 and F_6 is not precise any more, because F_5 and F_6 are not exact representation of x by involving uncertainties decided by δ. As we can see from the Fig. 3: the appropriateness degrees of using high level labels to describe low level concepts are depending on the uncertainty parameter δ. For example, given a data element m:

$$|\mu_{small}(F(\delta_1)) - \mu_{small}(m)| < |\mu_{small}(F(\delta_2)) - \mu_{small}(m)| < |\mu_{small}(F(\delta_3)) - \mu_{small}(m)|$$

So that:

$$\mu_{small}(m) = \lim_{\delta \to 0} \mu_{samll}(F(\delta))$$

where F is the function of the fuzzy label (a function of δ-either an interval, triangular fuzzy set or other type of fuzzy set) centered on m.

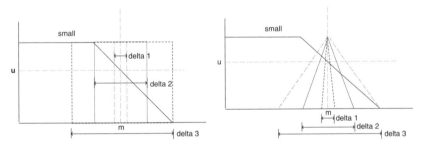

Fig. 3. The appropriateness degree of using *small* depends on the width of the vague concept of *about_m*.

4 Conclusions

In this paper, a methodology of using high level fuzzy labels to describe vague concepts or low level fuzzy labels is proposed based on label semantics framework. An example is given to show how to calcuate the mass assigments of high level fuzzy labels on a vague concept.

References

[1] J. F. Baldwin, T. P. Martin and B. W. Pilsworth, *Fril- Fuzzy and Evidential Reasoning in Artificial Intelligence*, John Wiley & Sons, Inc., 1995.

[2] George J. Klir and Bo Yuan, *Fuzzy Sets and Fuzzy Logic: Theory and Applications*, Prentice-Hall, Inc., Upper Saddle River, NJ, 1994.

[3] Jonathan Lawry, *Modelling and Reasoning with Vague Concepts*, Springer, 2006.

[4] Zengchang Qin and Jonathan Lawry, Decision tree learning with fuzzy Labels, *Information Sciences*, vol. 172: 91-129, June 2005.

[5] Zengchang Qin and Jonathan Lawry, Linguistic rule induction based on a random set semantics. *World Congress of International Fuzzy Systems Association* (IFSA-05), July 2005.

[6] Zengchang Qin and Jonathan Lawry, Prediction trees using linguistic modelling. *World Congress of International Fuzzy Systems Association* (IFSA-05), July 2005.

[7] Nicholas J. Randon and Jonathan Lawry, Classification and query evaluation using modelling with words, *Information Sciences* 176(4): 438-464, 2006.

[8] L. A. Zadeh, A new direction in AI - toward a computational theory of perceptions, *A.I. Magazine*, Spring 2001.

[9] L.A. Zadeh, Fuzzy logic = computing with words, *IEEE Trans. Fuzzy Systems*, Vol.2:103-111, 1996.

Part VII

Integrated Uncertainty Modelling in Applications

Possibilistic Channels for DNA Word Design

Luca Bortolussi[1] and Andrea Sgarro[2]

[1] Dept. of Mathematics and Informatics, University of Udine, Udine, Italia
   `luca.bortolussi@dimi.uniud.it`
[2] Dept. of Mathematics and Informatics, University of Trieste, and Consorzio Biomedicina Molecolare, Area Science Park, Trieste, Italia
   `sgarro@units.it`

We deal with DNA combinatorial code constructions, as found in the literature, taking the point of view of possibilistic information theory and possibilistic coding theory. The possibilistic framework allows one to tackle an intriguing information-theoretic question: what is channel noise in molecular computation? We examine in detail two representative DNA string distances used for DNA code constructions and point out the merits of the first and the demerits of the second. The two string distances are based on the reverse Hamming distance as required to account for hybridisation of DNA strings.

1 Introduction

Assume that an input word is chosen out of a list, or *codebook*, and is sent through a *noisy communication channel*; a distorted version of the input word is received at the output end of the channel. Several rational behaviours can be envisaged, according to the context. The observer can decode to the input word in the list whose *diversity* or *distortion* from the output is minimum; or: the observer of the output can *decode* to the input word whose *similarity* to the output is maximum; or: the observer can decode to the input word such that the *likelihood* of having observed the output is maximum; or: the observer can decode to the input word such that the *possibility* of having observed right that output is maximum. All these approaches, even if maximum likelihood is definitely less powerful, can be conveniently lodged under the shed of *possibilistic information theory* and *possibilistic coding theory*, as were put forward in [11], [12], [13], [14]. So far, this theoretical framework had been applied to devising error-correcting telephone keyboards and to the problem of ensuring safety in a telephone network with sensitive users [8], [2]; below, we show that it can consistently and conveniently lodge unusual forms of coding, as is *DNA word design* (for which cf. e.g. [4], [9], [10]), and it allows one to answer an intriguing question of a subtle information-theoretic nature: what is *noise* in molecular computation, or more specifically in DNA word design?

L. Bortolussi and A. Sgarro: *Possibilistic Channels for DNA Word Design*, Advances in Soft Computing **6**, 327–335 (2006)
`www.springerlink.com`

Below, Sections 2 and 3 deal with possibilistic coding up to the formulation of our main problem: what is channel noise in DNA word design? In Section 4 we deal with two representative cases, i.e. DNA code constructions based on checking the reverse Hamming distance between codewords, and those based on checking *both* Hamming distances, direct and reverse. Only this compound DNA string distance passes the "possibilistic control" (theorem 1), and allows a satisfactory definition of what channel noise may be in biological computation, while pure reverse Hamming distance fails (theorem 2). The last section contains a short biological reminder on DNA word design and a few concluding remarks. In this paper we consider only *crisp* codes, be it from the "soft" point of view of possibility theory; *fuzzy* DNA codebooks and their biological adequacy have been discussed in [3].

2 Channel Models

An input-word set $\mathscr{X}$ and an output-word set $\mathscr{Z}$ are given. A *codebook* $\mathscr{C}$, or simply a *code*, is a subset of the input set, and its elements are called *codewords*; $\mathscr{C} \subset \mathscr{X}$. The channel is specified through a *matrix* with rows headed to $\mathscr{X}$ and columns headed to $\mathscr{Z}$, and whose entries $\alpha(x,z)$ are non-negative real numbers. In this paper we shall cover explicitly two[3] cases:

Possibilistic matrix (the maximum entry in each row of the matrix is 1): its elements are *transition possibilities* from inputs to outputs.
Distortion matrix (the minimum entry in each row of the matrix is 0): its elements specify the *distortion* between input and output.

These matrices implicitly describe the *noise* which affects channel transmission. The *decoding strategies* are: in the case of transition possibilities decode to the input codeword in $\mathscr{C}$ which maximises the value in the matrix column headed to the output received; instead, one minimises the matrix value in the case of distortions. So, the first decoding strategy is a *maximising strategy*, while the second is a *minimising strategy*. The underlying implicit assumption is that the smaller the possibility (the higher the distortion, respectively), the less "likely" it is to occur during channel transmission, and this in a very uncommittal sense of the word "likely", cf. [11], [13], [14]. In case of ties one declares a *detected decoding error*; unfortunately, one can incur also into *undetected decoding errors*, even if the possibility of such events should be small for a good code.

Examples. In standard coding theory, one decodes by minimising Hamming distance, that is the number of positions in which the input sequence and the output sequence differ (both sequences are assumed to have the same length). The corresponding transition possibility is the frequency of positions where input and output

---

[3] We apologise for this redundancy: we might have stuck to possibilities only, as in [11], but we are confident that this redundancy will make the paper easily readable by a larger audience, inclusive of coding theorists and computational biologists.

sequences coincide; cf. below. In Shannon theory, instead, one minimises transition (conditional) probabilities, as arranged in a stochastic matrix. In [11], where the theoretical and practical bearing of the possibilistic approach to coding is discussed and vindicated, one mimics the Shannon-theoretic approach, after replacing transition probabilities by transition possibilities (probabilistic channels by possibilistic channels, probabilistic noise by possibilistic noise). An even more general possibilistic framework would be a "bayesian-like" generalisation: assume one has a possibility vector on the input set $\mathscr{X}$, and a transition possibility matrix from $\mathscr{X}$ to $\mathscr{Z}$. Form a *joint* possibilistic matrix by taking minima between prior possibilities and the corresponding transition possibilities: then one decodes to codewords which must have *both* a large prior possibility *and* a large transition possibility. In this case the matrix would be normal, but not all of its rows ("normal" means that the maximum entry is 1, cf. e.g. [6]).

Actually, what matters in these matrices is not the actual numeric values, but rather their mutual order. More specifically, two possibilistic matrices (two distortion matrices, respectively) are *equivalent* if there is a strictly increasing one-to-one correspondence $\beta(x,z) = f\big(\alpha(x,z)\big)$, between their entries $\alpha(x,z)$ and $\beta(x,z)$; a possibilistic matrix and a distortion matrix are equivalent when the one-to-one correspondence f is strictly *decreasing*.

Proposition. *Take a codebook $\mathscr{C}$. Decode according to two equivalent matrices. The first decoder makes a detected error or an undetected error, respectively, iff the second decoder makes an error, detected or undetected, respectively.*

(The straightforward proof is omitted). A distortion matrix can be soon converted to an equivalent possibilistic matrix. In practice, we shall have to deal only with words which are strings of the same length n, with the distortion spanning the integers from 0 to n, and so we shall *always consider jointly* a distortion matrix $d(x,z)$ and the possibilistic matrix defined by the transition possibilities:

$$\pi(z|x) = 1 - n^{-1} d(x,z) \tag{1}$$

In the case of decoding by maximum possibility we are ready to incur only into decoding errors which correspond to negligible transition possibilities π; if the decoding strategy is minimum distortion, we are ready to incur only into decoding errors which correspond to exorbitant distortions $d = n(1 - \pi)$.

3 Optimal Code Constructions for Possibilistic Channels

The *confusability* and the *distinguishability* between two input sequences x and y are defined, respectively, as:

$$\gamma(x,y) \;=\; \max_{z} \pi(z|x) \wedge \pi(z|y) \,, \quad \delta(x,y) \;=\; \min_{z} d(x,z) \vee d(y,z)$$

($\lor$ and $\land$ are alternative notations for *max* and *min*). The *maximum confusability* $\gamma_\mathscr{C}$ of the code $\mathscr{C}$ and its *minimum distinguishability* $\delta_\mathscr{C}$ are the maximum confusability and the minimum distinguishability between any two distinct codewords, respectively. Of course:

$$\gamma(x,y) = 1 - n^{-1}\delta(x,y), \ \ \gamma_\mathscr{C} = 1 - n^{-1}\delta_\mathscr{C}$$

The operational meanings of $\delta_\mathscr{C}$ and of $\gamma_\mathscr{C}$ from the viewpoint of coding are given by the following *reliability criterion*:

Reliability criterion ([13], [14]). *The maximum confusability $\gamma_\mathscr{C}$ is the highest possibility which is* not *always corrected when decoding by maximum possibility, while possibilities $> \gamma_\mathscr{C}$ are always corrected. Equivalently: the minimum distinguishability $\delta_\mathscr{C}$ is the lowest distortion which is* not *always corrected when decoding by minimum diversity, while distortions $< \delta_\mathscr{C}$ are always corrected.*

The classical *optimisation problem* of channel coding (*optimal code constructions*) is maximising the code's size subject to a specified reliability constraint, which in our case is that all "large" possibilities, or all "small" distortions, respectively, should be properly[4] corrected. Maximising the code size is the same as maximising its *transmission rate* $n^{-1}\log_2|\mathscr{C}|$, i.e. the number of information bits carried by each transmitted symbol (for this and for other basic notions of information and coding theory the reader is referred e.g. to [7]). We stress that the constraints with respect to which one optimises should *not* be expressed in terms of transition possibilities or distortions, but rather in terms of confusabilities, or distinguishabilities, as the reliability criterion makes it clear:

$$\gamma_\mathscr{C} \leq \rho \ \text{ or } \ \delta_\mathscr{C} \geq \lambda = n(1-\rho), \ 0 \leq \rho \leq 1$$

In the case when the distortion is the *Hamming distance* $d_H(x,y)$, the distinguishability and the confusability are soon found to be:

$$\delta_H(x,y) = \left\lceil \frac{d_H(x,y)}{2} \right\rceil, \ \ \gamma_H(x,y) = n^{-1}\left\lfloor n\frac{1+\pi(y|x)}{2} \right\rfloor$$

and so they are (weakly) *increasing* functions of the distortion and of the transition possibility, respectively. Given this monotonic dependence, one can construct reliable codes as one does in algebraic coding, that is with respect to reliability constraints expressed in terms of the minimum Hamming distance between distinct codewords, rather than constraints on confusabilities or distinguishabilities, as one may and should do in full generality. In other words, the Hamming distance acts also as a very convenient *pseudo-distinguishability* between codewords[5] and not

---

[4] Due to the maxitive nature of possibilities, this is the same as checking *decoding error possibilities*, exactly as one checks decoding error probabilities in Shannon theory, cf. [11].

[5] Since the monotonicity is only weak, by checking distances rather than distinguishabilities one ends up solving a more general combinatorial problem, and one found also "spurious" solutions. Such spurious solutions are made good use of for error detection rather than error correction, but we cannot deepen this point here; cf. [14].

only as a distortion between input sequences and output sequences. In the literature of DNA word design [4], [5], [9], [10] one is interested in code constructions where one checks suitable DNA string distances, out of which we shall consider two (representative) examples. Can one solve in a satisfactory way the *inverse problem* of constructing possibilistic channels (of describing suitable *channel noise*) which would "explain" those constructions, as one can do in the usual Hamming case of coding theory? Or: can these DNA distances be seen as suitable *pseudo-distinguishabilities* for a suitable noisy channel? As argued in next section, one of the two DNA distances will faithfully mirror the usual Hamming case, while the second will turn out to be totally unmanageable. It is no coincidence, we deem, that the second case is less justifiable also from a strictly biological point of view.

4 The Inverse Problem of Channel Noise in DNA Word Design

DNA word design is an "odd" form of coding used in molecular computation, where, based on biological facts (cf. section 5), one exhibits maximum-size code constructions relative to constraints of the form $d(x,y) \geq \lambda$ for a suitable DNA string distance d. An information-theoretic problem arises: what is the nature of the biological channel one is implicitly envisaging, or, equivalently: what sort of "biological noise" are we fighting against when we use these code constructions? Thinking of the above arguments, we can re-formulate the question as follows: can $d(x,y)$ be interpreted as a *pseudo-distinguishability*, i.e.: can one exhibit a transition possibility $\pi(z|x)$ between inputs and outputs such that the corresponding distinguishability function $\delta(x,y)$ is a non-trivial and non-decreasing function of $d(x,y)$? The possibilistic framework[6] allows one to answer these questions in a sensible way. We shall discuss two types of code constructions found in the literature: the answer will be positive in one case, which is better justified also from the biological point of view, and negative in the other.

We shall deal only with two DNA distortions, which however are very representative, the *reverse Hamming distance* and a variation thereof:

$$d_R(x,y) \quad \text{and} \quad d_{H \wedge R}(x,y) = d_H(x,y) \wedge d_R(x,y)$$

Here $d_H(x,y)$ is the usual Hamming distance, while the reverse Hamming distance is $d_R(x,y) = d_H(x,y^*)$, with y^* mirror image of y. In practice, in the case of d_R, codewords in a good code should have a large reverse Hamming distance, while they should have *both* a large Hamming distance *and* a large reverse Hamming distance[7] in the case of $d_{H \wedge R}$. We recall that $d_{H \wedge R}(x,y)$ is a *pseudometric*; one has

---

[6] Maximum likelihood is enough to account for usual code constructions, as those in [7], by referring to symmetric channels. The maximum likelihood approach, however, does not appear to be the right model for "odd" codes as the ones of DNA word design.

[7] The reason why we do not discuss explicitly complementarity and self-hybridisation are given in the last section.

$d_{H \wedge R}(x,y) = 0$ when $x = y$ or when x and y are mirror images of each other. Nothing so tame happens in the case of d_R, which violates the triangle inequality.

Below we shall try to "explain" the corresponding DNA code constructions by exhibiting a suitable possibilistic noisy channel and a suitable noise-fighting decoder. To achieve this, let us begin by the friendlier case, and let us compute the confusability $\gamma_{H \wedge R}$ and the distinguishability $\delta_{H \wedge R}$ corresponding to the string distance $d_{H \wedge R}$ taken as the distortion between inputs and outputs. We decode the output z by minimum distortion, and so we are implicitly assuming that it is "unlikely" (i.e. possible only to a small degree) that z has both a large Hamming distance and a large reverse Hamming distance from the codeword c actually sent over the channel.

Theorem 1. *Decode the output z by minimising $d_{H \wedge R}(c,z)$, $c \in \mathscr{C}$; the corresponding distinguishability and confusability functions are:*

$$\delta_{H \wedge R}(x,y) = \left\lceil \frac{d_{H \wedge R}(x,y)}{2} \right\rceil, \quad \gamma_{H \wedge R}(x,y) = n^{-1} \left\lfloor n \frac{1 + \pi(y|x)}{2} \right\rfloor$$

This is exactly the same situation as found with usual Hamming distances and the codes of algebraic coding (as for the straightforward proof, cf. similar computations to find distinguishability functions in [14]). In practice, this means that a possibilistic channel based on the transition possibilities $\pi(z|x) = 1 - n^{-1}d_{H \wedge R}(x,z)$ and the corresponding noise quite adequately "explain" the code constructions based on checking the pseudometric $d_{H \wedge R}$, as are those found in the literature.

Now, let us think of a DNA word design construction where one controls *only* the minimum reverse Hamming distance between codewords. The situation is less friendly, because if we decide to decode by minimum reverse Hamming distance, the corresponding distinguishability function turns out to be a non-decreasing function of the *usual* Hamming distance, and *not* of the reverse Hamming distance, as a simple computation shows. In other words, against this sort of noise one would need the *usual* codes of coding theory, and *not* those codes of DNA word design which we are trying to "explain". So, the following problem is relevant:

Problem: Exhibit a transition possibility $\pi(y|x) \geq 0$ whose distinguishability function $\Xi(x,y)$ is a non-decreasing and non-trivial function of the reverse Hamming distance $d_R(x,y)$. Equivalently, exhibit a distortion $\xi(x,y)$ joint with $\pi(x,y)$ as in (1).

Note that we are not even insisting that the distortion ξ used for decoding should be in any way "similar" to d_H. Unfortunately one has the following negative result (assuming $f(0) = 0$ as we do below is no real loss of generality):

Theorem 2. *Let $\pi(y|x)$ be a transition possibility such that its distinguishability function $\Xi(x,y)$ is a non-decreasing function f of the reverse Hamming distance $d_R(x,y)$, $f(0) = 0$. Under such hypotheses, Ξ is trivial, in the sense that $\Xi \equiv 0$ when n is even, while, when n is odd, $\Xi(x,y) = 0$ whenever $d_R(x,y) \leq n-1$. Moreover, if the joint distortion $\xi(x,y)$ is constrained to verify the triangle inequality, then $\Xi \equiv 0$ even for n odd.*

Proof. For all $x \in \mathcal{X}$, one has $d_R(x,x^*) = 0$, hence $\Xi(x,x^*) = f(0) = 0$. But this means that $\exists \tilde{z} \in \mathcal{X}$ such that $\xi(x,\tilde{z}) = \xi(x^*,\tilde{z}) = 0$. Therefore $\Xi(x,x) = \min_{z \in \mathcal{X}} \xi(x,z) = 0$, and so each string has distinguishability zero from itself. Now, if x is a string such that $d_R(x,x) = m$, one has $f(m) = f(d_R(x,x)) = \Xi(x,x) = 0$; as $d_R(x,x)$ can assume all even values from 0 to n, one may take $m = n$ for n even and $m = n - 1$ for n odd; so, the non-decreasing function f maps the integer interval $[0,n]$ to 0 for n even, and the integer interval $[0, n-1]$ to 0 for n odd. Binary examples with n as low as 3 (and with ξ symmetric, $\xi(x,x) = 0 \; \forall x$) show that one can have $\Xi(x,y) \neq 0$ for $d_R(x,y) = n$. Instead, if ξ is constrained to verify the triangle inequality, one has $\xi(x,y) \leq 2\Xi(x,y)$ (cf. [14]), and so $\xi(x,y) = 0$ whenever $d_R(x,y) \leq n - 1$; now, for $n = 3,5,\ldots$ one can always find a z at reverse Hamming distance $\leq n-1$ from both x and y; e.g. take z with the first digit equal to the last digit of x and with the last digit equal to the first digit of y; this implies $\Xi \equiv 0$ even for n odd. $\qquad\square$

In practice, the theorem means that within the possibilistic framework, ample as it is, code constructions based on checking reverse Hamming distances have *no* counterparts in terms of noisy channels and channel decoders; *no* possibilistic matrix $\pi(z|x)$ exists which would adequately support those constructions.

5 A Short Reminder on DNA Word Design

In the last ten years, a new computational paradigm emerged from a very uncommon place, i.e. wet labs of biologists. The fact that DNA contains all the basic information necessary to build very complex living organisms convinced Adlemann that it could also be used as a computational entity. In his milestone paper of 1994 [1], he proposed a computational model based on very simple manipulations of DNA that can be performed in a wet lab. This model is Turing-complete and bases its power on the massive parallelism achievable by using DNA. Moreover, one of the basic operations performed is the hybridisation of complementary DNA strings. Specifically, DNA strings are oriented strings over the alphabet $\Sigma = \{a, c, g, t\}$, where a-t and c-g are complementary letters. Two such strings are said to be complementary if they have the same length and if one can be generated by reversing the other and complementing each of its letters. Physically, complementary DNA strings can hybridise, i.e. they can attach one to the other, forming the famous double helix. Actually, hybridisation can occur also between strings that are not perfect complements, but close to it. In DNA computations, data is coded by short strings of DNA in such a way that hybridisations occurring determine the output of the "algorithm" [10]. Therefore, one of the main concerns is to avoid that "spurious" hybridisations occur, leading straight to the so-called *DNA word design* problem.

DNA word design (cf. [9, 4]) consists of identifying sets of DNA strings of a given length, called *DNA codes*, satisfying some constraints, usually related to distances between codewords. In particular, the main concern of DNA word design is to identify maximal set of strings satisfying the constraints, cf. Sections 3 and 4.

In the body of the paper complementarity has been forgotten out of simplicity, since it does not really change the mathematical problem, but makes notations and formulations heavier; cf. also [5]. Seemingly, we have forgotten also about self-hybridisation, i.e. we have forgotten to check that the codewords should not "resemble" their own mirror images. Notice however that in Section 4 we have *never* used the assumptions that the input space is made of *all* the strings of length n: the constraint on the reverse Hamming distance between a codeword and itself has a nature of its own, and may be conveniently used to restrict the input space for possible codewords only to those strings whose self-distance is large enough with respect to a prescribed threshold.

Acknowledgements

Supported by INdAM-GNCS and FIRB-RBLA039M7M_005 Project on Bioinformatics.

References

[1] L. Adlemann. Molecular computations of solutions of combinatorial problems. *Science*, 266:1021–1024, November 1994.

[2] L. Bortolussi and A. Sgarro. Fuzzy integrals and error correction in a telephone networks with sensitive users. *Proceedings of IPMU-2004*, Perugia, Italy, 1503–1508, 2004.

[3] L. Bortolussi and A. Sgarro. Fuzzy codebooks for DNA word design. *Proceedings of IPMU-2006*, Paris, France, July 2006.

[4] A. Condon and A. Brenneman. Strand design for bio-molecular computation. *Theoretical Computer Science*, 287(1):39–58, 2002.

[5] A. Condon, R.M. Corn, and A. Marathe. On combinatorial dna word design. *Journal of Computational Biology*, 8(3):201–220, November 2001.

[6] D. Dubois and H. Prade, eds. *Fundamentals of Fuzzy Sets*. Kluwer, 2000.

[7] J. van Lint. *Introduction to Coding Theory*. Springer Verlag, Berlin, 1999.

[8] F.L. Luccio and A. Sgarro, Fuzzy graphs and error-proofs keyboards. *Proceedings of IPMU-2002*, Annecy, France, 1503–1508, 2002.

[9] G. Mauri and C. Ferretti. Word design for molecular computing: A survey. In *Proceedings of 9th Int. Workshop on DNA Based Computers, DNA 2003*, pages 37–46, 2003.

[10] N. Pisanti. A survey on dna computing. *EATCS Bulletin*, 64:188–216, 1998.

[11] A. Sgarro. Possibilistic information theory: a coding-theoretic approach. *Fuzzy Sets and Systems*, 132-1, 11–32, 2002.

[12] A. Sgarro. Possibilistic time processes and soft decoding. In *Soft Methodologies and Random Information Systems*, ed. by M. Lopéz-Díaz, M.Á. Gil, P. Grzegorzewski, O. Hyrniewicz and J. Lawry, 249–256, Springer Verlag, 2004.

[13] A. Sgarro. Utilities and distortions: an objective approach to possibilistic coding. *International Journal on Uncertainty, Fuzziness and Knowledge-Based Systems*, 13-2, 139–161, 2005.

[14] A. Sgarro and L. Bortolussi. Codeword distinguishability in minimum diversity decoding. *Journal of Discrete Mathematical Sciences and Cryptography*, 2006 (to appear).

Transformation of Possibility Functions in a Climate Model of Intermediate Complexity

Hermann Held and Thomas Schneider von Deimling

Potsdam Institute for Climate Impact Research (PIK); PO Box 601203, 14412 Potsdam, Germany
held@pik-potsdam.de

Motivated by a preliminary series of expert interviews we consider a possibility measure for the subjective uncertainty on climate model parameter values. We consider 5 key uncertain parameters in the climate model CLIMBER-2 that represents a system of thousands of ordinary differential equations. We derive an emulator for the model and determine the model's mapping of parameter uncertainty on output uncertainty for climate sensitivity. Climate sensitivity represents a central climate system characteristic important for policy advice, however subject to huge uncertainty. While the ratio of output/input uncertainty induced by a single-parameter perturbation resembles the respective ratio when using a standard probability measure, we find the ratio qualitatively larger in the 5-dimensional situation. We explain this curse of dimension effect by a Gaussian analogue toy system.

1 Introduction

The climate modeling community faces a shift in statistical paradigm from frequentists' classical methods towards Bayesian updating. In particular, since 2001, Bayesian methods are being used when analyzing the key climate system or climate model characteristic, the climate sensitivity (CS). CS denotes the system's equilibrium response in global mean surface temperature when doubling the atmospheric (preindustrial) concentration of carbon dioxide. Over the last years, uncertainty analysis of CS has become a hot topic as uncertainty in CS comprises a major fraction of uncertainty in the mapping from greenhouse gas emissions to global warming impacts.

To our impression, Bayesian methods are used in this context for three reasons: (1) they allow for probability measures in CS, useful for further use in decision-analytic frameworks, (2) the Bayesian method allows for an elegant inclusion of subjective prior knowledge on model parameters, (3) Bayesian methods – operationalized through some sort of (Markov Chain) Monte Carlo-type techniques [5] – are more straightforward to implement than powerful classical tests on CS in multivariate nonlinear settings such as climate models.

H. Held and T.S. von Deimling: *Transformation of Possibility Functions in a Climate Model of Intermediate Complexity*, Advances in Soft Computing **6**, 337–345 (2006)
www.springerlink.com

While Bayesian learning on CS has been published for several implementations, unease with the structure of subjective knowledge and its adequate representation is growing. E.g. Frame et al. [6] highlight Bertrand's paradox: given the case of absent prior knowledge, the Bayesian school asks for "non-informative" priors, in general uniform distributions. Suppose we feel uninformed on the parameter x, then we are so on $y \equiv x^2$ as well and should assume a uniform distribution on y as well, incompatible with a non-uniform Jacobian between x and y.

Here we consider possibility measures (e.g. [1, 2, 4, 12]) as particular variant of imprecise probability measures [11] to represent subjective knowledge, for the following reasons: (1) They resolve Bertrand's paradox by avoiding the need of Jacobians. (2) The measure can be conveniently expressed through a possibility function in analogy to a probability density function (pdf) for the probability measure. (3) We elicited 7 experts in the climate modeling community[1] on the structure of their knowledge on uncertain parameters. Inter alia we asked to specify a subjective pdf for a given parameter. Then we proposed bets in order to test for consistency with the pdf specified. A larger fraction of the individuals said that in fact they could imagine to bet even higher on the center of their previously specified pdf than it would be appropriate in a fair bet. We will discuss below that possibilities can account for this empirical result. (4) In the same survey it became clear that it might be desirable to consider nonlinear transformation of a parameter rather than the parameter itself, while admittedly the choice of that transformation might be clear only in very fuzzy terms. As possibility functions do not need Jacobians when mapped (see also (1)), they respond in a much more robust way than pdfs to fuzziness in transformation.

In summary, we consider possibility functions in an explorative manner to represent model users' uncertain prior knowledge on model parameters. We ask what the effects of this more imprecise (as against a pdf) representation on the output quantity of interest would be in order to generate an impression on how "assuming" the expert really is. Furthermore, this also serves as a showcase for Bayesian learning with vanishing information content in the likelihood, a situation often faced in data-sparse decision situations.

2 Basic Properties of Possibility Measures

Any measure discussed in the remainder of this article may live on a σ algebra B on $\mathbb{R}^n, n \in \mathbb{N}$. A *possibility measure* Π [1, 2, 4, 12] represents a class $\mathscr{P}$ of probability measures P with [3]

$$\mathscr{P} = \{P | \forall_{A \in B} \, P(A) \leq \Pi(A)\} \tag{1}$$

and there exists always a (generating) possibility function $\pi : \mathbb{R}^n \to [0,1]$ with

$$\exists_{x \in \mathbb{R}^n} \, \pi(x) = 1 \quad \text{and} \quad \forall_{A \in B} \, \Pi(A) = \sup_{x \in A} \pi(x). \tag{2}$$

---

[1] Details will be published elsewhere.

For that reason, it is sufficient to study the properties of the function π.

We now ask how $\mathscr{P}$ transforms under a mapping $f : \mathbb{R}^n \to \mathbb{R}$.[2]

In [8] it is shown that under weak conditions[3] the transformed class $f(\mathscr{P})$, generated by transforming each member of that class according to the standard rules of probability theory, represents again a possibility measure π' with

$$\forall_{y \in f(\mathbb{R}^n)} \ \pi'(y) = \sup_{x \in f^{-1}(y)} \pi(x) \tag{3}$$

Note that no Jacobian of f is needed as it would be the case for pdfs. This simple transformation rule makes possibility measures particularly attractive.

Below we will need the following Lemma [8] that at the same time provides an attractive interpretation of possibility functions:

Lemma 1. *Let $\mathscr{P}$ the class of probability measures induced by the possibility function π according to Eqs. 1 and 2. Then*

$$\mathscr{P} = \{P \ | \ \forall_{a \in [0,1]} \ P(\{x | \pi(x) \le a\}) \ \le a\}.$$

This implies that for any member P of the class $\mathscr{P}$ the measure outside the α-cut of π must be α or less, consequently within the α-cut or more. This is the very feature which would be in line with the finding of our expert elicitation were experts would not exclude that the measure in the central areas of the parameter intervals specified could also be larger than their pdf would allow for.

3 Sampling and Emulating the Climate Model

The data used in this article where taken from a numerical experiment that was designed for standard Bayesian learning of the CLIMBER-2 climate model [9], [7]. The 5 model parameters that were regarded as most influential on CS were perturbed over a range that the authors of the model would find plausible. That way, a 5000 member ensemble along a Latin Hypercube Scheme was set up. In later experiments [10], 11 parameters were perturbed over wider ranges. However, for the present conceptual study we stay with the 5-parameter experiment as the pdf derived is more structured and in that sense more interesting. Among the 5 parameters were horizontal and vertical ocean diffusivity as well as cloud parameters. As some of them were varied over orders of magnitude, they were logarithmically transformed and then scaled by an affine transformation such that any of their sampling intervals would be mapped onto [0,1]. The vector of the so transformed parameters will be denoted by the vector $x = (x_1, ..., x_5)$ in the following. For any i, x_i was sampled according to a $\beta(7/4, 7/4)(x_i)$-distribution (see Fig. 1, upper graph) $\propto x_i^{3/4}(1 - x_i^{3/4})$, combined under a Latin Hypercube scheme.

---

[2] The following statement holds for arbitrary dimension, however, is not needed for this article in which we focus on the single output quantity CS.

[3] The statement given below holds e.g. if $\mathbb{R}^n$ where replaced by a finite approximation.

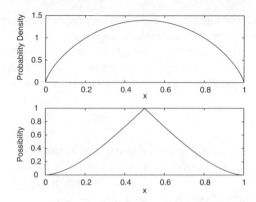

Fig. 1. $\beta(7/4, 7/4)$ distribution (top graph) and possibility substitute (bottom graph). The possibility function is chosen such that the β measure is contained in the class spanned by the possibility measure and that the β measure at the same time represents the "least localized" member (see text) of that class.

We can use 5000 realizations of $(x, CS(x))$. However, in order to implement Eq. 3, these data are not directly suited – no optimization algorithm could meaningfully operate on them.

Hence we emulate the model by a polynomial fit of 9th order, involving 2002 monomes. Here we intend to present a conceptual study and hence are satisfied if our emulator reproduces some gross features. We check that the histograms generated from the 5000 realizations of x look very similar for both the original model and the emulator (see Fig. 2).

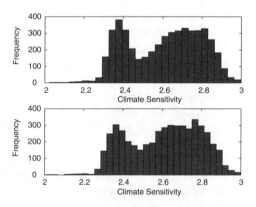

Fig. 2. Comparison of CS histograms for the original model (top graph) and the emulator (bottom graph). For this article, the agreement is sufficient, in particular the emulator reproduces the bimodality.

4 Mapping Single-Parameter Uncertainty

As multi-dimensional, nonlinear mapping is intricate to interpret, we start by analyzing the effects of a single parameter. For that we choose the "most influential" parameter on CS (which turns out to be the 4th component of our parameter vector) in also setting up a *linear* fit for f and asking for the component with the maximum gradient (in CS) modulus.

We study the influence of x_4 by fixing the remaining parameters to the maximum of their pdf, i.e. to 1/2. In Fig. 3, top graph, we display the transfer function from x_4 to CS that turns out to be markedly nonlinear and also bijective. We sample $x_4 \sim \beta$ and obtain a bimodal pdf for CS as a result (center graph). As the input pdf was unimodal, the bimodality is a direct consequence of the Jacobian being non-constant, i.e. the mapping being nonlinear.

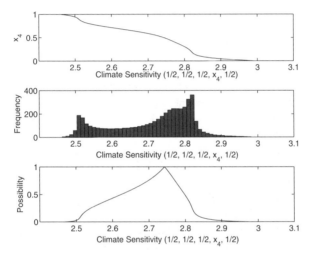

Fig. 3. Analysis of the effects of a single parameter according to the emulator transfer function. We fix $\forall_{i \neq 4} x_i := 1/2$, i.e. to the maximum of the input β-distribution. Then we sample x_4 according to the β-distribution. The top graph displays the nonlinear bijective mapping $x_4 \to$ CS. Due to extrema in the Jacobian, a bimodal structure is induced in the sampled pdf of CS. The related possibility function of CS displays roughly the same spread, however, no bimodality.

We now would like to use the same x_4-restricted transfer function in order to study the propagation of a possibility function. The point is to choose a possibility function for x_4 that suits the expert's knowledge better than β. On the one hand, the experts feel comfortable with a β-function. Hence $\mathscr{P}$ should contain β. On the other hand, according to the expert's betting behavior, $\mathscr{P}$ should also contain elements that allocate more probability measure to the mode of β. Both can be achieved if we define π such that $\mathscr{P}$ accommodates β as a limiting case according to Lemma 1:

$$\forall_{a \in [0,1]} \quad P(\{x | \pi(x) \le a\}) = a. \tag{4}$$

Given the symmetry of β w.r.t. 1/2, this is achieved by choosing

$$\forall_{x \in [0,1/2]} \; \pi_x(x) := 2 \int_0^x dx' \beta(x') \quad \text{and} \quad \forall_{x \in]1/2,1]} \; \pi_x(x) := 2 \int_x^1 dx' \beta(x'). \tag{5}$$

π_x is depicted in Fig 1, bottom graph. We calculate the possibility function for CS, π_{CS} by involving the transfer function f_4 in Fig. 3, top graph. As f_4 is bijective, Eq. 3 reduces to

$$\pi_{CS}(CS) = \pi_x(f_4^{-1}(CS)) \tag{6}$$

the result of which is displayed in Fig. 3, bottom graph. According to the last Eq., not involving a Jacobian, π_{CS} must be unimodal as π_x is unimodal.

5 The Multivariate Case

Our model represents a multi-dimensional mapping, as several parameters are found influential on model output CS: according to our linear fit (set up solely to rank parameters), the modulus of the gradient of CS, normalized by the largest component, reads $(1, > 0.80, > 0.25, > 0.04, > 0.01)$. Hence we regard it as necessary to set up a multi-dimensional possibility function.

The multivariate input pdf for x was chosen as $\beta(x_1) \cdot \ldots \cdot \beta(x_5)$. How could we generalize that to a possibility measure? It is not obvious how to generalize our single-parameter procedure for generating π to the multivariate case.

In order to transform the multivariate situation into an efficiently one-dimensional one, we map $\forall_i x_i$ to z_i such that the pdf of z_i is a standard normal distribution $N(z_i)$, and as a consequence, the pdf in $z := (z_1, \ldots, z_5)$ displays radial symmetry. Then we can apply the ideas of the previous Section to the radial coordinate $|| : \mathbb{R}^5 \to \mathbb{R}_0^+$, $|z| := \sqrt{\sum_{i=1}^5 z_i^2}$. The bijective mapping between x_i and z_i is then determined by

$$\forall_i \quad \int_{1/2}^{x_i} dx_i \, \beta(x_i) = \int_0^{z_i} dz_i \, N(z_i). \tag{7}$$

In analogy to Eq. 5 we then require

$$\forall_{z \in \mathbb{R}_5} \; \pi_z(z) := \int_{|z'| \ge |z|} dz' \, N(z'_1) \cdot \ldots \cdot N(z'_5) = \frac{\int_{|z|}^\infty dr \, r^4 \, e^{-\frac{1}{2}r^2}}{\int_0^\infty dr \, r^4 \, e^{-\frac{1}{2}r^2}}. \tag{8}$$

As the mapping from x on z is bijective, $\pi_x(x) = \pi_z(z(x))$. When we numerically implement Eq. 3 we obtain Fig. 4. Why is the multivariate case output possibility so much less informative than the output pdf?

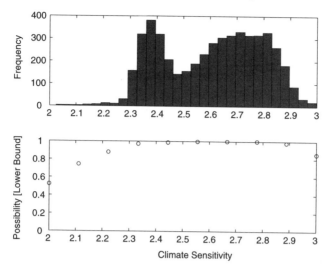

Fig. 4. Numerical results for π_{CS} in the multivariate case, generated from π_z. The result of a numerical optimization provides a lower boundary for the exact π_{CS}. Obviously the possibility function is much broader than the pdf. The upcoming Section provides an explanation.

6 The Curse of Dimension

We elucidate this phenomenon by considering the toy model $Y = \sum_{i=1}^{n} Z_i$, $Z_1, ...,$ Z_n iid $\sim N$. Then $\sqrt{<\text{var}(Y)>} = \sqrt{n}$. In analogy to the previous Section, we define π_Z. Then we define a width δY in the output possibility function π_Y by $\pi_Y(\pm\delta Y) = 1/2$ (bearing in mind that the maximum of any possibility function equals 1 by construction). How does δY scale with n? Let $y \in \mathbb{R}$ and $\forall_{z\in\mathbb{R}^n}\, z \equiv (z_1, ..., z_n)$. Then
$$\pi_Y(y) = \sup_{z\in\{z|y=\sum_{i=1}^{n} z_i\}} \pi_Z(z) = \pi_Z(\arg\inf_{z\in\{z|y=\sum_{i=1}^{n} z_i\}} |z|) = \pi_Z(y(1,...,1)/n) =$$
$$\int_{y/\sqrt{n}}^{\infty} dr\, r^{n-1}\, e^{-\frac{1}{2}r^2}$$
$$/\int_{0}^{\infty} dr\, r^{n-1}\, e^{-\frac{1}{2}r^2}.$$ If we request $\pi_Y(y = \delta Y) = 1/2$ then we can show that

$$\forall_{n\in\mathbb{N}>3} \exists_{c(n)\in[1,1.1]}\; \delta Y(n) = c(n)\sqrt{n(n-1)} \quad \text{and} \quad \lim_{n\to\infty} \frac{\delta Y(n)}{\sqrt{n(n-1)}} = 1. \quad (9)$$

In summary, $\sqrt{<\text{var}(Y)>} = \sqrt{n}$, while $\delta Y \approx n$.

Hence, the curse of dimension affects the width of possibility by a factor of $\sqrt{n}$ more than the standard deviation of the probability measure.

7 Conclusion

We have propagated climate model parameter uncertainty to climate sensitivity uncertainty. When making 1D uncertainty more imprecise by generalizing to a possibility measure, width in pdf and possibility are mapped quite similarly in the 1D case.

Quite the contrary, by assuming radially symmetric possibility measures in 5D parameter space, we find a markedly larger spread of uncertainty for the possibility measure (as against pdf). For a linear toy model we can attribute this to the fact that the possibility width scales with $\sim n$ as against $\sim \sqrt{n}$ for the pdf case.

Future work has to analyze whether our imprecise approach was too conservative or whether it in fact represents what the prior knowledge on an model output quantity is when influenced in a multi-variate manner.

Acknowledgement

We would like to thank E. Kriegler for drawing our interest to possibility theory and for helpful discussions. Furthermore, we would like to thank 7 Earth system modelers that volunteered in our survey on prior beliefs on uncertain model parameters. This work was supported by BMBF research grant 01LG0002, SFB555 and grant II/78470 by the Volkswagen Foundation.

References

[1] D. Dubois and H. Prade. *Possibility Theory: An Approach to Computerized Processessing of Uncertainty*. Plenum Press, New York, 1988.

[2] D. Dubois and H. Prade. Consonant approximations of belief functions. *International Journal of Approximate Reasoning*, 4:419–449, 1990.

[3] D. Dubois and H. Prade. When upper probabilities are possibility measures. *Fuzzy Sets and Systems*, 49:203–244, 1992.

[4] D. Dubois and H. Prade. Possibility theory: Qualitative and quantitative aspects. In Dov M. Gabbay and Philippe Smets, editors, *Quantified Representation of Uncertainty and Imprecision*, volume 1 of *Handbook of Defeasible Reasoning and Uncertainty Management Systems*, pages 169–226. Kluwer Academic Publishers, Dordrecht, 1998.

[5] C. E. Forest, P. H. Stone, A. P. Sokolov, M. R. Allen, and M. D. Webster. Quantifying uncertainties in climate system properties with the use of recent climate observations. *Science*, 295:113, 2002.

[6] D. J. Frame, B. B. B. Booth, J. A. Kettleborough, D. A. Stainforth, J. M. Gregory, M. Collins, and M. R. Allen. Constraining climate forecasts: The role of prior assumptions. *Geophysical Research Letters*, 32:L09702, 2004.

[7] A. Ganopolski, V. Petoukhov, S. Rahmstorf, V. Brovkin, M. Claussen, A. Eliseev, and C. Kubatzki. CLIMBER-2: a climate system model of intermediate complexity. Part II: model sensitivity. *Climate Dynamics*, 17:735–751, 2001.

[8] E. Kriegler and H. Held. Transformation of interval probability through measurable functions. *International Journal of Approximate Reasoning*, to be submitted.

[9] V. Petoukhov, A. Ganopolski, V. Brovkin, M. Claussen, A. Eliseev, C. Kubatzki, and S. Rahmstorf. CLIMBER-2: a climate system model of intermediate complexity. part i: model description and performance for present climate. *Climate Dynamics*, 16:1, 2000.

[10] T. Schneider von Deimling, H. Held, A. Ganopolski, and S. Rahmstorf. Climate sensitivity estimated from ensemble simulations of glacial climate. *Climate Dynamics*, DOI 10.1007/s00382-006-0126-8.

[11] P. Walley. *Statistical Reasoning with Imprecise Probabilities*. Chapman and Hall, London, 1991.

[12] L. A. Zadeh. Fuzzy sets as a basis for a theory of possibility. *Fuzzy Sets and Systems*, 1:3–29, 1978.

Fuzzy Logic for Stochastic Modeling

Özer Ciftcioglu and I. Sevil Sariyildiz

Delft University of Technology, Berlageweg 1, 2628 CR Delft, The Netherlands

Exploring the growing interest in extending the theory of probability and statistics to allow for more flexible modeling of uncertainty, ignorance, and fuzziness, the properties of fuzzy modeling are investigated for statistical signals, which benefit from the properties of fuzzy modeling. There is relatively research in the area, making explicit identification of statistical/stochastic fuzzy modeling properties, where statistical/stochastic signals are in play. This research makes explicit comparative investigations and positions fuzzy modeling in the statistical signal processing domain, next to nonlinear dynamic system modeling.

1 Introduction

The concept *computing with words* is a fundamental contribution of fuzzy logic [1] to the paradigm of artificial intelligence (AI). Computing with words became feasible via the utilization of linguistic variables, where the words can be interpreted as semantic labels in relation to the fuzzy sets, which are the basic conceptual elements of fuzzy logic. Consequently, comprehensible computer representation of the domain issues can be created. On one side, dealing with fuzzy qualities quantitatively is a significant step in AI. On the other side, due to the same fuzzy qualities, the interpretability issues arise [2]. While fuzzy logic contributes to science in dealing with domain related fuzzy issues, it is natural to anticipate that fuzzy logic associated with the probability theory and statistics can better deal with fuzziness of the domain issues, spanning the exact sciences and the soft sciences.

The statistical aspects of fuzzy modeling have received relatively less attention than *computing with words* or *soft computing*. In dealing with the latter two aspects Mamdani type of fuzzy models are more convenient [3], addressing soft issues especially in soft domains. In contrast to this, the Takagi-Sugeno (TS) type fuzzy model [4] is presumably more convenient in engineering systems where the fuzzy logic consequents are local linear models rather than fuzzy sets. In this way, the defuzzification process is greatly simplified making fuzzy logic more pragmatic approach

Ö. Ciftcioglu and I.S. Sariyildiz: *Fuzzy Logic for Stochastic Modeling*, Advances in Soft Computing **6**, 347–355 (2006)

in applications where data-driven modeling is a natural choice. In this research stochastic signals with TS fuzzy modeling are considered, since such signals are rich in probabilistic and statistical information that can be exploited by means of fuzzy logic. In particular, the fuzzy model is considered as the representation of a general nonlinear dynamic system.

2 Fuzzy Modeling

Takagi-Sugeno (TS) type fuzzy modeling [4] consists of a set of fuzzy rules as local input-output relations in a linear form thus:

$$R_i : If \quad x_l \quad is \quad A_{il} \quad and \dots x_n \quad is \quad A_{in} \tag{1}$$
$$Then \quad \hat{y}_i = a_i x + b_i, \quad i = 1, 2, \dots, K$$

where R_i is the i-th rule, $x = [x_1, x_2, \dots, x_n]^T \in X$ is the vector of input variables; A_{i1}, $A_{i2}, \dots, A_{in}$ are fuzzy sets and y_i is the rule output; K is the number of rules. The output of the model is calculated through the weighted average of the rule consequents, which gives

$$\hat{y} = \frac{\sum_{i=1}^{K} \beta_i(x) \hat{y}_i}{\sum_{i=1}^{K} \beta_i(x)} \tag{2}$$

In (2), $\beta_i(x)$ is the degree of activation of the i-th rule

$$\beta_i(x) = \pi_n^{j=1} \mu_{A_{ij}}(x_j), i = 1, 2, \dots K \tag{3}$$

where $\mu_{A_{ij}}(x_j)$ is the membership function of the fuzzy set A_{ij} at the input (antecedent) of R_i. To form the fuzzy system model from the data set with N data samples, given by $X = [x_1, x_2, \dots, x_N]^T, Y = [y_1, y_2, \dots, y_N]^T$ where each data sample has a dimension of n ($N \gg n$). First the structure is determined and afterwards the parameters of the structure are identified. The number of rules characterizes the structure of a fuzzy system. The number of rules is determined by clustering methods. Fuzzy clustering in the Cartesian product-space $X \times Y$ is applied for partitioning the training data. The partitions correspond to the characteristic regions where the system's behaviour is approximated by local linear models in the multidimensional space. Given the training data T and the number of clusters K, a suitable clustering algorithm [5] is applied. One of such clustering algorithms is known as Gustafson-Kessel (GK) [6]. As result of the clustering process a fuzzy partition matrix U is obtained. The fuzzy sets in the antecedent of the rules is identified by means of the partition matrix U which has dimensions $[N \times K]$, where N is the size of the data set and K is the number of rules. The ik-th element of $\mu_{ik} \in [0, 1]$ is the membership degree of the i-th data item in cluster k; that is, the i-th row of U contains the point wise description of a multidimensional fuzzy set. One-dimensional fuzzy sets A_{ij} are obtained from the multidimensional fuzzy sets by projections onto the space of the input variables x_j. This is expressed by the point-wise projection operator of the form $\mu_{Aij}(x_{jk})=proj_j\ (\mu_{ik})$ [7]. The point-wise defined fuzzy sets A_{ij} are then approximated by appropriate parametric functions. The consequent parameters for each rule are obtained by means of linear least square estimation. For this, consider the matrices $X=[x1,\dots,xN]T$, $X_e[X,1]$ (extended matrix $[N\times(n+1)]$) ; Λi (diagonal matrix dimension of $[N\times N]$) and $X_E=[(\Lambda_1 X_e);\ (\Lambda 2X_e);\dots\dots(\Lambda KX_e)]$ ($[N\times K(n+1)]$), where the diagonal matrix Λ_i consists of normalized membership degree as its k-th diagonal element

$$n\beta_i(x_k) = \frac{\beta_i(x_k)}{\sum\limits_{j=1}^{k}\beta_j(x_k)} \tag{4}$$

The parameter vector ϑ dimension of $[K\times(n+1)]$ is given by $\vartheta=[\vartheta_1 T\ \vartheta_2 T$ $\dots\dots\vartheta_k T]T$ where $\vartheta_i T=[aiT\ bi]$ $(1\leq i \leq K)$. Now, if we denote the input and output data sets as X_E and Y respectively, then the fuzzy system can be represented as a regression model of the matrix form $Y = X_E\ \vartheta+e$.

3 Dynamic System Modeling

For the investigation of fuzzy modeling with stochastic excitations, a nonlinear system

$$y(t) = 1 - e^{-x(t)/\tau} \tag{5}$$

is considered. Here $x(t)$ is the system variable. For a data driven fuzzy modeling approach, the system representation is cast into a recursive form as

$$y(t) = a(t)y(t-1) + u(t) \tag{6}$$

where the time varying AR coefficient $a(t)$ and the input $u(t)$ are given by

$$a = e^{-[x_2(t)-x_1(t)]/\tau}\quad and\quad u(t) = 1 - e^{-[x_2(t)-x_1(t)]/\tau} \tag{7}$$

For fuzzy modeling, first the system variable $x(t)$ is considered as band limited white noise and the system response is obtained from (5) for 200 samples. Based on this data the TS fuzzy model of the system is established for three clusters, i.e. three local models. The membership functions and the system performance are shown in Fig. 1. In the lower plot the true model output and the fuzzy model output is shown together. There is some difference between these outputs and this is constructive for the generalization capability of the model for unknown (test) inputs. In the model τ is taken as $\tau=20$. Figure 2 represents the model performance for the test data. The true and the estimated model outputs are shown together in the upper plot.

The model inputs corresponding to these outputs are shown in the figure as middle and lower plots, respectively. From Figs. 1 and 2 it is seen that the fuzzy model

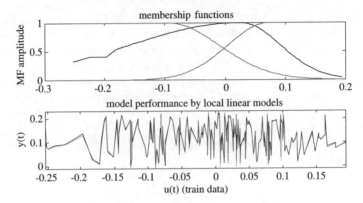

Fig. 1. Membership functions (upper) and the fuzzy model outputs (lower) as the true model outputs and the estimated counterparts

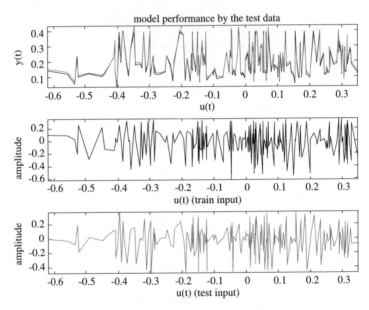

Fig. 2. True model output and its estimation by fuzzy modeling (upper); input to nonlinear system used for fuzzy model formation (middle); input to nonlinear system used for testing fuzzy model performance (lower)

has satisfactory performance for stochastic inputs. The data samples of system variable $x(t)$, which form the data-driven model, are from band-limited white noise. The system test input stems from perception measurements of a virtual agent reported elsewhere [8] where the present nonlinear system is representative of openness perception subject to measurement. It should be noted that, the test inputs to the system have wide frequency range. However, the nonlinear system behaves as a nonlinear low-pass filter so that three local models give satisfactory estimated system outputs, matching the true counterparts rather satisfactorily.

In order to investigate the pattern representation capabilities of fuzzy modeling a block of a time-series signal and its wavelet transform is considered. The time-series signal is a in particular band-limited white noise, and the number of clusters considered is five. For this case, the membership functions and the fuzzy model representation of the wavelet coefficients are shown in Fig. 3. Membership functions (upper) and the model outputs as true outputs and their estimated counterparts are also shown in the figure. The difference is significant due to the low number of fuzzy sets used for approximation. The above reported computer experiments show that TS fuzzy modeling is effective in modeling nonlinear dynamic systems and representation of patterns. In the nonlinear dynamic system representation, since the system is restricted to the lower frequency region, the Gaussian-shaped membership functions are capable of representing the system adequately. However in the pattern representation, since the frequency band is wide as the time-series data is band-limited white, in place of shaped Gaussian membership functions, the membership functions obtained directly from the clustering process are used. Otherwise, the shaped Gaussians are not enough narrow to represent the local variations. In other words, the local variations can not be represented by a restricted number of local linear models defined by the number of clusters.

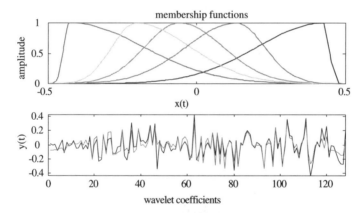

Fig. 3. Membership functions (upper) and model outputs as true outputs and their estimated counterparts involving five fuzzy sets

A similar situation obtains is the case of multivariable fuzzy modeling, where the membership functions are directly employed from the clustering. In this case the cause is different but the consequence is the same. Namely, there is an irrecoverable projection error due to projected and shaped membership functions from the clustered data, which prevents accurate representation of the dynamic model in multidimensional space [9]. By considering such basic features of fuzzy modeling, the fuzzy logic can be conveniently associated with the probabilistic entities, as this is stochastic signals and patterns, in this work.

4 Probability Density Functions

The probability density function (pdf) of the stochastic outputs of the fuzzy model can be computed from the pdf of the inputs. By studying both pdfs, one can obtain important information about the nature of the nonlinearity of the dynamic system. The pdf computations can be carried out as follows. Consider the nonlinear dynamic system given by $y = g(x)$. To find $f_y(y)$ for a given x we solve the equation $y = g(x)$ for x in terms of y. If $x_1, x_2, \ldots, x_n, \ldots$ are all its *real* roots, $x1 = g(y1)$ $x2 = g(y2)$ $= \ldots \ldots xn = g(yn) = \ldots$ Then

$$f_y(y) = \frac{f_x(x_1)}{|g'(x_1)|} + \ldots + \frac{f_x(x_2)}{|g'(x_2)|} + \ldots + \frac{f_x(x_n)}{|g'(x_n)|} + \cdots \tag{8}$$

According to the theorem above, we consider the nonlinear dynamic system given as (5).

$$y = g(x) = 1 - e^{-x/\tau} \quad \text{and} \quad f_y(y) = \frac{\tau}{1-y} f_x(x_1) \tag{9}$$

Then, if we assume uniform density between 0 and 1, for $f_x(x)$, the pdf of the system output is

$$f_y(y) = \frac{\tau}{1-y} \quad (0 \le y \le 1 - \exp(1/\tau)) \tag{10}$$

which satisfies $\int_0^{1-e^{1/\tau}} f_y(y)\,dy = 1$. The same computations for input with Gaussian pdf with a shift of x_0 yields

$$f_y(y) = \frac{\tau}{\sqrt{2\pi}\,\sigma} \frac{1}{1-y} \exp\left[-\frac{1}{2}(\ln\left(\frac{1}{1-y}\right)^\tau - x_0)/\sigma^2\right]^2 \tag{11}$$

The variation of $f_y(y)$ given in (10) and (11) are sketched in Fig. 4.

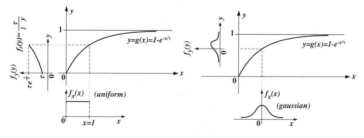

Fig. 4. Uniform probability and Gaussian density functions (pdf) at the model input and the ensuing pdfs at the model output

The pdf of $u(t)$ given by (7) is computed as follows.

$$u(t) = 1 - e^{-[x_2(t) - x_1(t)]/\tau} \tag{12}$$

We define a new variable w as $w = x_2 - x_1$.

$$f_w(w) = \int_{-\infty}^{+\infty} f_{x_1}(w - x_2) f_{x_2}(-x_2) dx_2$$

We assume x_1 and x_2 have uniform density, as this was the case in our research, $f_w(w)$ is obtained as seen in Fig. 5.

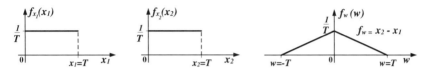

Fig. 5. Probability density function (pdf) of a random variable, which represents the difference of two other random variables with uniform density functions

From Fig. 5 we note that

$$f_w(w) = -\frac{w}{T^2} + \frac{1}{T} \ (w > 0) \quad and \quad f_w(w) = \frac{w}{T^2} + \frac{1}{T} \ (w < 0) \tag{13}$$

Using the theorem (8) we obtain for $u \leq 0$ and for $u \geq 0$, respectively

$$f_{-u}(u) = \frac{f_w(w_1)}{|g'(w_1)|} = \frac{\frac{\tau}{T^2} \ln(\frac{1}{1-u})^{\tau} + \frac{\tau}{T}}{1 - u} \tag{14}$$

$$f_{+u}(u) = \frac{f_w(w_1)}{|g'(w_1)|} = \frac{-\frac{\tau}{T^2} \ln(\frac{1}{1-u})^{\tau} + \frac{\tau}{T}}{1 - u} \tag{15}$$

are obtained, so that

$$\int_{1-e^{T/\tau}}^{0} f_{-u}(u) \, du + \int_{0}^{1-e^{-T/\tau}} f_{+u}(u) \, du = 1 \tag{16}$$

The input $u(t)$ to nonlinear system is seen in Figs. 2 and 3. The same calculations for the time varying autoregressive (AR) model coefficient a in (6) yields,

$$f_{a1}(a) = \frac{\tau}{Ta} \left[\frac{1}{T} \ln(\frac{1}{a})^{\tau} + 1 \right] \quad \text{for } a \leq 1, \text{ and} \tag{17}$$

$$f_{1a}(a) = \frac{\tau}{Ta} \left[-\frac{1}{T} \ln(\frac{1}{a})^{\tau} + 1 \right] \quad \text{for } a \geq 1$$

$$\int_{e^{-T/\tau}}^{1} f_{a1}(a) \, da + \int_{1}^{e^{T/\tau}} f_{a1}(a) \, da = 1 \tag{18}$$

The pdfs of u and a are shown in Fig. 6 for $\tau = 2$ and $T = 10$.

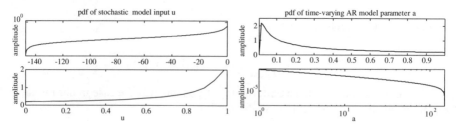

Fig. 6. *Pdf*s of model input u *(left)* and autoregressive model parameter a*(right)*

5 Discussion and Conclusions

TS fuzzy modeling is an essential means for the representation of nonlinear dynamic systems for identification, control etc. Such system dynamics are represented by a relatively small number of fuzzy sets compared to other approaches. For nonlinear dynamic system identification, the probability density of stochastic model inputs and outputs can reveal important information about the unknown system. In this respect, the capabilities of fuzzy modeling and its behavior with stochastic excitations are demonstrated in this work. Effective associations of probabilistic data can be made with fuzzy logic and these associations can be exploited in a variety of ways. Exemplary research can be seen in visual perception studies, where a theory of visual perception is developed as given in reference 8.

References

[1] Zadeh LA (1996) Fuzzy logic=computing with words. IEEE Trans. on Fuzzy Systems 4(2):103-111

[2] Casillas J, Cordon O, Herrara F, Magdalena L (eds.) (2003) Interpretability Issues in Fuzzy Modeling, Springer

[3] Mamdani EH (1974) Applications of fuzzy algorithms for control of a simple dynamic plant. In: Proceedings of the IEE, 121, 12:1585-1588

[4] Takagi T, Sugeno M (1985) Fuzzy identification of systems and its application to modeling and control. IEEE Trans. On Systems, Man, and Cybernetics 15:116-132

[5] Bezdek JC (1981) Pattern Recognition with Fuzzy Objective Function Algorithms. Plenum, NY

[6] Gustafson DE Kessel WC (1979) Fuzzy clustering with a fuzzy covariance matrix. In: Proc. IEEE CDC, San Diego, CA, pp 761-766

[7] Kruse R, Gebhardt J, Klawonn F (1994) Foundations of Fuzzy Systems. Wiley, NY

[8] Ciftcioglu Ö, Bittermann MS, Sariyildiz IS, (2006) Studies on visual perception for perceptual robotics. In: ICINCO 2006 3rd Int. Conference on Informatics in Control, Automation and Robotics, August 1-5, Setubal, Portugal

[9] Ciftcioglu Ö (2006) On the efficiency of multivariable TS fuzzy modeling. In: Proc. FUZZ-IEEE 2006, World Congress on Computational Intelligence, July 16-21, 2006, Vancouver, Canada

A CUSUM Control Chart for Fuzzy Quality Data

Dabuxilatu Wang

School of Mathematics and Information Sciences,
Guangzhou University, No. 230 Wai-Huan West Road ,
University Town, Panyu District, Guangzhou, 510006, P.R.China
dbxlt0@yahoo.com

Summary. Based on the concept of fuzzy random variables, we propose an optimal representative value for fuzzy quality data by means of a combination of a random variable with
a measure of fuzziness. Applying the classical Cumulative Sum (CUSUM) chart for these
representative values, an univariate CUSUM control chart concerning *LR*-fuzzy data under
independent observations is constructed.

Key words: *statistical process control; Cumulative sum chart; representative
values; fuzzy sets.*

1 Introduction

Cumulative Sum (CUSUM) control chart proposed by Page [9] is a widely used tool
for monitoring and examining modern production processes. The power of CUSUM
control chart lies in its ability to detect small shifts in processes as soon as it occurs and to identify abnormal conditions in a production process. For example, for
a given sequence of observations $\{X_n, n = 1, 2, \ldots\}$ on normal population, the monitored parameter of interest is typically the process mean, $\mu_n = E(X_n)$, the purpose
is to detect a small change in the process mean, one might specifies the levels μ_0
and $\mu_1 > \mu_0$ (or $\mu_1 < \mu_0$) such that under normal conditions the values of μ_n should
fall below (or above) μ_0 and the values of μ_n above (or below) μ_1 are considered
undesirable and should be detected as soon as possible. The CUSUM chart can be
used to monitor above process with the test-statistics $S_n = \max\{0, S_{n-1} + X_n - K\}$
(or $T_n = \min\{0, T_{n-1} + X_n + K\}$) and signal if $S_n > b$ (or $T_n < -b$), where b is the
control limit derived from a confidence interval assuming a Gaussian distributed observation, X_n $(n \geq 1)$ are the sample means at time t_n and $S_0 = T_0 = 0$, and K is the
reference value. It is well-known that CUSUM chart is more sensitive than Shewhart
chart ($\overline{X}$-chart) for detecting certain small changes in process parameters.

The random processes encountered in industrial, economic, etc. are typically
quality monitoring processes. J.M. Juran, an authority in international quality control circles, has pointed out that quality to customers, is its suitability rather than its

D. Wang: *A CUSUM Control Chart for Fuzzy Quality Data*, Advances in Soft Computing **6**, 357–364
(2006)
www.springerlink.com

conformity to certain standards. End-users seldom know what standards are. Customers' appraisal on quality is ways based on whether the products they have bought are suitable or not and whether that kind of suitability will last [1].

This is a kind of quality outlook which attaches primary importance to customers' feeling, so vague attribute of quality appraisal criterion and appraising customers' psychological reactions should, by no means, be ignored.

The fuzzy set theory [16] may be an appropriate mathematical tool for dealing with vagueness and ambiguity of the quality attribute. So it is very natural to introduce the concept of fuzzy set to the concept of quality and thus fuzzy quality is formed. As regards fuzzy quality, its "suitability" quality standard is expressed in the form of a fuzzy set. Also an outcome of the observation on quality characteristics may be appropriately represented by a fuzzy set because it is difficult to obtain a precise quality description of the inspected item in some case.

There are some literature on constructions of control charts based on fuzzy observations by [14], [11], [6], [4, 5], [13], [12] and [2]. Basically, the works mentioned above include two kinds of controlling methods , one of which is utilizing probability hypotheses testing rule for the representative values of fuzzy quality data, and the other is using a soft control rule based on possibility theory. However, how to deal with optimally both randomness and fuzziness of the process quality data is still a problem.

2 *LR*-fuzzy Data and the Measure of Fuzziness

2.1 *LR*-fuzzy Data

A fuzzy set on $\mathbb{R}$, the set of all real numbers, is defined to be a mapping $u : \mathbb{R} \to [0,1]$ satisfying the following conditions:

(1) $u_\alpha = \{x | u(x) \geq \alpha\}$ is a closed bounded interval for each $\alpha \in (0,1]$.
(2) $u_0 = supp u$ is a closed bounded interval.
(3) $u_1 = \{x | u(x) = 1\}$ is nonempty.

where $supp u = cl\{x | u(x) > 0\}$, cl denotes the *closure* of a set. Such a fuzzy set is also called a *fuzzy number*. The following parametric class of fuzzy numbers, the so-called *LR*-fuzzy numbers, are often used in applications:

$$u(x) = \begin{cases} L(\frac{m-x}{l}), & x \leq m \\ R(\frac{x-m}{r}), & x > m \end{cases}$$

Here $L : \mathbb{R}^+ \to [0,1]$ and $R : \mathbb{R}^+ \to [0,1]$ are fixed left-continuous and non-increasing function with $L(0) = R(0) = 1$. L and R are called left and right shape functions, m the central point of u and $l > 0$, $r > 0$ are the left and right spread of u. An *LR*-fuzzy number is abbreviated by $u = (m,l,r)_{LR}$, especially $(m,0,0)_{LR} := m$. Some properties of *LR*-fuzzy numbers for operations are as follows:

$$(m_1, l_1, r_1)_{LR} + (m_2, l_2, r_2)_{LR} = (m_1 + m_2, l_1 + l_2, r_1 + r_2)_{LR}$$

$$a(m, l, r)_{LR} = \begin{cases} (am, al, ar)_{LR}, & a > 0 \\ (am, -ar, -al)_{RL}, & a < 0 \\ 0, & a = 0 \end{cases}$$

$$(m_1, l_1, r_1)_{LR} - m_2 = (m_1 - m_2, l_1, r_1)_{LR}$$

For further properties of LR-fuzzy numbers the readers are refereed to [3].

Let $L^{(-1)}(\alpha) := \sup\{x \in R | L(x) \geq \alpha\}, R^{(-1)}(\alpha) := \sup\{x \in R | R(x) \geq \alpha\}$. Then for $u = (m, l, r)_{LR}$, $u_\alpha = [m - lL^{(-1)}(\alpha), m + rR^{(-1)}(\alpha)]$, $\alpha \in [0, 1]$.

An useful approach has been summarized by Cheng [2] for generating a fuzzy number based on a group experts' scores on a fuzzy quality item in a quality control process. By this approach, we may assign a fuzzy number for each outcome of a fuzzy observation on quality monitoring process. In this paper, we assume that the quality data collected from the fuzzy observation process can be assigned LR-fuzzy numbers. Such data is also called LR-fuzzy data. For example, the color uniformity of a TV set [1] under user's suitability quality view is with a fuzzy quality standard which could be expressed in a form of LR-fuzzy data $(d_0, 5, 5)_{LR}$, where $L(x) = R(x) = \max\{0, 1 - x\}$ is the shape function of a triangular fuzzy number, and d_0 is the designed value of the color uniformity. For the operational simplicity and a better description of fuzziness for fuzzy quality items, the triangular fuzzy number are often used.

LR-*fuzzy random variable* $X = (m, l, r)_{LR}$ has been defined by Körner [7], where m, l, r are three independent real-valued random variables with $P\{l \geq 0\} = P\{r \geq 0\} = 1$. Considering the fuzzy observations on a quality monitoring process, it is obvious that the LR-fuzzy data can be viewed as realizations of an LR-fuzzy random variable. Assuming the observational distribution is approximately normal, then the central variable m of an LR-fuzzy sample $X = (m, l, r)_{LR}$ obtained by method in [2] from the fuzzy observation process can be viewed as a Gaussian variable, and the spread variables l, r may be evenly distributed. The ith sample X_i is assumed to be a group mean of size n_i, $\{(x_{i1}, b_{i1}, c_{i1})_{LR}, \ldots, (x_{in_i}, b_{in_i}, c_{in_i})_{LR}\}$, i.e.

$$X_i = (\frac{1}{n_i} \sum_{j=1}^{n_i} x_{ij}, \frac{1}{n_i} \sum_{j=1}^{n_i} b_{ij}, \frac{1}{n_i} \sum_{i=1}^{n_i} c_{ij})_{LR} = (\bar{x}_i, \bar{b}_i, \bar{c}_i)_{LR},$$

and simply denoted by $X_i = (m_i, l_i, r_i)_{LR}$. By the central limit theorem, if the group size is relatively large, then l_i, r_i are approximately Gaussian variables.

2.2 The Measure of Fuzziness

The LR-fuzzy quality data are not easy to be plotted on a control chart directly. Therefore, it is necessary to convert a fuzzy data (a fuzzy sample) into a scalar (a real random variable) for plotting, such scalar (random variable) would be an optimal representative of the fuzzy data (the fuzzy sample). Some approaches for determining the representing value of a fuzzy data have been proposed in [14] and [6], etc.. In general, there are no absolute criteria for choosing methods on determining the representative value. However, we usually expect a method for determining one which is not only with lower complexity in computation but also with an optimal representativeness.

Recalling the concept of fuzzy random variables [7],[8], [10], we are aware of that fuzzy random variables are devoted to deal with the inherent randomness and fuzziness of samples simultaneously. Thus, we emphasize that a representative value should properly represent the main characteristics, randomness and fuzziness, of a fuzzy quality sample. Such features can be abstracted easier in the case of LR-fuzzy sample than that of other fuzzy sample because we are able to represent the randomness by the central variable and to represent the fuzziness of the fuzzy quality data by employing the concept of a measure of fuzziness.

A number used for measuring the fuzziness of a fuzzy set is a very important index when we deal with fuzzy concepts and fuzzy information. Fuzziness level of a fuzzy set is usually determined by the fuzziness level of each possible elements of the fuzzy set. For example, if the membership degree of one element is near 1, then the affirmation level with respect to the element must be high, and thus fuzziness level of the element becomes low; if the membership degree of one element is around 0.5, then its belongingness is extremely unsteady, and thus the fuzziness level of the element becomes high, and so on. Various measuring methods have been proposed based on the concept of measure of fuzziness [3] [17], for instance, Minkowski's measure of fuzziness $D_p(A)$ for a fuzzy set A on a discrete finite domain is as follows:

$$D_p(A) = \frac{2}{n^{1/p}} (\sum_{i=1}^{n} |A(x_i) - A_{0.5}(x_i)|^p)^{1/p},$$

where $p > 0$, $A_{0.5}$ denotes the 0.5-level set of the fuzzy set A, and $A_{0.5}(x)$ denotes the indicator of the non-fuzzy set $A_{0.5}$, i.e.,

$$A_{0.5}(x) := I_{A_{0.5}}(x) = \begin{cases} 1, & x \in A_{0.5} \\ 0, & x \notin A_{0.5} \end{cases}$$

When $p = 1$, $D_1(A)$ is said to be Hamming's fuzziness measure, and when $p = 2$, $D_2(A)$ is called Euclid's fuzziness measure. We employ an extension of Hamming's fuzziness measure to define a measure of fuzziness $D(X)$ for the LR-fuzzy quality sample $X = (m, l, r)_{LR}$, i.e.

$$D(X) = \int_{-\infty}^{+\infty} |X(x) - X_{0.5}(x)| dx$$

Theorem 1. *Let $X = (m, l, r)_{LR}$ be a fuzzy quality sample, then it holds that*

$$D(X) = l \left[L^{(-1)}(0.5) + \int_{L^{(-1)}(0.5)}^{L^{(-1)}(0)} L(x) dx - \int_{0}^{L^{(-1)}(0.5)} L(x) dx \right]$$
$$+ r \left[R^{(-1)}(0.5) + \int_{R^{(-1)}(0.5)}^{R^{(-1)}(0)} R(x) dx - \int_{0}^{R^{(-1)}(0.5)} R(x) dx \right].$$

Proof. It is obvious that

$$X_{0.5}(x) = \begin{cases} 1, & x \in \left[m - lL^{(-1)}(0.5), m + rR^{(-1)}(0.5)\right] \\ 0, & x \notin \left[m - lL^{(-1)}(0.5), m + rR^{(-1)}(0.5)\right] \end{cases}$$

$$\begin{aligned}
D(X) &= \int_{-\infty}^{+\infty} |X(x) - X_{0.5}(x)| dx \\
&= \int_{m-lL^{(-1)}(0.5)}^{m} (1 - L(\frac{m-x}{l})) dx + \int_{m}^{m+rR^{(-1)}(0.5)} (1 - R(\frac{x-m}{r})) dx \\
&+ \int_{m-lL^{(-1)}(0)}^{m-lL^{(-1)}(0.5)} L(\frac{m-x}{l}) dx + \int_{m+rR^{(-1)}(0.5)}^{m+rR^{(-1)}(0)} R(\frac{x-m}{r}) dx \\
&= l\left[L^{(-1)}(0.5) + \int_{L^{(-1)}(0.5)}^{L^{(-1)}(0)} L(x) dx - \int_{0}^{L^{(-1)}(0.5)} L(x) dx \right] \\
&+ r\left[R^{(-1)}(0.5) + \int_{R^{(-1)}(0.5)}^{R^{(-1)}(0)} R(x) dx - \int_{0}^{R^{(-1)}(0.5)} R(x) dx \right].
\end{aligned}$$

This completes the proof.

Example Let $u = (m_0, l_0, r_0)_{LR}$ be a triangular fuzzy data, where $L(x) = R(x) = \max\{0, 1-x\}$, then

$$D(u) = \frac{l_0 + r_0}{4}.$$

3 Construction of a CUSUM Chart for *LR*-fuzzy Data

3.1 A Representative Value

We now define a representative value denoted by $Rep(X)$ for fuzzy sample $X = (m, l, r)_{LR}$ as follows:

$$Rep(X) = m + D(X)$$

Let

$$\beta_1 := L^{(-1)}(0.5) + \int_{L^{(-1)}(0.5)}^{L^{(-1)}(0)} L(x) dx - \int_{0}^{L^{(-1)}(0.5)} L(x) dx$$

$$\beta_2 := R^{(-1)}(0.5) + \int_{R^{(-1)}(0.5)}^{R^{(-1)}(0)} R(x) dx - \int_{0}^{R^{(-1)}(0.5)} R(x) dx$$

then

$$Rep(X) = m + l\beta_1 + r\beta_2$$

Here, the central variable m just represents the randomness of the *LR*- fuzzy quality sample $X = (m, l, r)_{LR}$ extremely, since by its membership $X(m) = 1$ it implies no fuzziness , and it also largely determine the location of the *LR*-fuzzy quality sample. On the other hand, random variable $l\beta_1 + r\beta_2$ properly represents the fuzziness level of the *LR*-fuzzy quality sample because it is derived from a standard measure of fuzziness of a fuzzy set, which is well defined with a theoretical supporting. A kind of combination of the randomness with fuzziness of the fuzzy quality sample is realized simply by arithmetic addition, thus the related computation for obtaining the representative value becomes much easier. For a given fuzzy quality data $u = (m_0, l_0, r_0)_{LR}$, then its representative value is $Rep(u) = m_0 + l_0\beta_1 + r_0\beta_2$, which is a fixed scalar. The present methods for calculating representative values in the case of *LR*-fuzzy quality data somewhat have an advantage over that proposed in [14] and [6]. For instance, calculating representative values were done in five ways in [14] and [6]: by using the fuzzy mode as $f_{mode} = \{x|A(x) = 1\}, \quad x \in [a,b]$; the α-level fuzzy midrange as $f_{mr}(\alpha) = \frac{1}{2}(\inf A_\alpha + \sup A_\alpha)$; the fuzzy median as f_{med}, which satisfies $\int_a^{f_{med}} A(x)dx = \int_{f_{med}}^b A(x)dx = \frac{1}{2}\int_a^b A(x)dx$; the fuzzy average as $f_{avg} = \int_a^b xA(x)dx / \int_a^b A(x)dx$;and the barycentre concerned with Zadeh's probability measure of fuzzy events as $Rep(A) = \int_{-\infty}^\infty xA(x)f(x)dx / \int_{-\infty}^\infty A(x)f(x)dx$. Where A is a fuzzy set on some interval $[a,b] \subset \mathbb{R}, a < b$. In general, the first two methods are easier to calculate than the last three as well as our method, however, they only took account of the randomness of the fuzzy sample , e.g. $f_{mode} = m$ when the fuzzy sample is $X = (m,l,r)_{LR}$, which obviously may lead to a biased result. The third method used a non-standard measure of fuzziness , thus the fuzzy median may also be a biased representative of a fuzzy sample. The last two methods are reasonable, but the representative values derived from the methods are not easy to calculate. We can easily check that our method is easier to calculate than the fuzzy average and barycentre methods in the case of *LR*-fuzzy quality sample. We would like to point out that the accuracy of the representative for the given fuzzy sample is more important for constructing a representing control chart devoted for monitoring fuzzy quality, an inaccurate representative will lead to more false alarm or a wrong control scheme deviated from the original reality of fuzzy data . Also it is a common sense that every fuzzy data is characterized by the both randomness and fuzziness. Our proposed representative value is considerably accurate and simply and very comprehensive because we fully take the randomness as well as fuzziness measured by a standard fuzziness measure into account.

3.2 Construction of a CUSUM Chart

Using the classical CUSUM interval scheme [15], we can design a corresponding chart for the representative values of *LR*-fuzzy quality samples. As that mentioned in Subsection 2.1, the spread variable of a sample might be evenly distributed though the central variable is Gaussian, we need sampling in groups of varying number n_i of observations each, and n_i is relatively large, for instance, $n_i \geq 25$. Let the observation is:

$$X_{ij} = (m_{ij}, l_{ij}, r_{ij})_{LR}, \quad i = 1,2,\ldots,k; j = 1,2,\ldots,n_i$$

then $x_{ij} := Rep(X_{ij}) = m_{ij} + l_{ij}\beta_1 + r_{ij}\beta_2$. The representative value of the samples mean (group mean), denoted by $\bar{x}_i$, can be worked out by the following two ways: (1). $\bar{x}_i = \frac{1}{n_i}\sum_{j=1}^{n_i} x_{ij} = \overline{m}_i + \overline{l}_i\beta_1 + \overline{r}_i\beta_2$. (2).$\bar{X}_i = \frac{1}{n_i}\sum_{j=1}^{n_i} X_{ij} = (\overline{m}_i, \overline{l}_i, \overline{r}_i)_{LR}$, then $\bar{x}_i = Rep(\bar{X}_i) = \overline{m}_i + \overline{l}_i\beta_1 + \overline{r}_i\beta_2$. The standard error of mean for the representative values in group i, denoted by s_i, is:

$$s_i = \left(\frac{1}{n_i - 1}\sum_{j=1}^{n_i}\left[(m_{ij} - \overline{m}_i) + (l_{ij} - \overline{l}_i)\beta_1 + ((r_{ij} - \overline{r}_i)\beta_2)\right]^2\right)^{1/2}.$$

Then the standard error of samples mean can be estimated by

$$\hat{\sigma}_e = \left(\sum_{i=1}^{k}(n_i - 1)s_i^2 / \sum_{i=1}^{k}(n_i - 1)\right)^{1/2}.$$

Thus, we are able to construct a CUSUM control chart for the representative values of the samples as follows:

(1) Choose a suitable reference value T, here we assume that it is the overall mean $\hat{\mu}$ of the past observations.
(2) Use the standard scheme $h = 5, \quad f = 0.5$.
(3) Calculate the CUSUM S_n (Here S_n is with respect to the representative values of samples) with reference value $K_1 = T + f\hat{\sigma}_e = \hat{\mu} + 0.5\hat{\sigma}_e$. Keep it non-negative. Calculate the CUSUM T_n (Here T_n is with respect to the representative values of samples) with reference value $K_2 = T - f\hat{\sigma}_e = \hat{\mu} - 0.5\hat{\sigma}_e$. Keep it non-positive.
(4) Action is signalled if some $S_n \geq h\hat{\sigma}_e = 5\hat{\sigma}_e$ or some $T_n \leq -h\hat{\sigma}_e = -5\hat{\sigma}_e$.

This obtained CUSUM control chart is an appropriate representative CUSUM chart for the *LR*-fuzzy quality data involved process.

Conclusions

We have proposed an optimal representative value for fuzzy quality sample by means of a combination of a random variable with a measure of fuzziness. For *LR*- fuzzy data this kind of representative values are more accurate and easier to calculate, so by which the fuzzy control charts derived from using representative values methods could be improved to some sense. An accurate representative CUSUM chart for *LR*-fuzzy samples is preliminarily constructed. Likewise one could construct other control charts such as EWMA,P-chart and so on. The proposed representative value is expected to be extended to a general case where the normal, convex and bounded fuzzy quality data are monitored.

References

[1] Cen Y, 1996, Fuzzy quality and analysis on fuzzy probability, Fuzzy sets and systems, 83, 283-290.

[2] Cheng CB, 2005, Fuzzy process control: construction of control charts with fuzzy numbers, Fuzzy sets and systems, 154, 287-303.

[3] Dubois D, Prade H, 1980, Fuzzy sets and systems: Theory and Applcation. New York: Academic press.

[4] Grzegorzewski P, 1997, Control charts for fuzzy data, Proc. Fifth European congress on intelligent techniques and soft computing EUFIT'97, Aachen, pp. 1326-1330.

[5] Grzegorzewski P, Hryiewicz O, 2000, Soft methods in statistical quality control, Control and cybernet. 29, 119-140.

[6] Kanagawa A, Tamaki F, Ohta H, 1993, Control charts for process average and variability based on linguistic data, International Journal of Production Research, 2, 913-922.

[7] Körner R, 2000, An asymptotic α-test for the expectation of random fuzzy variables, Journal of Statistical Planning and Inferences. 83, 331-346.

[8] Kruse R, Meyer KD, 1987, Statistics with vague data. D. Reidel Publishing Comany.

[9] Page ES, 1954, Continuous inspection schemes, Biometrika, 41, 100-114.

[10] Puri MD, Ralescu D, 1986, Fuzzy Random Variables, Journal of Mathematical Analysis and Applications, 114, 409-422.

[11] Raz T, Wang JH, 1990, Probabilistic and membership approaches in the construction of control charts for linguistic data, Production Plann.Cont. 1, 147-157.

[12] Tannock TDT, 2003, A fuzzy control charting methods for individuals, International Journal of Production Research, 41(5), 1017-1032.

[13] Taleb H, Limam M, 2002, On fuzzy and probabilistic control charts, International Journal of Production Research, 40(12), 2849-2863.

[14] Wang JH and Raz T, 1990, On the construction of control charts using linguistic variables, International Journal of Production Research, 28, 477-487.

[15] Wetherill GB, Brown DW, 1991, Statistical Process Control, Chapman and Hall, London.

[16] Zadeh LA, 1965, Fuzzy sets, Information and control, 8: 338-353.

[17] Zimmermann HJ, 1985, Fuzzy set theory and applications, Kluwer-Nijhoff Publishing.

A Fuzzy Synset-Based Hidden Markov Model for Automatic Text Segmentation

Viet Ha-Thuc[1], Quang-Anh Nguyen-Van[1], Tru Hoang Cao[1] and Jonathan Lawry[2]

[1] Faculty of Information Technology, Ho Chi Minh City University of Technology, Vietnam
viettifosi@yahoo.com
nvqanh2003@yahoo.com
tru@dit.hcmut.edu.vn
[2] Department of Engineering Mathematics, University of Bristol, UK
j.lawry@bristol.ac.uk

Summary. Automatic segmentation of text strings, in particular entity names, into structured records is often needed for efficient information retrieval, analysis, mining, and integration. Hidden Markov Model (HMM) has been shown as the state of the art for this task. However, previous work did not take into account the synonymy of words and their abbreviations, or possibility of their misspelling. In this paper, we propose a fuzzy synset-based HMM for text segmentation, based on a semantic relation and an edit distance between words. The model is also to deal with texts written in a language like Vietnamese, where a meaningful word can be composed of more than one syllable. Experiments on Vietnamese company names are presented to demonstrate the performance of the model.

1 Introduction

Informally speaking, text segmentation is to partition an unstructured string into a number of continuous sub-strings, and label each of those sub-strings by a unique attribute of a given schema. For example, a postal address consists of several segments, such as house number, street name, city name, zip code, and country name. Other examples include paper references and company names. As such, automatic segmentation of a text is often needed for further processing of the text on the basis of its content, namely, information retrieval, analysis, mining, or integration, for instance.

The difficulty of text segmentation is due to the fact that a text string does not have a fixed structure, where segments may change their positions or be missing in the string. Moreover, one word, in particular an acronym, may have different meanings and can be assigned to different attributes. For example, in a paper reference, its year of publication may be put after the author names or at the end, and the publisher name may be omitted. For dealing with that uncertainty, a probabilistic model like HMM has been shown to be effective, for general text ([5]) as well as specific-meaning phrases like postal addresses or bibliography records ([1], [3]).

V. Ha-Thuc et al.: *A Fuzzy Synset-Based Hidden Markov Model for Automatic Text Segmentation*,
Advances in Soft Computing **6**, 365–372 (2006)
www.springerlink.com

However, firstly, the above-mentioned HMMs did not consider the synonymous words in counting their common occurrence probabilities. That affects not only the performance of text segmentation, but also information retrieval later on. Secondly, the previous work did not tolerate word misspelling, which is often a case. The segmentation performance would be better if a misspelled word could be treated as its correct one, rather than an unknown word. Besides, the segmentation task is more difficult with a language like Vietnamese, where a meaningful word can be composed of more than one syllable. For example, "*công ty*" in Vietnamese means "*company*".

In this paper, we propose an HMM that overcomes those limitations. Firstly, words having the same meaning are grouped into one synonym set (synset), and the emission probability for each state is distributed over those synsets instead of individual words. Secondly, the probability is fuzzified by using a string matching distance measure such as edit distance to deal with the word misspelling noise. Thirdly, the standard Viterbi algorithm is extended to group syllables into words for Vietnamese or an alike language.

The paper is organized as follows. Section 2 summarizes the basic notions of HMM and its application to text segmentation. Section 3 present our proposed fuzzy synset-based HMM and its extension for multi-syllable words. Experimental results are presented in Section 4. Finally, Section 5 concludes the paper with some remarks and suggestion for future work.

2 HMMs for Text Segmentation

2.1 Hidden Markov Models

An HMM is a probabilistic finite state automaton ([8]), consisting of the following parameters:

- A set of one start state, one end state, and n immediate states
- An $n \times n$ transition matrix, where the ij^{th} element is the probability of making a transition from state i to state j.
- A vocabulary set V_s for each immediate state s, containing those words that can be emitted from s.
- An emission probability p distributed over V_s for each immediate state s, where $p(w|s)$ measures the probability for s emitting word w in V_s.

These four parameters are learned from data in the training phase. Then in the testing phase, given a text, the most probable path of states, from the start state to the end, can be computed, where each state emits and corresponds to a word of the text in that sequence.

2.2 Learning Parameters

Learning the HMM parameters requires only a single pass over the training data set ([3]). Each training instance is a sequence of state-word pairs. The learned set of immediate states simply comprises all states appearing in the training data. The vocabulary set of each state can also be learned easily as the set of all words paired with that state in the training data.

Let N_{ij} be the number of transitions made from state i to state j, and N_i be the total number of transitions made from state i, according to the training data. The transition probability from state i to state j is learned as follows:

$$a_{ij} = N_{ij}/N_i$$

For the emission probability distribution of state s, suppose that the vocabulary set $V_s = \{w_1, w_2 \ldots w_M\}$ and the raw frequency, i.e., number of occurrence times, of each w_i in state s in the training data is f_i. Then, the probability that s emits w_i is computed as below:

$$p(w_i|s) = f_i / \sum_{j=1,M} f_j$$

The above formula would assign probability of zero to those words that do not appear in training data, causing the overall probability for a text string to be zero. So the model would not be applicable to a text string containing one or more unknown words. To avoid this, a non-zero probability is assigned to an unknown word with respect to a state, and the emission probability distribution of that state is adjusted accordingly. Such a smoothing technique is the Laplace one, as follows:

$$p(\text{``}unknown\text{''}|s) = 1/(\sum_{j=1,M} f_i + M + 1)$$

$$p(w_i|s) = (f_i + 1)/(\sum_{j=1,M} f_i + M + 1)$$

One can see that $\sum_{j=1,n} a_{ij} = 1$ and $\sum_{i=1,M} p(w_i|s) + p(\text{``}unknown\ word\text{''}|s) = 1$, satisfying the normalized conditions.

2.3 Text Segmentation

For text segmentation, given an input string $u = w_1, w_2 \ldots w_m$ and an HMM having n immediate states, the most probable state sequence, from the start state to the end state, that generates u can be obtained by the Viterbi algorithm as follows ([8]). Let 0 and $(n+1)$ denote the start and end states, and $Pr_s(i)$ be the probability of the most probable path for $w_1, w_2 \ldots w_i$ ($i \leq m$) ending at state s (i.e., s emits w_i). As such, $Pr_0(0) = 1$ and $Pr_j(0) = 0$ if $j \neq 0$.
Then $Pr_s(i)$ can be recursively defined as follows:

$$Pr_s(i) = \text{Max}_{t=1,n}\{Pr_t(i-1) \times a_{ts}\} \times p(w_i|s)$$

where a_{ts} is the transition probability from state t to state s, and the maximum is taken over all immediate states of the HMM. The probability of the most probable path that generates u is given by:

$$Pr(u) = \text{Max}_{t=1,n}\{Pr_t(m) \times a_{t(n+1)}\}$$

This probability function can be computed using dynamic programming in $O(mn^2)$ time.

3 Fuzzy Synset-Based HMMs

3.1 A Case Study: Vietnamese Company Names

For testing the performance of our proposed model as presented in the following sections, we have chosen the domain of Vietnamese company names. Figure 1 shows the HMM learned from our training data, where each immediate state corresponds to a field that a company name may contain:

- *Kind of Company* such as "*công ty*" (company), "*nhà máy*" (factory), ...
- *Kind of Possession* such as "*TNHH*" (Ltd.), "*cổ phˋân*" (stock), "*tu nhˋân*" (private), "*liên doanh*" (joint-venture), ...
- *Business Aspect* such as "*xăng dˋâu*" (petroleum), "*du lịch*" (tourism), "*yt*" (medical)...
- *Proper Name* such as "*Sài Gòn*", "*Microsoft*", "*Motorola*", ...

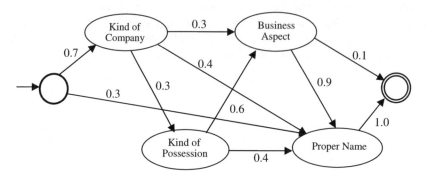

Fig. 1. An HMM for Vietnamese Company Names

3.2 Synset-Based HMMs

As mentioned above, synonymous words should be treated as the same in semantic matching as well as in counting their occurrences. For example, in Vietnamese, "*trách nhiêm hữu hạn*", "*TN hữu hạn*", and "*TNHH*" are full and acronyms of the same word meaning "*Ltd.*". So we propose synset-based HMMs in which words having the same meaning are grouped into a synset. Each training instance is a sequence of state-word-synset triples created manually. The probability of a synset emitted by a state is defined as the sum of the probabilities of all words in that synset emitted by the state, as exemplified in Table 1. Then, the model would operate on a given text as if each word in the text were replaced by its corresponding synset.

Since one ambiguous word may belong to different synsets, the one with the highest emission probability will be chosen for a particular state. For example, "*TN*"

Table 1. Emission Probabilities in a Synset-Based HMM

Word w	$p(w\mid$ state $= Kind\ of$ Possession)	Synset W	$p(W\mid$ state $= Kind\ of$ Possession)
trách nhiệm hữu hạn	0.05	trách nhiệm hữh hạn	0.4
TNHH	0.25	TNHH	
TN hữu hạn	0.1	TN hữu hạn	
cổ phần	0.15	cổ phần	0.3
CP	0.15	CP	
tu nhân	0.2	tu nhân	0.3
TN	0.1	TN	

in the two following company names has different meanings, where in the former it is an abbreviation of "*tu nhân*" (private) and in the latter of "*thiên nhiên*" (natural):

Công ty	*TN*	*Duy Lợi*
company	private	proper name

Cty	*nuóc khoáng*	*TN*	*La Vie*
company	mineral water	natural	proper name

Figure 2 illustrates the most probable paths of the two names in the proposed synset-based HMM, found by using the Viterbi algorithm. We note that using synsets also helps to fully match synonymous words emitted from the same state, such as "*Công ty*" and "*Cty*" in this example.

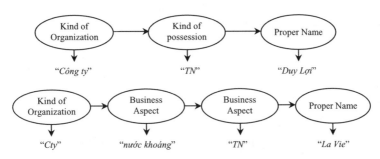

Fig. 2. A Synset-Based HMM Resolves Ambiguous Words

3.3 Fuzzy Extension

The conventional HMM as presented above does not tolerate erroneous data. If a word is misspelled, not in the vocabulary set of a state, it is treated as an unknown word with respect to that state. For instance, "*cômg ty*" and "*Micorsoft*" are misspellings of "*công ty*" and "*Microsoft*", respectively. The idea of our proposed fuzzy HMMs is that, if a word w is not contained in the vocabulary set V_s of a state s, but its smallest edit distance ([2]) to the words in V_s is smaller than a certain threshold, then it is considered as a misspelling. In the synset-based case, the distance of w to a synset in V_s is defined as the minimum of the distances of w to each word in that synset. Therefore, the fuzzy emission probability of w with respect to s is computed as follows:

if $(w \in W \text{in } V_s)$ **then**
        $fp(w|s) = p(W|s)$
else {

        $W_0 = Argmin_{W \in V_s} distance(w, W)$
                        // $distance(w, W) = Min_{x \in W} editDist(w, x)$
           if $(distance(w, W_0) < threshold_s)$ **then**
           $fp(w|s) = p(W_0|s)$ // w *might be misspelled from* W_0
           else
           $fp(w|s) = p(\text{"unknown"}|s)$ // w *is an unknown word*

   }

3.4 Extension for Vietnamese

The fact that a Vietnamese word may comprise more than one syllable makes word segmentation and part-of-speech tagging difficult ([6], [7]), as compared to English where words are separated by spaces. For example, the Vietnamese words "*xuòsng*", "*công ty*", "*tông công ty*" contain one, two and three syllables respectively. Therefore, in order to segment company names, for instance, using the HMM present above, one would have to pre-process it to group syllables into words first.

Here we propose to do both steps in one HMM, by modifying the Viterbi algorithm as follows. Assume that the maximal number of syllables that form a word is K, in particular 4 for Vietnamese. The probability $Pr_s(i)$ of the most probable path for a syllable sequence $e_1e_2...e_i$ ending at state s, among n immediate states of the HMM, is defined by:

$$Pr_s(i) = Max_{j=1,K}\{Max_{t=1,n}\{Pr_t(i-j) \times a_{ts}\} \times p(e_{i-j+1}... e_{i-1}e_i|s)\}$$

That is $j(1 \leq j \leq K)$ syllables may form a word ending at state s, which maximizes $Pr_s(i)$. The time complexity of the algorithm is $O(Kmn^2)$, for a syllable sequence of length m.

4 Experimental Results

The accuracy of a name segmentation method is defined to be the percentage of names correctly segmented in a testing set. We have evaluated the proposed synset-based HMM over a set of company names randomly extracted from several Vietnamese websites. Figure 3 shows that its accuracy is about 5% higher than the

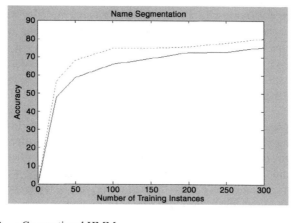

——————   : Conventional HMM
- - - - - - -.   : Synset-Based HMM

Fig. 3. Comparison between Conventional and Synset-Based HMMs

conventional HMM, being over 80% with 300 training instances. In a domain where the vocabulary sets contain many synonymous words, the improvement could be higher.

To obtain noisy data sets, we use a tool that randomly generates misspelled words from their original correct ones. The noise level of a data set is defined as the ratio of the number of erroneous characters per the total number of characters of a word, for every word in the set. Figure 4 compares the performances of a synset-based HMM and its fuzzy extension over three data sets with different noise levels.

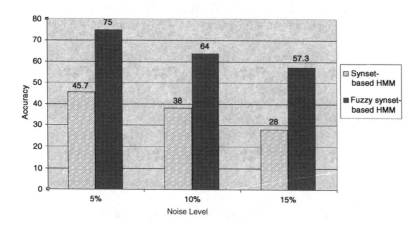

Fig. 4. Comparison between a Synset-Based HMM and its Fuzzy Extension

The experiment results show that a synset-based HMM works quite well on clean data, but its accuracy decreases with noisy data. Introducing fuzzy emission probabilities helps to reduce the misspelling effect.

5 Conclusion

We have presented an enhancement of HMMs for text segmentation whose emission probabilities are defined on synsets rather than individual words. It not only improves the accuracy of name segmentation, but also is useful for further semantic processing such as name matching. For dealing with misspelled words, we have introduced the notion of fuzzy emission probability defined on edit distances between words. Lastly, we have modified the Viterbi algorithm to segment text in Vietnamese and alike languages, where a meaningful word may comprise more than one syllable.

Conducted experiments have shown the advantage of the proposed fuzzy synset-based HMM. Other string distance measures are worth trying in calculating fuzzy emission probabilities. The model is being applied in VN-KIM, a national key project on Vietnamese semantic web, to automatically recognize named-entities in web pages and matching them for knowledge retrieval ([4]). These are among the topics that we suggest for further work.

References

[1] Agichtein E., Ganti V., (2004), Mining Reference Tables for Automatic Text Segmentation, Procs of ACM Conference on Knowledge Discovery and Data Mining (SIGKDD), pp. 20-29.

[2] Bilenko M., Mooney R., Cohen W., Ravikumar P., Fienberg S., (2003), Adaptive Name Matching in Information Integration, IEEE Intelligent Systems, Vol. 18, No. 5, 16-23.

[3] Borkar V., Deshmukh K., Sarawagi S., (2001), Automatic Segmentation of Text into Structured Records, Procs of the ACM SIGMOD Conference.

[4] Cao T.H., Do H.T., Pham B.T.N., Huynh T.N., Vu D.Q, (2005). Conceptual Graphs for Knowledge Querying in VN-KIM, Contributions to the 13th International Conference on Conceptual Structures, pp. 27-40.

[5] Freitag D., McCallum A.K., (2000), Information Extraction with HMM Structure Learned by Stochastic Optimization, Procs of the 18th Conference on Artificial Intelligence, pp. 584-589.

[6] Ha L.A., (2003), A Method for Word Segmentation in Vietnamese, Procs of Corpus Linguistics Conference, pp. 17-22.

[7] Nguyen Q.C., Phan T.T., Cao T.H., (2006), Vietnamese Proper Noun Recognition, Procs of the 4th International Conference on Computer Sciences, pp. 145-152.

[8] Rabiner L.R., (1989), A Tutorial on Hidden Markov Models and Selected Applications in Speech Recognition, Procs of the IEEE, Vol. 77, No. 2, 257-286.

Applying Fuzzy Measures for Considering Interaction Effects in Fine Root Dispersal Models

Wolfgang Näther[1] and Konrad Wälder[1]

TU Bergakademie Freiberg, Institut für Stochastik, D-09596 Freiberg
naether@math.tu-freiberg.de
waelder@math.tu-freiberg.de

Summary. We present an example how fuzzy measures and discrete Choquet integrals can be used to model interactivities between trees within a stochastic fine root dispersal model.

1 Introduction

Fine roots are roots with diameter smaller than 2 mm which are responsible for the soil water reception of trees. Investigations of the spatial dispersion of the fine root biomass can help to improve the knowledge about effects trees impose on soil resources. One point within this research are dispersal models where interaction between trees says something about their rivalry, for example with regard to water resources. In this paper the multi-tree case with trees from two different species is discussed. In this multi-tree case the total biomass of fine roots consists of the contributions of the individual trees. Here, interactions can affect the root biomass. Therefore, we want to describe the total mean of the fine root biomass by a weighted sum of the individual biomass contributions where the weights depend on the interactivities. From fuzzy theory it is known that fuzzy integrals are flexible tools of aggregation considering interaction ([2]). Especially, so-called discrete Choquet integrals can be applied for the aggregation of interacting critieria which in our case are given by the individual tree biomasses.

2 Fuzzy Measures for Modelling Interactivities

We restrict ourselves to a finite universe of discourse, say $\mathcal{N} = \{1,...,N\}$. Denote $\mathcal{P}(\mathcal{N})$ the power set of $\mathcal{N}$. A fuzzy measure v is a set function $v : \mathcal{P}(\mathcal{N}) \to [0,1]$ with

$$v(\emptyset) = 0, \ v(\mathcal{N}) = 1, \ v(A) \leq v(B) \text{ for } A,B \in \mathcal{P}(\mathcal{N}) \text{ and } A \subseteq B, \qquad (1)$$

see [1]. In general, a fuzzy measure v is a non-additive set function. The 'degree' of nonadditivity expresses the 'degree' of interaction between two subsets A and B from

W. Näther and K. Wälder: *Applying Fuzzy Measures for Considering Interaction Effects in Fine Root Dispersal Models*, Advances in Soft Computing **6**, 373–381 (2006)
www.springerlink.com

$\mathcal{N}$. A sub-additive v with $v(A \cup B) \leq v(A) + v(B)$ for $A \cap B = \emptyset$ models negative synergy or redundancy whereas a super-additive v with $v(A \cup B) \geq v(A) + v(B)$ for $A \cap B = \emptyset$ describes positive synergy.

Often the elements of $\mathcal{N}$ are interpreted as criteria. Then sub-additivity for example says that the evaluation $v(A \cup B)$ of the 'sum' $A \cup B$ of criteria is less than the sum of the single evaluations $v(A) + v(B)$. Using the interpretation as criteria, the evaluation of a single criterion $A = \{i\}, i \in \mathcal{N}$, is of special interest. Let us introduce the so-called importance index of criterion i. Note that for i being unimportant it is not enough that $v(\{i\})$ is small. If it happens that for some $A \subset \mathcal{N}$ the value $v(A \cup \{i\})$ is much greater than $v(A)$, then i may be important although $v(\{i\})$ is small. Considering these effects, the importance index or Shapley value is defined by

$$\Phi_i(v) := \sum_{A \subset \mathcal{N} \backslash \{i\}} \frac{(N - |A| - 1)! |A|!}{N!} [v(A \cup \{i\}) - v(A)], \qquad (2)$$

see [3]. Analogously to the concept of the importance index the interaction index between two criteria i and j is defined by

$$I_{i,j}(v) = \sum_{A \subset \mathcal{N} \backslash \{i,j\}} \frac{(N - |A| - 2)! |A|!}{(N-1)!} \Delta_{i,j}(A, v) \qquad (3)$$

$$\Delta_{i,j}(A, v) := v(A \cup \{i, j\}) - v(A \cup \{i\}) - v(A \cup \{j\}) + v(A).$$

If v reduces to a probability measure μ we always have $\Delta_{i,j}(A, \mu) = 0$, i.e. additive set functions cannot model interaction.

Consider a feature variable x which takes values x_i for the criteria $i \in \mathcal{N}$. For global evaluation or for aggregation of the feature values on $\mathcal{N}$ suitable tools seem to be certain means of x_i, more generally: certain integrals of x over $\mathcal{N}$. Classical integrals are linear operators with respect to a given measure. A much more powerful tool for a suitable aggregation are Choquet integrals with respect to a given fuzzy measure v, see [2]. For the ordered feature values $x_{(1)} \leq ... \leq x_{(N)}$ the discrete Choquet integral with respect to a fuzzy measure v is defined by

$$C_v(x) := \sum_{i=1}^{N} w_{[i]} x_{(i)} \qquad (4)$$

$$w_{[i]} := v(A(i)) - v(A(i+1)), \ i = 1, ..., N; \ A(i) := \{(i), (i+1), ..., (N)\}.$$

The set $A(i)$ collects the indices of the $N - i + 1$ largest feature values. Especially, it is $A(1) = \mathcal{N}, A(N+1) = \emptyset$.

An ordered weighted average (OWA) of the feature values is given by

$$\text{OWA}(x) = \sum_{i=1}^{N} w_i x_{(i)}, \sum_{i=1}^{N} w_i = 1, w_i \geq 0. \qquad (5)$$

Formally, $\text{OWA}(\boldsymbol{x}) = C_v(\boldsymbol{x})$ with respect to the special fuzzy measure

$$v(A) = \sum_{j=0}^{|A|-1} w_{N-j}$$

i.e. $w_{[i]}$ with respect to v from (4) coincides with w_i. For any fuzzy measure v with

$$v(T) = v(S) \text{ for all sets } T \text{ and } S \text{ with } |T| = |S| \qquad (6)$$

$\text{OWA}(\boldsymbol{x})$ coincides with the corresponding Choquet integral. In this simple case, the interaction index (3) is the same for any pair (i, j), given by

$$I_{i,j}(v) = \frac{w_1 - w_N}{N-1}, i, j \in \mathcal{N}, i \neq j, \qquad (7)$$

see [3].

One-parametric families of fuzzy measures where the parameter controls interaction in a transparent way are of special interest. Let us mention here the Yager family

$$v_q(A) = \left(\frac{|A|}{N}\right)^{\frac{1}{q}}, q > 0, \qquad (8)$$

where $q > 1$ models negative synergy and $q \in (0,1)$ positive synergy. The Yager family obviously satisfies (6) and leads to the weights

$$w_{[i]} = \left(\frac{N-i+1}{N}\right)^{1/q} - \left(\frac{N-i}{N}\right)^{1/q}; i = 1, ..., N; q \in (0, \infty). \qquad (9)$$

This fuzzy measure contains all possible types of interaction though the corresponding OWA is not too far away from the arithmetic mean ($q = 1$) which is natural for many dispersal effects in forests.

3 A Stochastic Model for Root Dispersal and Estimation of the Model Parameters

To describe a real root dispersal situation, a number of soil cores (with diameter 2,65 cm and volume 440 cm^3) is placed in the neighbourhood of the trees which collect a random number of root mass units (1 unit = 1 mg). At first let us consider a single tree and M soil cores, each of area a and fixed depth and with distance r_j from the tree, $j = 1, ..., M$. For the random number n_j of root mass units in soil core j we use a special nonlinear regression model:

$$E(n_j) = amp(r_j, \vartheta) =: \rho(r_j; m, \vartheta), j = 1, ..., M, \qquad (10)$$

m mean total mass of fine roots of the tree

$p(r, \vartheta)$ probability density for the location of a single root mass unit at distance r from the tree

ϑ unknown parameter.

The justification of (10) comes from theory of stochastic point processes which is suppressed here (see, e.g. [4]). An often used model for $p(r, \vartheta)$ assumes log-normality, i.e. with a normalizing constant c we have

$$p(r; \mu, \sigma^2) = \frac{c}{\sigma r} \exp\left(-\frac{(\ln r - \mu)^2}{2\sigma^2}\right), \sigma > 0, \mu \in R. \tag{11}$$

Much more interesting, especially with regard to interaction, is the multi-tree case. Firstly, we restrict ourselves to the case of N trees of the same species. Consider the model of N additive overlapping trees, i.e.

$$E(n_j) = \sum_{i=1}^{N} \rho(r_{ij}; m_i, \vartheta) =: \rho_j^N(m_1, ..., m_N, \vartheta), \tag{12}$$

where r_{ij} is the distance of soil core j from the tree i and let m_i be the total root mass of tree i. To avoid too much parameters we use an empirical relation between m_i and dbh_i, the stem *diameter at breast height* of tree i:

$$m_i = m \left(\frac{dbh_i}{30}\right)^\beta. \tag{13}$$

This relation is often used in forest sciences (see e.g. [4]) and expresses m_i by the mass m of a standard tree of $dbh = 30$ cm and an allometry parameter β, i.e. the N parameters m_i reduce to two parameters m and β. Now, (12) can be written as

$$E(n_j) = \sum_{i=1}^{N} \rho(r_{ij}; m, \beta, \vartheta) = \sum_{i=1}^{N} \rho_{ij}^N(m, \beta, \vartheta) =: \rho_j^N(m, \beta, \vartheta). \tag{14}$$

Note that (14) is given by the unweighted sum of the root masses of the N trees. But from ecological point of view this is not realistic for all cases of root dispersion. In some cases it seems to be more realistic to prefer the 'most intensive' or 'strong' trees, e.g. the trees closest to soil core j and to put (more or less) the remaining 'weak' trees at a disadvantage. For example, the strong tree takes up the total soil volume at some location and forces the roots of weaker trees to use other soil regions. On the other hand, it is conceivable that a strong tree with a number of fine roots can afford to accept roots of other trees, maybe from the same species, at some locations. These remarks lead in a natural way to a discrete Choquet integral of the root intensities. Consider the ordered intensities $\rho_{(ij)}^N(m, \beta, \vartheta)$ of soil core j, i.e. $\rho_{(1j)}^N(m, \beta, \vartheta) \leq ... \leq \rho_{(Nj)}^N(m, \beta, \vartheta)$ and aggregate them by (see (4))

$$E(n_j) = N \sum_{i=1}^{N} w_{[i]} \rho_{(ij)}^N(m, \beta, \vartheta) = NC_v(\rho_j^N(m, \beta, \vartheta)), \tag{15}$$

where the weights $w_{[j]}$ are defined in (4) and the vector $\boldsymbol{\rho}_j^N(m, \beta, \vartheta)$ contains as elements $\rho(r_{ij}; m, \beta, \vartheta), 1 \leq i \leq N$.

Let us explain the fuzzy measure v in (15). Let $A \subset \mathcal{N}$ be a subset of trees. Then $v(A)$ stands for the overall root mass if the trees of A produce alone, without any contribution of trees from $\mathcal{N} \setminus A$.

Now, let us justify the use of the discrete Choquet integral. In (15) all trees contribute at least $\rho_{(1j)}^N(m, \beta, \vartheta)$ root mass units to the total root mass at j. This results in a total root mass at least equal to $N\rho_{(1j)}^N(m, \beta, \vartheta)v(\mathcal{N})$ with $v(\mathcal{N}) = 1$. Each tree in $\mathcal{N} \setminus A_1$ contributes at least $\rho_{(2j)}^N(m, \beta, \vartheta)$ additional root mass units where A_1 collects the tree with the smallest individual contribution. Therefore, the increment of total root mass is at least equal to

$$N\left(\rho_{(2j)}^N(m, \beta, \vartheta) - \rho_{(1j)}^N(m, \beta, \vartheta)\right) v(\mathcal{N} \setminus A_1).$$

And so on. Summing up all these increments of total root mass units results exactly in the expression of the Choquet integral (15), see (4). Note that the weight of the smallest contribution is equal to $v(\mathcal{N}) - v(\mathcal{N} \setminus A_1)$ given by $v(A(1)) - v(A(2))$ in (4).

For a symmetric fuzzy measure fulfilling (6) (15) reduces to

$$E(n_j) = N\,\text{OWA}(\boldsymbol{\rho}_j^N(m, \beta, \vartheta)). \tag{16}$$

Obviously, if the chosen weights w_i of the OWA operator (see (4)) are increasing with i than - with regard to a given soil core - 'strong' trees suppress 'weak' trees, which expresses negative synergy. In the opposite case, if 'weak' trees contribute above the average, i.e. if the w_i's are decreasing in i, we have positive synergy, see (7).

Now, consider the more general case that root masses of trees from two species are given. Let N_1 be the number of trees from species 1 and N_2 the number of trees from species 2. We will propose a two-step approach for the total root mass consisting of the masses of the two species. At the first step, the mean of the root mass $n_j^{(l)}$ in a soil core j coming from species $l = 1, 2$, can be expressed following (15):

$$E(n_j^{(l)}) = N_l C_{v_l}(\boldsymbol{\rho}_j^{N_l}(m_l, \beta_l, \vartheta_l)) =: \rho_j^{N_l}(m_l, \beta_l; \vartheta_l),$$

where v_l is the specific fuzzy measure of species l. m_l is the total root mass of a standard tree from species l, β_l is the corresponding allometry parameter for this species l. The distributional parameters for species l are given by ϑ_l. Obviously, v_l controls the type of interaction inside species l, the so-called intra-specific interaction of species l. But in ecological context, interaction between species - the so-called inter-specific interaction - is also of great interest. We will model such effects at the second step. Proceeding from $E(n_j^{(1)})$ and $E(n_j^{(2)})$ for a given soil core j we can describe the mean of the total root mass n_j at j by an additional discrete Choquet integral:

$$E(n_j) = 2C_{v_{12}}\left(\left(\rho_j^{N_1}(m_1, \beta_1; \vartheta_1)\right), \rho_j^{N_2}(m_2, \beta_2; \vartheta_2)\right)$$

$$=: \rho_j^{N_1 N_2}(m_1, \beta_1, \vartheta_1, \gamma_1; m_2, \beta_2; \vartheta_2, \gamma_2; \gamma_{12}) \qquad (17)$$

where the inter-specific interaction is controlled by the fuzzy measure v_{12}.

The unknown model parameters in (17) are

m_l mean total mass of a standard tree from species l with $dbh = 30$ cm

β_l allometry parameter for species l

ϑ_l vector of distribution parameters in $p(r)$ for species l

γ_l vector of the parameters in the fuzzy measure v_l for species l controlling intra-specific interactivities

γ_{12} vector of the parameters in the fuzzy measure v_{12} for the total mass controlling inter-specific interactivities.

Now, we have to estimate the parameters by the use of soil core results n_j, $j = 1, ..., M$. The simplest way is a least squares approximation

$$\sum_{j=1}^{M} \left(n_j - \rho_j^{N_1 N_2}(m_1, \beta_1, \vartheta_1, \gamma_1; m_2, \beta_2; \vartheta_2, \gamma_2; \gamma_{12})\right)^2 \to \min.$$

Denote the estimated parameters by $\widehat{m_l}, \widehat{\beta_l}, \widehat{\vartheta_l}, \widehat{\gamma_l}$ and $\widehat{\gamma}_{12}$. As usual, the goodness of model fit can be expressed by the (mean) sum of squared residuals

$$S_M^2 := \frac{1}{M} \sum_{j=1}^{M} \left(n_j - \rho_j^{N_1 N_2}(\widehat{m}_1, \widehat{\vartheta}_1, \widehat{\gamma}_1; \widehat{m}_2, \widehat{\beta}_2; \widehat{\vartheta}_2, \widehat{\gamma}_2; \widehat{\gamma}_{12})\right)^2. \qquad (18)$$

In our case it is also useful to regard the sum of mean squared residuals for species l, i.e.

$$S_{M,l}^2 := \frac{1}{M} \sum_{j=1}^{M_l} \left(n_j^{(l)} - w_{[l]}\rho_j^{N_1}(\widehat{m}_l, \widehat{\vartheta}_l, \widehat{\gamma}_l)\right)^2. \qquad (19)$$

For further details and remarks see [5].

4 A Real-case Study

The study was carried out in a mixed spruce and beech stand consisting of 11 beech (species 1) and 17 spruce trees (species 2) in Germany (Saxony) near to Dresden. The study site is part of a greater nearly homegenous spruce stand. In 2003 soil cores were taken at 226 given sampling points with collections of the fine root biomass from the forest floor organic and mineral horizon.

At first, fine root biomass dispersion was modelled without considering interaction, i.e. inter- and intra-specific aggregations were carried out as additive sums of the contributions of the trees. The allometry parameters β_1 and β_2 were taken constant with value 2. Assuming a lognormal model, see (11), this leads to the following estimates

$$\widehat{m}_1 = 3.32 \cdot 10^6, \widehat{\mu}_1 = 2.45, \widehat{\sigma}_1 = 1.11 \qquad (20)$$
$$\widehat{m}_2 = 1.64 \cdot 10^6, \widehat{\mu}_2 = 1.85, \widehat{\sigma}_2 = 1.2$$

with $S_M^2 = 2365$, $S_{M,1}^2 = 1809$ for species 1 (beech) and $S_{M,2}^2 = 1126$ for species 2 (spruce), see (18) and (19). The empirical standard deviation of the root mass of species 1 s_1 is equal to 73.04. For species 2 $s_2 = 64.88$ holds. We denote

$$r_i := \frac{\sqrt{S_{M,i}^2}}{s_i} \qquad (21)$$

and obtain $r_1 = 0.58$ and $r_2 = 0.52$. From statistical point of view these values are not very good with respect to model fitting. But let us refer to some problems connected with fine root data. Fine roots are characterized by a high spatial and temporal variability depending for example on the changing availability of resources. Therefore, discussing investigations based on one spatio-temporal sample we have to accept some remaining variability of the residuals.

Now, interaction effects are considered applying an OWA operator with one-parametric fuzzy measures from the Yager family, see (8). The number of model parameters increases to 9. We obtain the following estimates:

$$\widehat{m}_1 = 4.67 \cdot 10^6, \widehat{\mu}_1 = 2.73, \widehat{\sigma}_1 = 1.3, \widehat{q}_1 = 0.74 \qquad (22)$$
$$\widehat{m}_2 = 9.27 \cdot 10^5, \widehat{\mu}_2 = 1.71, \widehat{\sigma}_2 = 0.94, \widehat{q}_2 = 1.37; \ \widehat{q}_{12} = 1.38$$

with $S_M^2 = 1903$, $S_{M,1}^2 = 1328$ and $S_{M,2}^2 = 951$. This leads to $r_1 = 0.54$ and $r_2 = 0.47$, see (21). Considering interactions results in visible improvement of the mean squared error of both species and the total root mass. To sum it up it can be said that the two species suppress each other, whereas fine root dispersal of the beeches is characterized by positive synergy. Considering that the study site comprises more spruces than beeches the supposition that positive synergy between the beeches enables their survival against the superiority of spruces is quite logical in ecological sense. In contrast to this the spruces are able to develop without intra-specific support or even with intra-specific suppression.

If the inter-specific interaction is modelled by an OWA operator, it is not possible to decide which species the other suppresses. This can be seen regarding the interaction index from (3). We obtain $\Phi_{spruce} = \Phi_{beech} = 0,5$. In ecological sense it is not satisfactory that the effect of suppression is equally distributed over the two species. Therefore, if we want to model that spruces suppress beeches we need a discrete Choquet integral as introduced in (17). Fortunately, it is easy to define a non-symmetric fuzzy measure for a set of two criteria. For negative synergy a sub-additive fuzzy measure is necessary. By

$$v(\{spruce, beech\}) = 1, v(\{beech\}) = w_1, v(\{spruce\}) = w_2, v(\emptyset) = 0 \qquad (23)$$

with $w_1, w_2 \leq 1$ and $w_1 + w_2 > 1$ such a fuzzy measure is given. In this case

$$\Phi_{spruce} = \frac{1}{2} + \frac{1}{2}(w_1 - w_2) \text{ and }, \ \Phi_{beech} = \frac{1}{2} + \frac{1}{2}(w_2 - w_1)$$

with $I_{12} = 1 - w_1 - w_2 < 0$ holds.

Now, the discrete Choquet integral (17) can be evaluated. Model fitting leads to

$$\widehat{m}_1 = 1.04 \cdot 10^7, \widehat{\mu}_1 = 2.33, \widehat{\sigma}_1 = 1.15, \widehat{q}_1 = 0.73$$
$$\widehat{m}_2 = 1.40 \cdot 10^6, \widehat{\mu}_2 = 2.18, \widehat{\sigma}_2 = 1.21, \widehat{q}_2 = 1.82 \tag{24}$$
$$\widehat{w}_1 = 0.42, \widehat{w}_2 = 0.8$$

with $S^2 = 2050, S^2_{M,1} = 1345$, $S^2_{M,2} = 997$, $r_1 = 0.54, r_2 = 0.49$ and

$$I_{12}(v) = -0.22, \Phi_{beech} = 0.31, \Phi_{spruce} = 0.69.$$

Obviously, negative synergy is given for inter-specific interactions. Further, the spruces are more important than the beeches with respect to fine root biomass dispersal, i.e. the spruces suppress the beeches. Analogously to the OWA case (22) the beeches support each other whereas interaction within the spruces is shaped by negative synergy.

5 Conclusions

The paper presents only some first results and shows that modelling of interaction effects by fuzzy measures leads to ecologically meaningful results. In a future project, we will analyze further ecologically interesting parameters, e.g. humus thickness and quality or the shape of the tree-tops, and we expect much more clear interaction effects.

Acknowledgement

This research was supported by grants from the Deutsche Forschungsgemeinschaft (DFG). The authors are grateful to S. Wagner for useful discussions and to A. Zeibig, who collected the root data used in Section 4.

References

[1] M. Grabisch. Fuzzy integral as a flexible and interpretable tool of aggregation, in: B. Bouchon-Meunier (Ed.), Aggregation and Fusion of Imperfect Information Physica Verlag, Heidelberg, 1998 51-72.
[2] M. Grabisch. Modelling data by the Choquet integral, in: V. Torra (Ed.), Information Fusion in Data Mining Physica Verlag, Heidelberg, 2003 135-148
[3] J.-L. Marichal and K. Roubens, Determination of weights of interacting criteria from a reference set, Faculte d'economie, de gestion et de sciences sociales, universite de Liege, working paper 9909 (1999).

[4] W. Näther and K. Wälder, Experimental design and statistical inference for cluster point processes- with applications to the fruit dispersion of anemochorous forest trees, Biometrical Journal 45 (2003) 1006-1022.

[5] W. Näther and K. Wälder, Applying fuzzy measures for considering interaction effects in root dispersal models, submitted to Fuzzy Sets and Systems.

Scoring Feature Subsets for Separation Power in Supervised Bayes Classification

Tatjana Pavlenko[1] and Hakan Fridén[2]

[1] TFM, Mid Sweden University
tatjana.pavlenko@miun.se
[2] TFM, Mid Sweden University
hakan.friden@miun.se

Summary. We present a method for evaluating the discriminative power of compact feature combinations (blocks) using the distance-based scoring measure, yielding an algorithm for selecting feature blocks that significantly contribute to the outcome variation. To estimate classification performance with subset selection in a high dimensional framework we jointly evaluate both stages of the process: selection of significantly relevant blocks and classification. Classification power and performance properties of the classifier with the proposed subset selection technique has been studied on several simulation models and confirms the benefit of this approach.

1 Introduction

There has been a recent explosion in the use of techniques for collecting and statistically analysing very high-dimensional data. A typical example is microarray gene expressions data: these allow the measurements of thousands of genes simultaneously and each observation is usually equipped with an additional categorical class variable, e.g. indicating tissue type. An important problem with statistical analysis of this type of data is to study the relationship between gene expression and tissue type, and to develop classification techniques that show good performance properties in these high dimension low sample size situations. Such problems can partly be reduced by the identification of strongly related functional subsets of genes whose expression levels are highly discriminative for specified tissue types. The genes can be merged into a single abstract subset (block) thereby reducing the problem of dimensionality in statistical estimation and generating some biological insights into gene co-regulation processes.

In light of this biologically motivation, we focus on the high-dimensional statistical model for supervised classification where the covariance structure of class-conditional distributions is sparse. This means that only a few underlying feature variables, or feature blocks are strongly associated with a class variable and account for nearly all of its variation - that is, determine the class membership. Several approaches have been considered to identify these blocks, e.g. unsupervised clustering (see [1] and references therein) and the self organizing maps technique [2], where the

T. Pavlenko and H. Fridén: *Scoring Feature Subsets for Separation Power in Supervised Bayes Classification*, Advances in Soft Computing **6**, 383–391 (2006)
www.springerlink.com

sets of feature variables with similar behavior across the observed data were identified. However, there are two important drawbacks in these approaches: on one hand, they identify thousands of blocks in a tree-like structure which makes them very difficult to interpret; secondly, the variables, which are blocked by similarities only, usually fail to reveal the discriminative ability in the classification problem since no information about the response variable was used.

We focus here on the multivariate approach suggested in [3] and [4] where the identification of highly relevant blocks is controlled by the information about the response variable and thereby reveal the blocks that are of special interest for class separation. We present a technique for measuring the block separation score which relies on calculating the cross-entropy distance between classes. We then explore the asymptotic distribution of the separation score in a high-dimensional framework, which in turn makes it possible to quantify the significance of each block and perform the subset selection by an appropriately specified threshold. Further, we combine the suggested subset selection technique with partial least squares (PLS) procedure which also directly incorporates the prior knowledge of class identity into a predictive model.

Finally, we illustrate the power and utility of our technique with experimental results using the Toeplitz matrix of order one to describe the within-block dependence structure of the class-conditional densities.

2 Supervised Bayes Classification Model

2.1 Problem Formalization

We focus here on a supervised classification model, given a training set of continuous feature variables $\mathbf{x} = (x_1, \ldots, x_p)$ as well as their associated class membership variable $\mathbf{Y} = 1, \ldots, \mathscr{C}$. The task is to build a rule for assessing the class membership of an observed vector $\mathbf{x}_0$.

Let the class conditional probability densities at feature vector $\mathbf{x}$ be $f(\mathbf{x}, \theta^j)$, where θ^j specifies a parametric model $\mathscr{F}_{\Theta_j}$, $j = 1, \ldots, \mathscr{C}$. According to Bayes decision theory, we use the following rule

$$\mathbf{Y} = j \quad \text{if} \quad \Pr(\mathbf{Y} = j | \mathbf{x}) = \max_k \Pr(\mathbf{Y} = k | \mathbf{x}), \tag{1}$$

where $\Pr(\mathbf{Y} = j | \mathbf{x}) \propto \pi_j f(\mathbf{x}; \theta^j)$ is a discriminant score that indicates to which the degree $\mathbf{x}$ belongs to the class j. $\Pr(\mathbf{Y} = j) = \pi_j$ are class prior probabilities, $j = 1, \ldots, \mathscr{C}$ and $\propto$ denotes proportionality. This approach can be combined with plug-in estimation of $f(\mathbf{x}, \theta^j)$ and is straightforward in the cases when the number of observations is greater than that of the dimensionality, i.e. when $n > p$ and p remains fixed. However, it becomes a serious challenge with very high dimensional data, where n/p and a plug-in classifier will be far from optimal.

To describe the high-dimensional framework asymptotically, we allow the dimension p to go to infinity together with n subject to the following constraints:

$\lim_{n\to\infty} p/n = c$, where $c \in (0,\infty)$ and the case when c is above one, i.e. $p > n$ makes standard classification methods degenerate. To overcome this problem, we assume that the true dependence among the features has a sparse structure such that not all p feature variables independently, but rather a few functional subsets of features, or *blocks of features*, are strongly associative with the response variable $\mathbf{Y}$. The natural goal in high-dimensional classification is to identify these blocks and thereby reduce the dimensionality by selecting the most informative blocks. More precisely, we aim to find $\mathcal{M}$ whose transpose maps an p-dimensional data vector $\mathbf{x}$ to a vector in q-dimensional space: $\mathcal{M} : R^{p\times 1} \to R^{q\times 1}$ such that such $(\mathbf{x}_1,\dots,\mathbf{x}_q)$ with $q l p$ are independent functional blocks, which constitute a disjoint and complete partition of the initial set of variables: $\{\cup_{i=1}^q \mathbf{x}_i\} \subseteq \{x_1,\dots,x_p\}$ and $\mathbf{x}_i \cup \mathbf{x}_j = \emptyset$ for any $i \neq j$. Since we seek a transform that optimally preserves class structure of the data in the reduced dimension space it natural to apply a searching technique which directly incorporates the response variable $\mathbf{Y}$ into the blocking process.

There are several ways to find $\mathcal{M}$ and determine the blocks of variables having high discriminatory power for a given class structure. We in this paper restrict the class conditional distributions to parametric models $\mathscr{F}_\Theta$ and adapt the supervised feature merging method (see [3] and [4]), which identifies the desirable blocks assuming that a block is highly discriminative if it is tightly clustered given Y, but well separated from the other blocks. Using this approach, we obtain q disjoint, non-empty independent m_q-dimensional blocks so that $p = \sum_{i=1}^q m_i$ with $m_i l p$. Observe that the asymptotics within the block is standard, meaning that n goes to infinity while m remains fixed.

The block structure of $\mathbf{x}$ reflects the notion of sparsity for high-dimensional data, meaning that each feature variable is expected to have a small number of direct neighbors. These blocks then can be considered as potential explanatory feature variables for a classification model revealing simultaneously those blocks which contain most relevant information about class separability.

2.2 Separation Score

Our method of revealing highly discriminative blocks for a given transform $\mathcal{M}$, relies on the distance-based separation score. To introduce this measure in classification framework, we first notice that given $\mathcal{M}$, each class conditional density can be factorized into a product of local interaction models as $f^j(\mathbf{x};\theta_i) = \prod_{i=1}^p f_i^j(\mathbf{x}_i;\theta)$, where the local m-dimensional density $f_i^j(\mathbf{x}_i;\theta_i^j)$ belongs to a specific parametric family $\mathscr{F}_{\mathcal{M},\Theta_j}$ which depends on a finite set of parameters $\theta_i^j \in \Theta_j$, $i = 1,\dots,q$, $j = 1,\dots,\mathscr{C}$.

We define the ith block separation score as a cross-entropy distance between two local probability densities given by

$$\mathscr{D}_i^{(j,k)} = \mathsf{E}\left[\ln \frac{f(\mathbf{x}_i;\theta_i^j)}{f(\mathbf{x}_i;\theta_i^k)}\bigg| f(\mathbf{x};\theta^j)\right] - \mathsf{E}\left[\ln \frac{f(\mathbf{x}_i;\theta_i^j)}{f(\mathbf{x}_i;\theta_i^k)}\bigg| f(\mathbf{x};\theta^k)\right] \tag{2}$$

which is a linear combination of Kullback-Leibler divergences (relative entropies) for i-the block, $i = 1,\dots,q$, $j,k = 1,\dots,\mathscr{C}$. This distance has also been known as entropy loss in a Bayesian framework. It is clear that the cross-entropy distance is a measure of discrepancy between two densities and this distance is minimized if we set $f(\mathbf{x}; \theta^j) = f(\mathbf{x}; \theta^k)$ for all $\mathbf{x}$. For the detailed properties of this distance see [5].

To give a theoretical justification of this definition in supervised classification, we relate the cross-entropy distance to the Bayesian misclassification risk. Since the primary role of the feature evaluation and selection is to improve the classification performance, the Bayesian risk is a natural measure of the block relevance. We turn to the special case of $\mathscr{C} = 2$, so that the response variable $\mathbf{Y}$ is binary and notice that typically, $\mathscr{C}$-class classification rule tends to be easier to construct and investigate for $\mathscr{C} = 2$ than for $\mathscr{C} > 2$ since only one decision boundary is analysed given that the new observations fit in the model validity domain. A generalization of the classification rule to the multiclass case can be given by so-called "max-wins" rule: solve each of the two-class problems and then, for an observed vector $\mathbf{x}_0$ combine all the pairwise classifiers to form a $\mathscr{C}$-classification rule; see details of this approach in [6]. This rule is quite intuitive: we assign to the class that wins the most pairwise comparison. Based on this pairwise classification criterion, we can then generalize the two-class distance based separation score to the multiclass case.

First, we observe that from the block independence it immediately follows that the classifier (1) allows factorization $\Pr(\mathbf{Y} = j|\mathbf{x}) \propto \pi_j \prod_{i=1}^{q} f_i^j(\mathbf{x}_i; \theta)$, $\quad j = 1, \dots, \mathscr{C}$. Furthermore, using plug-in estimates $f_i^j(\mathbf{x}_i; \hat{\theta})$ we get a block additive Bayesian classifier of the form

$$\mathscr{G}^{\mathscr{M}}(\mathbf{x}; \hat{\theta}^j, \hat{\theta}^k) = \sum_{i=1}^{q} \mathscr{G}_i^{\mathscr{M}}(\mathbf{x}_i; \hat{\theta}_i^j, \hat{\theta}_k^2) = \sum_{i=1}^{q} \ln \frac{f_i(\mathbf{x}_i; \hat{\theta}_i^j)}{f_i(\mathbf{x}_i; \hat{\theta}_i^k)} \lessgtr \ln \frac{\pi_k}{\pi_j}, \qquad (3)$$

where estimates $\hat{\theta}_i$ satisfies the standard set of "good" properties such as asymptotic unbiasedness and efficiency, uniformly in i as $n \to \infty$, $i = 1, \dots, q$, $\pi_j = \Pr(\mathbf{Y} = j)$ and $j, k = 1, 2$; see details in [7]. Classifier (3), given in terms of the decision boundary preserves the ordering of class posterior probabilities equivalent to the classifier (1). Furthermore, given that the dimensionality of $\mathbf{x}_i$ is fixed to m_i and $m_i \ln$, so that $q \to \infty$, we can get a closed form expression for asymptotic Bayesian misclassification risk $\mathscr{R}_{\mathscr{G}\cdot\mathscr{M}(\mathbf{x}; \hat{\theta}^j, \hat{\theta}^k)}$ which is the average value of the misclassification probabilities of the classifier $\mathscr{G}^{\mathscr{M}}(\mathbf{x}; \hat{\theta}^j, \hat{\theta}^k)$. Indeed, by the growing dimensions asymptotic assumption the classifier (3) can be seen as a sum of growing number of independent random variables and asymptotic normality of this sum an be stated under proper regularity conditions imposed on the family $\mathscr{F}_{\mathscr{M}, \Theta}$. This in turn gives

$$\mathscr{R}_{\mathscr{G}\cdot\mathscr{M}(\mathbf{x}; \hat{\theta}^j, \hat{\theta}^k)} = \Phi\left(-\frac{1}{2}\left[\mathscr{D}^{(j,k)}\right]^{1/2}\left[1 + \frac{2m\rho}{\mathscr{D}^{(j,k)}}\right]^{-1/2}\right), \qquad (4)$$

where $\Phi(x) = \frac{1}{\sqrt{2\pi}} \int_{-\infty}^{x} \exp(-z^2/2)dz$, $\mathscr{D}^{(j,k)} = \sum_{i=1}^{q} \mathscr{D}_i^{(j,k)}$ is the total cross-entropy distance between classes j and k and $\rho = \lim_{n \to \infty} \frac{q}{n}$; see detailed proof of the normal convergency in [7].

The message of the result (4) is that, the relevance of the ith block about classification can be estimated by its input towards the total cross-entropy distance $\mathscr{D}^{(j,k)}$, since the Bayesian misclassification risk is proved to be a monotone, strictly decreasing function of this distance.

3 Asymptotic Distributions and Significance Analysis

Numerical values of the separation score attached to each block mean a lot more if they come with statistical significance figures. In fact, comparison across blocks and selection of the most informative ones are essentially impossible without a uniform figure of merit or a selection threshold.

To give a suitable technique for comparing block scores and thresholds we establish the limiting distribution of $\hat{\mathscr{D}}_i^{(j,k)}$, $i = 1,\ldots,q$. First, we notice for each i and a fixed block dimension m, the block score allows the representation $\hat{\mathscr{D}}_i^{(j,k)} = (\hat{\theta}_i^j - \hat{\theta}_i^k)'I(\theta_i)(\hat{\theta}_i^j - \hat{\theta}_i^k) + \mathcal{O}(n^{-3/2})$, where $I(\theta_i)_{jk} = \int \frac{\partial \ell(\mathbf{x},\theta^j)}{\partial \theta_i^j} \frac{\partial \ell(\mathbf{x},\theta^j)}{\partial \theta_k^j} f(\mathbf{x},\theta^j)\,d\mathbf{x}$ is the ith block's information matrix, $\ell_i(\mathbf{x}_i;\theta_i^j) := \ln f_i(\mathbf{x}_i;\theta_i^j)$, $\theta_i = \frac{\theta_i^j + \theta_i^k}{2}$ and $n_j = n_k = n$. Furthermore,

$$(\hat{\theta}_i^j - \hat{\theta}_i^k)'I(\theta_i)(\hat{\theta}_i^j - \hat{\theta}_i^k) = n^{-1}\langle(\omega_i + T_i^j - T_i^k),(\omega_i + T_i^j - T_i^k)\rangle \tag{5}$$

where $\langle\cdot,\cdot\rangle$ denotes the scalar product, $\omega_i = \sqrt{n}[I^{1/2}(\theta_i)]'(\theta_i^j - \theta_i^k)$ and $T_i^j = n^{1/2}(\hat{\theta}_i^j - \theta_i^j)'I^{1/2}(\theta_i)$ represents the standardized bias of the estimate $\hat{\theta}_i^j$. According to the property of asymptotic efficiency of $\hat{\theta}_i^j$, we assume that uniformly in i, T_i^j converges to m-dimensional standard normal distribution $\mathcal{N}_m(0,I)$ as $n \to \infty$, $i = 1,\ldots,q$, $j,k = 1,2$.

Observe that the distribution of the random variable $\frac{T_i^j - T_i^k}{\sqrt{2}}$ also approaches $\mathcal{N}_m(0,I)$ uniformly with respect to i since T_i^j and T_i^k are independent random vectors, whose distributions are asymptotically normal. Therefore the distribution of $\langle\omega_i + T_i^j - T_i^k, \omega_i + T_i^j - T_i^k\rangle$, as well as $\frac{n\hat{f}_i(n)}{2}$, approaches non-central χ^2 distribution with m degrees of freedom and non-centrality parameter $\omega_i/\sqrt{2}$, $i = 1,\ldots,q$.

The asymptotic distribution of the normalized separation score gives us means to compute the selection threshold. We compute the plug-in separation score for the ith block using the observations from classes j and k and, viewing the score $\hat{\mathscr{D}}_i^{(j,k)}$ as χ^2-distributed test statistics order then blocks according to their potential significance for the class separation.

Another way to select significantly informative blocks is to test them against random data. More precisely: we test the hypothesis of whether and how many blocks achieve high separation score by chance only. We do this by performing permutation-based empirical analysis and estimate the probability of a block scoring better than some fixed threshold τ in randomly labeled data. This probability is the P-value, or the significance level corresponding to the scoring technique and the given threshold τ_P. Blocks with the separation score exceeding the threshold τ_P have low P-values and therefore contain the most useful information for classification. These blocks will be selected as the potential input feature variables for the reduced/pruned classification model. This selection technique will be studied in Section 4.

4 Empirical Study

In this section we evaluated utility of the distance based scoring technique by estimating the discriminative potential of the selected blocks. An easy to implement is a PLS model which in a supervised way constructs a weighted linear combination of feature variables that have maximal covariance with the response variable $\mathbf{Y}$ and can be used as a classifier.

We generated two populations as p-dimensional normal distributions $\mathcal{N}_p(\mu^j, \Sigma^j)$, where $\Sigma^j = (\sigma_{ls}^2)_{l,s=1}^p$ had the the block diagonal form reflecting the sparse structure of the data specified by $\mathcal{M}$ in Section 2, $j = 0, 1$. Assuming that the block size m is the same for both classes, we modeled Σ^j so that each diagonal element Σ_i^j represented the $m \times m$ covariance matrix of the ith block with Toeplitz structure of order one, having the same covariances σ^2 on the both upper and lower mid-off diagonals. In this case the cross-entropy distance $\mathcal{D}$ for $i$the block could be written as

$$\mathcal{D}_i^{(0,1)} = \frac{1}{2}\text{Tr}(\Sigma_{0_i}^{-1}\Sigma_{1_i} + \Sigma_{1_i}^{-1}\Sigma_{0_i} - 2I_m) + \frac{1}{2}(\mu_i^0 - \mu_i^1)'(\Sigma_{0_i}^{-1} + \Sigma_{1_i}^{-1})(\mu_i^0 - \mu_i^1). \quad (6)$$

A normally distributed population was generated by $N = 20,000$ observations with $q = 16$, $m = 3$ and covariance structure specified by $\sigma^2 = 0.3$ (for the first data set), and $\sigma^2 = 0.5$ (for the second data set), on the both mid-off diagonals, representing different dependence strength within the block. Variable average vectors μ_j for each block were simulated from independent normal distributions without mean shift for the first two blocks, but with a mean shift of 0.5 and 0.75 for the last two blocks and with the shift of 0.1 and 0.05 for the variables in between these blocks. All variables were autoscaled as the first step in the modelling and the binary response variable $\mathbf{Y} = \{0, 1\}$ was added to each data set to be used then in the class prediction phase.

Parameters μ_j^i and Σ_{i_j} where estimated for each j, by randomly drawing $n = 50$ observations with known $\mathbf{Y}$ from the population and were subsequently plugged into (6), showing considerably higher score values for the last two blocks: $\hat{\mathcal{D}}_{15} = 2.0562$, $\hat{\mathcal{D}}_{16} = 2.9711$ for $\sigma^2 = 0.3$ and $\hat{\mathcal{D}}_{15} = 1.8886$, $\hat{\mathcal{D}}_{16} = 2.7609$ for $\sigma^2 = 0.5$. These blocks therefore could be selected as the most informative. To judge whether the informative blocks selected from the original data set were of better quality than we would get by chance, we investigated the empirical distribution of $\hat{\mathcal{D}}_i$. We applied the bootstrap technique, generated $r = 500$ resampled data sets from both populations and estimated the ith block distance score for each sample. The results are summarized in Figure 1 (two upper plots of the left and right panels), clearly indicating top score values for the last two blocks for both data sets. Further, to specify more precisely the selection threshold and evaluate the statistical significance of our results, we explored the empirical distribution of the distance scores from the permuted data. We let $(y_1, \ldots, y_n)$ be the original set of responses specified by out

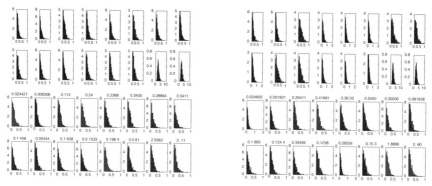

Fig. 1. Histograms displaying the original and permuted distributions for the block separation scores. The x-axis represents sample-based block scores and frequencies are on the y-axis. The two upper parts of the left and right panels summarize the bootstrap distributions ($r = 500$) of the block scores for $q = 16$ and $m_i = 3$ with $\sigma^2 = 0.3$ and $\sigma^2 = 0.5$ respectively. The two lower parts of both panels represent the permuted distributions ($r = 500$), where numbers on the top of each plot indicate the score values estimated by the original data sets. Values in the range of the score of the last two blocks estimated by the original data were never achieved with any of the permuted data.

training samples and supposed that $(\tilde{y}_1^r, \ldots, \tilde{y}_n^r)$ was a permuted set of responses constructed from the original set by a random permutation for each $r = 1, \ldots, 500$. We then assigned the element of the permuted response to each of the observation $\mathbf{x}$ from the training sample, which gave a set of pairs $\{(\mathbf{x}_1, \tilde{y}_1^r), \ldots, (\mathbf{x}_n, \tilde{y}_n^r)\}$ for each $r = 1, \ldots, 500$ and calculated the $\hat{\mathscr{D}}_i^{(0,1)}$ using permuted data. The histogram in Figure 1 (the two lower plots of both panels) displays the permuted distributions for both data sets, showing that the scores of the two top blocks estimated by the original data set considerably exceeded the 99% quantile of each of the permuted scores of the other blocks, reaching very high significance level. This meant that the 99% quantile, $\tau_{0.99}$ averaged over permuted block distributions for $j = 1, \ldots, 14$ could be used as the selection threshold. Moreover, these results confirmed that it was quite unlikely that the blocks we selected had such a high separation score by chance. To evaluate the discriminative potential of the selected blocks we applied the PLS technique and contrasted the classification results for the full PLS model to the reduced one, which was based on the selected blocks only. We varied the initial population parameters μ_i^j and σ^2, drew new samples and classified them by both PLS models assuming the same sample size and using the decision boundary of 0.5. The results are demonstrated in Figure 2 showing quite satisfactory predictive accuracy of the reduced model. The estimator $\hat{\mathbf{Y}} = \hat{\mathbf{Y}}(\mathbf{x}_1, \ldots, \mathbf{x}_{\tilde{q}})$ based on $\tilde{q}$ selected blocks is the predicted value of the class membership of the observed vector $\mathbf{x}$. The left hand side of both panels show a slight increase in false positives and false negatives when classifying by the selected blocks only. The percentage of false positives by $\hat{\mathbf{Y}}$ increased from 10% to 14% for the model with $\mu_i^1 = 0.75$ and $\mu_i^0 = 0.1$ (uppper plot of the left panel) and from 0% to 8% for the model with $\mu_i^1 = 0.5$ and $\mu_i^0 = 0.05$ (upper plot of the right panel). This increase was due to the considerable dimensionality re-

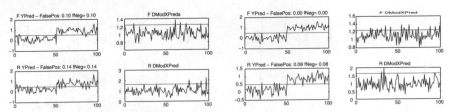

Fig. 2. Comparison of the predictive potential of the full and reduced PLS classifiers. Parameters μ_j^i and σ^2 were set as in Figure 1. The two upper parts of the left and right panels indicate the percentage of false positives and the distance to the model when using all $q = 16$ blocks. The two lower parts of both panels correspond to the reduced classifier. The dashed lines correspond to the decision boundary of 0.5 in the graphs, summarizing prediction potential, and to the 95% confidence limit indicating the consistency of the classified observation **x** with the training data set.

duction: instead of two highly informative blocks we had eight times as many single feature variables in the non-reduced PLS classifier. So far, we can conclude that the suggested distance-based scoring and selection technique really identified the blocks which had high discriminative ability.

The stability of our predictive results of the reduced PLS classifier is an important issue. To validate the output of the PLS classifier given in Figure 2, we tested if it remained unchanged for similar input data. 500 new samples were generated and classified by the reduced PLS model with subsequent comparison of the original class membership and the percentage of false positives and false negatives. False positive and false negative rates showed larger ranges for smaller distances (see left panel of Figure 3), indicating better effect of subset selection when classes were close to one another. However we conjecture that the reduced PLS model discriminates classes with a small number of variables only and the prediction ability of this model was just about 5% lower than for the full model.

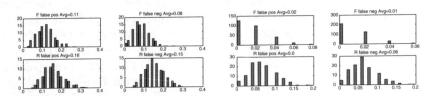

Fig. 3. Histograms showing the empirical distribution ($r = 500$) of the fraction of false positives and false negatives for the full PLS model (four plots on the left panel) and reduced PLS model (four plots on the right panel).

Conclusion

In summary, this paper presents the distance-based scoring measure for evaluating the separation power of the feature variables and is especially designed for the sparse data structure in a high-dimensional framework. The idea behind the distance-based criterion is theoretically justified by the straightforward relationship between the classification accuracy and the cross-entropy distance. We propose two approaches that allow a significance measure to be attached to the separation power: one is based on the asymptotic distribution of the sample-based block score which is shown to approach the non-central χ^2 distribution; the other is based on the permutation test and empirically evaluates a potential selection threshold. We also apply our scoring technique to randomly permuted data which justifies that the blocks showing high discriminative power are more than just an artifact. To examine the predictive properties of selected blocks, we embed the selection technique into the PLS classifier and show empirically that our approach makes it possible to reduce immensely the dimensionality while just slightly decreasing the predictive accuracy of the classifier.

Acknowledgements

Work was supported in part by the Research Center of the Västernorrland County Council under grant FoU: JA-2004-035 RSD and by ERDF (European Regional Development Fund), Objective 1, "Södra skogsänen", project Y3041-1013-03.

References

[1] Liu H, Yu L: Towards integrating feature selection algorithms for classification and clustering. *IEEE trans knowl and data eng 2005*, **3**:1-11.

[2] Tamayo P, Slonim D, Mesirov J, Zhu Q, Kitareewan S, Dmitrovsky E, Lander E, Golub, T: Interpreting patterns of gene expression with self-organizing-maps: methods and application to hematopoietic differentiation. *Proc Natl Acad Sci USA 1999*, **96**:2907-2912.

[3] Dettling M, Buhlman P: Finding predictive gene groups from microarray data. *J Multivariate Anal 2004*, **90**:106-131.

[4] Park P, Pargano M, Bonetti M: A nonparametric scoring algorithm for identifying informative genes from microarray data. *Pac Symp Biocomput 2001*, **6**:52-63.

[5] Wang X, Zidek J: Deviation of mixture distributions and weighted likelihood function as minimizers of KL-divergence subject to constraints. *Ann Inst Statist Math 2005*, **57**: 687-701.

[6] Hastie T, Tibshirani R: Classification by pairwise coupling. *Annls Statistics 1998*, **26**:451-471.

[7] Pavlenko T, von Rosen D: On the optimal weighting of high-dimensional Bayesian networks. *Adv and appl in stat 2004*, **4**:357-377.

Interval Random Variables and Their Application in Queueing Systems with Long–Tailed Service Times

Bartłomiej Jacek Kubica[1] and Krzysztof Malinowski[1][2]

[1] Warsaw University of Technology, Institute of Control and Computation Engineering, ul. Nowowiejska 15/19, 00–665 Warsaw, Poland
bkubica@elka.pw.edu.pl
[2] Research and Academic Computer Network (NASK), ul. Wkawozowa 18, 02–796 Warsaw, Poland
K.Malinowski@ia.pw.edu.pl

Key words: interval random variables, queueing, long–tailed distributions, Laplace transform

1 Introduction

The theory of interval random variables has been introduced by the authors in [12]. Up to now, it has been used in a few applications connected with Internet servers admission control and queueing systems.

The theory may be considered to be remotely related to the p–bound (i.e. "probability bound") concept (see e.g. [2], [4]), using bounds on the CDF of a random variable. For example, [13] and [15] consider some links between these theories.

Likewise, the approach based on the Evidence Theory (e.g. [7]) exhibits several similarities.

Actually, the interval random variables theory may also be considered to be the antitype of all approaches using set–valued probabilities – it operates on events having certain, non–interval probabilities. Nevertheless, values of random variables, assigned to these events are uncertain.

Earlier papers, e.g. [12], concentrated on an application connected with the estimation of uncertain parameters. Suitable notions and propositions, including the analog of Kolmogorov's theorem were introduced.

This paper is devoted to another problem, for which the developed theory is useful – numerical computation of the Laplace transform of random variable's PDF, which is useful e.g. in queueing theory with long–tailed distributions of service (or interarrival) times, e.g. [5], [18].

2 Basic Notions of Probability Theory

One of the most fundamental notions is the random variable.

B.J. Kubica and K. Malinowski: *Interval Random Variables and Their Application in Queueing Systems with Long–Tailed Service Times*, Advances in Soft Computing **6**, 393–403 (2006)
www.springerlink.com   © Springer-Verlag Berlin Heidelberg 2006

Definition 1. *Let the probability space* (Ω, S, P) *be given, where* Ω *is the set of elementary events,* S *– the* σ-*field of events and* $P \colon S \to [0, 1]$ *– the (*σ-*additive) probability measure.*

A mapping $X \colon \Omega \to \mathbb{R}$ *is called a (real) random variable, if the inverse images of Borel subsets of* $\mathbb{R}$ *are events (i.e. the elements of* S*).*

This condition is necessary to define a probability distribution of the variable.

By the *probability distribution* we mean a function $P_X(X_0) = P(\{\omega \colon X(\omega) \in X_0\})$ where X_0 is a Borel subset of $\mathbb{R}$.

Several important (and well–known) notions of the probability theory (like the expected value of a random variable) are beyond the scope of this paper. What we have to mention is the Laplace transform of the PDF.

For continuous random variables, with the PDF $f(x)$, the Laplace transform of the PDF may be defined as follows (see e.g. [5], [18]):

$$\widetilde{f}(s) = \int_0^\infty e^{-sx} f(x)\, dx \, . \tag{1}$$

Equation (1) may be transformed to the form of the integral w.r.t. (with respect to) the measure $P(\cdot)$. Then, in the case when the random variable is non–negative, it takes the form:

$$\widetilde{f}(s) = \int_\Omega e^{-sX(\omega)}\, dP(\omega) \, . \tag{2}$$

3 Basics of Interval Computations

To define the notion of interval random variable, we need to have some basic notions of intervals and their arithmetic. We follow a wide literature, like the article [8] or books [6], [9], [17], to name just a few.

We define the (closed) interval $[\underline{x}, \overline{x}]$ as a set $\{x \in \mathbb{R} \mid \underline{x} \leq x \leq \overline{x}\}$. We denote all intervals by brackets; open ones will be denoted as $]\underline{x}, \overline{x}[$ and partially open as: $[\underline{x}, \overline{x}[$, $]\underline{x}, \overline{x}]$. (We prefer this notation than using the parenthesis that are used also to denote sequences, vectors, etc.)

We also use boldface lowercase letters to denote interval variables, e.g. x, y, z. Following [10], $\mathbb{IR}$ denotes the set of all real intervals and $\mathbb{IC}_{\mathrm{rect}}$ – the set of "rectangular" complex intervals (i.e. pairs of intervals for real and imaginary part).

We design arithmetic operations on intervals so that the following condition was fulfilled: if $\odot \in \{+, -, \cdot, /\}$, $a \in a$, $b \in b$, then $a \odot b \in a \odot b$. We omit the actual formulae for arithmetic operations; they can be found in a wide literature e.g. ([6], [8], [9], [17]).

Now, we define a notion to set links between real and interval functions.

Definition 2. *A function* $f \colon \mathbb{IR} \to \mathbb{IR}$ *is an* inclusion function *of* $f \colon \mathbb{R} \to \mathbb{R}$, *if for all intervals* x *within the domain of* f *the following condition is satisfied:*

$$\{f(x) \mid x \in x\} \subseteq f(x) \, . \tag{3}$$

The definition is analogous for functions $f \colon \mathbb{R}^n \to \mathbb{R}^m$.

Remark 1. There is a confusion in interval community; some researchers use terms "interval enclosure" or "interval extension" instead of "interval inclusion function". We also use the term "interval enclosure" below, but in a slightly different case: a function assigning intervals to non–interval arguments (see Definition 4).

4 The Notion of an Interval Random Variable

Now, we define a "random variable" that maps the events not to real numbers, but rather to intervals of real numbers.

Definition 3. *Let the probability space* (Ω, S, P) *be given (as in Definition 1). Let us define a partition of* Ω *into sets* $A(x)$ *of the form:*

$$A_X(x) = \{\omega \in \Omega \mid X(\omega) = x\} \ , \ where \ x \in \mathbb{I}_X \ . \tag{4}$$

Any mapping $X: \Omega \to \mathbb{I}_X \subseteq {}^*\mathbb{IR}$, *satisfying the condition that for each* $x \in \mathbb{I}_X$ *the set* $A_X(x)$ *is an event, is called an* interval random variable.

According to [10], ${}^*\mathbb{IR}$ denotes the set of intervals, the endpoints of which may be not only finite real numbers, but also $-\infty$ or $+\infty$.

The definition of an interval random variable differs from the definition of a real random variable not only in the set of values. We omit here the condition about the reverse images of the Borel sets, replacing it by a simpler one. Why ? To formulate a relevant condition it would be necessary to define a <u>reverse image</u> first. And this notion is not explicitly defined.

There are several possible definitions of a reverse image of an interval valued function (see e.g. [11] and Section 4.1). In this paper we consider (and so we do in the earlier works, e.g. [12], [13], [15]) only those random variables that have a finite set $\mathbb{I}_X$ of intervals as its possible values. This assumption allows us to define the probability function only for interval arguments from the set $\mathbb{I}_X$:

$$P_X(x) = P\Big(A_X(x)\Big) ,$$

for any interval $x \in \mathbb{I}_X$.

Papers [12], [13], [15] consider also several other notions (e.g. the expected value of an interval random variable) that may be important in many applications, but are of no importance for the considered problem.

Now, we define some notions that will allow to associate interval variables with real variables, namely an *interval enclosure* and an *interval discretization* of a real random variable.

Definition 4. *Suppose, we have a real random variable* X.
The interval random variable X *that fulfills the condition:*

$$X(\omega) \in \mathbf{X}(\omega) \qquad \forall \omega \in \Omega \ , \tag{5}$$

will be called an interval enclosure *of the random variable* X.

Definition 5. *Suppose, we have a real random variable X. Let the values of X be contained in the interval $[a, b]$, where $a \in \mathbb{R} \cup \{-\infty\}$ and $b \in \mathbb{R} \cup \{+\infty\}$.*

Let us divide the interval $[a, b]$ into n subintervals. We denote their endpoints by x_i. We obtain the sequence $(x_i)_{i=0}^n$, where:

$$a = x_0 < x_1 < \ldots < x_{n-1} < x_n = b .$$

The interval random variable $\mathbb{X}$ will be called an interval discretization *of the random variable X, if the following conditions are fulfilled:*

- $\mathbb{X}$ *is an interval enclosure of X,*
- *the set of values of $\mathbb{X}$ is equal to $\mathbb{I}_X = \left\{ [x_{i-1}, x_i] \quad i = 1, \ldots, n \right\}$.*

Remark 2. In recent papers (e.g. [12], [13], [15]) a less restrictive condition was used in the definition of an interval discretization. The interval discretization was supposed to take the value $x_i = [x_{i-1}, x_i]$ with probability:

$$p_i = \int_{x_{i-1}}^{x_i} f(x)\, dx = F(x_i) - F(x_{i-1}) ,$$

for each $i = 1, \ldots, n$.

Defining the interval discretization by the sets of elementary events instead of the values of probability measure seems more appropriate.

Property 1. If $\mathbb{X}$ is an interval discretization of X, then:

$$P(\{X \in x\}) = P\Big(A_X(x)\Big) \forall x \in \mathbb{I}_X .$$

This property does not hold for interval enclosures that are not interval discretizations.

Remark 3. Precisely, Property 1 is fulfilled for interval discretizations of continuous random variables. Nevertheless, it can be generalized for the case of discrete random variables, relatively simply.

Namely, we have to consider only disjoint intervals, which means we cannot use closed ones only. If the probability that a random variable X takes a single value x_1 is nonzero, then computing probabilities: $P\Big(\{X \in [x_0, x_1]\}\Big)$ and $P\Big(\{X \in [x_1, x_2]\}\Big)$, we add $P\Big(\{X = x_1\}\Big)$ to both these quantities. We have to use either intervals $[x_0, x_1[$ and $[x_1, x_2]$ or $[x_0, x_1]$ and $]x_1, x_2]$.

It is well known that the distribution does not determine the random variable uniquely. Different random variables may have exactly the same distribution, but associate different values with different elementary events.

Obviously, the same holds for interval–valued random variables. We shall introduce now a notion to represent the distribution of an interval random variable.

Definition 6. *Consider a finite subset of $\mathbb{R}$, $\mathbb{I}_X = \{x_1, \ldots, x_n\}$.*
A generalized histogram *is a mapping $P \colon \mathbb{I}_X \to \mathbb{R}_+ \cup \{0\}$, such that:*

$$\sum\nolimits_{i=1}^n P(x_i) = 1 .$$

Remark 4. What is the difference between a generalized histogram and an ordinary one ? Only such that the intervals x_i, $i = 1, \ldots, n$ do not have to be pairwise disjoint.

Obviously, each interval random variable defines a generalized histogram of the form:

$$P(x) = P\Big(A_X(x)\Big) \quad x \in \mathbb{I}_X \ .$$

Remark 5. In many cases we are more interested in the distribution of a random variable than in the assignment of values to specific elementary events. Hence, researchers sometimes do not distinguish between the random variable and its distribution when it is not important. Also, we shall use notions "interval random variable" and "generalized histogram" alternately, when elementary events are not explicitly considered.

4.1 Interval Random Variables or Random Sets ?

Yet one more important question has to be answered: what is the relation between the theory of interval random variables and the more general theory of *set–valued random variables*, also known as *random sets* (e.g. [11], [16]).

A random set is a measurable mapping from the space Ω of elementary events to some family of sets. Measurable means that all sets $\{\omega : X(\omega) \cap x \neq \emptyset\}$ are events.

Though the interval random variables' theory was developed independently from the theory of random sets ([12]), it is obvious that interval random variables are a particular case of set–valued random variables.

Nevertheless, they are an important specific case, because they are computationally far more tractable and their theory is simpler. It is especially worth noting, that for interval–valued random variables it is reasonably simple to consider unbounded random variables (which we actually do in this paper), while papers on set–valued random variables usually assume, they are bounded and compact (see e.g. [16].

5 TAM – Transform Approximation Method

The $\mathscr{L}$–transform is well–defined and finite for the PDF of each random variable. Unfortunately, for some probability distribution functions the Laplace transform does not have an analytic form. According to e.g. [18], this is the case for all power–tailed distributions (e.g. Pareto distribution) and most other long–tailed ones (including lognormal and Weibull distributions).

The Laplace transforms, useful e.g. in $M/GI/1$ and $GI/M/1$ queueing systems analysis (see e.g. [1]), have to be approximated somehow. Below, we present a popular method to approximate such transforms.

5.1 Transform Approximation Method

TAM (Transform Approximation Method) is described e.g. in [5], [18].

Let us consider a random variable X with the PDF $f(x)$ and CDF $F(x)$.

The essence of TAM is very simple: we discretize the domain of the random variable X (at least the set $[0, +\infty[$), obtaining n points: $x_1 < x_2 < \ldots < x_n$.

Let us denote the CDF's values in these points in the following way:

$$y_i = F(x_i) \qquad i = 1, \ldots, n .$$

We associate some probability masses with these points:

$$
\begin{aligned}
p_1 &= \frac{y_1 + y_2}{2} , \\
p_n &= 1 - \frac{y_{n-1} + y_n}{2} , \\
p_i &= \frac{y_{i+1} - y_{i-1}}{2} \qquad i = 2, \ldots, n-1 .
\end{aligned}
\tag{6}
$$

Then we can approximate the $\mathscr{L}$–transform $\tilde{f}(s)$ of the PDF of X by a finite sum:

$$\check{f}(s) = \sum_{i=1}^{n} p_i \cdot e^{-s \cdot x_i} . \tag{7}$$

The above description does not specify how to choose points x_i (or y_i). There are a few approaches to do it (see below), but how to do it optimally remains an open problem.

Possible parameterizations.

The method was first developed in 1998 by Gross and Harris. The formula $\check{f}(s) = \frac{1}{n} \cdot \sum_{i=1}^{n} e^{-s \cdot x_i}$, where $x_i = F^{-1}\left(\frac{i}{n+1}\right)$ was used there. Such an approach is called *uniform–TAM*, or shortly UTAM.

Currently, more widely used is the GTAM (*geometric–TAM*), which sets: $y_i = 1 - q^i$ (for some q such that $0 < q < 1$) and $x_i = F^{-1}(y_i)$.

6 Interval Transform Approximation Method

To introduce the interval analog of TAM, let us consider a real–valued random variable X with the PDF $f_X(x)$ and CDF $F_X(x)$. Consider an interval discretization X of X.

We can formulate the interval inclusion function for the Laplace transform of the PDF of X. It is the function $\check{f}_X : \mathbb{IC}_{\text{rect}} \to \mathbb{IC}_{\text{rect}}$ of the form:

$$\check{f}_X(s) = \sum_{i=1}^{n} p_i \cdot e^{-s \cdot x_i} ,$$

where $x_i = [x_{i-1}, x_i]$ and $s = [\underline{s}, \overline{s}]$ is an interval complex variable, i.e. $\underline{s}$ and $\overline{s}$ are complex numbers.

We want to use this approximation for Laplace transforms of the PDFs of services time in queueing systems. It should be especially useful when the distribution of service time is long–tailed. In such a case (and actually each case when the time is unbounded) one of the intervals $[x_i, x_{i+1}]$ will be of the form: $[x_{n-1}, +\infty]$.

So, the interval extension of the $\mathscr{L}$–transform will be finite only for values of the argument s, satisfying $\mathrm{Re}\, s > 0$.

Now, let us prove that $\check{f}_X(s)$ is indeed an inclusion function of $\widetilde{f}_X(s)$.

Theorem 1. *Let an interval random variable X be interval enclosure of a real random variable X.*

Then, for each complex s such that $s \in s$, the following condition is fulfilled:

$$\widetilde{f}_X(s) \in \check{f}_X(s) \ .$$

The theorem is holds specifically for $s = [s, s]$.

Proof.

According to (2), we obtain:

$$\widetilde{f}_X(s) = \int_\Omega e^{-s \cdot X(\omega)} \, dP(\omega) \ .$$

Using the partition of Ω into sets $A(x_i)$, defined by equation (4), we can reformulate the above integral into the form of the following sum:

$$\widetilde{f}_X(s) = \sum_{i=1}^n \int_{A(x_i)} e^{-s \cdot X(\omega)} \, dP(\omega) \ . \tag{8}$$

From the definition of an interval enclosure we have that $X(\omega) \in X(\omega)$. The rules of interval computations (e.g. [6], [8], [9], [17]) imply that for each $x \in x_i$ and $s \in s$ we have: $e^{-s \cdot x} \in e^{-s \cdot x_i}$.

Hence, we obtain:

$$\left(\int_{A(x_i)} e^{-s \cdot X(\omega)} \, dP(\omega) \right) \in \left(e^{-s \cdot x_i} \int_{A(x_i)} dP(\omega) \right) \ .$$

The right side simply reduces to the form:

$$e^{-s \cdot x_i} \cdot p_i \ .$$

So:

$$\left(\sum_{i=1}^n \int_{A(x_i)} e^{-sX(\omega)} \, dP(\omega) \right) \in \left(\sum_{i=1}^n e^{-s \cdot x_i} \cdot p_i \right) \ . \tag{9}$$

Then, from (8) and (9) we obtain:

$$\widetilde{f}_X(s) \in \sum_{i=1}^n e^{-s \cdot x_i} \cdot p_i = \check{f}_X(s) \ .$$

QED

The Essence of the Method

Let us now refer to TAM, described in Section 5.

Having defined the notion of interval discretization and the interval inclusion of $\mathscr{L}$–transform of a random variable, it was simple to develop an interval analog of TAM. We may call it "Interval TAM" or ITAM for short.

This approach is similar to classical TAM, except for using interval discretization instead of a traditional one and computing the interval inclusion of the $\mathscr{L}$–transform basing on this interval discretization.

The advantages of such approach in comparison with the traditional TAM are obvious:

- we use correct probabilities associated with the intervals, not probability masses quite arbitrarily associated with chosen points, as in (6),
- we can naturally bound the discretization error and truncation error,
- as in other interval methods, we can bound the numerical error (see e.g. [6], [8], [9], [17]).

7 Laplace Transform for Queueing Systems

In case of $M/GI/1$ systems the $\mathscr{L}$–transform of the sojourn time is given by the so–called Pollaczek–Khinchin formula (see e.g. [1]):

$$\widetilde{w}(s) = \frac{(1-\rho) \cdot \widetilde{b}(s) \cdot s}{s + \lambda \cdot \left(\widetilde{b}(s) - 1\right)} , \qquad (10)$$

where $\widetilde{b}(s)$ is the $\mathscr{L}$–transform of PDF of the service time B, λ is the arrival rate and $\rho = \lambda \cdot \mathbb{E}B$.

Assume the service time to be Pareto–distributed; this is a typical power–tailed distribution, commonly used to model various levels of computer network traffic. The most commonly encountered form uses two parameters: the shaping parameter $\alpha > 0$ and the location parameter $\beta > 0$. A Pareto–distributed variable X has the CDF $F_X(x) = 1 - \left(\frac{\beta}{x}\right)^{\alpha}$ (for $x \geq \beta$; otherwise $F_X(x) = 0$) and PDF $f_X(x) = \frac{\beta^{\alpha}}{x^{\alpha+1}}$ (also for $x \geq \beta$).

As it was mentioned before, PDF of a Pareto–distributed random variable posses an $\mathscr{L}$–transform (as PDFs of all random variables do), but that transform does not have a closed analytical form. Hence, some approximation of $\widetilde{b}(s)$ has to be used, in particular in formula (10), to get an approximation $\widecheck{w}(s)$ of $\widetilde{w}(s)$.

So, we can now approximate the Laplace transform of the sojourn time. Where can we use such an approximation ? Obviously, we can invert it numerically, to obtain the distribution of the sojourn time. But the next subsection describes a different application.

7.1 Optimization

Consider the following problem: we want to find the arrival and service rate of a queueing system, to optimize some performance measure for users waiting for the completion of their tasks. It can be set as the following optimization problem:

$$\max_{\lambda,\mu} \left(Q = V(\lambda) - \lambda \cdot G(\lambda,\mu) - C(\mu) \right)$$

s.t.

$$0 \leq \lambda \leq \Lambda \,,$$

$$\lambda - \mu \leq -\varepsilon \,.$$

The meaning of the above notions is as follows:

- $V(\lambda)$ – increasing, concave and strictly differentiable is the aggregated utility of the users,
- $G(\lambda,\mu)$ – is the delay cost of the user,
- $C(\mu)$ – is the capacity cost (usually a linear structure is assumed $C(\mu) = c \cdot \mu$,
- ε is a small positive number used to avoid a strict inequality $\lambda < \mu$.

What about the delay cost G ? In [3] a few measures are proposed: linear cost, polynomial cost, etc. However in the case of a Pareto–distributed service time (with $\alpha < 2$) most of them are useless: they are infinite regardless the values of parameters λ and μ (proof given in [15], Subsection 4.1.4). The only useful measure of the delay cost is the exponential one (see [3]), expressed as:

$$G = \frac{v}{k} \cdot \left(1 - \widetilde{w}(k) \right) \,,$$

where $v > 0$ and $k > 0$ are some real–valued parameters, estimation of which is beyond our interest (interval random variables might be useful there too, though; [12].

So, $\mathscr{L}$–transform of the sojourn time is explicitly used here to measure performance of the queueing system. More details may be found in [15].

Numerical Experiments

The lack of space makes the authors to present only a limited number of experiments. They are presented in Tables 1, 2 and 3.

Table 1 presents the results for Erlang distribution. Obviously, this is not a long–tailed distribution and TAM does not have to be used here. We present it, however, to show the failure of traditional real–valued TAM, which provides incorrect values there. Intervals computed by ITAM are somewhat wide, but correct.

Table 2 presents the results for Pareto distribution. We do not know the actual values of the $\mathscr{L}$–transform, so we can only use some kind of TAM.

Finally, Table 3 shows the performance of an interval optimization algorithm, setting the parameters of a queueing system with the Pareto service time (i.e. solving the problem from Subsection 7.1).

Table 1. Approximate values of the Laplace transform of the Erlang–distributed random variable's PDF ($r = 2$, $\mu = 100.000000$); TAM with 100 points

s	$\mathscr{L}$–transform	real–valued TAM	ITAM
(5.000000, -3.000000)	(0.951203, 0.027846)	(0.789720, 0.089621)	([0.579440, 1.000000],[4.705679E-009, 0.295521])
(5.000000, -2.000000)	(0.951543, 0.018568)	(0.797220, 0.060250)	([0.594440, 1.000000],[3.163486E-009, 0.198670])
(5.000000, -1.000000)	(0.951746, 0.009286)	(0.801750, 0.030276)	([0.603500, 1.000000],[1.589685E-009, 0.099834])
(5.000000, 0.000000)	(0.951814, 0.000000)	(0.803265, 0.000000)	([0.606530, 1.000000],[-0.000000, 0.000000])
(5.000000, 1.000000)	(0.951746, -0.009286)	(0.801750, -0.030276)	([0.603500, 1.000000],[-0.099834,-1.589685E-009])
(5.000000, 2.000000)	(0.951543, -0.018568)	(0.797220, -0.060250)	([0.594440, 1.000000],[-0.198670,-3.163486E-009])
(5.000000, 3.000000)	(0.951203, -0.027846)	(0.789720, -0.089621)	([0.579440, 1.000000],[-0.295521,-4.705679E-009])

Table 2. Approximate values of the Laplace transform of the Pareto–distributed random variable's PDF ($\alpha = 1.1$, $\beta = 1.0$; TAM with 1000 discretization points

s	real–valued TAM	ITAM
(0.100000, 0.000000)	(0.733017, 0.000000)	([0.732661, 0.757252],[0.000000, 0.000000])
(0.200000, 0.000000)	(0.593593, 0.000000)	([0.593003, 0.602140],[0.000000, 0.000000])
(0.500000, 0.000000)	(0.343648, 0.000000)	([0.342789, 0.344802],[0.000000, 0.000000])
(1.000000, 0.000000)	(0.157934, 0.000000)	([0.157144, 0.158725],[0.000000, 0.000000])
(2.000000, 0.000000)	(0.040318, 0.000000)	([0.039914, 0.040722],[0.000000, 0.000000])
(5.000000, 0.000000)	(0.001082, 0.000000)	([0.001054, 0.001109],[0.000000, 0.000000])
(10.000000, 0.000000)	(4.183039E-006, 0.000000)	([3.974061E-006,4.392017E-006],[0.000000, 0.000000])

Table 3. Results of the interval branch–and–bound for the single $M/P/1$ queue, capacity cost $c = 1.0$ and exponential delay cost with different values of v and k; ITAM with 100 discretization points

v	k	execution time	function evaluations	number of boxes that can contain a solution
10	10	0.66 sec.	73	7
10	2	25.94 sec.	3491	121
5	2	2.78 sec.	374	68
0.1	0.4	1.7 sec.	290	16

8 Conclusions

The proposed ITAM is an efficient way to approximate the Laplace transform of PDFs of random variables. Its computation may be a bit more costly than in the case of traditional, real–valued TAM, but it is significantly more precise and safe. It seems to be another useful application of the presented interval random variables theory.

References

[1] I. Adan, J. Resing, "*Queueing Theory*", 2001, available at: http://www.cs.duke.edu/~fishhai/misc/queue.pdf .

[2] D. Berleant, "Automatically Verified Arithmetic on Probability Distributions and Intervals", [in:] *Applications of Interval Computations* (ed. R. B. Kearfott and V. Kreinovich), Applied Optimization, Kluwer, Dordrecht, Netherlands, 1996.

[3] S. Dewan, H. Mendelson, "User Delay Costs and Internal Pricing for a Service Facility", *Management Science*, Vol. **36**, No. **12** (1990).

[4] *Engineering Reliability Design Handbook* (ed. E. Nikolaidis, D. Ghiocel, S. Singhal), CRC Press, 2004.

[5] M. J. Fischer, D. Gross, D. M. B. Masi, J. F. Shortle, "Analyzing the Waiting Time Process in Internet Queueing Systems With the Transform Approximation Method", *The Telecommunications Review*, No. **12** (2001), pp. 21-32.

[6] E. Hansen, "*Global Optimization Using Interval Analysis*", Marcel Dekker, New York, 1992.

[7] C. Joslyn, V. Kreinovich, "Convergence Properties of an Interval Probabilistic Approach to System Reliability Estimation", *International Journal of General Systems*, Vol. **34**, No. **4** (2002), pp. 465-482.

[8] R. B. Kearfott, "Interval Computations: Introduction, Uses, and Resources", *Euromath Bulletin*, Vol. **2**, No. **1** (1996), pp. 95-112.

[9] R. B. Kearfott, "*Rigorous Global Search: Continuous Problems*", Kluwer, Dordrecht, 1996.

[10] R. B. Kearfott, M. T. Nakao, A. Neumaier, S. M. Rump, S. P. Shary, P. van Hentenryck, "Standardized notation in interval analysis", available at: http:// www.mat.univie.ac.at/~neum/ software/int/notation.ps.gz .

[11] E. Klein, A. C. Thompson, "*Theory of correspondences. Including applications to mathematical ecomomics*", Wiley, New York, 1984.

[12] B. J. Kubica, "Estimating Utility Functions of Network Users – An Algorithm Using Interval Computations", *Annals of University of Timisoara, Mathematics & Informatics Series*, Vol. **40** (2002), pp. 121–134, available at: http://www.math.uvt.ro/anmath/ issues/anuvt2002_3/ kubica_synasc02.ps .

[13] B. J. Kubica, K. Malinowski, "*Interval Approximations of Random Variables and Their Applications*" (in Polish), KZM'04 (Conference of Mathematics' Applications), 2005; the abstract available at: http://www.impan.gov.pl/ ~zakopane/34/Kubica.pdf .

[14] B. J. Kubica, K. Malinowski, "An Interval Global Optimization Algorithm Combining Symbolic Rewriting and Componentwise Newton Method Applied to Control a Class of Queueing Systems", *Reliable Computing*, Vol. **11**, No. **5** (2005).

[15] B. J. Kubica, "*Optimization of Admission Control for Systems with Uncertain Parameters*", PhD Thesis, 2005, Faculty of Electronics and Information Technology, Warsaw University of Technology.

[16] S. Li, Y. Ogura, "Convergence of Set Valued Sub- and Supermartingales in the Kuratowski–Mosco Sense", *The Annals of Probability*, Vol. **26**, No. **6** (1998), pp. 1384-1402.

[17] H. Ratschek, J. Rokne, "*Interval Methods*", [in:] *Handbook of Global Optimization* (ed. R. Horst, P. M. Pardalos), Kluwer, 1995.

[18] J. F. Shortle, M. J. Fischer, D. Gross, D. M. B. Masi, "Using the Transform Approximation Method to Analyze Queues with Heavy–Tailed Service", *Journal of Probability and Statistical Science*, Vol. **1**, No. **1** (2003), pp. 17-30.

Online Learning for Fuzzy Bayesian Prediction

N.J. Randon[1], J. Lawry[1], and I.D. Cluckie[2]

[1] Artificial Intelligence Group, Department of Engineering Mathematics, University of
Bristol, Bristol, BS8 1TR, United Kingdom
Nick.Randon, J.Lawry@bris.ac.uk
[2] Water and Environmental Management Research Centre, Department of Civil Engineering,
University of Bristol, Bristol, BS8 1TR, United Kingdom
I.D.Cluckie@bris.ac.uk

Summary. Many complex systems have characteristics which vary over time. Consider for
example, the problem of modelling a river as the seasons change or adjusting the setup of
a machine as it ages, to enable it to stay within predefined tolerances. In such cases offline
learning limits the capability of an algorithm to accurately capture a dynamic system, since
it can only base predictions on events that were encountered during the learning process.
Model updating is therefore required to allow the model to change over time and to adapt
to previously unseen events. In the sequel we introduce an extended version of the fuzzy
Bayesian prediction algorithm [6] which learns models incorporating both uncertainty and
fuzziness. This extension allows an initial model to be updated as new data becomes available.
The potential of this approach will be demonstrated on a real-time flood prediction problem
for the River Severn in the UK.

1 Introduction

Many data modelling approaches in Artificial Intelligence (AI) rely on an offline
learning strategy where a static model is learned from historical data. This type of
modelling is appropriate if the underlying dynamics of the system under consideration does not change over time. However, often this is not the case as the behaviour of
a system varies and evolves over time. In this situation an offline learning approach
cannot account for these changes unless the model is completely re-learned.

2 Fuzzy Bayesian Methods

The fuzzy Bayesian learning algorithm proposed in Randon and Lawry [5] allows for
the induction of prediction models that incorporate both uncertainty and fuzziness
within an integrated framework. In the following we give a brief exposition of this
approach.

Consider the following formalization of a prediction problem: Given variables
$x_1, \ldots, x_{n+1}$ with universes $\Omega_1, \ldots, \Omega_{n+1}$, each corresponding to a compact interval

N.J. Randon et al.: *Online Learning for Fuzzy Bayesian Prediction*, Advances in Soft Computing **6**, 405–
412 (2006)
www.springerlink.com

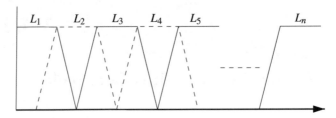

Fig. 1. Trapezoidal fuzzy sets discretizing a continuous universe

of $\mathbb{R}$, suppose that x_{n+1} is dependent on $x_1,\ldots,x_n$ according to some functional mapping $g : \Omega_1 \times \cdots \times \Omega_n \to \Omega_{n+1}$ (i.e. $x_{n+1} = g(x_1,\ldots,x_n)$). In fuzzy Bayesian algorithms fuzzy labels are used to partition both input and output and probability values are then estimated for the corresponding imprecise regions of attribute space. Figure 1 illustrates how trapezoidal fuzzy sets can be used to discretize a continuous universe. Such a discretization can be generated from a crisp partition where we now identify a fuzzy label L_i with each partition interval (i.e. a bin) such that L_i applies to those elements in the interval to degree 1 and also applies to some points in neighbouring intervals to a non-zero degree. Examples of labels include *small*, *medium*, *large*, *tall*, *high* etc. and in the fuzzy Bayesian model the membership of x in L is interpreted as the probability that L is a valid or appropriate label given value x. i.e.

$$\forall x \in \Omega \ \mu_L(x) = P(L|x)$$

Now unlike crisp partitions the labels L_i overlap so that more than one label may be appropriate to describe a particular x value, and hence we cannot directly define probability distributions on the set $LA = \{L_1,\ldots,L_n\}$. Instead, we must base our analysis on the set of atoms generated from $LA = \{L_1,\ldots,L_n\}$, each identifying a possible state of the world and taking the form:

$$\alpha = \bigwedge_{i=1}^n \pm L_i \text{ where } +L_i = L_i \text{ and } -L_i = \neg L_i$$

For example, in the case that we have only two labels L_1 and L_2 then there are 4 atoms; $\alpha_1 = L_1 \wedge L_2$, $\alpha_2 = L_1 \wedge \neg L_2$, $\alpha_3 = \neg L_1 \wedge L_2$ and $\alpha_4 = \neg L_1 \wedge \neg L_2$. In general, if we have n labels then there are 2^n atoms, however, if as in figure 1 at most two labels can be applied to any x then only $2n - 1$ atoms can have non-zero probability. For example, the atoms generated by the fuzzy labels in figure 1 are shown in figure 2. Let $\mathscr{A}$ denote the set of atoms with non-zero probability for at least some $x \in \Omega$.

For a given $x \in \Omega$ the distribution on atoms $P(\alpha|x) : \alpha \in \mathscr{A}$ can be represented by a mass assignment $m_x : 2^{LA} \to [0,1]$ on the power set of LA as follows:

$$\forall T \subseteq LA \ P(\alpha_T|x) = m_x(T) \text{ where } \alpha_T = \left(\bigwedge_{L \in T} L\right) \wedge \left(\bigwedge_{L \notin T} \neg L\right)$$

For example, if $LA = \{L_1,\ldots,L_n\}$ then:

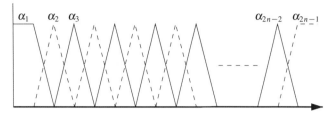

Fig. 2. Probability functions $P(\alpha|x)$ for atoms $\alpha \in \mathscr{A}$ where $\alpha_1 = L_1 \wedge \neg L_2 \wedge \neg L_3 \wedge \cdots \wedge \neg L_n$, $\alpha_2 = L_1 \wedge L_2 \wedge \neg L_3 \wedge \cdots \wedge \neg L_n$, $\alpha_3 = \neg L_1 \wedge L_2 \wedge \neg L_3 \wedge \cdots \wedge \neg L_n$ etc

$$P\left(L_1 \wedge L_2 \wedge \bigwedge_{i=3}^{n} \neg L_i|x\right) = m_x\left(\{L_1, L_2\}\right)$$

Intuitively $m_x(T)$ is the probability that the set of all labels appropriate to describe x is T. A consequence of this translation process is that the current algorithm can be embedded in the label semantics framework as proposed by Lawry ([2] and [3]). Now under certain circumstances label semantics can be functional, allowing a mapping from the fuzzy label definitions $\mu_L(x) : L \in LA$ to the conditional probabilities on atoms $P(\alpha|x) : \alpha \in \mathscr{A}$. One such possibility, as discussed in Lawry [2], is that for each x there is a natural ordering on the appropriateness of labels and that the values of m_x are evaluate so as to be consistent with this ordering. This means that the mass assignment m_x is consonant or nested and consequently can be determined from $\mu_L(x) : L \in LA$ as follows: If $\mu_{L_1}(x) \geq \mu_{L_2}(x) \geq \ldots \geq \mu_{L_n}(x)$ then:

$$P\left(\bigwedge_{j=1}^{n} L_j|x\right) = m_x\left(\{L_1, \ldots, L_n\}\right) = \mu_{L_n}(x)$$

$$P\left(\bigwedge_{j=1}^{i} L_j \wedge \bigwedge_{j=i+1}^{n} \neg L_j|x\right) = m_x\left(\{L_1, \ldots L_i\}\right)$$

$$= \mu_{L_i}(x) - \mu_{L_{i+1}}(x) : i = 1, \ldots, n-1$$

$$P\left(\bigwedge_{j=1}^{n} \neg L_j|x\right) = m_x(\emptyset) = 1 - \mu_{L_1}(x)$$

and $P(\alpha|x) = 0$ for all other atoms α.

In the case where for any $x \in \Omega$ at most two labels have non-zero probability (as in figure 1) then the above consonant mapping is simplified further so that if for a given $x \in \Omega$ $\mu_{L_i}(x) \geq \mu_{L_j}(x) > 0$ and $\mu_{L_k}(x) = 0 : k \notin \{i, j\}$ (for the labels in figure 1 either $j = i+1$ or $j = i-1$) then only two atoms have non-zero probability given by:

$$P\left(L_i \wedge L_j \wedge \bigwedge_{k \notin \{i,j\}} \neg L_k | x\right) = m_x\left(\{L_i, L_j\}\right) = \mu_{L_j}(x)$$

$$\text{and } P\left(L_i \wedge \bigwedge_{k \neq i} \neg L_k | x\right) = m_x\left(\{L_i\}\right) = \mu_{L_i}(x) - \mu_{L_j}(x)$$

For the membership functions shown in figure 1 the probability functions on atoms inferred in this way are shown in figure 2 across all values of x.

In the fuzzy naïve Bayes algorithm each input universe is partitioned using trapezoidal fuzzy sets (as in figure 1) and the probability function for the atoms generated as in figure 2. Let $\mathscr{A}_i$ denote the atoms generated for variable x_i where $i = 1, \ldots, n$. Then for output atom $\alpha_{n+1} \in \mathscr{A}_{n+1}$ and input atom $\alpha_j \in \mathscr{A}_j$ we infer the conditional probability $P(\alpha_j | \alpha_{n+1})$ from the training database $DB = \{\langle x_1(i), \ldots, x_n(i), x_{n+1}(i)\rangle : i = 1, \ldots, N\}$ as follows:

$$P(\alpha_j | \alpha_{n+1}) = \frac{\sum_{i \in DB} P(\alpha_j | x_j(i)) P(\alpha_{n+1} | x_{n+1}(i))}{\sum_{i \in DB} P(\alpha_{n+1} | x_{n+1}(i))}$$

From this we can use Jeffrey's rule [1] (an extension of the theorem of total probability) to infer a marginal density conditional on α_{n+1} such that:

$$f(x_j | \alpha_{n+1}) = \sum_{\alpha_j \in \mathscr{A}_j} P(\alpha_j | \alpha_{n+1}) f(x_j | \alpha_j)$$

Where assuming a noninformative uniform prior distribution on the input universe Ω_j, from Bayes' theorem we have:

$$\forall x_j \in \Omega_j, \ \alpha_j \in \mathscr{A}_j \ f(x_j | \alpha_j) = \frac{P(\alpha_j | x_j)}{\int_{\Omega_j} P(\alpha_j | x_j) \, dx_j}$$

From this we can apply Bayes' theorem together with the naïve Bayes conditional independence assumption as in the standard Bayesian model [4] to obtain the conditional probability $P(\alpha_{n+1} | x_1, \ldots, x_n)$ of each output atom given a vector of input values, as follows:

$$P(\alpha_{n+1} | x_1, \ldots, x_n) = \frac{P(\alpha_{n+1}) \prod_{j=1}^{n} f(x_j | \alpha_{n+1})}{\sum_{\alpha_{n+1} \in \mathscr{A}_{n+1}} P(\alpha_{n+1}) \prod_{j=1}^{n} f(x_j | \alpha_{n+1})}$$

Hence, given an input $\langle x_1, \ldots, x_n\rangle$ we can now obtain a density function on output values using Jeffrey's rule as follows:

$$f(x_{n+1} | x_1, \ldots, x_n) = \sum_{\alpha_{n+1} \in \mathscr{A}_{n+1}} P(\alpha_{n+1} | x_1, \ldots, x_n) f(x_{n+1} | \alpha_{n+1})$$

A single output prediction can then be obtained by taking the expected value so that:

$$\hat{x}_{n+1} = \int_{\Omega_{n+1}} x_{n+1} f(x_{n+1} | x_1, \ldots, x_n) \, dx_{n+1}$$

3 Online Updating

In a dynamically changing prediction problem the functional mapping $g_t : \Omega_1 \times \ldots \times \Omega_n \to \Omega_{n+1}$ from input to output variables is time dependent. In this section we introduce a version of the fuzzy naïve Bayes algorithm which learns incrementally in an online manner, updating the prediction model at each step.

Suppose we now receive the data as a series where our current data index is $i-1$. Given a new training example $\mathbf{x}(i) = \langle x_1(i), \ldots, x_n(i), x_{n+1}(i) \rangle$ we update the conditional probabilities for each output atom as follows:

$$P'(\alpha_j | \alpha_{n+1}) = \frac{|\alpha_{n+1}| P(\alpha_j | \alpha_{n+1}) + wP(\alpha_j | x_j(i)) P(\alpha_{n+1} | x_{n+1}(i))}{|\alpha_{n+1}| + wP(\alpha_{n+1} | x_{n+1}(i))}$$

Here $P(\alpha_j | \alpha_{n+1})$ is the current probability estimate obtained through updating on the first $i-1$ examples and $P'(\alpha_j | \alpha_{n+1})$ denotes the updated probability taking into account example i . $|\alpha_{n+1}|$ indicates the degree to which output atom α_{n+1} has been previously encountered during learning given by:

$$|\alpha_{n+1}| = \sum_{k=1}^{i-1} P(\alpha_{n+1} | x_{n+1}(k))$$

w is a learning parameter controlling the updating impact of a new training example and is typically assumed to be a decreasing function of $|\alpha_{n+1}|$ with limit 1 (see figure 3). For example, one possibility is $w(|\alpha_{n+1}|) = \frac{c}{|\alpha_{n+1}|} + 1$ where c is a constant controlling the level of initial updating. Note that if $c = 0$ then after updating on all N training examples the conditional probabilities are identical to those obtained using the offline algorithm as described in section 2. In the absence of any data concerning the atom α_{n+1} conditional probabilities are a priori assumed to be uniform so that:

$$P(\alpha_j | \alpha_{n+1}) = \frac{1}{|\mathscr{A}_j|} : \alpha_j \in \mathscr{A}_j$$

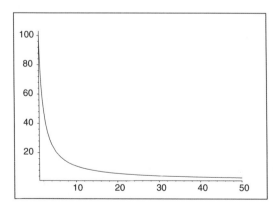

Fig. 3. Weight function $w = \frac{100}{|\alpha_{n+1}|} + 1$

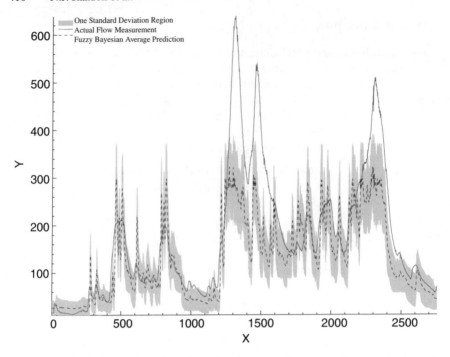

Fig. 4. Fuzzy Bayesian model using offline learning to predict flow at Buildwas 33 hours in advance from upstream flow measurements at Abermule

4 River Flow Modelling

The River Severn is situated in the west of the U.K. and is about 350km long. The source of the river is in the Cambrian Mountains in Wales and its mouth is in the the Bristol Channel. In this paper we look at one section of the River called The Upper Severn which runs from Abermule near Powys, a small village in mid Wales, to Buildwas in Shropshire. The data used in this experiment was taken from flow gauges situated upstream at Abermule which has a catchment area of 580 km^2, and downstream at Buildwas which has a catchment area of 3717 km^2. The flow data for these gauges was obtained from level measurements by applying the rating curve conversion. See [7] for a more detailed description of this catchment.

The offline version of fuzzy naïve Bayes was trained on 1 hourly data consisting of 13119 examples between 01/01/1998 and 01/07/1999. The aim of the learning process was to infer a model to predict flow levels 33 hours ahead [3] at Buildwas from earlier flow data both at Buildwas and upstream at Abermule. Hence, the functional mapping was assumed to take the form $x^B_{t+33} = g(x^A_t, x^B_t)$ where x^A_t and x^B_t denote the flow levels at time t for Abermule and Buildwas respectively. The offline model was then tested on 1 hourly data between 07/09/2000 and 30/12/2000. As well as the actual flow values at Buildwas, figure 4 shows the predicted value for x^B_{t+33}

---

[3] A 33 hour lead time was selected so as to be consistent with the study reported in [7]

taken as the expected value of the conditional output distribution $f\left(x^B_{t+33}|x^A_t,x^B_t\right)$. Also shown is a region of uncertainty associated with each prediction corresponding to one standard deviation on either side of the mean.

The three large peaks in flow during this period resulted in a major flood event and are not representative of peak flow in the training data. This results in relatively poor performance of the offline algorithm at these peak values. The online algorithm was then applied directly to the year 2000 data, with learning parameter $c = 100$ so that the updating weight function corresponded to $w = \frac{100}{|\alpha_{n+1}|} + 1$ as shown in figure 3. The results for the online learning algorithm on the year 2000 data are then shown in figure 5. Clearly, the overall performance is significantly improved from that of the offline approach, with the first and the third peak being captured, albeit with some time delay. Notice, however, that the second peak is still being significantly underestimated. This may be due to the influence of an upstream tributary not included in the model, but further research is required to resolve this issue.

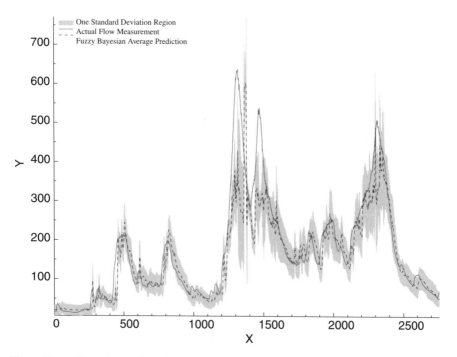

Fig. 5. Fuzzy Bayesian model using online learning to predict flow at Buildwas 33 hours in advance from upstream flow measurements at Abermule

5 Conclusion

An online updating version of the fuzzy naïve Bayes algorithm [5] has been introduced to model systems, the characteristics of which change over time. The potential of this approach has been demonstrated on a river flow modelling problem for the River Severn U.K.

Acknowledgements

This research has been funded by the EPSRC, Flood Risk Management Research Consortium.

References

[1] R. C. Jeffrey, *The Logic of Decision*, Gordon and Breach,New York, 1965.
[2] J. Lawry, A Framework for Linguistic Modelling, *Artificial Intelligence*, Vol. 155, pp 1-39, 2004.
[3] J. Lawry, *Modelling and Reasoning with Vague Concepts*, Springer, 2006.
[4] D.D. Lewis, Naïve (Bayes) at Forty: The Independence Assumption in Information Retrieval, *Machine Learning ECML-98*, LNAI 1398, 1998, pp 4-15.
[5] N.J. Randon, J. Lawry, Linguistic Modelling using a Semi-naïve Bayes Framework, In *Proceedings of IPMU 2002*, pp 1243-1250, 2002.
[6] N.J. Randon, J. Lawry, D. Han, I.D. Cluckie, River Flow Modelling based on Fuzzy Labels, In *Proceedings of IPMU 2004*, 2004.
[7] R.J. Romanowicz, P. C. Young, K. J. Beven, Data Assimilation in the Identification of Flood Inundation Models: Derivation of On-line Multi-step ahead predictions of Flows, In *Hydrology: Science and Practice for the 21st Century*, Vol. 1, pp 348-353, 2004.

Index

absolute quantifier   174
alpha-cut   176

cardinal comparative quantifier   174
cardinality coefficients   179

data summarization   173
database   173
determiner fuzzification scheme   176
DFS   *see* determiner fuzzification scheme

exception quantifier   174

fuzzy data summarization   173
fuzzy median   176
fuzzy quantifier   174

generalized quantifier   174
granularity scaling   180

histogram of a fuzzy set   179

illusion of precision   179
imprecise data   173
induced fuzzy connectives   176

linguistic adequacy   176
linguistic data summarization   173
linguistic fitting   179
linguistic quantifier   174
linguistic summary   173

model of fuzzy quantification   176

optimal quantifier   175
optimal quantifier selection   178
optimality criterion for quantifier selection
       175

propagation of fuzziness   176
proportional quantifier   174

QFM   *see* quantifier fuzzification
       mechanism
quality indicator   175
quantifier fuzzification mechanism   174
quantifier selection   173
quantity in agreement   173

relative proportion coefficient   180

semi-fuzzy quantifier   174
semi-fuzzy truth function   175
standard DFS   176
supervaluation   177

TGQ   *see* theory of generalized quantifiers
theory of generalized quantifiers   174
three-valued cut   177
truthfulness score   *see* validity score
types of linguistic quantifiers   174

validity score   173, 175

Printing: Krips bv, Meppel
Binding: Stürtz, Würzburg